International Conference on Engineering Science and Technology (ICEST 2019)

IOP Conference Series: Materials Science and Engineering
Volume 807

Lampung, Indonesia
29 - 30 August 2019

ISBN: 978-1-7138-3158-7
ISSN: 1757-8981

Printed from e-media with permission by:

Curran Associates, Inc.
57 Morehouse Lane
Red Hook, NY 12571

Some format issues inherent in the e-media version may also appear in this print version.

Institute of Physics
Dirac House, Temple Back
Bristol BS1 6BE UK

Phone: 44 1 17 929 7481
Fax: 44 1 17 920 0979

techtracking@iop.org

Additional copies of this publication are available from:

Curran Associates, Inc.
57 Morehouse Lane
Red Hook, NY 12571 USA
Phone: 845-758-0400
Fax: 845-758-2633
Email: curran@proceedings.com
Web: www.proceedings.com

TABLE OF CONTENTS

IOP Conf. Series: Materials Science and Engineering **807** (2020) 011001 doi:10.1088/1757-899X/807/1/011001

Preface The 1st ICEST2019

The International Conference on Engineering Science and Technology (ICEST 2019) is the first international conference organized by Malahayati University Lampung Indonesia. The goal of the conference is to facilitate researchers around the world to publish and to share their current research. The 2019 edition of the conference held in Lampung, Indonesia on August 29, 2019 at Malahayati University, Lampung Indonesia. These proceeding contains the selected scientific manuscripts submitted to the conference. It is with great pleasure to welcome you to the "1st International Conference on Engineering Science and Technology (ICEST 2019) " that is held at Lampung, Indonesia. The call for papers attracted submissions of over more than 50 abstracts from both local and international universities. Thanks to Universiti Malaysia Pahang (UMP), Abulyatama University Aceh, IST AKPRIND Yogyakarta, Batanghari University and Semarang State University to become as Co-Organizer for this conference.

The scientific papers published in these proceeding have been revised and approved by the technical committee of the 1st ICEST2019. All of the papers exhibit clear, concise, and precise expositions that appeal to a broad international readership interested in science, technology and engineering. The topics of the Conference were: (1) Mechanical and Industrial Engineering; (2) Civil Engineering; (3) Computer and Informatics Engineering; (4) Development and Environmental Engineering; (5) Electrical Engineering; and (6) Agricultural Engineering.

The reports present original ideas or results of general significant supported by clear reasoning and compelling evidence, and employ methods, theories and practices relevant to the research. The authors state clearly the questions and the significance of their research to theory and practice, describe how the research contributes to new knowledge, and provide tables and figures that meaningfully add to the narrative. In this edition of representatives coming from academia, industry, governmental and private sectors. The plenary and invited speakers present, discuss, promote, and disseminate research in all fields of mechanical engineering in a constructive manner. Numerous social events that provide opportunities for the participants to renew old contacts and establish new ones were scheduled to accompany and follow the scientific activities.

Malahayati University would like to thank the Organizing Committee, Co-Organizer and Sponsors supporting the conference and everyone who contributed along the conference and publication of these proceeding.

Dr Eng Rina Febrina, S.T, M.T
The Chairman of ICEST2019

Assoc. Prof. Dr Mohd Shahrir Mohd Sani
The Chief Editor of ICEST2019

IOP Conf. Series: Materials Science and Engineering **807** (2020) 011002 doi:10.1088/1757-899X/807/1/011002

ORGANIZING COMMITTEE ICEST 2019

Advisor	:	Prof. Dr. Rizalman Mamat
		DR. Muhammad Kadafi, SH., MH
		R. Agung Efriyo Hadi, Phd
		DR. Ir. Toto Rusianto, MT
Chairman	:	DR. Eng. Rina Febrina, ST., MT
Vice Chairman	:	Ellyawan S Arbintarso, ST., Msc, Phd
		Marcelly Widya Wardhana, ST., MT
Treasure & Finace	:	Fertina Candra, SE
Publication	:	Associate Prof. Dr. Mohd Shahrir Bin Mohd Sani
		Rani Ismiarti Ergantara, SP., MIL
		Dr. Januar Parlaungan Siregar
Secretariat	:	Hazman Bin Abdul
		Dewi Fadilasari, ST.,MT
Website & Media	:	Fredy
		Diah Ayu Wulandari, ST., MSI

IOP Conf. Series: Materials Science and Engineering **807** (2020) 011003 doi:10.1088/1757-899X/807/1/011003

Peer review statement

All papers published in this volume of *IOP Conference Series: Materials Science and Engineering* have been peer reviewed through processes administered by the proceedings Editors. Reviews were conducted by expert referees to the professional and scientific standards expected of a proceedings journal published by IOP Publishing.

IOP Conf. Series: Materials Science and Engineering **807** (2020) 012001 doi:10.1088/1757-899X/807/1/012001

Preliminary Study of Melting Basalt Rock As A Raw of Advanced Material

Kusno Isnugroho[1], David C Birawidha[1], Yusup Hendronursito[1], Muhammad Amin[1], Muhammad Al Muttaqii[1], R Agung Efriyo Hadi[2], Muhammad Fadil Fazma[2]

[1] Research Unit for Mineral Technology, Indonesian Institute of Sciences, Tanjung Bintang, Lampung, Indonesia
[2] Mechanical Engineering, Malahayati University, Bandar Lampung, Lampung, Indonesia

Corresponding e-mail: kusn005@lipi.go.id

Abstract. The aim of this research is to conduct initial research on the process of melting basalt rocks using an upright shaft furnace. The preliminary process of melting basalt rock has been carried out as a raw of advanced material. The raw material of basalt was obtained from East Lampung, Lampung Province, Indonesia. Coke is used as fuel for the smelting process. Laboratory scale of shaft furnace which is designed and made by Research Unit for Mineral Technology is used for the smelting process. The basalt smelting process lasts for 4 hours, starting from the charging materials to the tapping process, with a process temperature of 1500-1600 °C. The yield of this process can reach 68,7%, while the rest was trap inside the bottom of the furnace due to lack of pressure at the end of the smelting process and the other was lost during the process. The product was basalt cast with the following dominant composition is MgO 5.244%, Al_2O_3 21.295%, SiO_2 48.750, CaO 10.138%, Fe_2O_3 11.019%, TiO_2 1.361% and Na_2O 2.132%. However, the products obtained are still brittle. Therefore, further research related to the crystallization process needs to be done to obtain products that are not brittle.

1. Introduction

Materials that are utilized in high-technology (or high-tech) applications are sometimes termed advanced materials. By high technology we mean a device or product that operates or functions using relatively intricate and sophisticated principles; examples include electronic equipment, computers, fiber-optic systems, spacecraft, aircraft, and military rocketry. These advanced materials are typically either traditional materials whose properties have been enhanced or newly developed and become high-performance materials. Furthermore, they may be of all materials types (e.g. metals, ceramics, polymers) and normally relatively expensive [1].

The need for advanced material continues to increase along with the enhancement in technological developments. The industrial revolution continued without being prevented. The order of human life in various fields is demanded to continue to adjust to the conditions of technological development, no exception in the field of material science and engineering. Various studies have been being carried out in an effort to create material that can support technological development. One of the materials that can be utilized as an advanced material is basalt rock. Processed basalt rocks can be in the form of basalt fiber and basalt cast, which are can be used as advanced material.

Continuous basalt fiber is a new type of green, high-tech inorganic material that has been extensively used in civil engineering, transportation, energy, environmental engineering, automobiles, ships, aerospace, petrochemical industry, and military because of its excellent mechanical property, resistance to high temperature, resistance to chemical erosion and environmental resistance. Basalt fiber, using basalt as the only source material, is formed by drawing the basalt melt (melting temperature of 1450-1500 °C) with a wire drawing machine. Basalt melting process is the most important step in the production of basalt fiber, because it will affect the quality of the basalt melt significantly, and they will affect the quality and performance stability of the basalt fiber [2-8].

For basalt casting technology, the basic principle is using the same technology as basalt fiber, where the difference lies only in the final product. Basalt rock must be melted in the furnace at a temperature of 1430 - 1450 ° C then basalt melted need to be cast under the influenced of heat treatment. In the basalt casting process, the products are in the form of cast basalt according to the desired product form. The casting process can use sand and resin as a casting medium and permanent mold casting. Basalt rock reserves in Indonesia have a high potential deposit which reaches ± 12,000 million tons and the possibility of its use can become a reference for basalt processing to become more valuable. The purpose of this study is to conduct initial research on the process of melting basalt rocks using an upright shaft furnace so that the conditions of the smelting process can be obtained in accordance with the characteristics of local raw materials.

2. Materials and Methods

2.1. Raw Material and Equipment

Basalt rocks as raw material were originating from Lampung province, Indonesia. Raw materials preparation was done through a jaw crusher to obtain a size of 2-4 cm of basalt rock. As an energy source, coke with a size of 5 cm was used. The equipment used to melt basalt is a shaft furnace type. This furnace was designed and made at BPTM-LIPI. The furnace has the following dimension, which is for the inside diameter is 40.5 cm and 156 cm for the high of the furnace. The inner walls in the furnace section are using fire bricks, and the outer wall is coated with mild steel with 6 mm thickness. The dimension of inner diameter of furnace is 450 mm and the height of the furnace is 1300 mm, which is equipped with four tuyeres as the air inlet into the furnace and a tap hole. Tap hole serves as the exit for the melting liquid and tuyere serves as a channel for air to flowing into the furnace with adequate volume and pressure so that the combustion process go on stable. The blower is used as an air source during the smelting process. The picture of the furnace can be seen in Figure 1. Material testing was carried out using X-ray Fluorescence (XRF) and X-Ray Diffraction (XRD) methods.

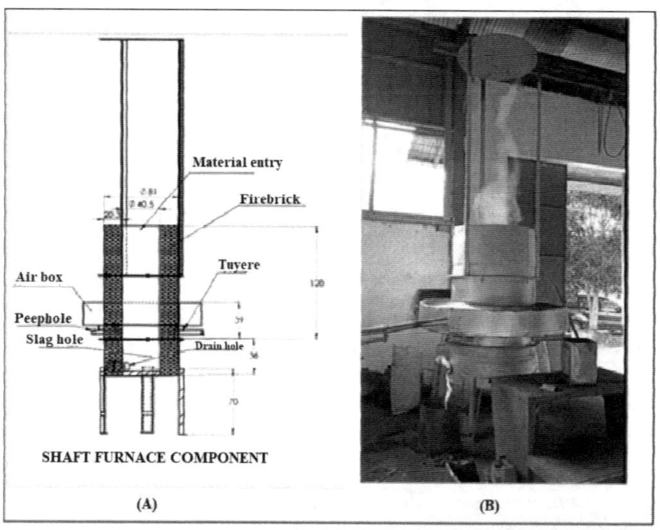

Figure 1. (a) Shaft Furnace component, (b) Basalt rock smelting furnace

2.2. Smelting Process of Basalt

The smelting process in basalt melting is almost the same as the operational procedure in iron smelting using an upright furnace. The furnace preparation process begins with the preparation of the bed furnace using sand casting with a thickness of 4 cm. For the initial combustion is carried out by wood charcoal combustion in the lower layer of the furnace. Gradually followed by adding coke as fuel until the coke burns evenly in the furnace. In order to maintain the combustion process which is taken place inside the furnace, combustion air is blown up through the tuyere hole. Air is blown into the furnace with the help of a blower.

The raw material is entered gradually into the furnace with the first order was coke, then basalt rock, and then coke again, and then basalt rock and so on. The smelting process in the furnace was carried out in a temperature range of 1500-1700 ° C. The temperature observation process is carried out by using a thermocouple which is installed in the melting furnace zone. After the basalt rock has melted, the tap hole was open to harvest the basalt liquid. The series of smelting processes last for 4.5 hours/batch.

3. Results and Discussion

An analysis of the chemical composition of basalt rock as a raw material has been carried out, including analysis for the composition of elements and compounds contained in basalt rock. The examination has been carried out before and after smelting. The results of chemical composition analysis using XRF can be seen in Table 1.

Table 1. The result of chemical composition analysis of basalt rock

Compounds	Percentage (% wt)
SiO_2	47..234
Na_2O	2.998
MgO	4.046
Al_2O_3	18.043
SO_3	0.111
K_2O	0.592
CaO	12.348
TiO_2	1.284
MnO	0.253
Fe_2O_3	12.751
Total	99.66

Based on Table 1, the dominant element found in basalt rocks is silica with a percentage reaching 47.234%. Continued with aluminum oxide and iron oxide which is affect the smelting process in the furnace, this results was similar to Isnugroho et al., [9]. During at basalt smelting process, the visual analysis was carried out. The process temperature can reach 1700 ° C. The amount of raw materials and fuel usage during the process can be seen in Table 2. The total basalt rock which is melted is 65.5 Kg. While the product obtained have a total weight of 45 Kg as the actual smelting process. The obtained percentage level is 68.7%. This condition indicates that the smelting process still not been optimal. Based on table 2, wood charcoal was feed as initial fuel into the shaft furnace in a total of 28 kg, then coke is feed into the furnace as much as 117 kg.

Table 2. Raw Material charge

No. Feed	Time (WIB)	Char (Kg)	Coke (Kg)	Basalt rock (Kg)	Information
1	13.00	Preheating (firewood)			
2	13.53	5	-	-	
3	13.59	5	-	-	
4	14.05	6	-	-	
5	14.11	6	-	-	

IOP Conf. Series: Materials Science and Engineering **807** (2020) 012001 doi:10.1088/1757-899X/807/1/012001

6	14.17	6	-	-	
7	14.23	-	10	-	Blower on
8	14.24	-	10	-	
9	14.25	-	11	-	
10	14.26	-	13	-	
11	14.27	-	11	-	
12	14.50	-	11	-	
13	14.51	-	11	-	
14	14.54	-	-	20	
15	14.55	-	-	23	
16	15.00	-	-	22,5	
17	15.24	-	10	-	
18	15.25	-	10	-	
19	15.26	-	9	-	
20	15.44				Tapping
TOTAL		28	117	65.5	

Based on Table 3, the furnace temperature range during the smelting process is 314-1700 ° C. Only in 57 minutes, shaft furnace temperature tends to increase rapidly, its because the blowing air from the blower through the 4 tuyeres around the shaft furnace wind box stimulate the combustion in the furnace and the relatively short height of the furnace also play a role in that process. The quickly increasing temperature causes some of the material contained in the basalt rock to evaporate in the form of gas. The rapid rise in temperature also results in a sudden decrease in temperature because the fuel has burned out when the tapping process still in progress. Which in result the liquid has not flowed completely and some of the liquid settles on the inner walls of the furnace and agglomerate at the bottom of the furnace, as shown in Figure 2.

Table 3. Temperature process

Time (WIB)	Shaft furnace temperature (^oC)	Information
13.00	30	Precombustion using firewood
13.53	314	Precombustion using firewood
14.23	1090	Precombustion using coke
14.26	1116	44 Kg coke consumption
14.54	$\pm$1700	Starting basalt rock melting
15.44	$\pm$1500	Basalt rock melting
16.56	$\pm$1300	Finished smelting process

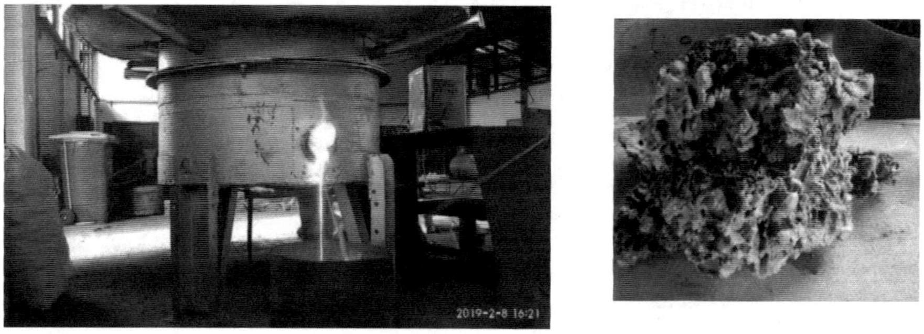

(a) (b)

Figure 2. (a) Tapping process, (b) Agglomerate of basalt melting

The XRD results from basalt rock melting can be seen in Figure 3. To obtain the phases, which is formed in the basalt rock melting, the sample was analyzed using Match! 3 Software. From the results, the sample has an amorphous structure. It is because from diffractogram result does not have a specific or typical peak at a certain angle, so the distance between each atom does not have a pattern and tends to be random. From XRD results cannot reveal the real distance between each atom. Because there is no crystallization process happened in the cooling process, it resulted in fragile characteristic in the product. According to Birawidha et al., [10] the crystal structure is unchanged from the parent material but the mechanical characteristic of material changed.

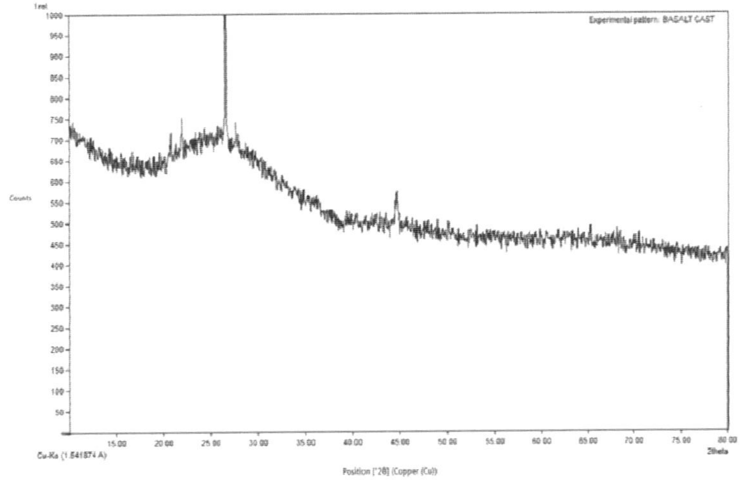

Figure 3. The result of XRD analysis after basalt rock smelting process

4. Conclusions

Smelting process was carried out at temperatures of 1500 to 1600 ° C. Based on the XRD, the sample has an amorphous structure and fragile characteristics. The main product of basalt cast such as Al_2O_3 21.295%, SiO_2 48.750, CaO 10.138%, and Fe_2O_3 11.019%. In this work, Basalt rock from Lampung Province can be utilized as an advanced material in the form of basalt fiber and basalt cast, which is through spinning and heat treatment process. The process is carried out as a series of processes in engineering material.

Acknowledgments

The authors would also like to acknowledge Research Unit for Mineral Technology, Indonesian Institute of Sciences for fund and facility support, so that this research is possible to do.

References

[1] Callister D W 2001 *Fundamentals of Materials Science and Engineering* Fifth Edition (New York: John Wiley & Sons, Inc)
[2] Chen X, Zhang Y, Hui D, Chen M, Wu Z 2017 *Composite Part B* **116** 53-60
[3] Dhand V, Mittal G, You R K, Park S J, Hui D 2015 *Composite Part B* **73** 166-80
[4] Wu Z, Wang X, Wu G 2012 6th International Conference on FRP Composite in Civil Engineering
[5] Yesilkaya S S, Guven O Z, Yilmaz C N, Yildiz A, Icelli O, Okutan M 2106 *Composite Part B* **9** 488-92
[6] Zhu H, Wu Gang, Zhang L, Zhang J, Hui D 2014 *Composite Part B* **60** 680-7
[7] Fiore V, Scalici T, Di Bella G, Valenza A 2015 *Composite Part B* **74** 74-94
[8] Singha K 2012 Int J Text Sci 1(**4**) 19-28
[9] Isnugroho K, Hendronursito Y, Birawidha D C 2018 *IOP Conf. Ser.: Mater. Sci. Eng.* **285** 012014
[10] Birawidha D.C., Isnugroho K, Hendronursito Y, Amin M, Al Muttaqii M 2019 *Multitek Indonesia: Jurnal Ilmiah* 13 (**2**) 6-14

IOP Conf. Series: Materials Science and Engineering **807** (2020) 012002 doi:10.1088/1757-899X/807/1/012002

Comparative Analysis the Addition of Natural and Artificial Fibres in Concrete

Devi Oktarina[1], Dewi Fadilasari[1],Yan Juansyah[1]

[1]Civil Engineering Department, Universitas Malahayati, Bandar Lampung

Corresponding e-mail: oktarina_sipil@yahoo.co.id

Abstract. Concrete structurally has a fairly large compressive strength, but has a lack of tensile strength and is brittle. This weakness can be corrected by adding natural or artificial fibres to the concrete mix evenly with a random orientation. The purpose of this study was to investigate about the effect of adding natural fibres and artificial fibres on compressive strength and tensile strength of concrete. Test specimens used are cylindrical concrete measuring 150 mm x 300 mm with natural fibres from bamboo Betung and artificial fibres from local Bendrat wire. The variation of the percentage value of the fibre fraction volume is 0%, 0.2%, 0.4%, and 0.6% and the test is carried out when the concrete is 28 days old. The results showed that the addition of natural fibres and artificial fibres to the concrete mix can increase the compressive strength and tensile strength of normal concrete at a fraction volume of 0.4%. Increased compressive strength of normal concrete is 24.31% for bamboo Betung fibre and 20.81% for Bendrat wire fibre. Whereas the increased in tensile strength of normal concrete for Betung bamboo fibre increased by 77.12% and for Bendrat wire fibre increased by 3.24%.

1. Introduction

Concrete is one building material that is widely used in everyday life. Concrete has a strength that is greatly influenced by the composition and quality of the ingredients mixing. The strength of concrete lies in its compressive strength, while its weakness lies in its low tensile strength. The value of concrete tensile strength has range from 9% - 15% of the compressive strength. In line with the development of today's concrete technology, various attempts were made to improve the unfavorable character of concrete. The improvement method is by adding fibre into the concrete mixture. The main objective of adding fibre into concrete is to increase the tensile strength of the concrete. Very low tensile strength causes the concrete to crack easily, which in turn reduces the durability of the concrete. The intercalation of fibre in the concrete mixture is done evenly with random orientation. Types of fibres used can be natural fibres or artificial fibres. This research uses natural fibres from Betung bamboo peels (*Dendrocalamus Asper*) and artificial fibres from steel fibres in the form of Bendrat wire.

1.1. Steel Fibre

There are several types of steel fibres that are commonly used [1] which are straight, hooked, crimped, double duo form, ordinary duo form, paddled, enlarged ends, irregular, or indented. The cross section of the steel fibre can be round or wire, rectangular or sheet, or irregular or melt extract. The types of steel fibres glued together can be seen in Figure

1.1.1. Fibre glued together into a bundle
The bending and deflection load characteristics for each types of steel fibres and the illustration of the fibres can be seen in Figure 1 and Figure 2 respectively below.

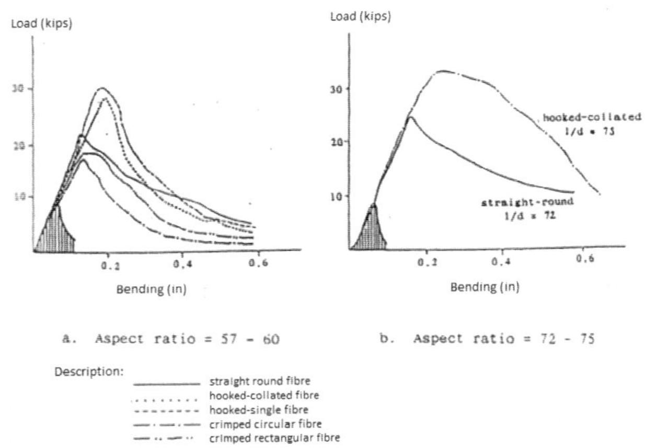

Figure 1. Bending – deflection load characteristic of steel fibre concrete with different fibre types [1]

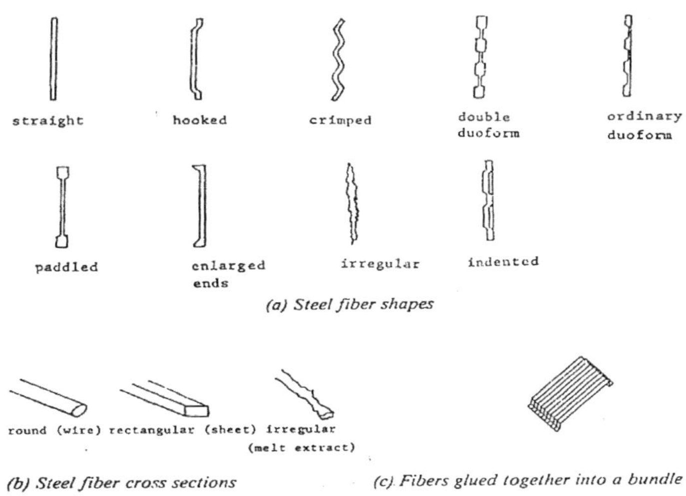

Figure 2. Various types of steel fibre shapes

1.2 Bamboo Fibre
Bamboo is a plant of the order of *bamboo idae* which grows quickly and can be harvested at the age of around 3 years. During the growth period, bamboo can grow vertically 5 cm per hour or 120 cm per day. [2] states, the presence of sklerenkin fibres in bamboo stems caused bamboo to have strength and can be

used for building materials. The strength of bamboo is generally influenced by the amount of sklerenkin fibre and cellulose in bamboo. The strength of bamboo on the outside is much higher than inside the bamboo.

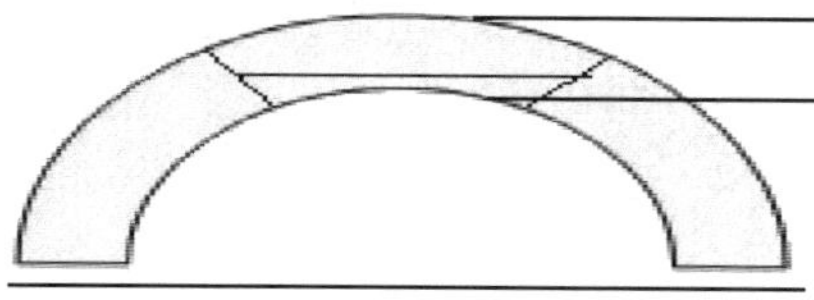

Figure 3. Intake of Bamboo Specimens

Bamboo Betung (*Dendrocalamus Asper*) is a very strong and large bamboo with a diameter of 10 – 15 cm. Bambu Betung has a short internodes and thick walls and can grow very tall up to 10 – 20 meters. This bamboo tensile strength is 441 MPa [3]. This type of bamboo is usually used as the main structure of buildings, namely columns and beams.

Bamboo is an ingredient which is ductile so that it can be become immaterial for the house of earthquake proof. The level of high ductile, it is because bamboo is material who is a lightweight and the frame system work as a hinge. As natural material that can be renewed for 3-5 year, bamboo can be environmentally friendly. Excellence and this is what other unparalleled other materials.

2. Methods

The study was conducted in a laboratory to obtain actual data from the results of the study [4, 5]. Mixture comparation plan for concrete with fibres refers to regulations made by the ACI Committee 544, 1982. Trial concrete mix based on volume of weight was first made to obtain the desired concrete [6]. The specimens to be made consist of cylinders with a diameter of 150 mm with a height of 300 mm, a beam measuring 750 mm x150 mm x 150 mm. Each specimen was made of 12 specimens consisting of four variations in the percentage of fibre content, namely 0%; 0.2%; 0.4%; 0.6%. The tests conducted in this research consisted of Slump testing, Compressive Strength testing, Tensile Strength testing.

3. Result and Discuss

3.1 The results of testing Workability

The severity of the fibre concrete mix was measured using the slump test and VB-test apparatus with vibration in the test. The slump value and the VB-time value of each mixture are presented in Table 1 below.

Table 1. Slump and VB-Time Values

Code	Fraction volume (%)	Cement Water Factor (fas)	VB-time		Slump Value (cm)		Comparation (cm)
			Bamboo fibres	Bamboo fibres	Bamboo fibres	Bamboo fibres	
BN-0%	-	0.5	-	13	18	8	10
BF-0,2 %	0,2	0.5	24	22	8	5	3
BF-0,4 %	0,4	0.5	26	23	1.5	5	-3.5

BF-0,6 %	0,6	0.5	25	25	3	4	-1

Based on Table 1 above, it is known that the addition of fibre in a concrete mixture adds VB-Time value and decrease the value of Slump. The addition of inserted fibre into the concrete mix makes the concrete mixture become thicker and speediness smaller. This makes the concrete mixture more difficult to work with.

3.2 Compression Strength Calculation Results

The calculation results of the average compressive strength of three test pieces for each fibre concrete mixture are presented in Table 2 below.

Table 2. Concrete Compressive Strength

Code	Fraction volume (%)	Bamboo Fibre	Compressive Strength (MPa)		Influence (%)	Comparison (MPa)
			Influence (%)	Fibre Wire		
BN-0%	-	26.33	-	27.65	-	1.32
BF-0,2 %	0,2	23.78	-3,072	26.80	-9,68	3.02
BF-0,4 %	0,4	32.73	+20,81	33.40	+24,31	0.67
BF-0,6 %	0,6	17.37	-37,18	17.37	-34,03	0

Based on Table 2 above, it is known that the highest compressive strength value of concrete is in the volume of 0.4% fibre fraction both using artificial fibres (Bendrat wire fibres) which increase the compressive strength of normal concrete by + 24.31% and natural fibres (bamboo Betung fibres) increased compressive strength than normal concrete by + 20.81%. Comparison of compressive strength of concrete with artificial fibres (Bendrat wire) is higher than using natural fibres (bamboo Betung).

Research ananta ariatama (2005) of the fibres hooked wire obtained strong press increased 14,67 % and strong pliable increased 48,06 % and for research by edwar bamboo abraham said 2016) obtained strong press increased 7,00 % and 42,86 %.

3.3. Tensile Strength Calculation Results

The results of the calculation of the average tensile strength of three test pieces for each concrete fibre mixture are presented in Table 3 below.

Table 3. Concrete Tensile Strength

Code	Fraction volume (%)	Bamboo Fibre	Tensile Strength (MPa)		Influence (%)	Comparison (MPa)
			Influence (%)	Fibre Wire		
BN-0%	-	1.53	–	2,83	-	1.3
BF-0,2 %	0,2	2.17	+41,83	1,93	-31,67	-0.24
BF-0,4 %	0,4	2.71	+77,12	2,92	+3,24	0.21
BF-0,6 %	0,6	2.37	+54,9	2,32	-18,14	-0.05

Based on Table 3 above, it is known that the highest value of concrete tensile strength is in the volume of 0.4% fibre fraction both using artificial fibres (Bendrat wire fibres) which increase the tensile strength of normal concrete by + 3.24% and natural fibres (bamboo Betung fibres), increased the tensile strength of normal concrete by + 77.12%.

4. Conclusion

Addition of natural fibres and artificial fibres to the concrete mix can increase the compressive strength and tensile strength of normal concrete at a fraction volume of 0.4%. the increasing of compressive strength of normal concrete is 24.31% for Betung bamboo fibre and is 20.81% for Bendrat wire fibre, while for normal concrete tensile strength concrete is 77.12% for Betung bamboo fibre and 32.45% for Bendrat fibre.

References

[1] Soroushian P, Bayasi Z 1991 Concept of Fibre Reinforced Concrete *Proceding of The International Seminar on Fibre Reinforced Concrete* (Michigan State University, Michigan, USA)
[2] Morisco. 1999 Rekayasa Bambu (Yogyakarta)
[3] Subyakto 2009 Proses Pembuatan Serat Selulosa Berukuran Nano dari Sisal (Agave Sisalana) dan Bambu Betung (Dendrocalamus Asper) *Berita Selulosa* **4**(2)
[4] Devi O, Olyvia A, Edi P 2018 Pengaruh Penambahan Serat Kawat Bendrat Terhadapat Kuat Tekan Dan Kuat Tarik Beton *Prosiding SNAST AKPRIND* (Yogyakarta)
[5] Edi P, Devi O, Siti H 2017 Pengaruh Penambahan Serat Bambu Betung Terhadap Kapasitas Ultimit Beton Bertulang *Jurnal RTS Fak. Teknik Univ. Malahayati* **1(2)** 87-93
[6] Edwar H, 2016 Pengaruh Penambahan Serat Kulit Bambu Terhadap Sifat Mekanik Beton *Jurnal JomFTEKNIK* **3(1)**

IOP Conf. Series: Materials Science and Engineering **807** (2020) 012003 doi:10.1088/1757-899X/807/1/012003

Analysis of Aluminium Basalt Particulate Composite Using Stirring Casting Method Through Taguchi Method Approach

Yusup Hendronursito[1*], Tumpal Ojahan Rajagukguk[2], Rizal Nur Safii[2], Achmad Sofii[1], Kusno Isnugroho[1], David Candra Birawidha[1], Muhammad Amin[1], Muhammad Al Muttaqii[1]

[1] Research Unit of Mineral Technology, Indonesian Institute of Science, South Lampung, Indonesia

[2] Mechanical Engineering Department, Faculty of Engineering, Malahayati University, Indonesia

Corresponding e-mail: yusu016@lipi.go.id

Abstract. Aluminium matrix composites (AMCs) becomes a choice for automotive and aerospace industries due to its tenable mechanical properties such as high hardness and good strength to weight ratio. Reinforcement like particulate alumina, silicon carbide, graphite, fly ash, etc can easily be incorporated in the melt using cheap and widely available stirring casting method. This research investigates the optimum hardness of AMCs containing basalt powder as reinforcement by setting varying parameter. The design of experiments (DOE) approach using Taguchi method orthogonal arrays L9 (3^3) factors 3 runs 9, was employed to predict the hardness of material casting with variable parameters such as matrix composition, stirring time, and particle size (mesh) of basalt. ANOVA and S/N ratio for determining optimum hardness (HRb) material and P-value confirms most significant parameter that affects the hardness at optimum level parameters to improve the mechanical properties of AMCs. Analysed Taguchi design observed that particle size (mesh) of basalt has a significant contribution on hardness (72.71%) followed by concentration of basalt (18.62%) and stirring time (8.45%) with an error about 0.22%. Based from previous study and field experiment, this result is match especially from different particle size will affect most of mechanical properties of composite materials.

1. Introduction

In this few past years, the global needs for low cost, high performance and good quality materials had turn the interest of research from the uses of monolithic materials to composite materials. Composite is a material that are made from two or more mixture of materials. In other words, it is a combination of two or more nano, micro or macro constituents with an interface separating them that vary in form and chemical composition and they are essentially insoluble for each other [1]. In case of metal matrix composites (MMCs), aluminium matrix due to their high strength to weight ratio, low cost and high wear resistance are widely manufactured and use in structural applications along with aerospace and automotive industry [2]. AMCs can be classified into four types depending on the type of reinforcements which are the particle reinforce AMCs (PAMCs), the whisker or short fibre-reinforced AMCs (SFAMCs) the continuous fibre reinforced (CFAMCs) and the monofilament reinforced AMCs (MFAMCs) [1-2].

AMCs has a lower density, wear resistance, high strength, and high elasticity so that mechanical properties which are desired can be determined depends on a combination of the matrix, reinforcement, and interface. Those properties of AMCs are the main attention for the researcher to make AMCs as a substitute for the current conventional materials [3]. It was found that most of the researches do studies about AMCs that used the aluminium composite with a particle as reinforcement [4], but seldom found the uses of recycle aluminium waste as the matrix in AMCs. Recycle aluminium waste to make AMCs became an interest due to the performance, economic and environmental benefits. Even in the country of United States of America (USA), the production of aluminium is also yield from the recycling process [5]. In Lampung, Indonesia, the aluminium waste such as dross are occasionally found to be recycled as the waste are observed littered in the surrounding environment. Aluminium waste still possesses the characteristic of the normal aluminium, but with a higher aluminium oxide content due to the aluminium melting process. When this material is used as a matrix component in the composite material, it could produce a mix characteristic that would overcome the weakness of each of the material used as raw composite. With the used of aluminium waste as component in the composite material, it could eventually solve the environmental problems as well as added a higher value to the aluminium waste.

2. Modelling of design research

2.1. Stir casting

In the stir casting process, usually, the particulate reinforcement is distributed into the molten aluminium by mechanical stirrer. The important parameters in the process of stir casting determine by the physical or mechanical properties. The parameters are the temperature of casting, the stirring time and the velocity of stirring [6]. Based on Raja et al. [7], the study uses basalt particle in order to increase the hardness of the material and resistance of wear where the value is directly proportional to the percentage of basalt as the reinforcement. In this research, there are three levels selected for three parameters, the percentage of weighing basalt particle, the stirring time, and the particle size (mesh) of basalt.

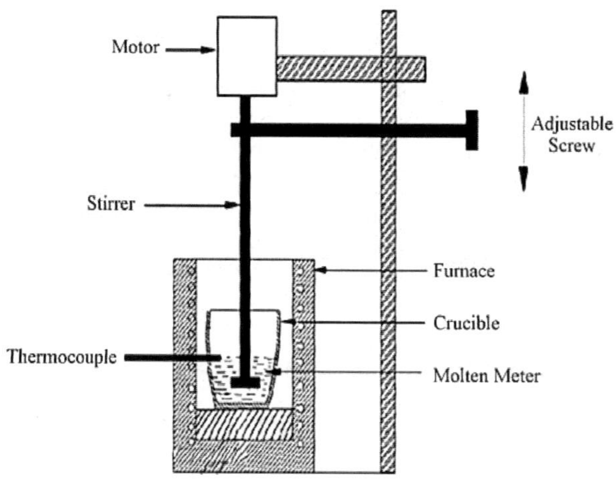

Figure 1. Schematic diagram of stirring casting process [4]

2.2. Taguchi Method

Taguchi design is the most powerful approach in order to achieve a high quality of work at a reduced cost. It is the statistical analysis tool by Dr Taguchi that was introduced which is focused on three concepts adequate to achieve the desired quality of products; system design parameters, and tolerance

IOP Conf. Series: Materials Science and Engineering **807** (2020) 012003 doi:10.1088/1757-899X/807/1/012003

[8]. The selection of materials and methods for the experimentation according to Taguchi method L9 orthogonal array selected for three levels of three parameters, are shown in Table 1.

Table 1. Taguchi orthogonal array design used L9 (3^3) factors 3 runs 9

Factors	Level 1	Level 2	Level 3
% w/t Basalt	5	10	15
Stirring time (minutes)	5	10	15
Particle size (mesh)	100	200	325

Table 2. Level and factor in stir casting

Sample No.	Basalt (%)	Stirring time (minutes)	Particle size (mesh)
1	5	5	100
2	5	10	200
3	5	15	325
4	10	5	200
5	10	10	325
6	10	15	100
7	15	5	325
8	15	10	100
9	15	15	200

For experimental procedure, the total of nine samples runs are taken as shown in Figure 2.

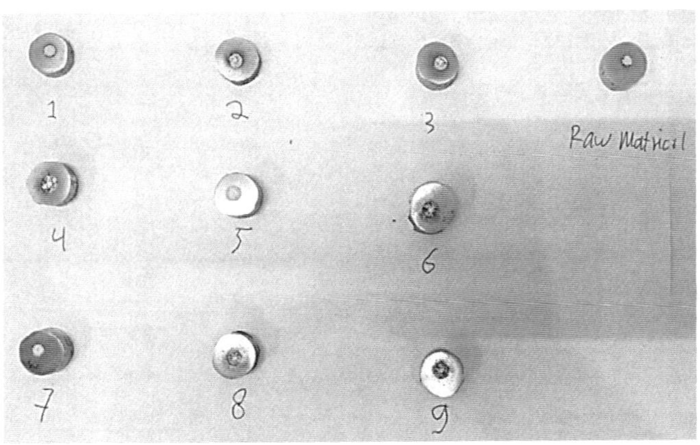

Figure 2. Samples of experiment runs

3. Analysis Based on Taguchi Method

Analysis of signal-to-noise ratio ($\frac{S}{N}$) and analysis of variance (ANOVA) are given as follow.

3.1. Analysis of signal to noise ratio

Signal-to-noise ratio indicates the output desirable value that are mean of output characteristics and noise indicates undesirable value such as squared deviation of output characteristics. S/N ratio denoted by η and the unit is dB. The equation for η for higher the best is given as equation 1.

IOP Conf. Series: Materials Science and Engineering **807** (2020) 012003 doi:10.1088/1757-899X/807/1/012003

$$\frac{s}{n} = -10 \times log \sum \frac{\left(\frac{1}{y^2}\right)}{n} \tag{1}$$

Table 3. L9 orthogonal array for experiment values

Experiment No.	Basalt (%)	Stirring Time (minutes)	Particle size (mesh)	Hardness (HRb)	S/N value (dB)	Mean
Raw				54.167		
1	5	5	100	141	42.9844	141
2	5	10	200	148	43.4052	148
3	5	15	325	151	43.5795	151
4	10	5	200	145	43.2274	145
5	10	10	325	145	43.2274	145
6	10	15	100	125	41.9382	125
7	15	5	325	148	43.4052	148
8	15	10	100	101	40.0864	101
9	15	15	200	149	43.4637	149

Analysis of variance (ANOVA) is a statistical analysis tool of data analysis. ANOVA includes the design parameters that significantly affect the output result. In the ANOVA method, the sum of square (SS), mean square (MS), and P values are calculated for deciding significant factors which affect the process and percentage contribution are calculated. The average effect response table for hardness and the rank of parameters as shown in Table 4 and Figure 3-4.

Table 4. Response table for signal to noise ratios (larger is better)

Level	A	B	C
1	43.32	43.21	41.67
2	42.80	42.24	43.37
3	42.32	42.99	43.40
Delta	1.00	0.97	1.73
Rank	2	3	1

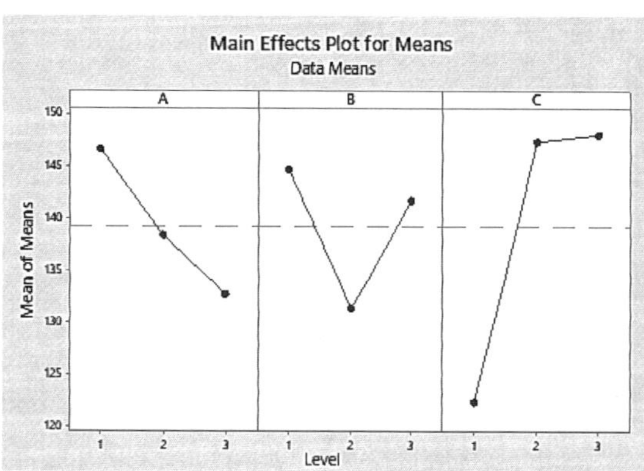

Figure 3. Main effect plot for means

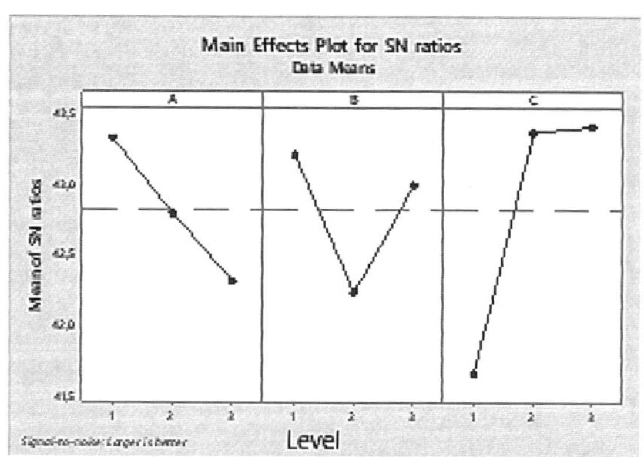

Figure 4. Main effect Plot for S/N Ratio

IOP Conf. Series: Materials Science and Engineering **807** (2020) 012003 doi:10.1088/1757-899X/807/1/012003

3.2. ANOVA

The percentage of contribution by each of the process parameters in the total sum of the squared deviation can be used to evaluate the importance of the process parameter change on the quality characteristic. In addition, P-value used in order to determine the accepted H_0 or rejected. Optimum process parameter and the effect of a factor on hardness material based on Taguchi analysis are the factor of percentage of basalt level one, stirring time level one and particle size (mesh) basalt level three. The most significant parameter according to ANOVA table is the percentage of basalt which affect the performance of material hardness. Basalt itself besides of high in silica content, the composition also high in other minerals such as Fe and Al which also can improve the mechanical properties of the composites [9]. Besides of the percentage of basalt, the significant parameter is particle size (mesh) of basalt and next is the stirring time. The percentage contributed by A (Basalt); B (stirring time) and C (particle size) are 18.62%, 8.45%, and 72.71% respectively. From the calculation prediction, particle size variable would highly affect in the hardness value of the composite. From the experiment, when the particle size is smaller, the hardness value is higher than the other samples with different variation. This is because if the particle size is smaller, the spreading of the particle is more homogenic towards the total surface area of the AMCs with stirring method. This result corresponds with the statement in Israa [10] and Jaya et al. [11], which is when the particle size is smaller, the mechanical properties of the composites are better.

Table 5. Analysis of variance for transformed response

Source	DF	Seq SS	Contribution (%)	Adj SS	Adj MS	F-Value	P-Value
A	2	278.19	18.62	278.19	139.096	86.09	0.011
B	2	126.17	8.45	126.17	63.083	39.05	0.025
C	2	1086.14	72.71	1086.14	543.071	336.14	0.003
Error	2	3.23	0.22	3.23	1.616		
Total	8	1493.73	100				

4. Conclusions

The Taguchi design can be used to analyse AMCs hardness test result which is made by stir casting method. ANOVA table determines the significance of stir casting parameters and according to ANOVA, the particle size do highly affect the hardness value which is the percentage of basalt. The second parameter that affects the hardness is the stirring time which it does less affect the hardness value. The experimental results correspond with the Taguchi method in term of the good materials quality and the hardness parameters in AMCs is optimized. Basalt particle would enhance the mechanical properties of AMCs, but the optimum value for enhancement especially in hardness can be achieved when the added particles is smaller.

Acknowledgments

The authors acknowledged financial and facility supports from Research Unit of Mineral Technology, Indonesian Institute of Sciences and thanks to the support from Mechanical engineering department, Malahayati University, Lampung, Indonesia. All author in this paper contribute equally to produce this article.

References

[1] Kala H, Mer K K S and Kumar S 2014 *Procedia Mater. Sci.* **6** 1951
[2] Lee W B and London W 1997 *Journal of Materials Processing Technology*. **63** 339
[3] Anam S 2012 *Kekerasan Dan Kekuatan Bending Komposit Aluminium Yang Diperkuat Serbuk Besi Produk Proses Stirring Casting* Diponegoro University Semarang. Indonesia
[4] Srivatsan T S, Ibrahim I A, Mohamed F A, and Lavernia E J 1991 *J. Mater. Sci.* **26** 5965
[5] Surappa M K 2003 Sadhana. **28** 319

[6] Arifin A and Juanidi 2017 *J. Tek. Mesin Untirta* **3** 21
[7] Raja M A, Manikandan V, Amuthakkannan V P, and Rajesh S 2018 *J Aust Ceram Soc* ***54*** 119
[8] Parihar R and Jathar S 2015 International Journal of Mechanical Engineering and Technology (IJMET) **6** 194
[9] Isnugroho, K., Hendronursito, Y. and Birawidha, D.C. 2018. Characterization and utilization potential of basalt rock from East-Lampung District. IOP Conf. Ser. Materials Science and Engineering 285(1):012014.
[10] Gh Israa 2017 Study the Effect of the Particle Size on Mechanical Properties of Particulate Natural Composite Materials.
[11] Jaya et al. (2016). "Sawdust polyethylene composites," BioResources 11(3), 6489 – 5504

The activity of modern agricultural technique, mina paddy, on salt water and the income of farmers in west matang raya-north aceh regency

D Aziz[1]

[1] Geography Education Department, Faculty of Teacher Training and Education-Syiah Kuala University

Corresponding e-mail: hda.unsyiah@gmail.com

Abstract. This is a study of activity in agricultural modern technique, mina paddy, on salt water and the income of Kena group farmer in West Matang Raya, North Aceh district. This study focuses on identifying the activity of modern agricultural technique on salt water and the income of farmers in Kena group. The major objective of this research is to find out the activity of modern agricultural technique on salt water and the income of Kena group. This study was conducted on February, 2017 with 20 members of Kena group. The data were obtained through library studies, documentations, observation, questionnaires and interviews. The data were analyzed descriptively. Based on the result of this study, the proper activity of modern agricultural technique, mina paddy, can increase the productivity of agricultural products and also increase the farmers' income compared to the agricultural activities in general. It is suggested that the government can encourage the farmers to do the mina paddy technique on the proper land and help them by providing the fertilizer, seed, pesticide and feed. Therefore, the farmers will have no difficulties and get the motivation to do the mina paddy technique which can improve their prosperity and ensure their food security for the future, especially for the farmers in Kena group, West Matang Raya.

1. Introduction

Mina paddy is one of the methods implemented by farmers from the monoculture system to the agricultural diversification system. The minapaddy cultivation business is an integrated business that can increase the productivity of paddy fields, gain more income for farmers' prosperity and keep the existence of food security [1]. The implementation of the modern technique ofminapaddy is one of the foundations of food, many efforts and breakthroughs have been made to increase the agricultural productivity. One of them is integrated farming, known as the minapaddy system, which is a method of fish and paddy farming in one site of paddy field.

In agricultural land, the weeds will automatically become the food for Vannamei Shrimp and Milkfish. Consequently, it becomes fertile because the presence of fish' feces which is containing various nutrients. Therefore, farmers can save the fertilizer because the land that has been mixed with dirt and fish food can already be an organic fertilizer that minimizes the use of chemicals. Fish can also limit the growth of other plants that are competent to rice in the utilization of nutrients, as well as to reduce the cost of weeding wild plants. The minapaddy technique has long been developed. Besides providing food sources of carbohydrates, this technique also provides protein, hence it is good enough to improve the food quality for the population. The properminapaddy techniquecan alsoincrease the income if compared to the usual techniques that are often applied by farmers over the years. The benefits obtained from the minapaddy farming business are rice and fish production. Vannamei Milkfish and Shrimp chosen by the Kena group for the mina paddy system because they are not susceptible to disease andits market price tend to rise and stable.

The difficulty in managing the minapaddy system happens when there isa large number of birds that eat rice in the period leading up to harvest, when ensuring the irrigation runs well and also when feedingthe fish regularly. However, with the various advantages such as saving fertilizers, minimizing weeding and also the benefits of milkfish and vannamei shrimp, all of the disadvantages are not a big deal. The application of this minapaddy technique is good enough in increasing agricultural products as well as improving community' productivity. For that reason, the researcher was interested in conducting a study entitled "The Activity of Modern Agricultural Technique, Mina Paddy, on Salt Water and The Income of Farmers (Kena Group) in West Matang Raya—NorthAceh Regency".

Based on the background of the problem, this study focused on identifying the activity of modern agricultural technique on salt water and the income of farmers in Kena group. The aim of this research was to find out the activity of modern agricultural technique, mina paddy, on salt water and the income of Kena group farmer in West Matang Raya.

2. Methodology and Location

The study was conducted on February, 2017 in the West Matang Raya Village in North Aceh regency, especially for farmers in Kena group who implement the mina paddy technique. The data were obtained through library studies, documentations, observation, questionnaires and interviews. The data were analyzed descriptively.

The astronomical location of the West Matang Raya Village is 5010'20 "LU-509'20" LU and 970220 "BT-97022'50" BT. It is located in a low land, with an area of 275 Ha and a distance to the district capital as far as 7 km. As a lowland area and located near the coastal sea, it is very suitable for planting rice, ponds and coconuts. The geographical conditions are very supportive. However, the facilities and infrastructure are inadequate so the natural potential cannot be managed maximally.

3. Result and Discussion

3.1. *The Activity of Modern Agricultural Technique Mini Paddy*

According to Rahim [2] agriculture is an activity to develop (reproduce) plants and animals to grow better due to fulfill human needs, such as farming, breeding, and sailing. Agriculture is also a type of business or economic activity in the form of planting crops or farming (food, horticulture, plantations, and forestry), breeding (livestock) and fisheries (aquaculture). The mina paddy cultivation system is a method of maintaining fish on the sidelines of rice plants in order to make the alternation between two seasons of rice plants and/or to replace the crops in rice fields by breeding the fish. The presence of fish to this integrated farming business can enrich the planting media, because it develops organic fertilizers and produces plankton, which is the source of fish feed.

Montazeri [3] explained that mina paddy is one of the agricultural technologies for improving environmental quality as the anticipation of climate anomalies. Due to its' integrated cultivation system, it can increase the productivity of paddy fields as well as increasing farmers' income through rice production up to 10%; increasing the diversity of agricultural products because it produces fish; increasing soil and water fertility (reducing fertilizer up to 30%); and also reducing the pests that caused disease in rice plants. The mina paddy farming system is classified into [1,4]:

a. Aquaculture as Alternation of Rice Plant

Growing fish as an alternation is done while the farmers work on the field where the paddy will be planted. The duration of farming usually takes 20-30 days until the seeds are ready to be planted. In this system, the seeds dispersal is only done with 1-3 cm size seeds for the purpose that after 20-30 days, the seeds will turn into fish that are ready to be spread in the pond (3-5 cm size or seedlings).

b. Aquaculture with Plants
Aquaculture with plants means that the farmers are farming fish in the paddy fields. The length of farming is from the time the rice seeds are planted, continued to the first weeding, the second weeding, and until the rice flowering, even until drying. The harvest result can be a fish amounting to 100 grams/tail.
c. Aquaculture as Substitute for Crops
This aquaculture is carried out as a substitute for secondary crops in the pattern of rotation of rice crops. The purpose of this method is to restore the fertility of the rice fields. In general, fish farming as secondary crops is carried out after two consecutive rice planting periods.

The purpose of growing fish with rice is to increase farmers' income, in addition to the results of rice cultivation, the fish production also brings many profits. Moreover, families' nutritional value can be fulfilled and the risk of crop failure can be reduced.

3.1.1 The Process of Managing Mina Paddy Technique
The mina paddy technique done by Kena group [5] is the cultivation of fish in the paddy fields. The length of farming is from the time the rice seeds are planted, continued to the first weeding, the second weeding, and until the rice flowering, approximately when the paddy aged 50 days. The length of the farming is adjusted to the rice planting plan, usually takes 30-40 days. After the soil has been processed, then it is fertilized and irrigated to a 5-10 cm surface height. Next, after being left alone for 4-7 days, then the fish seeds are ready to be spread. The seeds are 1-3 cm (aged 20 days-1 month) with density levels ranging from 100-120 thousand tails/hectare. The farmers can also spread the fish seeds that have reached a size of 5-7 cm with a density ranging from 50-60 thousand tails per hectare. The harvest is done after the fish seeds have been kept for a month reaching a length of between 8-12 cm. The height of the water level in the plot of rice fields should not be too low. For the 1-3 cm fish seeds, 5 cm of water level is already enough. While the spreading of fish seeds which having 5-7 cm in size, the height of the water surface is set between 5-10 cm. The activity of maintaining fish with rice should be adjusted to the growth of rice plants. In one rice planting several types of fish can be carried out, according to the purpose of maintenance, whether for seed size only or for the size of consumption fish. The cultivation of fish with plants should be adjusted to the growth of rice plants. In one planting several kinds of fish farming can be carried out based on the farming purpose, whether for seed size only or for the size of fish to be consumed.

3.1.2 Land Preparation -The Making of Rice Field Dike
The paddy fields owned by farmers in Kena group are made to be high enough and strong enough to hold the water. The height of the dike should be between 25-40 cm, depending on the height of the water surface. The width of the base dike is not less than 50 cm, while the width of the top dike is only 25 cm. The making of dike does not use the materials derived from plants, because this material is rotten as a result, it caused leaks in the dike. The dike is made from soil that is compacted by stepping on it until it forms the expected result. The holes along the dikes are patched with soil to avoid seepage from the fields. If the hole is too large, the dike around the hole should be dismantled first and then rebuilt.

3.1.3 Land Preparation -The Making of Canal
The making of canal is intended to protect fish from pest attacks, such as birds, snakes or weasels. The dangers of drought are often caused by high seawater evaporation and increased water temperatures due to the heat of the sun. This canal is made across or parallel to the dike. Its width is around 50 cm with a depth of not less than 30 cm. During the farming, the water in the canal should be controlled.

3.1.4 Land Preparation -The Making of Drainage
The drainage is made with the aim of regulating the water level in the rice fields so that it is not too deficient or too excessive. Drainages for the water-entry and water-exit are made from bamboo or pipes that are planted in rice fields. There are two drainages made by the farmers, one serves to drain water contained in the canal hence it will facilitate fishing at the time of harvest. While the other one functions to regulate the desired water level. Farmers only need one water-entry-drainage and it must be located

higher than the water-exit-drainage, so that the water that has been flowed does not flow back. To prevent the entry of wild fish or rubbish to the farming fish, it is better to install a filter made from bamboo or wire netting on the drainages.

3.1.5 Land Preparation -The Making of Reservoir
The reservoir is useful for accommodating fish during the harvesting so that fish can be easily caught. This shelter is made around the drainage channel. The size of the shelter depends on the available rice fields so that it can accommodate all the fish that are farmed. The reservoir is deeper than the canal, so when the canal is dry the reservoir is still filled with water to hold the fish.

3.1.6 Land Cultivation
The cultivation on the Kena's group mina paddy land provides a good medium for supporting the growth of rice plants and fish food organisms. First, the land is hijacked 20 cm deep. Second, the water is drained so the soil becomes slightly muddy. Third, the urea fertilizer is sprinkled evenly across the ground with a dose of 100-200 kilograms for each hectare. After the seeds are planted, then the water is poured back until the surface of the water reaches 20 cm in height, and left it alone for 4-7 days to provide an opportunity for fish food organisms to grow. Finally, fish seeds are ready to be spread based on a certain density.

3.1.7 Fish Seeds' Density
To get the satisfactory results, the fish seeds that are going to be spread should meet the following requirements:
- Fish seeds do not have body striking-colors, because they can attract the attention of predatory animals
- Omnivorous fish
- Seeds come from superior, healthy and preferred types of fish
- Able to live in shallow water and heat-resistant

Seeds density at the time of spreading the size 1-3 cm is 40-60 thousand tails/hectare, with the total 4 weeks of farming. To get a larger fish seeds, farming is carried out until the second weeding. The spreading density reaches 30-50 thousand tails/hectare. The spreading of fish to be consumed: (a) For 3-5 cm size, the density is 2000 tails/hectare with 60 days of farming and 40 grams/fish will be obtained; (b) For sizes of 5-8 cm, the density is 1000-2000 tails/hectare with 50 days farming, 40 grams/fish will be obtained; (c) For sizes 8-11 cm, the densitiy is 1,000-1500 tails/hectare with a farming period takes 50 days and 60 grams /fish will be obtained.

3.1.8 How to Farm Fish with Plants: Fish Farming as the Alternation of Rice Plants
Growing fish as an alternation is done while the farmers work on the field where the paddy will be planted. The length of the farming is adjusted to the rice planting plan, usually takes 30-40 days. After the soil has been processed, then it is fertilized and irrigated to a 5-10 cm surface height. Next, after being left alone for 4-7 days, then the fish seeds are ready to be spread. The seeds are 1-3 cm (aged 20 days-1 month) with density levels ranging from 100-120 thousand tails/hectare. The farmers can also spread the fish seeds that have reached a size of 5-7 cm with a density ranging from 50-60 thousand tails per hectare. The harvest is done after the fish seeds have been kept for a month reaching a length of between 8-12 cm. The height of the water level in the plot of rice fields should not be too low. For the 1-3 cm fish seeds, 5 cm of water level is already enough.

3.1.9 Fish farming with Plants
The cultivation of fish with plants should be adjusted to the growth of rice plants. In one planting several kinds of fish farming can be carried out based on the farming purpose, whether for seed size only or for the size of fish to be consumed. There are two forms of seed farming, namely: 1) the dispersal of 1-3 cm fish seeds, carried out 5 days after the rice is planted and harvested during the first weeding of rice. The spreading density is 40-60 thousand tails/ hectare. The length of farming for 24 days, will producing

3-5 cm fish seeds; 2) the spreading of 1 cm size of fish seeds is carried out 5 days after the rice is planted and harvested during the second weeding. The spreading density is 10-15 thousand tails/hectare. At the time of the first weeding, the farmers tried to keep the troughs at the bottom of ponds or rice fields with water, so that the fish continued to be cultivated until the second weeding. On this second weeding, the 5-8 cm fish seeds will be obtained.

The farming of fish for the purpose consumption starts by spreading as much as 3000 fish seeds (2-3 cm in size)/hectare. The dispersal is done 5 to 7 days after the rice is planted. For fish seeds that are 5-8 cm in size, they are spread around 1000-1500 fish/hectare. The spreading is done after the first weeding. For fish seeds that have size from 8-12 cm, they are spread around 1000 fish/hectare. The spreading is carried out after the second weeding and harvested during the rice harvest with a farming period up to 50 days.

3.1.10 Rice Harvest Period and How to Harvest Fish

The farmers in Kena group state that the length of the rice harvest period is 3 months whereas the length of the fish harvest period is 4 to 5 months. How to harvest the fish by using mina paddy technique are mentioned below:

- The filter is installed on the drainages. The water-entry-drainage is closed and the water-exit-drainage is opened, so that the water level drops.
- The fish are collected in the canals and then the water is lowered again until the water stays in the ditch.
- The fish in the canals are taken to the reservoir and ready to be caught by using scoop-net and placed in the clean water.

3.1.11 Advantages and Disadvantages of Mina Paddy Technique

Minapaddy has several advantages [6-9]:

- Increasing the income of paddy farmers who experience the crop failure because the attack from planthopper pests due to climate change. The presence of fish in this field will help the farmer because the fish will consume planthopper pests that fall into the water.
- Increasing the rice production about 10-20% and at the same time also increasing fish production at least 1 ton/ha for each planting season.
- Helping to accelerate environmental improvement because the minapaddysystem will reduce methane gas which is discarded from the remaining fertilization.
- Saving the government's budget on fertilizer subsidiesbecause with mina paddy technique the fertilizer can be reduced up to 30%.
- Increasing fish consumption in order to improve family nutrition. It can be seen that the data of fish consumption in the main-rice-producing province has the lowest per capita fish consumption (18-23 kg/capita while the national average is 30 kg/capita).
- Developing the growth of rural fish industries as well as rice industries.
- Increasingthe family income due to two types of production at once.
- Monitoring field becomes one of the most important things to do because the farmers are required to check the flow of water that irrigate their fields every day and check the filters that exist so that the fish in the fields are not easily released.
- Improving soil structure because fish often eat by flicking through the mud.
- Fish will help eat small animals that act as the pests of paddy.
- Reducing the dependency on the imported-meat because the farmers can produce fish with a cheaper price than meat.
- Food security that has been supported by rice will be reduced and at the same time the excess rice produced will be able to fill the world's food needs.
- Increasing the potential of existing paddy fields.
- Increasing the diversity of agricultural products, besides being able to produce organic rice,it can also produce fish.
- Increasing income due to both rice and fish businesses.

- Harvestingthe good quality of organic rice, because the use of chemicalsalready replaced by the fish dirtwhich acted as organic fertilizer.
- Harvesting fish by saving the feed, because fish can eat Azolla and moss in the fields as an extra feed.
- Increasing fish production with pond size and water level.
- Helping to reduce pest and penile attacks.
- Breeding fish can be done based on the age of paddy plant, when the rice is ready to be harvested, the fish also ready to be consumed.

According to Anwar (2012), the disadvantage of the minapaddy system is because of the excessive pesticides are given to paddy itself. It affects the fish life in the pond and pests such as snakes, frogs, and birds can be easily entered the rice fields.

3.1.12 Mina Paddy on Salt Water by Kena Group

Saltwater intrusion is the movement of salt water into freshwater that can contaminate groundwater' sources. On the paddy field of farmers in Kena group, the salt water intrusion can occur naturally because the location of the field is not too far from the coastal sea, thus it is a certain degree on most coastal aquifers due to the hydraulic connection between groundwater and sea water. Since salt water has a higher mineral content than freshwater, seawater has a higher density and greater water pressure. Therefore, salt water moves towards freshwater. The activity of mina paddy farmers, especially in pumping the groundwater from coastal aquifers can improve sea water intrusion because groundwater pressure decreases and becomes relatively smaller than the pressure from sea water. Another cause of saltwater intrusion is the navigation and drainage canal which creates a gap for sea water to move ashore across the surface and through tides. Sea water intrusion can also occur in extreme weather conditions such as storms and big waves.

Agriculture done by farmers in Kena group is located close to the shore; the groundwater flows from deeper field meets with salty-groundwater from the sea. Freshwater flows due to high differences. Since salt water has a higher level of dissolved minerals, seawater is denser than fresh water, thus it has a higher hydraulic height than fresh water. The hydraulic height refers to the pressure of the liquid shown in the water column; columns that have a higher hydraulic height will move to the water column which has a lower hydraulic height if both columns are connected.

3.2. *The Activity of Modern Agricultural Technique Mini Paddy*

Income can be interpreted as wages obtained by a person for the business or activity he/she does. The size of the income depends on the type of activities and capabilities of each individual, education also can be one of the factors between one's income and others. In general, income is the output from business. According to Skousen and Stice (2010: 161) "Revenue is an inflow or settlement (or a combination of both) from delivering or producing goods, providing services or carrying out other activities which are activities it self".

3.2.1 Venture Capital and Farmers' Expenditure in Using Mina Paddy

The venture capital and other expenses for mina paddy technique is commonly range from Rp. 2.000.000, - Rp. 4.000.000, -. It means that it is the capital spent by farmers in working on the agricultural sector after being summed up for both of rice and fish. The expenditures of Kena group are slightly different compared to other groups. It happens because the size of lands cultivated by each farmer are different. From the explanation above, it is known that in the activity of Mina paddy, the venture capital is higher than the expenditure.

3.2.2 Venture Capital and Farmers' Expenditure in Using Mina Paddy

Based on the data from this study, the total sales of rice production in one harvest range from Rp. 13.000.000, - to Rp. Rp. 15.000.000, - with 3 months of harvest period. For fish production, most of the respondents answered that the total of fish production in one harvest (4-5 months) up to Rp. 4.000.000, - Rp. 5.000.000.

3.3 Activity of Mina Paddy and Income of Farmers "Kena" Group

The study of mina paddy technique on salt water used by a group of farmers, called Kena, has been implemented since 2013 in West Matang Raya, Baktiya subdistrict. The ownership status of the land where mina paddy agricultural activity implemented is not the farmers' own, as a consequence, they use a profit-sharing system.

The capital donation such as the availability of water sources that flow through agricultural land from the irrigation, the pumping equipment and other expenditure costs in implementing the mina paddy activity are obtained from the farmers themselves. This technique requires up to 6 workers. Unfortunately, the farmers' problems in applying mina paddy technique are that the agriculture in their village does not use electricity, they also never get help in getting seeds, fertilizers, pesticides and fish feed—they use their own money to provide all of the equipment needed. However, they often get counseling about mina paddy agriculture from government agencies and several private institutions as well as the support from the agricultural service, the fisheries services, the irrigation services and the Development Planning Agency at Sub-National Level (BAPPEDA) of North Aceh.

Kena group explains that by using this technique, the rice-harvest period was 3 months, while the fish-harvest period took 4 to 5 months. The total of rice production in one harvest is 1.750 kg up to 2.750 kg, while for fish production is 150kg - 200 kg. The average income from the sale of rice production in one harvest amounting to Rp. 14.000.000, - whereas the income for fish production in one harvest amounting to Rp. 4.950.000, -. If both of the income is added, then the productivity of rice and fish once harvested within 4 months is Rp. 18.950.000, - then it can be said that the income of the farmers' family reaches an average of Rp. 4.737.500, - per month. Moreover, the mina paddy technique used by this Kena group is also supported by the fisheries services, the irrigation services and the Development Planning Agency at Sub-National Level (BAPPEDA) of North Aceh.

3.4 Discussion

Modern agricultural technique, mina paddy, on salt water is an activity done by a group of farmers, named Kena, in West Matang Raya, Baktiya subdistrict, North Aceh regency. This activity turned into a business that has been running since 2013 until today. Mina paddy is one of the techniques where the rice planting and fish farmingcarried out simultaneously in one land. The purpose of this business is to increase the productivity of agricultural products so that farmers' income will increase. Besides planting the rice, the products from fish farming are also very useful to fulfill the food needs and nutritional value of the community who live there as well as can reduce the risk of harvest-failure in an agricultural business. Farmers who apply the mina paddy technique show that this is a breakthrough to increase the productivity in agricultural. Through this technique, the process of fertilizing and farming fish in salt water can be harvested after 3 months 5 days. As a result, the farmers in West Matang Raya village receive an average income from rice production in a single harvest as much as IDR 14.000.000, - while the average sale of fish production in one harvest is IDR 4.950.000, -.

Mina paddy agriculture is one of the new strategies implemented by the farmers from the monoculture system to the agricultural diversification system. The mina paddy cultivation business is an integrated business that can increase the productivity of paddy fields, gain more income for farmers' prosperity and keep the existence of food security for the future.

4. Conclusion and Suggestion

Based on the results of the study that have been conducted, it can be concluded:

1. The people in West Matang Raya village are mostly works as farmers. Mina paddy farmers get higher profits/income compared to ordinary farmers. The income from the sale of rice production in one harvest amounting to Rp. 14.000.000, - whereas the income for fish production in one harvest amounting to Rp. 4.950.000, -. If both of the income is added, then the productivity of rice and fish

IOP Conf. Series: Materials Science and Engineering **807** (2020) 012004 doi:10.1088/1757-899X/807/1/012004

once harvested within 4 months is Rp. 18.950.000, - then it can be said that the income of the farmers' family reaches an average of Rp. 4.737.500, - per month.

2. The products from mina paddy's technique in the West Matang Raya indeed brings the improvement for the farmer in Kena group, it can be seen from the higher incomes obtained from both rice and fish business, which are mutually beneficial. Caterpillars in rice roots that usually hold up the growth of rice now becomes the fish food. The farmers do not bother with the fish feeds and the paddy can grow better without the interfere from caterpillars. The integrated-mina paddy technique can reduce the costs that farmers have to spend and do not require many workers because it can be managed simultaneously.

As a breakthrough in developing the agricultural technique, it is suggested that the government can encourage the farmers to do the mina paddy business and help them by providing the fertilizer, seed, pesticide and feed. Therefore, the farmers will have no difficulties and get the motivation to do the mina paddy business which can improve their prosperity and ensure their food security for the future, especially for the farmers in Kena group, West Matang Raya.

References

[1] Anonim, 1985. P*etunjuk Budidaya Ikan di Sawah, Proyek Peningkatan Produksi Perikanan*. Jawa Barat; UPP BudidayaAir Tawar.
[2] Rahim. 2007 "Ekonomika Pertanian (Pengantar, Teori, dan Kasus). Jakarta : *Penerbit Penebar Swadaya*.
[3] Montazeri, M. (2012). Inovasi Teknologi Minapadi Dalam Mengurangi Pemanasan Global.
[4] Sudjana. 2005. *Metode Statistika*. Bandung.
[5] Arikunto,Suharsimi.2006. *Prosedur Penelitian*. Jakarta; Rineka Cipta.
[6] Soekartawi,dkk. 1998. *Ilmu Usaha Tani dan Penelitian untuk Pengembangan Petani Kecil* . Jakarta; Universitas Indonesia.
[7] Suntanto., R. 2002. *Penerapan Pertanian Organik*. Yogyakarta: Kanisius.
[8] https://id.wikipedia.org/wiki/Intrusi_air_asin, diakses; kamis 7 februari 2017
[9] Kementrian Kelautan dan Perikanan Republik Indonesia. 2016. *Petunjuk Teknis Sarana Budidaya Minapadi*. Jakarta.

IOP Conf. Series: Materials Science and Engineering **807** (2020) 012005 doi:10.1088/1757-899X/807/1/012005

Factors Influencing The Greenhouse Gas Emission Disclosure on Manufacturing Firms in Indonesia

Erna Listyaningsih[1], Natalina[2]

[1] Economics Faculty of Universitas Malahayati
[2] Engineering Faculty of Universitas Malahayati

Corresponding e-mail: ernatya04@yahoo.com

Abstract. Greenhouse gas (GHG) emission disclosure study is interesting to be investigated in recent years. Nevertheless, the results are still varying. This research investigated the influence of firm size, Return on Asset (ROA), leverage and environmental performance on GHG emission disclosures of listed manufacturing firms of Indonesia Stocks Exchange (IDX). Multiple regression methodology employed in this study. Sample used was 36 firms that consisted of manufacturing companies in the ranking of PROPER and IDX during 2015-2017. It was found that firm size and environmental performance had an effect on the GHG emission disclosure while ROA and leverage did not have effect. Furthermore, large companies tend to carry out information on GHG emissions disclosures voluntarily than the small one. It indicates that the ranking of PROPER motivated large firms to give information of GHG emission. This finding supported the previous studies in which firm size influenced the GHG emission disclosure.

1. Introduction

The issue of climate change and its effects has caused emerging the new environment rule namely the alleviation of the GHG around the world such as carbon pricing [1]. Previously, it was made the convention in Kyoto, Japan in 1997 which is known as The Kyoto Protocol which is an international community to alleviation the GHG emission [2]. Additionally, through the Kyoto Protocol, the international community has recognize the possible high cost associated with reducing GHG emission and as a result, allowed ratifying countries to use flexible mechanisms including emission trading, joint implementation, clean development mechanism, and joint fulfillment [3]. The other research found that the companies which in countries under the GHG law have higher emission disclosure levels [4].

Indonesia is one of the countries that has ratified the Kyoto Protocol certainly has consequences for implementing the contents contained in the protocol. However, Indonesia is including in developing countries therefore they are not required to reduce emissions but can do it voluntarily and are asked to implement sustainable development that is cleaner and more climate friendly. The Protocol Kyoto addressed on obligation of firm to admit, quantify, note, provide and expose carbon emission [5].

Figure 1 presents that electricity and heat production is 25% of 2010 global GHG emission. Electricity and heat production have the biggest contribution on GHG emission, the second and third is agriculture, forestry & other land use and industry respectively. It indicates that sector such as electricity, agriculture and industry support the bigger GHG emission than the other sector. Additionally, size of manufacturing firms are also influence to disclose the GHG emission [6].

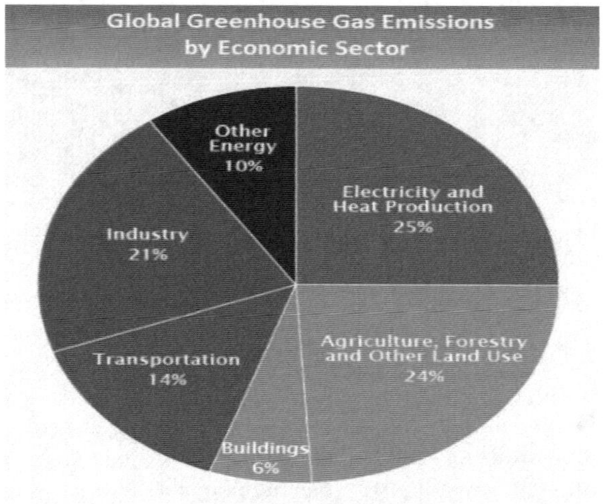

Figure 1. Global GHG Emission by Economis Sector

Previous studies have investigated the factors which influenced on GHG gas emission [1,5–13]. However, the findings are still varying. Those factors which influenced the GHG emission disclosure was such as business performance (profitability, leverage), size of firm, industry sector, media exposure, corporate governance and environmental performance. While other study investigated the effect of foreign direct investment inflows, economic development, and energy consumption on GHG emissions [14]. Therefore, this study attempts to investigate the factors which influencing the GHG emission disclosure by using manufacturing firms which is listed in Indonesia Stock Exchange (IDX). Furthermore, The firm that is not good in performance will take less emission alleviation, however to quantify the performance itself is not easy due to no single measurement clearly [10]. The GHG emission disclosure is one of the actions to gain the legitimacy from stakeholder[6,11,15]. In the context of Indonesia, a firm with a high ranking companies based on PROPER classification which is compiled by the Ministry of Environmental of Republic Indonesia have a good environmental performance. With this assessment, the company tries to inform its environmental performance in this term is GHG emission disclosure to gain stakeholder support and legitimacy of the community.

The objectives of this study is to investigate whether size of firm, business performance namely return on assets (ROA) and leverage as performance measurement, and environmental performance which is measured from the ranking list of PROPER influence GHG emission disclosure. In addition, to measure the extent of GHG emission disclosure, this study employed the Disclosure Index adopted from Choi et al. (2013). This study conducted multiple regressions for analysis data. To the authors' best knowledge, previous studies on this issue in Indonesia only focused on the effect of environmental performance and GHG emission disclosure on firm value, while the other just measured the extend of GHG emission disclosure without including environmental performance. Therefore, this study conducted by adding environmental performance which is employing PROPER proxy toward GHG emission disclosure. The results of this study could provide the latest empirical insight regarding factors which influence GHG emission disclosure on manufacturing firms which is listed in IDX which would be useful for organization and investors.

2. Methodology

This study used purposive sampling method to collect the data. The data year used was 2015-2017. All data retrieved from IDX and sustainability report. To measure GHG emission disclosure, this study employed the Choi index on Carbon Disclosure [5,12].

This study used multiple regression to investigate the influence of factors to GHG emission disclosure. To measure the GHG disclosure, the index methodology is conducted by using some GHG reporting frameworks [9].

3. Literature Review

3.1 Legitimacy Theory
According to legitimacy theory, firm makes sure that stakeholders look what they have done was legitimate [15]. The GHG emission disclosure is one of the actions to gain the legitimacy [6,11,15]. However, the other study found that environmental legitimacy significantly negatively influences the likelihood of corporate carbon disclosure [16].

3.2 GHG emission disclosure
In the context of Indonesia, the disclosure of GHG is non-mandatory. However, firm which conducting the GHG emission disclosure will obtain usefulness such as legitimacy from investors, reducing operating cost, fines and penalty [15].

Furthermore, larger firms tend to conduct voluntary GHG emission disclosure to decrease the potential risks [13]. In addition, they tend to present more credible the GHG emission disclosure [6]. Furthermore, profitability and leverage influenced the GHG emission disclosure, in where unprofitable firms using the GHG emission disclosure for legitimacy goal, and the more leverage used from creditor the bigger possibility voluntary disclosure [6,13].

The PROPER rating also becomes signal that even though they play a role for changes in extreme weather, however they have made various efforts which is in accordance with the norm and tested by the independent party or government through the program PROPER to restore the environment so as to provide support for company in gaining legitimacy from the surrounding community until finally the business they are not hindered by environmental issues [17].

If the ownership of firm is hold broadly, then managers tend to disclose information regarding GHG emission voluntarily [11,15]. Therefore, this study posits the hypothesis. The hypothesis is , there are influence of firm size, ROA, leverage, and ranking of PROPER on GHG emission disclosure

4. Results

4.1. The influences of size, ROA, Leverage and environmental performance on GHG disclosure
Table 1 shows that the biggest mean is firm size, while the lowest mean is dummy gold. Additionally, the biggest standard deviation is ROA and the lowest one is dummy gold.

Table 1. The Descriptive Statistics

Variables	Mean	Deviation Standard
GHGDisc	4.565	3.703
Firm Size	15.316	1.509
ROA	7.321	10.251
Leverage	0.460	0.192
DummyGold	0.02	0.141
DummyGreen	0.10	0.303
DummyBlue	0.78	0.418

Table 2 shows that the firm size, dummy gold and dummy green have effect on the GHG emission disclosure while others do not have effect on the GHG emission disclosure, in where firm size has significance at 1% level, dummy gold and dummy green have significance at 5% levels respectively. In addition, the value of adjusted R square is 28.6%. It means that the contribution of firm size, ROA, leverage and environmental performance on GHG emission disclosure is 28.6 % and the remaining explained by other variables beyond those variables. This result support the previous study in which firm size influenced the GHG emission disclosure [6,13]. Additionally, regarding dummy gold and dummy green are significant at 5% level on GHG emission disclosure, this findings support the previous study about legitimate theory [6,11,15].

Table 2. Multiple Linear Regression Test

Variables	β Coefficient	T	p-value	Adjusted-R square
Firm Size	0.981	4.356	***0.000	0.286
ROA	0.037	1.135	0.259	
Leverage	2.660	1.612	0.110	
Dummy Gold	5.127	2.083	**0.040	
Dummy Green	3.104	2.076	**0.041	
Dummy Blue	1.180	1.098	0.275	

5. Discussion

This research found that firm size and environmental performance influence the GHG emission disclosures. The company can provide guarantees with the existence of environmental information for the public that even though they contribute in climate change, but they have also minimized it. Big firms tend to do so due to they have been able to. However, there were a number of companies that have good PROPER ratings, apparently not publishing GHG disclosure information. It can be explained that it is possible due to the company has limitation resources to participate in PROPER because there are various requirements from two categories the main thing that must be owned by the company, namely the criteria for obedience and aspect assessment environmental management [17]. Nevertheless, this study supported the previous study in which there were influence firm size and environmental performance on the GHG emission disclosure [6,13].

6. Conclusion

This study found that firm size and environmental performance influenced on the GHG disclosure, while other independent variables did not influence on the GHG disclosure. These findings indicate that the firm that has big size and has a good ranking as well tends to conduct disclosure information due to they are able to do so. These findings supported the previous studies that firm size affected the GFG emission disclosure. Additionally, environmental performance which is measured from the ranking list of PROPER affected the GHG emission disclosure as well in which this study employed dummy variables for environmental performance.

References

[1] Borghei-Ghomi Z and Leung P 2013 An Empirical Analysis of the Determinants of Greenhouse Gas Voluntary Disclosure in Australia *Account. Financ. Res.* **2** 110–27
[2] Setyaningsrum W 2015 Analisis Yuridis Implementasi Protokol Kyoto Di Indonesia Sebagai Negara Berkembang *J. Komun. Huk.* **1** 185–98
[3] Donehower J 2008 Analyzing carbon emissions trading: a potential cost efficient mechanism to reduce carbon emissions *Environ. Law.* **38** 177–208
[4] Reid E M and Toffel M W 2009 Responding To Public And Private Politics : Corporate Disclosure Of Climate Change Strategies *Strateg. Manag. J.* **1178** 1157–78
[5] Irwhantoko I and Basuki B 2016 Carbon Emission Disclosure: Studi pada Perusahaan Manufaktur Indonesia *J. Akunt. dan Keuang.* **18** 92–104
[6] Rankin M, Windsor C and Wahyuni D 2010 An investigation of voluntary corporate greenhouse gas emissions reporting in a market governance system *Accounting, Audit. Account. J.* **24** 1–36

[7] Hermawan A, Aisyah I S, Gunardi A and Putri W Y 2018 Going Green : Determinants of Carbon Emission Disclosure in Manufacturing Companies in Indonesia **8** 55–61
[8] Jannah R and Muid D 2014 Analisis Faktor-Faktor Yang Mempengaruhi Carbon Emision Disclosure Pada Perusahaan Di Indonesia (Studi Empiris pada Perusahaan yang Terdaftar di Bursa Efek Indonesia Periode 2010-2012) *Diponegoro J. Account.* **3** 2337–3806
[9] Tauringana V and Chithambo L 2015 The effect of DEFRA guidance on greenhouse gas disclosure *Br. Account. Rev.* **47** 425–44
[10] Broadstock D C, Collins A, Hunt L C and Vergos K 2018 Voluntary disclosure, greenhouse gas emissions and business performance: Assessing the first decade of reporting *Br. Account. Rev.* **50** 48–59
[11] Kılıç M and Kuzey C 2018 The effect of corporate governance on carbon emission disclosures: Evidence from Turkey *Int. J. Clim. Chang. Strateg. Manag.*
[12] Bahari N A S, Alrazi B and Husin N M 2016 A Comparative Analysis of Carbon Reporting by Electricity Generating Companies in China, India, and Japan *Procedia Econ. Financ.* **35** 74–81
[13] Halimah N P and Yanto H 2018 Determinant of Carbon Emission Disclosure at Mining Companies Listed in Indonesia Stock Exchange *ICE-BEES 2018* vol 2018 pp 127–41
[14] Sarkodie S A and Strezov V 2019 Science of the Total Environment Effect of foreign direct investments , economic development and energy consumption on greenhouse gas emissions in developing countries *Sci. Total Environ.* **646** 862–71
[15] Berthelot S and Robert A-M 2011 Climate Change Disclosures: An Examination of Canadian Oil and Gas Firms *Issues Soc. Environ. Account.* **5** 106
[16] Li D, Huang M, Ren S, Chen X and Ning L 2018 Environmental Legitimacy, Green Innovation, and Corporate Carbon Disclosure: Evidence from CDP China 100 *J. Bus. Ethics* **150** 1089–104
[17] Anggraeni D Y 2015 Pengungkapan Emisi Gas Rumah Kaca, Kinerja Lingkungan, Dan Nilai Perusahaan *J. Akunt. dan Keuang. Indones.* **12** 188–209

Optimization of Fuel Savers Preheater Water System (PWS) for Diesel Fuel

A Pranoto[1*] and A Purwanto[1]
[1]Mechanical Engineering Department, Institut Sains & Teknologi AKPRIND, Yogyakarta, 55222, Indonesia

Corresponding author: aji_pranoto@akprind.ac.id

Abstract. The present work aims to design and investigate the performance of diesel fuel heating system called Preheater Water System (PWS). The PWS is made of stainless-steel container, copper spiral pipe and electric heater. The container has a diameter of 140 mm and height of 170 mm. The tube has a diameter of 8.5 mm. This system heats the diesel fuel before entering the fuel injector pump. The system utilizes heat from the heat from radiator cooling water and also uses additional electrical heating. The system is equipped with temperature sensor to measure the diesel fuel entering the fuel injector pump. The results show that at engine speed of 800 rpm (idle condition), temperature of the fuel increases 55°C, fuel consumption is 26% more efficient. At medium speed of 1200 rpm, the temperature increases up to 55°C and fuel consumption is 15% less. Meanwhile at the engine speed of 2200 rpm, the temperature rises to 60°C and fuel consumption increases up to 15%. In can be conclude that the PWS works well and the optimum temperature of the diesel fuel is 50°C.

1. Introduction

1.1. Combustion Process and Effect of Fuel Temperature on Diesel Engines

Many previous ways have been done as an effort to improve diesel engine combustion, such as fuel heating method, booster methods, magnetic and current power method. The advantages of those methods are engine performance increases, fuel consumption and damage risk reduce. This method able to prevent the engine overheating, over vibration, over noise and the most severe result in the engine breaking [1-3]. Another alternative that can be applied to all vehicles is by increasing the temperature of the fuel before be sprayed into the combustion chamber. From the working process of the engine, 30% of the heat produced by combustion is absorbed by the engine cooling system and wasted to the atmospheric air [4]. In the water-cooling system, it is known that the cooling water temperature is maintained between 83-93°C. While the temperature of diesel fuel accommodated in the tank is close to the temperature of the outside air which is around 20°C. Thus, the heat of engine cooling water can be used to raise the temperature of the fuel. This technology is useful for increasing the temperature of the fuel before being injected into the combustion chamber. Increasing the heat on the fuel is expected to facilitate the fuel evaporation process so that it makes the fuel easier to reach the flash point, in turns the combustion process getting better. The better the combustion process, the higher the energy produced and the less fuel is consumed [5, 6].

Figure 1 shows the combustion process in a diesel engine and the effect of temperature on spontaneous combustion. The combustion process of the diesel engine itself is divided into four periods which is delayed burning (AB), flame propagation (BC), direct combustion (CD), and continued combustion (DE). From gas equation of state, it is known that the higher the pressure, the higher the temperature produced. The diesel engine combustion occurs better at higher pressure and temperature.

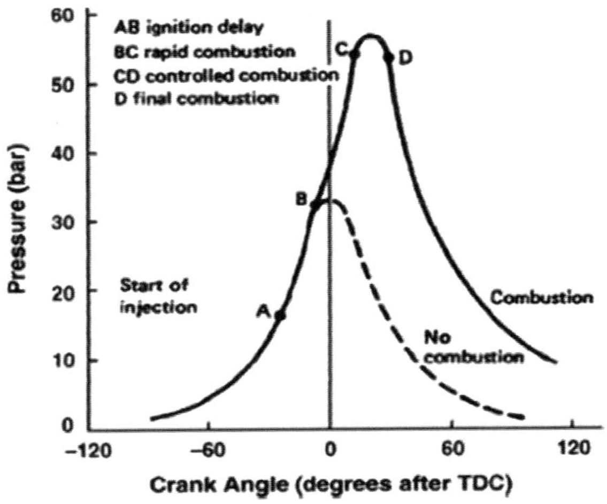

Figure 1. Diesel Burning Diagram [7]

Figure 2 indicates an effect of temperature on the power output of the diesel engine. Several problems often encountered in diesel engines are improper the amount of fuel in the combustion chamber with the load, poor injection process, and poor mixing between the fuel and air which causes the incomplete combustion process. To overcome these problems, the size of fuel spray into the combustion chamber have to be very tiny in order to obtain a more homogeneous mixture [8, 9].

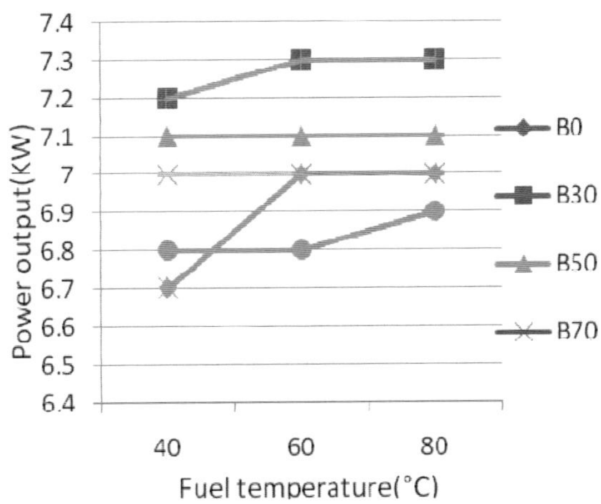

Figure 2. Effect of Temperature on Power Output [10]

1.2. Fuel and Diesel Heating Systems

Figure 3 shows the flow of in-line diesel fuel system. The systems consist of fuel tank, feed pump, filter, injection pump, high pressure line and nozzle where the tank is used as a reservoir or fuel provider to meet fuel needs while the engine is working, the feed pump, functioned to suck fuel from the tank then pump it into the filter to filter fuel from impurities contained in the fuel so as to prevent the occurrence of blockages in the fuel system channel, while injection pump, used for creating high-pressure fuel which is flowed towards the nozzle through the high pressure line to be sprayed into the combustion chamber by converting the pressurized fuel from the injection pump into fuel mist using nozzle.

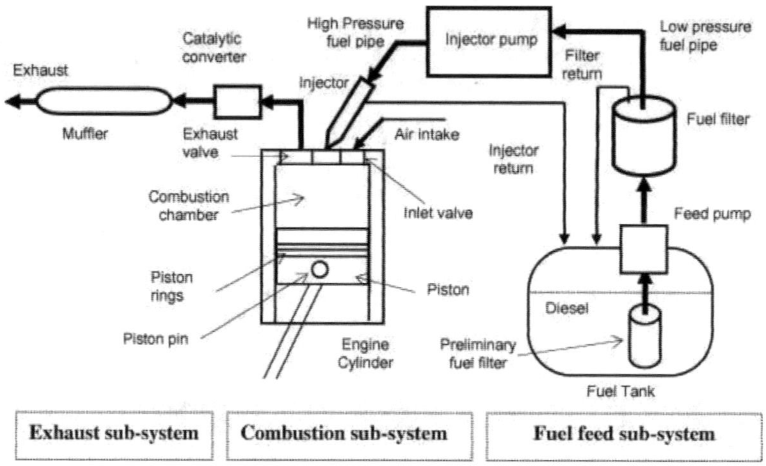

Figure 3. Fuel System of Diesel Engine [11]

Stages of combustion of diesel engines can be divide into 3 stages, namely heating, evaporation, and burning. The temperature of the fuel temperature affects the diameter of the spray. The smallest the diameter has the fastest ignition delay times. The ignition delay time are influenced by the size of the fuel spray size and by the fuel jet velocity and specific gravity of the fuel. Meanwhile, spontaneous combustion in diesel engines is influenced by fuel temperature and pressure in the cylinder. This is in line with that spontaneous combustion is influenced by cetane numbers, effective pressure, also influenced by fuel traction.

Previous work which comparing the effect of solar and biodiesel temperature on the performance of a constant-speed direct injection diesel engines concluded that the increase in temperature of biodiesel and diesel fuel resulted in 8% fuel consumption, 4% engine power, and diesel engine thermal efficiency of 23.7% [12]. Meanwhile other work on the performance of diesel motors with the addition of solar heaters shows that the changes in fuel diesel temperature injected in the combustion chamber affected the torque and power by 4.1%, fuel consumption by 7%, thermal efficiency by 23.4% [13, 14]. However, those two works still have the problem which the solar heaters produced are still less effective to produce the temperature needed in each condition, because it is very dependent on the heat of the radiator to the heater. To overcome the lack of the systems, an effective and efficient diesel fuel heating system has to be designed.

1.3. Heat transfer

Heat is transferred from high temperatures to lower temperatures. The heat transfer rate will depend on the mode of transfer (conduction, convection, radiation) and the heat transfer medium.

a. Conduction heat transfer [15]

$$q = kA\frac{\Delta T}{L} \tag{1}$$

Where:
q = Conduction heat transfer rate (W)
k = thermal conductivity (W /mK)
A = cross-sectional area (m^2)
ΔT = Temperature difference (^{o}K)
L = Distance between hot and cold temperature (m)

b. Convective heat transfer [15].

$$q_c = hA\Delta T \quad (2)$$

Where:
q_c = Convective heat transfer rate (W)
h = Convection coefficient (W / $m^{2o}K$)
A = Heat transfer contact area (m^2)
ΔT = Temperature difference (^oK)

Fuel viscosity decreases as fuel temperature increases which causes the fuel flash point is also getting lower. So that it can make the fuel volatile and flammable. Diesel fuel that has been heated is sprayed into the pressurized and high temperature combustion chamber, result in better and faster combustion. Better combustion leads to increase the engine's efficiency.

1.4. Novelty and Purpose
The present work aims to design an effective and a low-cost Diesel Fuel Preheating System using the heat from radiator's heat via cooling water of the radiator. The system is called Preheating Water System (PWS). No works on utilization of heat of radiator for heating the diesel fuel have been reported so far.

2. Material and Method
Figure 4 displays the flow diagram of the present work.

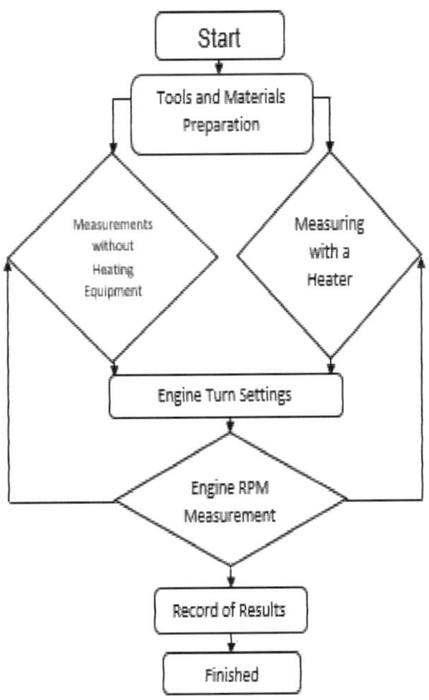

Figure 4. Flow diagram of the present work.

Meanwhile, Figure 5 shows the schematic diagram of the experimental setup of the preheater water system (PWS). The nozzle on the diesel fuel system will work if there is a pressure from the fuel injection pump both in-line models and distributors. Fine grains and whether or not spraying (injecting) the nozzle into the combustion chamber is influenced by the quality of the fuel.

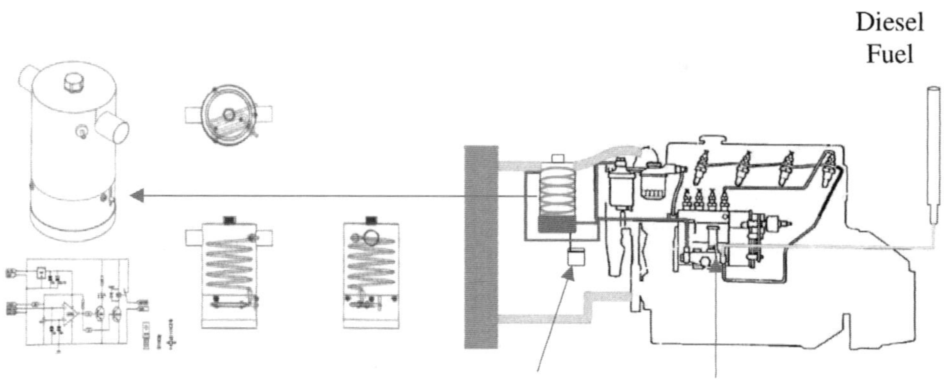

Figure 5. Installation and design of PWS tools.

The PWS is installed before the injection pump. Fuel is transfer to the injection pump via copper tube through the cooling water Stainless Steel container. Heat from radiator is transfer to the fuel via radiator cooling water. As a result the fuel that enters the injection pump will become hot, so that the viscosity will decrease as a result of the fuel spray size through the nozzle decreases and homogeneous.

Figure 6 shows the design and a photograph of the PWS. Heating container is made of stainless steel. The reason for using stainless steel is due to its corrosive resistant and has high heat conductivity which allows to maintain at high temperatures. Stainless steel used is 0.5 mm thick.

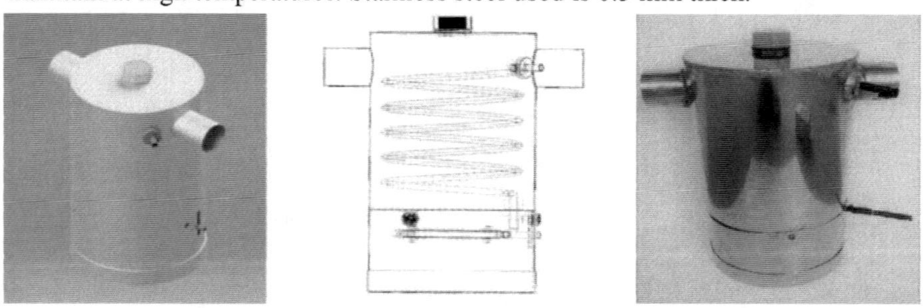

Figure 6. Heating Container

Whereas, the spiral copper tube is shown in Figure 7. Copper is selected due to its high heat conductivity, Heat transfer rate between the hot water to the fuel is high. This heat is used to heat diesel fuel that will be used for combustion. The spiral copper tube has a diameter of 6.5 mm and a length of 750 mm.

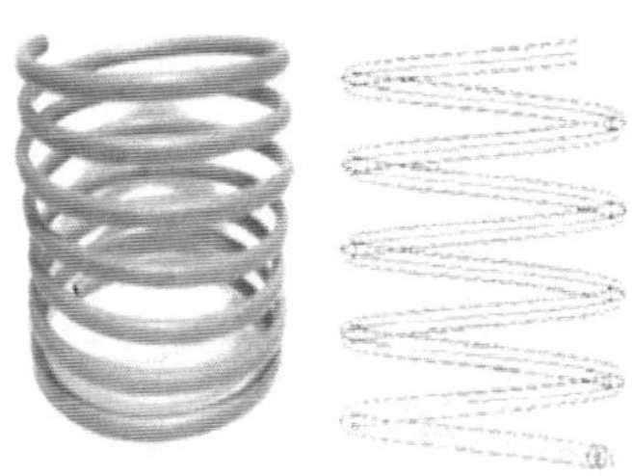

Figure 7. Spiral Copper Tube

Electric heater tube is also made of Stainless Steel. The use of the electric heater aims to prevent the heat sensor readings from disturbance of air temperatures outside the specified system. The dimensions of this electric heater tube are 50 mm in height and 140 mm in diameter. This dimension is accordance to the size of the PTC heater which is used as an electric heater when the system temperature condition has not yet worked.

Figure 8. Electric Heaters and Trolleys

3. Result and Discussion

Figure 9 to Figure 11 presents the results of the present work. The graph show that a very significant difference between diesel engines without and using a fuel heater (PWS) is observed.

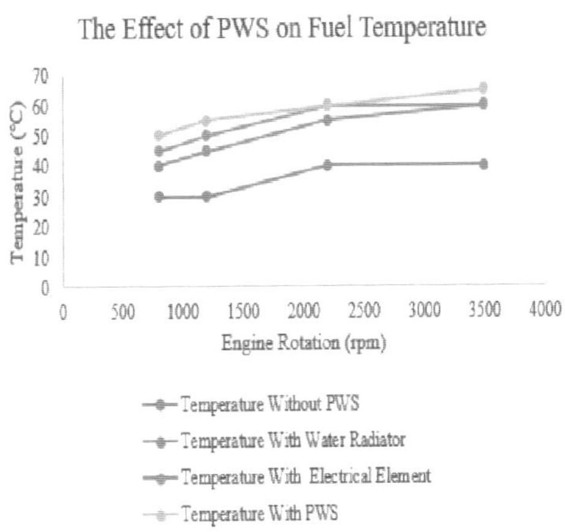

Figure 9. The structure with bolted joints.

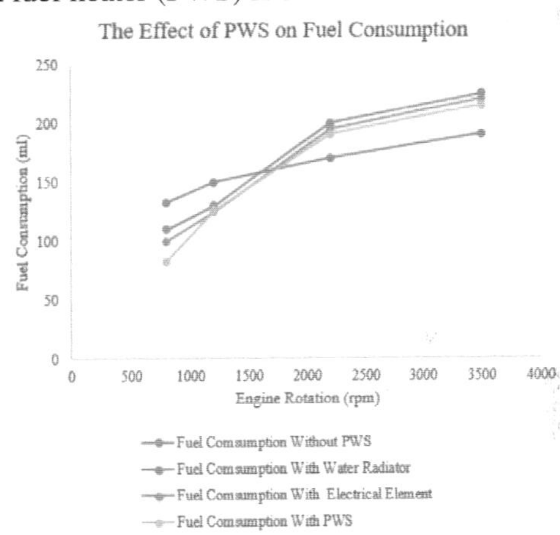

Figure 10. Cross-section of the bolted joints.

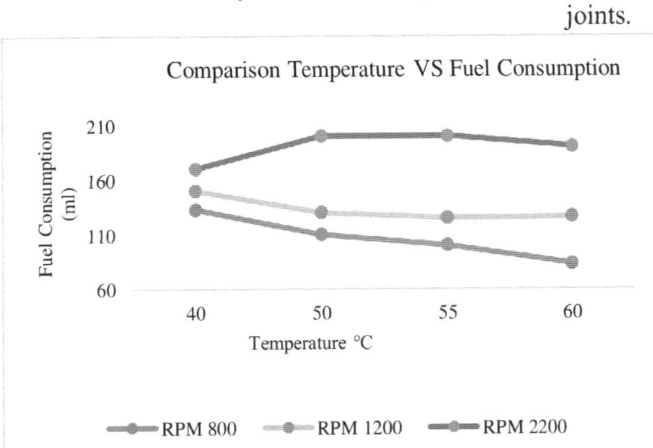

Figure 11. The Effect of Temperature of Diesel Fuel on Fuel Consumption

This fuel heater is a device to increase the temperature of diesel fuel. With higher diesel fuel temperatures, better combustion occurs. This can minimize fuel consumption, because with less diesel fuel and in one attempt to get the same amount of energy. From the results of the experimental data, it can be seen that the PWS able to save the fuel consumption. This can be known by the following calculation [15];

$$E = \frac{C_{start} - C_{finish}}{C_{start}} x100 \quad (3)$$

Where: E = Efficiency (%)
C = Fuel consumption

The data shows that there is a relation between fuel temperature and viscosity and also fuel flash points. If the fuel temperature increases, the viscosity of the fuel will decrease where the fuel flash point is reached. For high fuel viscosity, the combustion process will be more difficult. Hence, by increasing the temperature of the fuel, the fuel viscosity decreases that make the fuel combustible [16-17]. As a result of the heated fuel has been sprayed into a pressurized and high temperature combustion chamber, it will produce better and faster combustion.

Increasing the amount of fuel efficiency in each engine RPM condition is influenced by the movement of the piston, where the greater the RPM, the higher the fuel injection [18]. With faster flow, fuel circulation is also faster, this affects the difference in fuel temperature, so the fuel increases at higher temperature. The system proven to produce higher power and less fuel consumption. The system able to safe a fuel 9-16%.

4. Conclusion

The design of the Preheating Water System in this study can be made easily and work well. The system has two chambers, namely heaters with water radiators and controlled electric heating elements. In can be concluded that the PWS works well and the optimum temperature of the diesel fuel is 50°C. At engine speed of 800 rpm (idle condition), temperature of the fuel increases 55°C, fuel consumption is 26% more efficient. At medium speed of 1200 rpm, the temperature increases up to 55°C and fuel consumption is 15% less. Meanwhile at the engine speed of 2200 rpm, the temperature rises to 60°C and fuel consumption increases up to 15%.

Acknowledgements

The authors would like to thank profusely to Institut Sains & Teknologi AKPRIND that has given funding of the conference activities, and to all member of Mechanical Engineering Department for all of their support during the project.

References

[1] Shuaijun Wang Chentao Liu and Yao Zhou 2017 Study on Waste Heat Utilization Device of High-Temperature Freshwater in the Modern Marine Diesel Engine *IOP Conf. Series: Earth and Environmental Science 111 (2018) 012009*

[2] S Pinzi MD Redel-Macías DE Leiva-Candia JA Soriano MP Dorado 2017 Influence of ethanol/diesel fuel and propanol/diesel fuel blends over exhaust and noise emissions *Energy Procedia 142 849–854*

[3] Stephen Busch, Kan Zha, Paul C. Miles, Alok Warey, Francesco Pesce, Richard Peterson and Alberto Vassallo 2015 Experimental and Numerical Investigations of Close-Coupled Pilot Injections to Reduce Combustion Noise in a Small-Bore Diesel Engine *SAE International Journal of Engines Vol. 8, No. 2 (April 2015), pp. 660-678.*

[4] Roonak Daghigh Abdellah Shafieian 2016 An investigation of heat recovery of submarine diesel engines for combined cooling, heating and power systems *Energy Conversion and*

Management Volume 108, 15 January 2016, Pages 50-59

[5] B Challen R Baranescu 1999 *Diesel Engine Reference Book* Butterworth Heineman New Delhi 263-304.

[6] Murat Karabektas, Gokhan Ergen, Can Hasimoglu, Ahmet Murcak 2016 Performance and Emission Characteristics of a Diesel Engine Fuelled with Emulsified Biodiesel-Diesel Fuel Blends *International Journal of Automotive Engineering and Technologies Vol. 5, Issue 4, pp. 176–185,* Academic @ Paper ISSN 2146-9067

[7] H. K. Imdadul H. H. Masjuki M. A. Kalam N. W. M. Zulkifli M. M. Rashed H. K. Rashedul I. M. Monirula and M. H. Mosarofa 2015 A comprehensive review on the assessment of fuel additive effects on combustion behavior in CI engine fuelled with diesel biodiesel blends *RSA Advance* doi: 10.1039/C5RA09563H

[8] V. Gnanamoorthia G. Devaradjaneb 2015 Effect of compression ratio on the performance, combustion and emission of DI diesel engine fueled with ethanol–Diesel blend *Journal of the Energy Institute Volume 88, Issue 1, February 2015, Pages 19-26*

[9] Marzena Dzida Piotr Prusakiewicz 2008 The effect of temperature and pressure on the physicochemical properties of petroleum diesel oil and biodiesel fuel *Journal Fuel Volume 87, Issues 10–11, August 2008, Pages 1941-1948* doi.org/10.1016/j.fuel.2007.10.010

[10] Pushpendra Kr. Singh Rathore M.P. Singh Sugandha Agnihotri 2013 Performance Analysis of 4-stroke Diesel Engine by using Mahua Methyl Ester (Bio Diesel) and its Blends with Diesel Fuel International *Journal of Systems, Algorithms & Applications Volume 3, Issue ICRASE13, May 2013, ISSN Online: 2277-2677*

[11] Jung, D., Dennis N., & Assanis 2001 *Multi-Zone DI Diesel Spray Combustion Model for Cycle Simulation Studies of Engine Performance and Emissions* The University of Michigan 2001-01-1246, 1-23.

[12] O.M I Nwafor 2003 The effect of elevated fuel inlet temperature on performance of diesel engine running on neat vegetable oil at constant speed conditions *Renewable Energy Volume 28, Issue 2, February 2003, Pages 171-181* doi.org/10.1016/S0960-1481(02)00032-0

[13] Jaime Gimeno Pedro Martí-Aldaraví Marcos Carreres Jesús E Peraza 2018 Effect of the nozzle holder on injected fuel temperature for experimental test rigs and its influence on diesel sprays *International Journal of Engine Research Volume: 19 issue: 3, page(s): 374-389*

[14] M. Nandeesh C. R. Rajashekar N. R. Banapurmath 2018 Effect of Injection Timing on the Utilization of B20 Blends of Dairy Scum Oil Methyl Esters (Dsome) Fulled Diesel Engine *European Journal of Sustainable Development Research, 2 (2)*

[15] Peter von Böckh Thomas Wet elz 2012 Heat Transfer Basics and Practice *Springer Heidelberg Dordrecht London New York* ISBN 978-3-642-19182-4 e-ISBN 978-3-642-19183-1 doi 10.1007/978-3-642-19183-1

[16] K Esmail, T Meisam, D Ayhan, A Mortaza 2017 Impacts of additives on performance and emission characteristics of diesel engines during steady state operation *Progress in Energy and Combustion Science Volume 59, March 2017, Pages 32-78*

[17] SamsudinAnis Galuh, Nur Budiandono 2019 Investigation of the effects of preheating temperature of biodiesel-diesel fuel blends on spray characteristics and injection pump performances *Renewable Energy Volume 140, September 2019, Pages 274-280*

[18] Y. Devarajan, D. Munuswamy, S. Radhakrishnan, A. Mahalingam, and B. Nagappan, Experimental Testing and Evaluation of Neat Biodiesel and Heptanol Blends in Diesel Engine *Journal of Testing and Evaluation 47, no. 2 (2019): 987-997.* doi.org/10.1520/JTE20170307

IOP Conf. Series: Materials Science and Engineering **807** (2020) 012007 doi:10.1088/1757-899X/807/1/012007

Development of a Household Solar Power Plant: System Using Solar Panels

M Suyanto[1], T Rusianto[2], Subandi[3]

[1,3]Electrical Engineering Department, Institut Sains & Teknologi AKPRIND Yogyakarta 55222 Indonesia
[2]Mechanical Engineering Department, Institut Sains & Teknologi AKPPRIND Yogyakarta 55222 Indonesia

Corresponding e-mail: myanto@akprind.ac.id

Abstract. The purpose of this study was to find a model system of power generation by using solar-cells for house. The research was a realization of concern in overcoming the electricity energy crisis. Solar panels, inverters, batteries, and solar charge controllers were used to support on this study. Polycrystalline solar panels with 4 × 50Wp power were chosen to generate energy. The output voltage ranges from 14.8 to 17.5 V DC (direct current) per panel with a capacity of 2 × 70 Ah. Battery capacity of 24 volts was connected in series with an inverter of 1000 watts. The result of the study showed the voltage output of the solar panels was at maximum average of 30.2 V. The distribution of batteries charging used a solar charger controller at a voltage of 24.5 volt DC. The measurement of voltage resulted at the inverter output from 24 volt DC to converse of 220 volt AC (alternating current) without load. Voltage conditions as outputs of solar panels reach maximum voltage during the day from 10:00 am to 1:00 pm Western Indonesia Time (WIB) and it begin down until 6:00 pm. The design greatly influenced the performance of solar power plants.

1. Introduction

The energy is needed to increase with the advancement of technology year to year. Excessive use of energy can lead to energy depletion such as non-renewable such as crude oil, coal, and natural gas. Human energy needs continue to increase. Humans are trying to find alternative energy that is environmentally friendly as well as solar energy. SEC (State Electricity Company) as a provider of electricity services uses fossil fuels to drive turbine generators. Until now there are still areas that have not been served electricity by SEC. so it needs to be sought by researchers to find suitable alternative energy. Therefore researchers effort to find an alternative energy source, making a system of generating electricity from solar panels. Thus dependence on SEC electricity sources can be overcome. With this phenomenon, the researchers try to use solar energy as an energy source to produce electricity. Solar cells (photovoltaic cells) are devices that can convert solar energy into electricity. The solar panels are semiconductor devices consisting of *p-n* (positive-negative) junction diodes. Intersections direct electrons transferred from *n*-type semiconductors to *p*-type semiconductors and vice versa for polar movements [1]. Based on the explanation, electrons and polar movements produce photogeneration as shown in Figure 1.

IOP Conf. Series: Materials Science and Engineering **807** (2020) 012007 doi:10.1088/1757-899X/807/1/012007

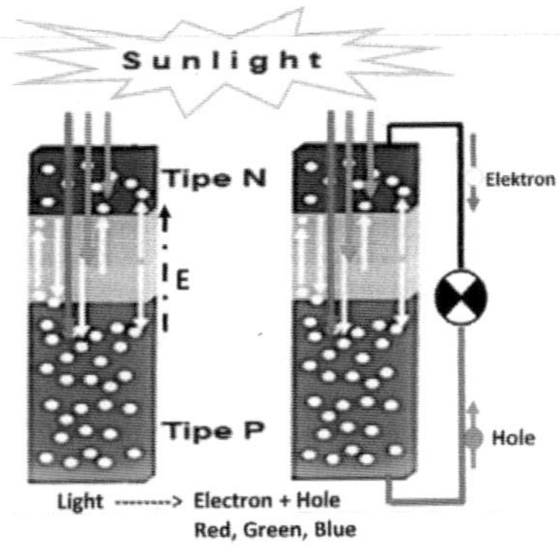

Figure 1. Photogeneration process [1].

The basic characteristics of solar panels influence parameters of solar panels or photovoltaics to produce optimal electrical energy - parameters as a reference to be used to determine the performance of solar panels in generating electricity [2]. The graph of the I-V (current-voltage) solar panel parameter is given in Figure 2.

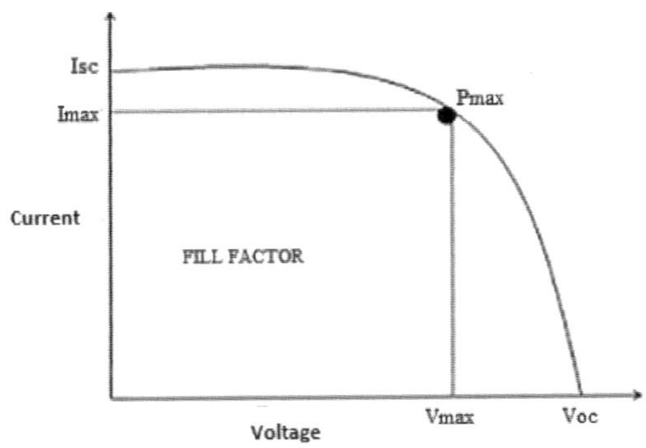

Figure 2. The relation between current (I) vs voltage (V) to get a maximum power output of a solar panel [2].

Short Circuit Current (I_{SC}) is defined as the maximum current output from a solar panel that is released without a hitch or a shortcut. On the other hand, Open Circuit Voltage (V_{OC}) is the maximum voltage that can be achieved by a solar panel without load (no current from the solar panel to load) [3]. The high temperature in solar panels, caused by waste heat generated due to absorption of solar radiation. From radiation falling to Photovoltaic (PV) panels, only 20% of the solar energy event is converted to electricity [4]. The rest of the department section is turned into heat. As a result, the accumulation of heat energy increases the operating temperature in the PV panel, due to a decrease in its electrical efficiency. PV panel conversion efficiency is reduced by approximately 0.40 to 0.50% for each increase in temperature degree [5-6].

However, all data rarely fulfil actual sun conditions because it depends on the specific climatic conditions of each location. This study looks at voltage, PV cells have an important role in controlling the output of temperature coefficients from V_{OC}, P_{max} and Fill Factor (FF) are recognized negatively, while positive for I_{SC} [7]. The maximum power point (V_{mpp} and I_{mpp}) on the I-V curve is an optimal operation where the optimal power is generated by solar panels when loaded. Maximum power output can be calculated by equation (1).

$$P_{max} = V_{oc} I_{sc} \mathrm{FF} \tag{1}$$

where P_{max} is the maximum power point (MPP), V_{oc} is the open circuit voltage, I_{sc} is the open circuit current and FF is the Fill Factor, which is a parameter that affects the maximum power of the solar panel and also the quality and can be calculated according to equation (2). The important characteristic of FF is written as equation (3).

$$\mathrm{FF} = \frac{P_{max}}{V_{oc} I_{sc}} \quad (2)$$

$$\frac{V_{mpp} I_{mpp}}{V_{oc} I_{sc}} \; FF = \frac{V_{mpp} I_{mpp}}{V_{oc} I_{sc}} \frac{P_{max}}{V_{oc} I_{sc}} \quad (3)$$

where V_{mpp} is the maximum voltage and I_{mpp} is the maximum current. The power can be calculated using equation (4).

$$\mathrm{P} = \mathrm{V} \cdot \mathrm{I} \quad (4)$$

where P is the power in Watt, V is the voltage in Volt, and I is the current in Ampere.

Solar panels are assembled devices of photovoltaic cells that convert sunlight into electricity. However, with advancements in commercial solar panel technology having a 15% efficiency, it is one of the main reasons why the solar energy industry still cannot compete with fossil fuels. Commercial solar panels rarely exceed the efficiency of 20%. Photovoltaic cells, such as batteries, produce direct current, which is generally used for small loads (electronic equipment). When DC from photovoltaic cells is used for remote home applications, then electric utilities use alternating current networks, in this case, must be converted to alternating current (AC) using an inverter, a solid-state device that converts DC power to AC [8-9].

Many solar cells system optimization efforts have utilized factors from performance solar cells. In this study, the focus is on the factors solar panels, inverters, batteries, and solar charge controllers that were used to support the power system. This study of the solar cells panels was designated at location and time/date that play a significant role in the performance of the solar cells panels. The purpose of the study is to find a model system of power generation by using solar-cells for a house and the optimum settings to improve the performance of the solar cells panel.

2. Methodology

Solar cells can convert sunlight energy directly into electricity, in this research was focused on the design of solar panels power plant for a household where the flowchart of research is shown in Figure 3. The power density emitted from the sun in the outer atmosphere is 1.373 kW/m^2 [10].

The last incident of sunlight on the earth's surface has a peak density of 1 kW/m^2 during the day in the tropics. These cells can be grouped to form a panel, as shown in Figure 4. These panels can be grouped to form a large array of solar cells. The term array is usually used to describe solar cells panels or groups of panels. Solar Cell is an active element that converts sunlight into electrical energy. Solar Cells generally have a minimum thickness of 0.3 mm, which is made of semiconductor material slices with positive poles and negative poles. The basic principle of making Solar Cells is to utilize the photovoltaic effect, which is an effect that can convert sunlight directly into electrical energy [11].

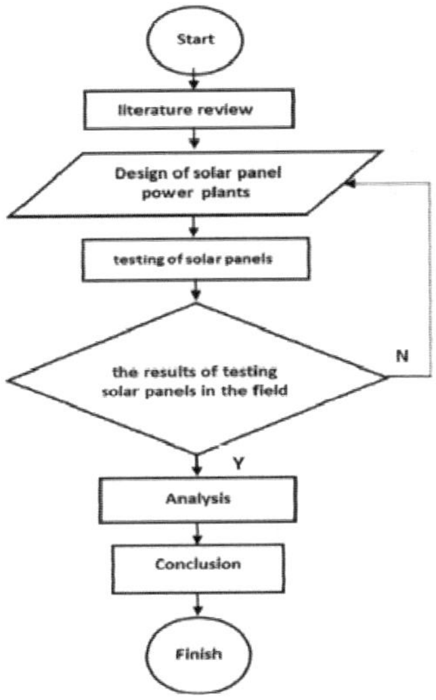

Figure 3. Research flowchart.

Figure 4. Solar cells panels..

The Solar Charge Controller is an electronic device used to adjust direct current which is charged to the battery and taken from the battery to the load [12-14], as shown in Figure 5. The solar charge controller regulates overcharging (excess charging - because the battery is 'full') and excess voltage from the solar panel. Overvoltage and charging will reduce battery life. The goal is to maintain the battery when given DC voltage supply correctly and safely for a long time. The controller fills the battery, prevents excess battery discharge, protects against excessive electricity, and displays battery status and power flow [15-16].

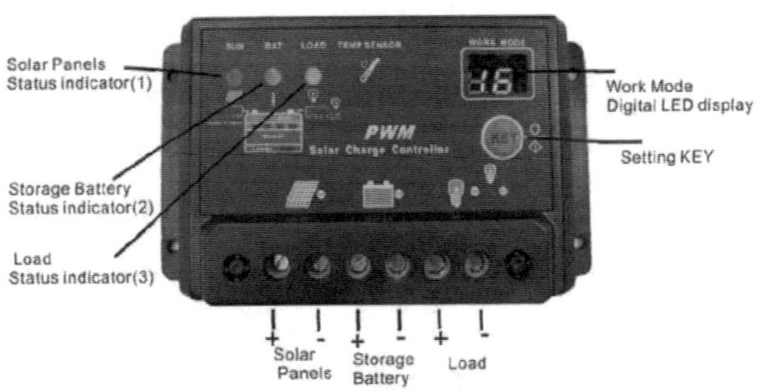

Figure 5. Solar charger controller.

Figure 6. Accumulator (battery).

Accumulators or batteries are one of the direct current (DC) source elements. The battery was used in the design is 2 x 70 Ah, including electrochemical elements. The electrochemical elements is a current source system that converts chemical energy into electrical energy.

Inverters were used in solar power plants for converting DC (direct current) to AC (alternating current) [17-19]. Inverters used in electrical devices, with a capacity of 1000 watts 24 volts, as shown in Figure 7.

Figure 7. Power inverter

2.1. Electric Motor

In this design, the load used is a 125 watt and 200 watt electric motor. The electric motor was used to operate a water pump. The specifications of the water pump motor are shown in table 1.

Table 1. Load specifications, in the form of 125 watts and 200-watt water pump motors.

Model	PWH137C	PWH236C
Electric Motor	1 phase induction	1 phase induction
Voltage Source	220 Volt AC/50 Hz	220 Volt AC/50 Hz
Power Output	125 watt	200 watt
Current input	1.55 Ampere	2.3 Ampere
Number of Poles	2	3
Suction Power	9 meter	9 meter
Maximum Flow Height	30 meter	30 meter
Maximum Water Capacity	30 litre/minute	45 litre/minute
Suction Pipe / Push	1 inci	1 inci
Size	210 x 170 x 225 mm	225 x 175 x 225 mm
Net / Gross Weight	5,3 kg/5,8 kg	7.2 kg/7.7kg

3. Results and Discussions

The results of the study show that solar cells produce a voltage by converting solar energy into electricity. The voltage produced by 17.5 Volt DC solar cells. The solar cells used four pieces the type of Polycrystalline (Poly-crystalline) with 50 WP per piece, so that it can fill the 2 × 70Ah battery connected in series.

3.1. Battery capability testing.

The current and voltage distribution of the installed solar panels produce a voltage of ±17.5 V per panel, but their distribution of charging batteries is very stable at an average of 13.5 V. The distributions are regulated by the solar charge controller (SCC). The flow will start to increase in the morning at 07.30WIB; then it will reach the maximum level at noon at 10:00-14.00 WIB, then begins to drop at 18.00 WIB. Figure 8 shows the ability of battery with AC 125-watt during the day and night.

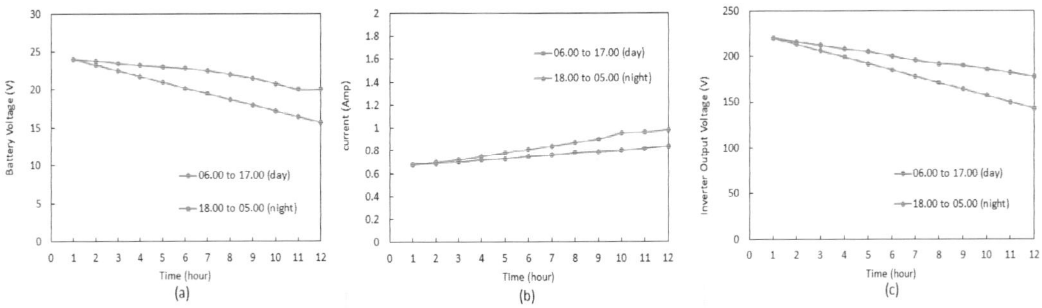

Figure 8. The ability of battery in AC 125-watt load during the day and night with measurement of (a) battery voltage (b) current of the battery and (c) Inverter output voltage.

The system designed is, electrical energy derived from solar energy sources can be used for solar power plants. Also, it can be used to drive a water pump motor, with electrical energy that has been converted through an inverter from 24.5 volts DC to 220 volts AC. The conversion of electric DC to AC could not achieve 100%, it may be lost to converse. Vinay (2015) reported, that AC power in overall the conversion DC to AC in actual field condition is an efficiency of about 82-89% [2]. The electrical energy from solar panels can be flowed and stored into batteries; the process of storing and using electricity is regulated by using a Solar Charge Controller (SCC). The stored electrical energy can be used to drive loads in the form of lighting and electric motors, through an installed inverter requiring a voltage of 220 volts AC with a frequency of 50 HZ. The ability of battery with 200-watt AC during the day and night is shown in Figure 9. At the day the measurement of voltage and electric current more than higher at the night, it corresponded to Amelia (2016). Amelia reported that high solar irradiation may increase output power [7].

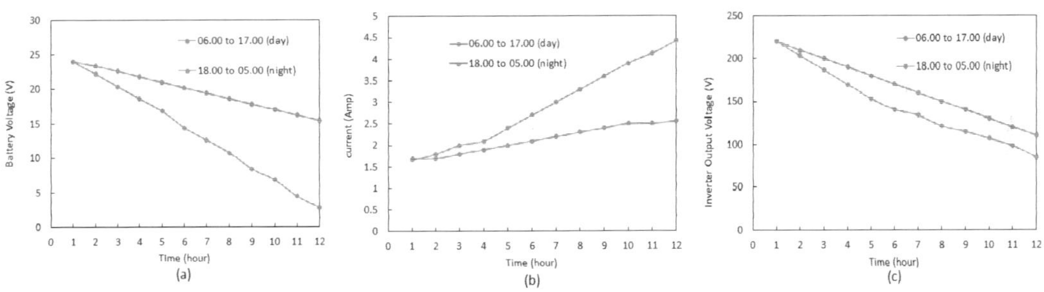

Figure 9. The ability of battery with AC 200-watt load during the day and night with measurement of (a) battery voltage (b) current of the battery and (c) Inverter output voltage.

4. Conclusion and recommendation

4.1. Conclusion

Based on the results of measurements and testing of data retrieval and the overall analysis that has been carried out, it can be concluded that:

i. The working principle of PLTS is to convert solar thermal energy into electrical energy using solar panels which are then stored in batteries.
ii. To change the 24 volt DC voltage from the battery/battery to a 220 volt AC voltage using an inverter circuit.
iii. In the current and voltage distribution of solar panel sources, the resulting ± 30.2 V is in series, but its distribution of charging the battery is very stable with a maximum average of 24.5 V where a solar charge controller regulates all distribution.

iv. The voltage and current will begin to increase in the morning at 06.00 WIB to 10.00 WIB, then will reach the maximum level at noon at 10:00 to 12:00 WIB, and begin to decrease until the afternoon.

4.2. Recommendation

The disadvantage of this Solar cells system is the lack of power efficiency in very changing weather conditions. If the weather is sunny, the filling process is very good, it all also depends on the characteristics of the type of solar cells that used.

For further development, to get a large amount of energy in the panel, it is better to install an automatic solar tracking system that can follow the direction of sunlight.

Acknowledgements

The authors would like to thank profusely to Institut Sains & Teknologi AKPRIND that has given funding of the conference activities, and to all member of the Mechanical Engineering Department for all of their support during the project.

References

[1] Suyanto M, Firman B, Jatmiko A W 2017 Micro Solar Power Plant with Capacity Of 900 Watt for Power Supply of Rural Village Information System *Int. J. Adv. Res. Technol.* **6** 182-185.

[2] Pandey V K 2015 Study on Installation and Operation of Solar Photovoltaic Power Plant as Non-Conventional Energy Resource *Int. J. Adv. Res. Technol.* 2278-0181

[3] Oliveira T P, Narvaez D I, Villalva M G 2019 Comparison of Irradiance Decomposition and Energy Production Methods in a Solar Photovoltaic System *International Journal of Energy and Power Engineering* **5** 13

[4] Duran D, Martínez I, Weber B, Rincón E, Juárez J 2013 Design Of A Mobile Photovoltaic Modulesystem For Demonstration And Experimentation *ISES Solar World Congress*, (Universidad Autónoma del Estado de México, Facultad de Ingeniería)

[5] Attou A, Massoum A, Saidi M 2016, Photovoltaic Power Control Using MPPT and Boost Converter *1st International Conference on Electrical Energy and Systems* (Djillali Liabes University of Sidi bel-Abbes)

[6] Dash P K, Gupta N C 2015 Effect of Temperature on Power Output from Different Commercially available Photovoltaic Moduls *Int. J. Eng. Res. Appl.* **5** 1

[7] Amelia A R, Irwan Y M, Leow W Z, Irwanto M, Safwati I, Zhafarina M 2016 Investigation of the Effect Temperature on Photovoltaic (PV) Panel Output Performance, *Int. J. Adv. Sci. Eng. Inf. Techno* **6** 5

[8] Barchowsky A, Parvin J P, Reed G F, Korytowski M J, Grainger B M 2012 A Comparative Study of MPPT Methods for Distributed Photovoltaic Generation *Conference Publications, Innovative Smart Grid Technologies (ISGT)*

[9] Rompicherla S M 2013 Solar Energy: The Future *International Journal of Engineering Trends and Technology (IJETT)* **4** 6

[10] Muraa P G, Baccolia R, Innamoratia R, Mariottia S 2015 Solar energy system in a small town constituted of a network of photovoltaic collectors to produce Electricity for homes and hydrogen for transport services of municipality *6th International Building Physics Conference* (Cagliari, Italy)

[11] Selvaganapathi S, Kumar A S 2014 Simulation and Analysis of Solar Cells based Boost Converter fed Electric Drives *Aust. J. Basic Appl. Sci.* **8(13)** 516-521

[12] Kolluru V R, Narne R, Patjoshi R K, Varghese G T 2017 Implementation of a novel P&O MPPT Controller for Photovoltaic System at Standard Test Conditions *International Journal of Applied Engineering Research* **12** 1

[13] Rahaman M A, Matin M A, Sarker A, Uddin M R 2015 A Cost Effective Solar Charge Controller *Int. J. Adv. Res. Technol* **04** 314

[14] Singh A K, Agrawal A K, Vohra S, Thakur S S, Patel G 2017 Solar charge controller *Int. J. Acad. Res. Dev.* **2(6)** 994-1001
[15] Al-Asadi S Q A, Kazem H A 2013 Feasibility Study of Photovoltaic/Wind/Battery Hybrid System for Oman *Engineering Science and Technology: An International Journal* **3** 3
[16] Xia S, Wu X, Zhang Z, Cui Y, Liu W 2019 Practical Challenges and Future, Perspectives of All-Solid-State Lithium-Metal Batteries *Chem.***5(4)** 753-785
[17] France R M, Geisz J F, Garcıa I, Steiner M A, McMahon W E 2015 Quadruple-Junction Inverted Metamorphic Concentrator Devices *IEEE Journal of Photovoltaics* **5 (1)** 432 - 43
[18] Najafi E, Yatim A H M 2012 Design and Implementation of a New Multilevel Inverter Topology *IEEE Trans. Ind. Electron.* **59(11)** 4148-4154
[19] Stala R 2011 Application of balancing circuit for dc-link voltages balance in a single-phase diode-clamped inverter with two three-level legs *IEEE Trans. Ind. Electron.* **58(9)** 4185-4195

IOP Conf. Series: Materials Science and Engineering **807** (2020) 012008 doi:10.1088/1757-899X/807/1/012008

Modification Compression Strength of Test Machine by Using Control System on Compaction Energy

Lusmeilia Afriani[1]*, Idharmahadi Adha[1], Yan Juansyah[2] and Rina Febrina[2]

[1] Department of Civil Engineering, Engineering Faculty, University of Lampung, Indonesia
[2] Department of Civil Engineering, Engineering Faculty, University of Malahayati, Indonesia

*Corresponding e-mail: lusmeilia.afriani@eng.unila.ac.id

Abstract. Soil compaction for specific construction is determined soil density tests. Several methods have been developed by previous researches and the prominent method is soil density. Several types of soil compaction method in the field, namely compaction method using vibration machine, impact, and pressure. In laboratories, soil compaction test was performed by using a hammer which uses human strength to solidify soil. An experimental investigation was conducted to explore a new modification related to compression strength test machine with energy compaction control system control system the energy of compaction. Researchers tried to design and make a soil compactor with a different mould measures from the standard Proctor method to modified method. The special of this new tool is that it can measure the compaction energy from each layer and dry density values. This research used different pressures as much as 1.67 N/mm^2 - 20 N/mm^2, every load conducted for sandy clay and silt samples. The values of dry density obtained from this research were between 1.59 kN/m3 - 1.68 kN/m3 and the value is higher compared with the value obtained from experiments using the Standard Proctor system. Meanwhile, the soil compaction results in the laboratory by using a new modification compression strength test machine with energy compaction control system can be an option of heavy equipment which can be used in the field.

1. Introduction

Since the construction cost of soil management works is quite high therefore soil play a very crucial role in supporting infrastructures such as road, dams and embankment. Moreover, the development of science in the field of civil engineering construction must follow the progress of field needs so that the infrastructures which are going to be built could have decent quality, economical value in terms of the planning and implementation of the work and all of which must meet the required qualifications. Soil compaction work becomes the second which require high costs, great deal of time and also selection for decent materials after cut and fill works.

Soil compaction is a process of soil density by reducing the distance between particles; consequently, there will be reduction in air volume and compressed soil grains. The soil would have reduction in air volume [1-2]. By applying compaction method, it is expected to obtain stable soil and not easy to get penetrated by water so that building constructions would last longer.

Research studies on soil compaction held in a laboratory with the proctor test tool which include the uses of proctor standard and proctor modified. Consequently, the results of this kind of experiment will eventually get a maximum soil density and optimum moisture content. The maximum soil density can also be achieved depending on the soil grains and soil water content. The importance of compacted soil is that it will have an impact on some soil characteristics such as the shear strength and permeability value [1-3].

Theoretically, it is much convenient to run the experiment of maximum soil compaction in a laboratory although the practicality of running the experiment in the field could be very challenging. In order to get a decent compaction, one must meet the standards of Bina Marga. The construction works would be suitable for a building in a flat area and has good hard soil; hence the implementation would not be going to be too difficult. However, when the construction works are undergoes at a non-flat area, therefore excavation and embankment works are needed.

The aim of this research is to develop a compaction tool which could be used often in soil compaction experiments in laboratories. This tool was made by modifying the proctor in the laboratory. The purpose of making this tool is not to replace the existing tools and become the standard tool in all soil mechanics laboratories. This soil compactor tool could obtain the right solution in order to observe the value of compaction energy and it could be distinguished by the amount of energy that could be used to compact soil on the field. Meanwhile as contractors or even the government, in this case the decision maker, should be able to quickly determine the use of compactor at various kinds of soil. Whereas the advantage is that it can shorten the compaction work time in the laboratory, and the energy used to condense a type of soil measured. During soil compaction tests, the energy used can be measured. For example, a soil compaction experiment is given a pressure of 5 N/mm^2 in each layer, after the end of the research it will produce its dry density value. It means that the value of dry density is obtained by using 5 N/mm^2 of pressure energy. The energy of soil compaction using the Standard Proctor method is using formula I. However, by using this new modification equipment, the amount of direct compaction energy can be obtained. Other advantages of using the modification tool is it can shorten the implementation time and energy as well as the accuracy of the results.

The process of soil compaction will increase the soil density by knowing the density between the soil particles so that it will reduce the volume of air and remove some of the groundwater. The effect of groundwater will act as a lubricant and can increase the compaction quickly when the load is placed by on the ground. It is essential because the results of experiments in the laboratory become a reference for implementation in the field.

At the beginning of the compaction process, with the fixed compaction energy, the value of soil density (γd) will increase by increasing the water content. But if the optimal water level reaches a maximum value and the water continue to flow then the dry density will gradually decrease [4]. This kind of soil conditions will approach saturated and weak soil conditions, thus, in the implementation of compaction energy or the burden given to compact the soil tends to exceed the strength of the soil and it is carrying capacity. This condition will produce excessive wheel slippage, and the curve of the wheel will damage the ground. This event often experienced by expansive soil conditions [5]. Therefore, it is essential to know the density of soil energy before engineers implement it in the field.

2. Experimental

Standard Proctor and Modified Proctor tests are performed to obtain optimal moisture content and optimal dry density. The relationship of parameters between the optimum moisture content and the maximum dry volume weight (γd_{max}) for various types of soil has been observed by several scientists and researchers in the field of soil mechanics such as Proctor [6],[7]. Until now the Proctor's tool was created by Proctor for soil compaction testing which produces both parameters.

Relationship between dry density (γd) with the weight of volume (γb) and moisture content (w), is expressed in the equation (1) [8]:

$$\gamma d = \frac{\gamma b}{1 + \omega} \quad (1)$$

Based on the Proctor test experiment, the researcher tries to complete the research by making other compaction test equipment, namely making soil compaction testing with pressure. The idea and purpose of creating a soil compactor with pressure is to convert human energy into a machine of power and make it easier to use. When the soil is compacted, the energy is used according to the research plan and has the same functions as a Proctor system. The hydraulic system uses a manometer to measure the pressure applied when the soil is under pressure and the same functions as a Proctor system. In addition, it uses a manometer to measure the pressure applied when the ground is under pressure (the pressure energy can be measured). This tool is called modelling the soil compression test by the pressure method, which is shown by Figure 1 and 2.

For this reason, a test will be conducted between the soil compression modelling test with the pressure method and comparing the standard and proctor modified test equipment to the compaction amount of the compaction energy meter.

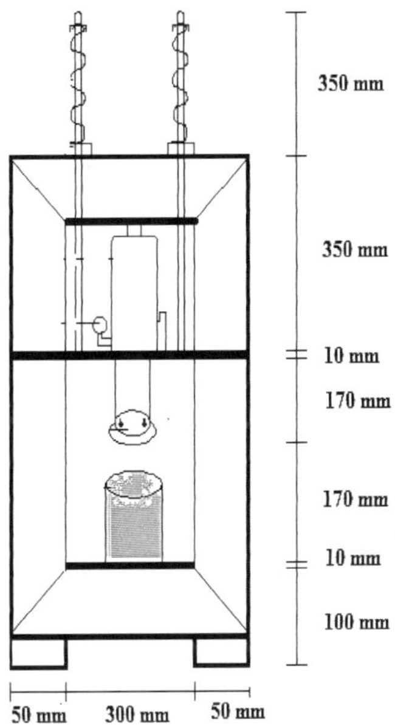

Figure 1. Prototype of modification compression test for soil compactor

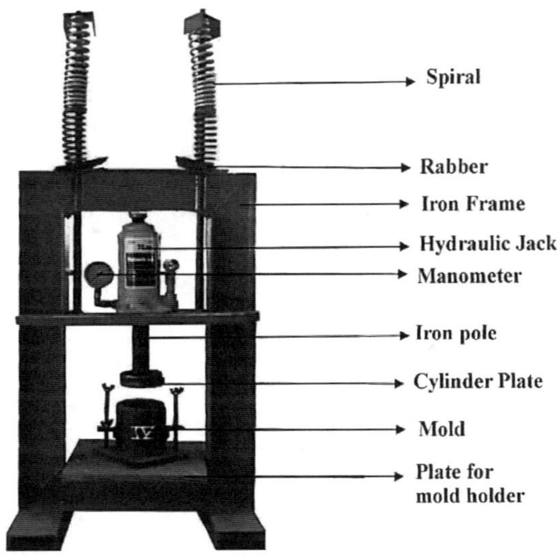

Figure 2. Modification compression testing machine .

The height of this tool is 110 cm, with a length of 40 cm, and a width of 30 cm. The size is not too big, and it can be used only for research material by using fine-grained soil samples until sandy clay soil samples.

The first test is to find the optimal moisture content and dry density using the Modified Proctor test and standard Proctor test. Optimal water content (w_{opt}) as well as maximum dry density (γd_{max}) that are found from the compacting curve, [7],[8]. After γd_{max} and w_{opt} are obtained, the optimal air content becomes the reference used in each subsequent soil test sample.

After the soil sample containing water content optimal is ready, the soil sample is then put into a mould with 10.20 cm in diameter and a height of 11.60 cm. Compaction system is divided into three layers or five layers. Each layer must be compacted by having the same energy and each layer is pressed with a present pressure. The pressure applied to each specimen is 1,67 N/mm^2 – 20 N/mm^2.

The energy required for compaction at standard compaction is formulated as in equation (2) [4],[8]:

$$E = \frac{N_b N_i W H}{V} \quad (2)$$

Notes:
E= Compaction energy (ft-lb/ft^3)
N b=Amount of pounding each layer
N i=Amount of layers
W=Weight of hammer (lb)
H=Falling height of hammer (ft)
V=Volume of mould (ft^3)

The working system of the ground press tool is to pump the hydraulic jack manually, at the end of the jack, there is an iron plate as thick as 5 mm thick. At the end of the jack is placed a 5 mm thick iron plate. When the hydraulic jack pushes the plate down, the iron plate will press the ground in the mould; then the soil becomes solid. When the plate has touched the soil and has begun to be compacted, the manometer will move, so that it can know the amount of pressure received by the ground when reading the value on the manometer. The initial test is applied testing the physical properties of the soil, for density, right before the soil is going to be tested. The data was taken from secondary data, [9].

3. Results and Discussion

The primary function of the soil layer is to support the building foundation in hence the required soil layers must be in a stable state. Thus, before building a construction, the base soil must be compacted first, adding soil to the ground from another place and then compacted is the solution for problems that often occur when the location of the building has a distant elevation difference. Soil compaction system usually has thickness of 30 cm for each layer. Additionally, soil compaction is needed so that the soil particles become solid and stable towards the load of the building structure.

Based on the background of the problem, this discussion will discuss about the physical properties of soil, density, and energy required when the soil is compacted. Sampled soils are from two different locations, the first sample was from Sukabumi Bandar Lampung and the second sample was from

Gedung Agung, South Lampung. These areas were chosen as an area of the study because the soil often used as a dam and other infrastructure works.

Soil compaction is a mechanical method for obtaining stable soil layers in support of building structural loads. It is necessary to study the amount of compaction energy since the result of soil compaction depends on the compaction energy. The amount of energy is measured by the making of a compactor model designed to conduct experiments of soil compaction. Compaction system uses a hydraulic jack as pressure or as a soil loading that will be compacted. This pressure has a function as compaction energy, while the instrument is called the soil density modification test with the control system on compaction energy.

The limitation problems in this study are sandy soil samples and clay samples. This study is limited to testing the physical characteristics of the soil which are tested by using standard proctor and modified proctor methods as well as design the testing tools for soil compaction with pressure conditions. This new pressure tool is expected to have the similar energy which is used in compaction experiment in the laboratory with compaction energy in the field.

Soil samples used in this study were from Sukabumi deposit and Gedung Agung deposit in South Lampung, Indonesia and these locations are less than 5 km apart. The reason these places were chosen was because these areas are often used for landfill of road bodies. The first research conducted was to find the value of physical properties and the value of w_{opt} and γd_{max}. [9],[10].

Based on the value of soil samples from Gedung Agung which was taken from the results of the test of the physical properties of the soil indicated that no more than 35% of material is larger than 200. It means, it is less than 35%. While for the liquid limit value (LL) was 39.42%, it means it did not exceed 40%. While, the plastic index value (PI) was 9.5%, it also did not exceed 10 %. The sieve graphic analysis can be seen in Figure 3. The results of soil analysis and plasticity index according to the AASHTO Classification System and USCS, (Standard procedures carried out in line to ASTM D698 and AASHTO T90) [7]. Meanwhile, samples from Sukabumi are categorized as sandy soil and samples from Gedung Agung soil are categorized in soil group A-2-4 sandy clay. In previous studies, both of this soil samples were used with different parameters [11],[12].

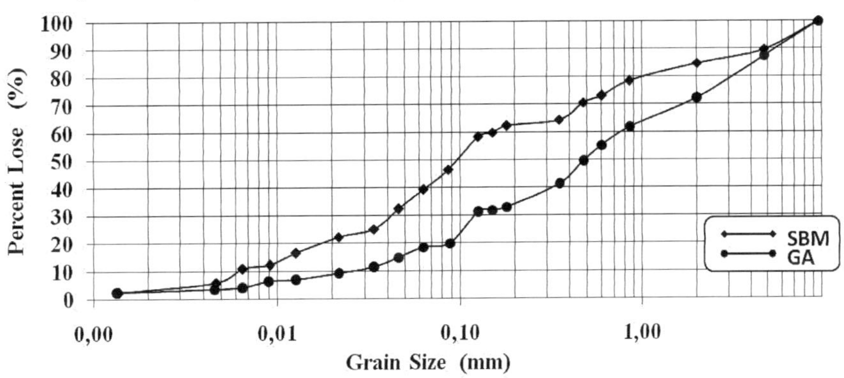

Figure 3. Graph of soil analysis [10]

The results of the study using a modified proctor for Gedung Agung soil were maximum dry volume weight of 1.70 gr/cm^3 and an optimum moisture content of 19.3%. While the energy needed for compaction according to equation (1) is equal to E = 56,250 (ft.lb/ft^3) = 2,6933,106 J/m^3 or equivalent to 2.7 N/mm^2. The amount of effort stress will be converted into pressure on the machine compactor. The compacting energy approach is for heavy equipment such as grid rollers at 6.2 MPa - 10 MPa, [12],[13]. The optimal moisture content obtained from the Proctor Modified testis used as a mixture of all samples in subsequent trials. The pressure used in this experiment using a new modification tool is 1.67 N/mm^2; 5.0 N/mm^2; 6.2 N/mm^2; 7.0 N/mm^2; 8.0 N/mm^2: 10.0 N/mm^2; 15.0 N/mm^2 and 20.0 N/mm^2. This research was also examined again by other researchers with different parameters by using the same amount of pressure and sample. From the results of the study as many as three parameters that were using a pressure test (research results using a machine compression tool modification compression testing machine) and modified proctor, the effect is observe as in Figure 4.

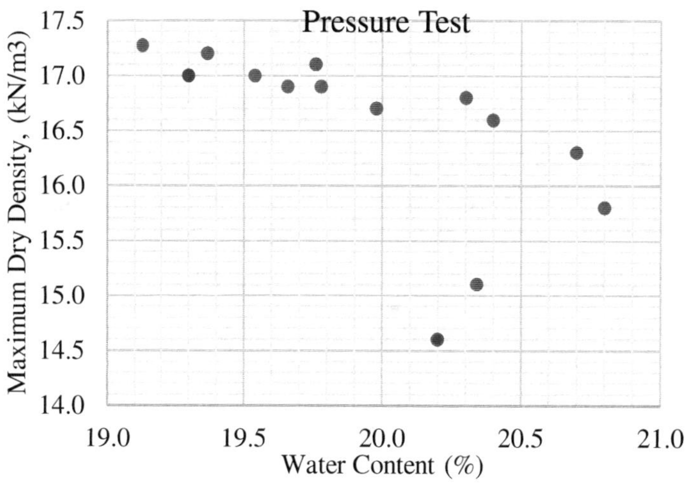

Figure 4. Water content against optimal dry density

The results of the study as seen in Figure 4 and henceforth are the optimal dry density and water content values of each experiment. Figure 4 shows that the greater the pressure for all samples, the higher the γd_{max} and the higher the water content will decrease the γd_{max} value. The higher the pressure to compact of soil, the denser the soil will be, and the water will quickly come out, even the soil becomes challenging to compact, as shown in Figure 5 and Figure 6.

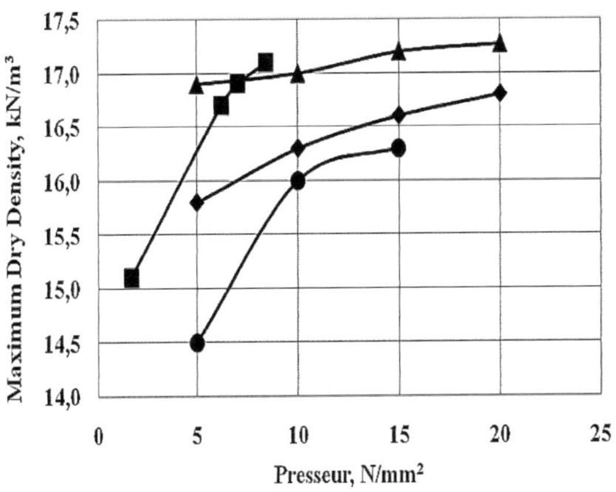

Figure 5. The value of maximum dry density against pressure

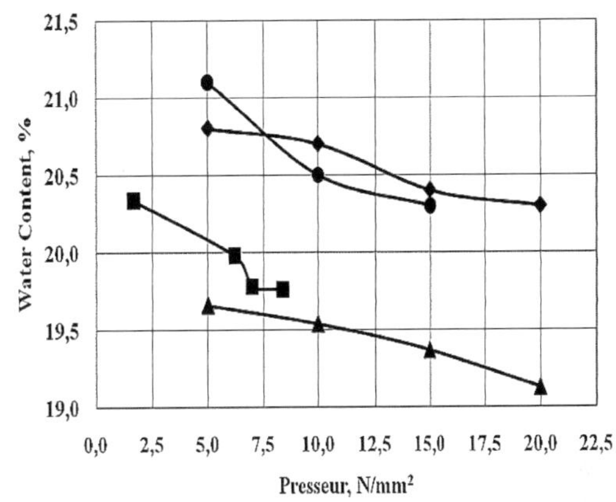

Figure 6. The value of water content against pressure

From graph in Figure 5 and Figure 6, it can be concluded that the compaction with a pressure of 1.6 N/mm^2 - 15 N/mm^2 is enough. It means that the soil density approaches the Zero Air Void (ZAV) value. In addition, the amount of water content used in this study is higher than the value of the optimal water content of the test results with modified proctor, it is still within the tolerance range of 4% - 8% of optimal water content.

Meanwhile, at a pressure of 20 N/mm^2, the soil density approaches the ZAV value, which means that the soil has no space for water and air. By converting from the optimal dry volume weight value obtained from the pressure experiment and the proctor modification (Figure 5 and Figure 6), the using of compaction energy pressure results in almost the same dry volume weight from the two tools. Therefore, the modified compactor can be used as a substitute for the modified proctor testing tool in compaction energy testing. Other researchers, such as Hilman and Rubianty [11],[12] get the same result. Testing with a pressure compaction tool using a hydraulic jack to reach a pressure of 1.67 N/mm^2 - 20 N/mm^2 then the dry density results obtained was between 14.5 kN/m^3 - 1.73 kN/m^3. This value is higher than the value obtained from experiments using the Standard Proctor system. Whereas, the values obtained for the pressure of 5 N/m^2 - 20 N/m^2 the dry density value was between 1.59 kN/m^3 - 1.68 kN/m^3. Since the soil samples were divided into several layers at the time of the research. Moreover, each layer is compacted until the mould full filled. Figure 6 shows the results of experiments were using three layers and five layers. The result is an experiment with five layers of denser compared to the experiment with three layers. It means that dry density is more considerable for five segments of denser soil.

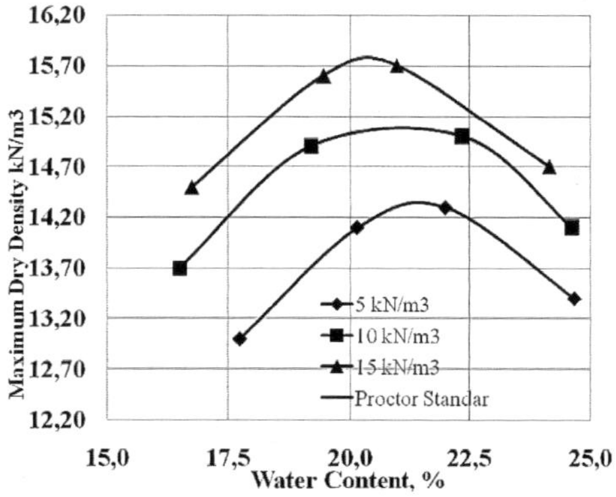

Figure 7. The results of experiments with 3 layers with a pressure between 5 N/mm^2 - 20 N/mm^2

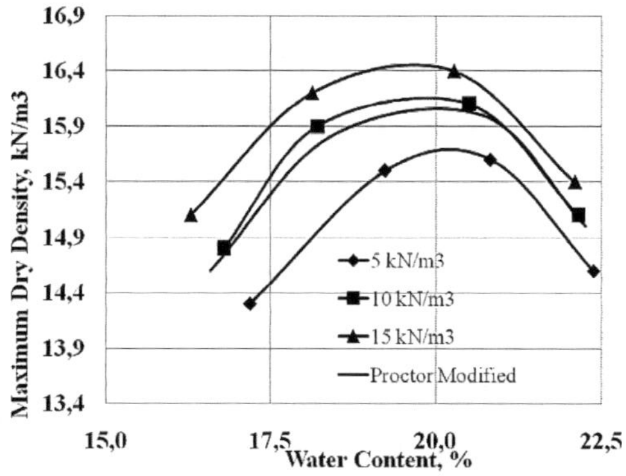

Figure 8. The results of experiments with 5 layers with a pressure between 5 N/mm^2 - 20 N/mm^2

Soil compaction work will obtain the results in which the state that compaction will reduce the void ratio and soil pores so that water is more difficult to flow into the soil. This condition is essential when the oil is used to collect water as needed for dams, dikes, and pavement structures so that it will reduce water loss. Compaction can also prevent the occurrence of high-water pressures and cause the soil to soften, shown in Figure 7 and Figure 8. It means that the importance of soil compaction is not only for compaction of road piles but also for the body of the dam.

Soil compaction is a process of pressure that is applied to the soil surface and causes the density toward the soil. Soil density occurs due to the presence of water and air coming out of the soil pores. If the area is small, it is enough to use a compacting or vibro tamper in the field. But if the area is large, the compaction machine is used. The choice of a compacting machine tool is depending on soil type and humidity conditions. The most used compactor is the sheep leg roller or vibro tamper. The speed and number of sheep leg roller trails depend on the type of soil and on the project requirements. Field compactors typically use a vibro machine for compacting small areas and this machine is suitable for compaction of all types of soil [14].

Contact pressures from sheep leg rollers are between 1400 and 7000 kPa or 6.9 N/mm^2, [11]. This study proves that the pressure value on the roller is a compacting machine almost close to the pressure on a modified compression test of 8 N/mm^2. It means that the new test equipment is a recommendation for the selection of heavy equipment in the field; this can be seen in Figure 9.

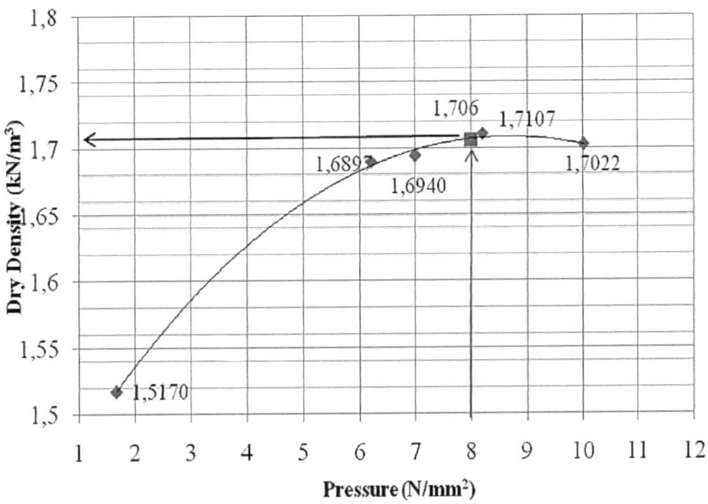

Figure 9. Relation of dry volume weight with pressure on modified compression test equipment and modified proctor test press, [10].

In the picture 7, it shows clearly the effect of compaction results with a new device and with the results using a modified proctor. That $\gamma d_{max}$of proctor modified by 17.01 kN/m^3 also can be seen from the picture showing the value of γd of 17.01 kN/m^3 could be able to be pressed with a machine pressure of 9 N/mm^2.

Compaction energy is influenced by the amount of clay content in the soil. If the compaction energy is low, the effect of the clay content has a significant impact. But when energy levels are high, the result of clay content is more striking, as found in many practical projects, this effect is less significant. When the clay content of the mixture in the soil increases, the optimum moisture content also increases [11]

In Figure 4, the higher the compaction energy it will reduce the value of soil water content. There is a supporting proof by Blotez [13]. Whereas, a research on nine types of soil had been carried out. In addition to study in the laboratory also showed using the equation (3), related the effect of compaction energy on maximum dry unit weight and optimum water content versus the logarithm of compaction energy. So, there is a linear relationship between the optimal dry unit weight and the general logarithm of compaction energy (log E) from equation (2). This verification is the result of the tests based on the observed behaviours of well-graded micaceous silty sand.

$$\gamma d_{max} = [3 - 1.23\log(\%clay)]\log E + 5.27\log(\%clay) + 10.59 \qquad (3)$$

Other researchers stated that compaction can increase shear strength, reduce voids ratio thereby reducing permeability and porosity, reducing volume changes and swelling-shrinkage. The results of the study found a change in the influence of the void ratio for the modified modifier test as high as 1.69. This is smaller than the void ratio of the Standard Proctor test, 1.85 at the same moisture content. This experiment was using compaction energy for the Standard Proctor is 594.8 KJ/m^3. While the energy compaction for Modified Proctor is 2698 kJ/m^3 [14].

Vinoid and others [16] researched the effect of compaction energy of soil with the bearing capacity of the soil (California Bearing Ratio, CBR Value). The results of the study state that when compaction energy can increase from 237 kJ/m^3 to 1197 kJ/m^3, CBR increase from 1% to 10.2%. The reason for the rise in CBR is like UCS of soil. Higher compaction energy reduces OMC, which results in flocculated soil structure with stronger inter-particle bonding so that it obtains a higher value of CBR. That compacting soils at the dry side of optimum reduce the permeability of the soil, resulting in a decrease in the swell potential of soil. It means that soil compaction energy very much needed in the process of building

Conclusions

This paper summarizes the result of laboratory analyses on all tests conducted. The first test was modelling of Modification Compression Strength Test Machine with Control System on Energy of Compactor. The result modelling the compaction method was using hydraulic jack as pressure and converted it into a very successful working tool. The models that best explain variability include compaction energy, moisture content, Atterberg limits, material passing not more than 200 sieve, and maximum dry density. The results showed that compaction energy was the main factor in determining the strength and stiffness parameters of the soil. Therefore, compaction work becomes an essential consideration during the planning stages of earthwork construction operation.

Acknowledgments

The authors would like to thank the University of Lampung for the financial support of the Research Grant through the DIPA UNILA 2017

Reference

[1] Afriani L 2014 *Soil Shear Strength* (Graha Ilmu Publisher) p 88
[2] Bowles, Joseph E. 1970.*Engineering Properties of Soils and Their Measurement* (McGraw-Hill Book) p 187
[3] Hardiyatmo H C 2002.*Soil Mechanics* (Jakarta: PT Gramedia Pustaka Utama Publisher) p 208
[4] Look B G 2007 *Handbook of Geotechnical Investigation and Design* p 356
[5] Hussain S 2017 Effect of compaction energy on engineering properties of expansive soil *J. Civil Engineering* **3** 8
[6] Craig R F 1991 *Soil Mechanics* (London: Taylor & Francis Group) p 256
[7] Das and Braja M 2010 Principles of geotechnical engineering **7** 662
[8] Afriani L, Adha I, Setyanto 2018.Innovative design of soil pressure modelling test apparatus to determine the amount of soil compaction energy to dry density value *Proc. Int. Coral Reef Sympsosium.***2** 12
[9] Gatot S B 2011 *Soil Testing in The Laboratory, Explanations and Guides* (Yogyakarta: Graha Ilmu Publisher) p 123

[10] Situmeang T P 2017 *Comparative Study of Proctor Modified with Modification Compaction Tools Using Contact Pressure of Soil Compression Equipment* Thesis of Civil Engineering University of Lampung p 78
[11] Rubianty E 2017 *Testing the Soil Compression Method Standard Proctor and Press Test Tool Modification Compactor* Thesis of Civil Engineering University of Lampung p 55
[12] Hilman F 2018 *Analysis of Soil Density in Sub grade Layers with Pressure Method Test Apparatus* Thesis of Civil Engineering University of Lampung p 62
[13] Blotez L, Benson C and Boutwell G 1998 Estimating optimum water content and maximum dry unit weight for compacted clays *J Geotechnology Engineering* **124** 9 907-12
[14] Jesmani M and Manesh N 2008 Optimum water content and maximum dry unit weight of clayey gravels at different compact efforts *Elec. J. Geotechnical Engineering* **3** 1-14
[15] Al-Khafaji S A A Effect of the different energy of compaction on subset course of roads *Saudi J. Eng. Tech.* **1** 3 86 – 91
[16] Vinod P P, Sridharan A and Soumya R J 2015 Effect of compaction energy on CBR and compaction behavior. *Proc. Inst. Civil Engineering* **168** 2 116-21.

IOP Conf. Series: Materials Science and Engineering **807** (2020) 012009 doi:10.1088/1757-899X/807/1/012009

Grain Refinement in Aluminium 1xxx Series as Effect of Vibration Torch Welding

S Huda[1], E S Arbintarso[1,*], and J Waluyo[1]
[1]Mechanical Engineering Department, Institut Sains & Teknologi AKPRIND, Yogyakarta, 55222, Indonesia

[*]Corresponding author: ellyawan@akprind.ac.id

Abstract. Grain refinement is the best way to improve the mechanical properties of fusion welds of the aluminium. The grain refinement was thereby found to be strongly dependent upon the aluminium chemical composition, where the pure aluminium has the highest grain refining efficiency. In order to increase the strength of the weld area is carried out by TIG welding with vibration on the torch which aims to obtain a smoother welded grain deposit so that its strength will increase. Theoretically, smoothing of the weld grain deposits occurs due to the welding along with with the torch vibrations arising from the peak pressure generation, the development of shear stress, the density power and the presence of laminar flow around the dendritic arm resulting in the intersection of the dendritic arms which smooths the grains. To observe the effect of the frequency and amplitude of the torch vibration in welding to refine the welded grain deposits, the research conducted with 5 Hz and 10 Hz torch vibration and torque vibration amplitude (0.25; 0.50; 0.75; 1.10; 1.60 and 2.00) mm and welding speed of 100 mm/minute and 200 mm/minute, respectively. As a comparison of the extent to which the effect or effectiveness of vibration on the torch on the weld deposit grain size produced, welding without vibration with welding speed of 100 mm/minute. Increase the welding speed will reduce grain size; on the other hand, increase the frequency have affected different grain size. Meanwhile, increase the amplitude to 0.75 mm will reduce the grain size then the grain size will unstable with increasing amplitude.

1. Introduction

The welding and connecting process is an essential part of the actual development of manufactured products. However, this process often appears as a high cost of production in the manufacturing process. Meanwhile, to obtain a welding process with low cost, and on the other hand, adequate quality is not an easy thing [1-3], some reasons that can explain this problem. First, welding is a work with many variations of process options (such as fastening, adhesive bonding, soldering, brazing, arc welding) as well as the broad disciplines needed for problem-solving (such as mechanics, material knowledge, physics, chemistry, and electronics) [3]. Second, the difficulty of welding or connection usually occurs deep into the manufacturing process, where the relative value of scrap is high enough [4]. Thirdly, a very large percentage of broken/failed products occur in welded joints because they are usually placed in high-voltage regions in an assembly, and also the weakest part of an assembly [5-6]. Careful attention to the connection process can provide economically sufficient results in the manufacturing process and can also produce reliable products.

However, in each case, the strength of the weld metal deposit is a critical part of the connection performance. The strength of welding is traditionally based on both tensile and yield strength of the welding deposit [7]. The engineers generally rely on the classification of the electrode to determine the strength of the weld/weld joint [8-10]. This classification provides some basic information on the nature

of the connection, such as yield strength and tensile strength. Although this classification is used for design purposes, the level of strength of the actual weld deposit may vary, this may, among others, be due to a modified or deviating process from a common procedure.
To obtain welding results at low cost but with adequate connection quality and strength, especially in the aluminium welding process, this study is carried out. The welding is carried out by providing vibrations at the torch in certain subsonic and amplitude frequencies, in this case, to obtain fine weld deposits. By refining of the weld deposit grains, then the strength of the weld bag of welding will be higher.

In this research, welding is done by vibrating the torch in the subsonic frequency on the aluminium work. Torch vibration is expected to occur the fracture in the secondary arm of dendrite growth, where this will produce fine weld deposits [11-13]. Observations were made by drawing a graph of the relationship between the grain size of the weld hail with the amplitude and frequency of the torch vibration at a particular welding speed. With the finest grain of the weld deposit, the hardness of the weld welding deposit will increase so that the tensile strength of the weld deposit will also increase [14]. The purpose of this research is to know the effect of frequency and amplitude of torch vibration on the large grain refinement produced in welding with the vibrated torch.

2. Experimental apparatus and procedure

2.1 Vibration Torch Welder

To conduct welding research with accompanying vibration on the torch, a tool that can support the research implementation is needed. This tool must have the ability to be controlled at the variable level set at the laboratory scale, which includes welding speed, vibration frequency, vibration amplitude, and the gap between the workpiece and the electrode (tungsten). For that purpose, a device has been made in the design and manufacture is done by trial and error. The design of the tool is not specific because, in this study, the orientation is on metallurgical analysis, not on the experimental tool itself. The tool has welding speed up to 300 mm/minute, vibration frequency and amplitude of torch are up to 15 Hz and 2 mm, respectively. The gap welding has also customized by the tool, and Figure 1 shows the schematic vibration torch welding.

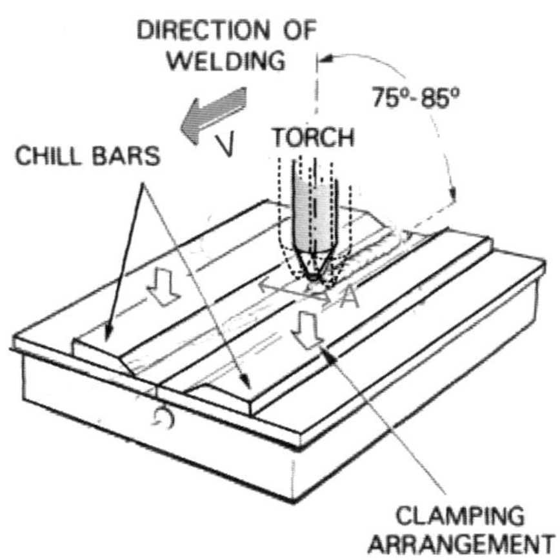

Figure 1. Schematic of vibration torch welding (adopted from [15] page 235)

2.2 Experimental Setup

The schematic diagram of the experimental setup is shown in Figure 2, consist of a vibratory torch welder which welding is done by using Tungsten Inert Gas (TIG) type Direct Current Straight Polarity (DCSP) with big current 100 A, with argon gas as a protective gas. The aluminium 1xxx series plate was used as a specimen with 3 mm thickness. The workpiece was not cut into two parts but remains intact, and then was welded in the middle, as shown in Figure 2. Welding is performed with the torch that vibrates at frequencies 5, and 10 Hz, both of the frequency test have the amplitude of 0.0; 0.25; 0.5; 0.75; 1.1; 1.6 and 2 mm, respectively. Two welding speed of 100 mm/minute and 200 mm/minute was used at all frequencies. Direction of the torch amplitude perpendicular to the direction of welding. The gap between the tungsten and the workpiece is 2 mm, with a tungsten diameter of 1/8 inch. Figure 2 also shows the performed analysis of microstructure using a microscope and ImageJ software processing.

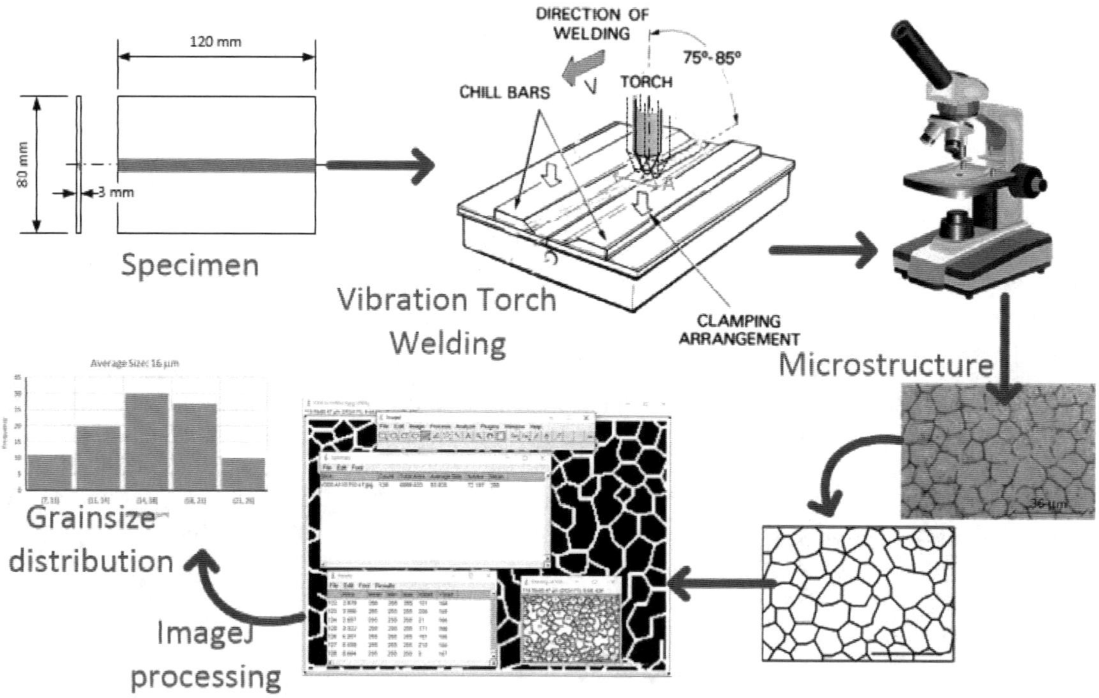

Figure 2. Schematic diagram of experimental setup.

TIG welding was used in this research because it was suitable for aluminium, and the design of the tool is simple. The welding can be done without a metal filler with a constant gap throughout the welding process. The study conducted several test variables that lead to obtaining the influence of vibration of the torch to the grain size of aluminium welding. In an attempt to obtain a larger grain of welded welder deposits, a study carried out by welding accompanied by vibrations on the torch. In this process, the main variables observed are the effect of frequency and amplitude of vibration on the torch to the smoothness of the weld deposit grain obtained. But, the other study said that the number of blowholes decreases with the decreasing frequency, and increases when the frequency is less than 15 kHz [16-17]. To observe the effectiveness of the effect of torch vibration on the refinement of the weld deposit grain, as a comparison is High-Frequency with variable welding speed, which also aims to refine the weld deposit grain. From these two studies will be obtained the answer of the effectiveness of vibration on the torch against refining weld deposit grain. Compared with the ordinary K-TIG (Keyhole TIG) weld, where the melting zone grains in High-Frequency K-TIG weld are so much refined [18]. On account of the larger welding current, the grain sizes and organizations of the welding zone in VPTIG (Variable Polarity power in TIG) welding were coarser than in DCEN (Direct Current Electrode Negative) A-TIG (Active agent TIG) welding [19]. In the welding process, HAZ (Heat Affected Zone) was subjected to higher temperatures, and thus growth of grains in HAZ was more evident than that in BM (Base Metal), and also in WM (Weld Metal) the grains become coarser with increasing TIG current and the density of the grain boundaries decreased [20-22].

3. Result and Discussion

In the results of this study, the material presented includes data and photo microstructure. The research is aimed to analyze the effect of vibration on the torch in welding to the yielded grains, where the variables used are frequency, amplitude, and welding speed. Grain size analysis was performed using ImageJ software [23-25]. Based on the microstructure images obtained, the results of grain size calculations were grouped on the influence of frequency, amplitude, and welding speed of the grain, as shown in Table 1.

Table 1. The average grain size (μm) of the vibrated torch of Aluminium 1xxx

A (mm)	V (mm/min) 100		V (mm/min) 200	
	f (Hz)		f (Hz)	
	5	10	5	10
0	36.47	20.64	33.75	36.39
0.25	28.33	29.11	19.91	17.46
0.50	18.74	19.80	15.37	18.39
0.75	16.12	10.77	13.48	9.41
1.10	16.17	17.99	17.98	13.10
1.60	27.59	26.11	16.47	20.60
2.00	20.17	19.27	13.32	16.02

The microstructure analysis was performed to confirm the grain sizes of the specimens. Figure 2 shows microstructure images of microstructure of vibrated torch welding and their distribution of grains at 0.75 mm amplitude. Figure 2 (a) shows the microstructure image with welding speed (V) 100 mm/min, and frequency (f) 5 Hz. The grain sizes spread between 2.65 to 27.94 μm and the distribution of grain size dominated by large grains. Increasing the welding speed (200 mm/min) as shown in Figure 2 (b), the average grain size decrease with the distribution of grain size center in the average range. Figure 2 (c) shows the microstructure image with welding speed (V) 200 mm/min, and frequency (f) 5 Hz. The grain sizes spread between 4.11 to 27.87 μm and the distribution of grain size also dominated by large grains. The average grain size decrease compares by the lower speed. Again, increasing the welding speed (200 mm/min) as shown in Figure 2 (d), the average grain size decrease with the distribution of grain size center in the average range. The average grain size reached the smallest size at this welding speed, with 10 Hz frequency.

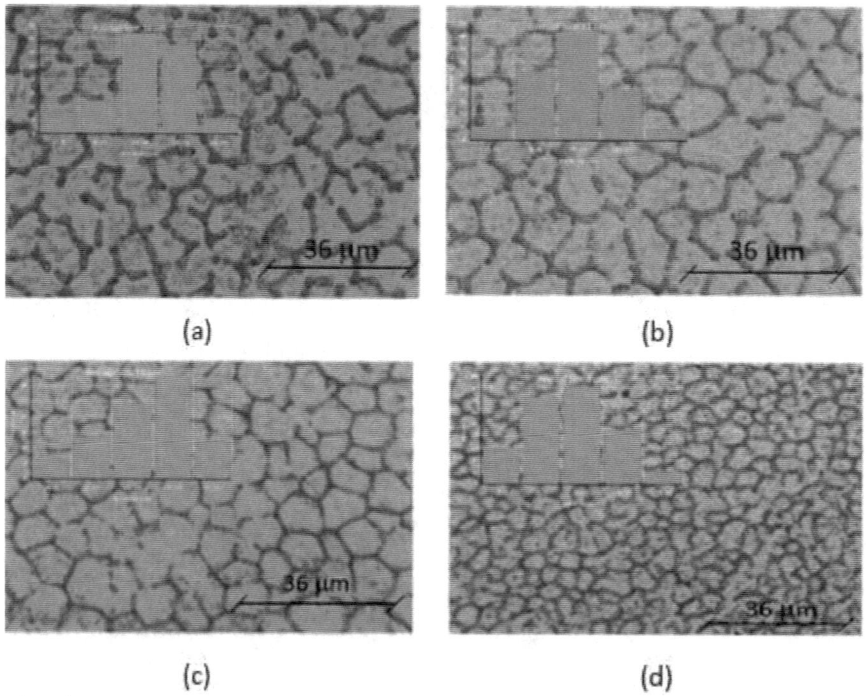

Figure 3. Microstructure of vibrated torch welding and their distribution of grains at 0.75 mm amplitude a) welding speed (V) 100 mm/min, frequency (f) 5 Hz, b) V: 100 mm/min, f: 10 Hz, c) V: 200 mm/min, f: 5 Hz, and d) V: 200 mm/min, f: 10 Hz.

The main variable in analyzing the effect of vibration on the torch in welding to grain refinement is welding speed, amplitude, and frequency. The effect of these variables can be described in detail based on the graph of the relationship between the amplitude and the grain size, as shown in figure 4.

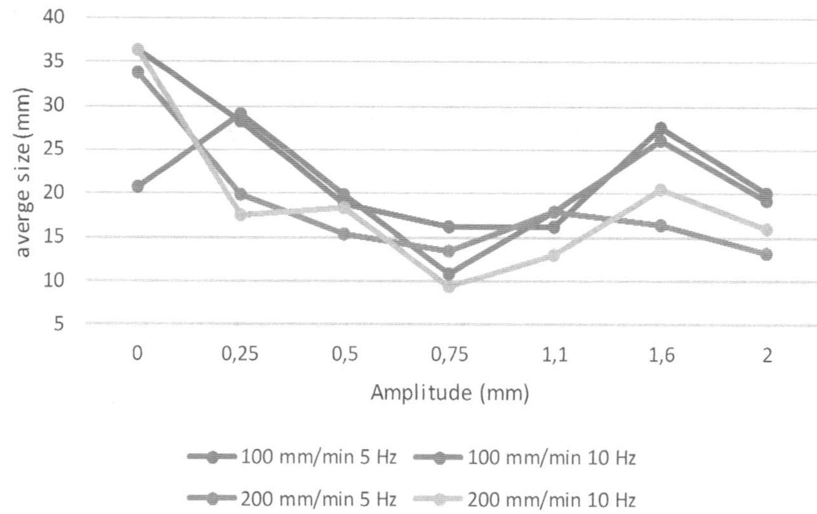

Figure 4. Average grain size as a function of the amplitude at different welding speeds and frequencies.

Figure 4 shows that the amplitude gives optimum effect to the grain refinement. While at the 0.75 mm amplitude with 200 mm/minute welding speed and the frequency of 10 Hz obtained the finest average grain size of 9.41 μm. From these results, it is seen that at higher frequencies obtained a finer grain deposit weld at the same welding speed (0.75 mm amplitude). In other words, an increase in frequency, to some extent, can improve the refining of the weld deposit grain. The analysis results obtained that the relationship between welding speed and grain size is non-linear. Based on Sivaraj et al. [26] and Sree Sabari et al. [27], the faster welding speed will be obtained finer grains. At higher welding speed, the heat input is lower, so the crystal grains are smaller because the energy for dendrite/grain growth is also smaller. In this research, it was found that the grains decrease as increasing amplitude up to 0.75 mm then the grains increase with increasing amplitude. In welding with torch vibration at the same welding speed, increasing the frequency will refine the grain size between 52 to 67%. Grain refinement occurs because of the generation of peak pressure, the development of power density shear stress, and the presence of laminar flow around the dendritic arm [13]. The results were in the intersection of the dendritic arm due to vibrations cause the finer grains.
Figure 5 shows the comparison of increasing welding speed at 5 Hz and 10 Hz, respectively. Both graphs have an anomaly of the average size of the grain. At 100 mm/minute, and frequency 5 Hz with 1.1 mm amplitude, the average grain size is around 16 μm which finer than average grain size at the frequency 10 Hz with the same amplitude. The distribution of grain size tent to higher than the average. Meanwhile, at frequency 10 Hz in beginning welding speed of 100 mm/minute, the value drops as the distribution small and large grain size has a significant gap, the average grain size should be more than 35 μm.

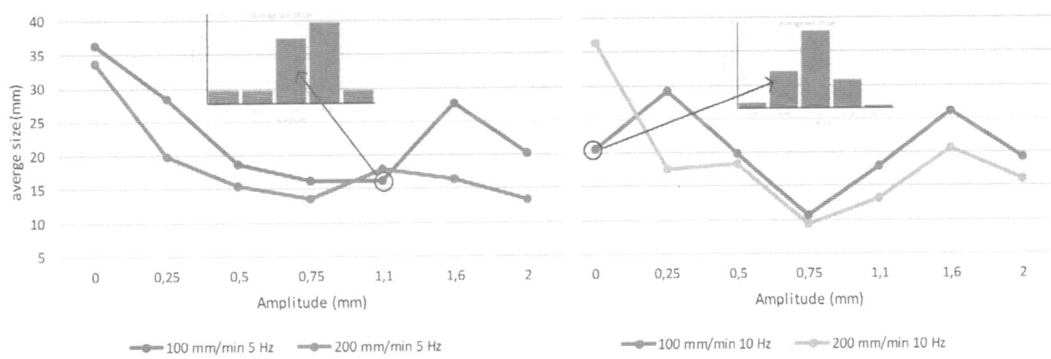

Figure 5. Increasing welding speed refining the average grain size at 5 and 10 Hz.

According to Chen et al. [28], the vibration of the welding will refine the grain because of the acceleration of the welding rate when compared to the welding speed in the direction parallel to the liquid-solid interface. The theory has limitations, where at a high enough frequency, the metal did not have time to melt. Higher frequencies will result in higher resultant speeds so that finer grains are obtained. Seen at higher frequencies the finer welding grain size of the weld is obtained at the same welding speed or in other words an increase in frequency within a certain limit can increase the refining of the weld deposit. The faster the welding rate, the finer the results will be obtained. Due to the higher welding speed, the lower heat input, so the crystal granules are smaller where the growth energy of automatic dendrite grains is smaller.

4. Conclusion

Welding speed gives the effect of grain refinement. Increase the welding speed will reduce grain size; on the other hand, increase the frequency has affected different grain size. The welding speed will smooth the grain size approximately 23% at both 5 Hz and 10 Hz. Meanwhile, increase the amplitude to 0.75 mm will reduce the grain size (approximately 52% and 67% at 5 and 10 Hz, respectively) then the grain size will unstable with increasing amplitude.

Acknowledgments

The authors would like to thank profusely to Institut Sains & Teknologi AKPRIND that has given funding of the conference activities, and to all member of the Mechanical Engineering Department for all of their support during the project.

References

[1] FG Armao 2011 Aluminum Welding: The Cost and Cure of Porosity in *Fabricating Metalworking* http://www.fabricatingandmetalworking.com/2011/12/aluminum-welding-the-cost-and-cure-of-porosity/.

[2] P Govindarao P Srinivasarao A Gopalakrishna A and MMM Sarkar 2012 Affect of Vibratory Welding Process to Improve the Mechanical Properties of Butt Welded Joints *International Journal of Modern Engineering Research* Vol. 1 Issue 4 pp 2766-2770 ISSN:2249-6645.

[3] M Olabode P Kah and J Martikainen 2013 Aluminium alloys welding processes: Challenged, joint types and process selection *Proceeding of the Institution of Mechanical Engineering, Part B: Journal of Engineering Manufacture* vol 227 pp 1129-1137. doi.org/10.1177/0954405413484015.

[4] MW Safeen and PR Spena 2019 Main Issues in Quality of Friction Stir Welding Joints of Aluminum Alloy and Steel Sheets in *Metals* Vol 9 Issue 5 610 doi.org/10.3390/met9050610.

[5] R Deekhunthod 2014 Weld Quality in Aluminium Alloys *Master Thesis* Uppsala Universitet Sweden ISSN: 1401-5773

[6] P Kah A Jibril J Martikainen and R Suoranta 2012 Process Possibility of Welding Thin Aluminium Alloys in *International Journal of Mechanical and Materials Engineering* Vol 7 No 3 pp 232-242 ISSN: 2198-2791.

[7] M Pedram and MR Khedmati 2014 The effect of welding on the strength of aluminium stiffened plates subject to combined uniaxial compression and lateral pressure in *Int J nav Archit Ocean Eng* Vol 6 pp 39-59 doi.org/10.2478/IJNAOE-2013-0162.

[8] J Peng S Fukumoto L Brown and N Zhou 2004 Image analysis of electrode degradation in resistance spot welding of aluminium in *Science and Technology of Welding and Joining* Vol 9 No 4 pp 331-336 doi.org/10.1179/136217104225012256.

[9] OI Oluwole and OJ Ajibade 2010 Effect of Welding Current and Voltage on the Mechanical Properties of Wrought (6063) Aluminium Alloy in *Materials Research* Vol 13 Issue 2 pp 125-128 ISSN: 1980-5373.

[10] SM Manladan F Yusof S Ramesh M Fadzil Z Luo and S Ao 2016 A review on resistance spot welding of aluminum alloys in *Int J Adv Manuf Technol* doi.org/10.1007/s00170-016-9225-9.

[11] S Wang J Kang X Zhang and Z Guo 2017 Dendrites fragmentation induced by oscillating cavitation bubbles in ultrasound field in *Ultrasonics* Vol 83 pp 26-32 doi.org/a0/a016/j.ultras.2017.08.004.

[12] E Vandersluis C Ravindran 2016 Comparison of Measurement Methods for Secondary Dendrite Arm Spacing in *Metallogr. Microstruct. Anal.* doi.org/10.1007/s13632-016-0331-8.

[13] RG Guan and D Tie 2017 A Review on Grain Refinement of Aluminum Alloys: Progresses, Challenges, and Prospects in *Acta Metall. Sin. (Engl. Lett.)* doi.org/10.1007/s40195-017-0565-8.

[14] OG Rivera PG Allison LN Brewer OL Rodriquez JB Jordon T Liu WR Whittington RL Martens Z McClelland CJT Mason L Garcia JQ Su and N Hardwick 2018 Influence of texture and grain

refinement on the mechanical behavior of AA2219 fabricated by high shear solid state material deposition in *Materials Science and Engineering: A* Vol 724 pp 547-558. doi.org/10.1016/j.msea.2018.03.088.

[15] Rolls-Royce Plc 1996 *The Jet Engine* 5th edition Renault Printing Co Ltd England ISBN: 0902121 235.

[16] Y Morisada H Fujii F Inagaki and M Kamai 2013 Development of high frequency tungsten inert gas welding method in *Materials and Design* Vol 44 pp 12-16. doi.org/10.1016/j.matdes.2012.07.054.

[17] H Li J Zou J Yao and H Peng 2017 The effect of TIG welding techniques on microstructure, properties and porosity of the welded joint of 2219 aluminum alloy in *Journal of Alloys and Compounds* Vol 727 pp 531-539 doi.org/10.1016/j.jallcom.2017.08.157.

[18] Y Fang Z Liu S Cui Y Zhang J Qiu and Z Luo 2017 Improving Q345 weld microstructure and mechanical properties with high frequency current arc in keyhole mode TIG welding in *Journal of Materials Processing Tech* Vol 250 No pp 280-288. doi.org/10.1016/j.jmatprotec.2017.07.026.

[19] H LI JS Zou JS Yao and HP Peng 2017 Uniform design and optimization of active agent and technology research for A-TIG welding of 2219 aluminum alloy in *Int j Adv Manuf Tecnol* Vol 92 pp 3435–3446 doi.org/10.1007/s00170-017-0356-4.

[20] K Devireddy V Devuri M Cheeou and B Kranthi Kumar 2018 Analysis of The Influence of Friction Stir Processing on Gas Tungsten Arc Welding of 2024 Aluminum Alloy Weld Zone in *International Journal of Mechanical and Production Engineering Research and Development* Vol 8 Issue 1 pp 243-252 ISSN: 2249-8001.

[21] Y Liu C Deng B Gong and Y Bai 2019 Effects of heterogeneity and coarse secondary phases on mechanical properties of 7050-T7451 aluminum alloy friction stir welding joint in *Materials Science & Engineering A* Vol 764 138223 doi.org/10.1016/j.msea.2019.138223.

[22] Y Liang J Shen S Hu H Wang J Pang 2018 Effect of TIG current on microstructural and mechanical properties of 6061-T6 aluminum alloy joints by TIG-CMT hybrid welding in *Journal of Materials Processing Technology* Vol 255 pp 161-174. doi.org/10.1016/j.jmatprotec.2017.12.006.

[23] M Terzi 2018 Particle size distribution analysis in aggregate processing plants using digital image processing methods in *Romanian Journal of Materials* Vol 48 Issue 4 pp 514-521 ISSN: 1583-3186.

[24] W Tian S Li J Liu M Yu and Y Du 2017 Preparation of bimodal grain size 7075 aviation aluminum alloys and their corrosion properties in *Chinese Journal of Aeronautics* Vol 30 No 5 pp 1777-1788 doi.org/10.1016/j.cja.2017.06.001.

[25] AF Ireti AE Omeiza KM Oluwasegun IS Adeyemi OA Adewale and E Gnozi 2018 Comparison of Imagej Analysis of Structure of Two Constructional Steel in *American Journal of Engineering and Applied Sciences* Vol 11 Issue 1 pp 318-326. doi.org/10.3844/ajeassp.2018.318.326.

[26] P Sivaraj P Hariprasath C Rajarajan and V Balasubramanian 2019 Analysis of grain refining and subsequent coarsening along on adjacent zone of friction stir welded armour grade aluminium alloy joints in *Materials Research Express* Vol 6 066566 doi.org/10.1088/2053-1591/ab0e37.

[27] S Sree Sabari S Malarvizhi and V Balasubramanian 2016 Influences of Tool Transverse Speed on Tensile Properties of Air Cooled and Water Cooled Friction Stir Welded AA2519-T87 Aluminium Alloy Joints in *Journal of Materials Processing Technology* Vol 237 pp 286-300 doi.org/10.1016/j.jmatprotec.2016.06.015.

[28] Q Chen S Lin C Yang C Fan and H Ge 2017 Grain fragmentation in ultrasonic-assisted TIG weld of pure aluminum in Ultrasonics Sonochemistry Vol 39 pp 403-413. doi.org/10.1016/j.ultronch.2017.05.001.

Optimization of information technology utilization based on geography information system as a marketing media for micro, small, and medium enterprises to improve competitiveness in Bantul Regency, Indonesia

Denny Hardiyanto[1], **Muhammad Sholeh**[2]* **and Rr Yuliana Rachmawati**[2]

[1] Department of Electrical Engineering, Faculty of Industrial Technology, Institut Sains & Teknologi AKPRIND Yogyakarta, Yogyakarta, Indonesia

[2] Department of Informatics Engineering, Faculty of Industrial Technology, Institut Sains & Teknologi AKPRIND Yogyakarta, Yogyakarta, Indonesia

*Corresponding author: muhash@akprind.ac.id

Abstract. Micro, small and medium enterprises (MSMEs) are the largest group of economic actors in the Indonesian economy. The growth of the micro industry grew by 8% every year, especially in Bantul Regency, Special Region of Yogyakarta. Various factors that led to the growth of the small micro industry in Bantul were caused by the increasing stretch of business in the creative economy sector, the majority of which were engaged in the craft sector. As support, the Government of Bantul provides policies that protect or secure raw materials for handicraft or industrial production properly and facilitate physical promotion by participating in various exhibition events through online media. In the era of information technology development, the support of online information system become one of the promotional media that can reach a large area. The research objectives are to build a website for MSME in Bantul Regency and to provide a media for MSME actors to advertise their goods and services. In this study, the MSME information system that was built was equipped with facilities for maps, images and videos of MSME products. This map facility makes it easy for users to go to business locations. The system was developed by utilizing the Application Programming Interface (API) of google map and database used to store all MSME data using MySQL with the PHP: Hypertext Pre-processor Programming language. The result is development of marketplace website (*https://wartacara.com/ukm*) that have many features to explore and advertise MSME actor in Bantul Regency.

1. Introduction

Today, the development of the digital era must be anticipated by MSME factors. One of the efforts that must be done is to utilize the information and communication technology to carry out promotions and conduct transactions by using social media. At present, the use of social media has become a necessity of the community. Dependency of information on social media is something that must be anticipated in

MSME factors. Social media has an important role in the business world especially in the field of marketing and it is one of the media that can rapidly spread information. Social media is not only a place to show or express oneslef to the outside world, but also as a place to get a lot of information. Various kinds of social media are useful for consumers to get information. [1]

When using internet or social media in the promotion process, MSME factors do not have to provide their own infrastructure. The facilities provided by social media include Facebook, Twitter and other media can be used as a promotion platform. Besides social media, the facilities offered by the government or the community can also be used to promote MSMEs. The use of promotional media, of course, will save costs in conducting promotions and does not require investment in managing promotional media on the internet. All promotional needs on the internet such as the availability of information systems, hardware, and other requirements, have been provided by the service providers. Some media on the internet that can be used include *Bukalapak, Tokopedia*, or other government websites that provide information about MSME factors. Online sales in Indonesia have experienced extraordinary growth reaching 20% consistently every year. In 2016, it was estimated that there were more than 180 million internet users and had reached approximately 40% of the 250 million total population. Based on the survey, there are four of the most popular online vendors in Indonesia, such as: *Lazada, Tokopedia, OnLine eXchange* and *Bukalapak*, with a strong customer base [2].

The availability of service managers who provide MSMEs promotion facilities already exists in several cities, for example there is a website – *http://umkm.jogjakota.go.id* in the city of Yogyakarta. The information presented is certainly beneficial for MSMEs actors or people who are seeking information on an MSMEs. With the existence of this website, promotion of MSMEs can be done easily, all UMKM facilities can be displayed on the website, such as location, potential, images and other information related to MSMEs while the weakness of the website is that there is no facilitation of map location and route to the location.

Research conducted by Meilani [3] showed that micro, small and medium enterprises (MSMEs) have great potential in driving community economic activities and at the same time become a source of income for most people in improving their welfare. Therefore, to improve the efficiency and effectiveness of independent and developing MSMEs, Information Technology (IT) is needed to promote enterprises and their products. For this purpose, Mobile Geographic Information System (MGIS) application was developed for android based mapping of MSMEs. In data collection process, the authors employed observation and interview as methods. The purpose of developing this application is as an alternative promotion media and to make it easy to find the location of MSMEs in Surabaya. Black box test was conducted to test the appropriateness of the application. In addition, testing was also conducted using a questionnaire involving 50 respondents with 15 questions representing 5 stakeholder needs. From the questionnaire results, it was concluded that 4 needs, namely, the need to promote products, to spread information, community service, and information of MSMEs have been highly fulfilled by the application, while the need to increase gross domestic product was sufficiently fulfilled by the application.

Kusuma's research [4] is developing an application that combines GIS technology with the Google Maps API to present information about tourism in Sidoarjo Regency as an effort to realize community empowerment strategies through developing tourism based on populist or community-based tourism development so that people are expected to be able to participate in developing tourism activities in their area, especially the people of Sidoarjo Regency.

Taryadi et al [5] has conducted research about Geographic Information System (GIS) application in mapping the potency of batik. Spatial model presented by GIS may directly be applied to understand the mapping of the potency of batik industry centre in Pekalongan city. System Development Life Cycle development methodology used in creating this application adopted Use-Case Diagrams tool, whereas for the database design The Entity Relationship Diagram was used. The test result with the Alpha test and the beta test were used to test the suitability of the application developed with the needs of users. The Alpha test result using black-box method indicates that all the function in the application of GIS is working well, suit to the design of the application. The result of the Beta testing can be summarized as follow which are the GIS application developed has a good interface and easy to use, the user can easily follow the provided instruction and the system is able to provide the information required by user, such as the mapping of the business field, the mapping of the amount of business in one district, the potency of the business type per village and displayed suit to the map location. The presentation of the results of this research with the use of GIS technology can facilitate Describe the results of the research in the form of data visualization. Use of GIS technology in more detail of information visualization so you need information about SMEs can be very complete.

In addition, Susanti et al [6] have conducted marketing research on batik based on GIS. This study has two purpose. First, this study aims to analyse the requirements of batik industry on effective and powerful software to select their supplier effectively. Second, this study aims to design, build, and test a web-based decision support system that using Geographic Information System (GIS) to select the best supplier for raw material. A prototype approach is used as a method for developing a web-based decision support system and several SMEs and supplier of batik industry in Central Java and Jogjakarta is used as a preliminary of object for proposed web-based decision support system. Thus, after the initial prototype of an information system was developed, this study uses acceptance sampling with alpha and beta testing and use system testing with positional accuracy testing as a tool for revising and enhancing the initial prototype. The result of all of those testing indicate that application of supplier selection based on GIS in the batik industry has been successfully used and did not experience an error, satisfied the user need and there was no differences between the coordinate of selected starting and ending point from GIS with the coordinate of selected starting and ending point from Global Position System (GPS). The other researches related to the use of GIS are as in reference [7].

2. Research Method

2.1. Research Location

The object of this research is MSMEs in Bantul regency. The website that has been built is able to display the map of Bantul regency with Google Map coordinates, with latitude -7.8866666666667 and longitude 110.3275. Besides the main coordinates of Bantul regency, the coordinate objects needed are sub-district, villages, and coordinates of MSME locations. Table 1, displays the coordinates of the sub-districts in Bantul regency and Table 2, showing the coordinates of village in the Kasihan sub-district.

Table 1. GPS and Google Map Coordinates of Sub-district in Bantul Regency

Sub-district	GPS Coordinates						Google Map Coordinates	
	BT			LS				
	DD	MM	SS	DD	MM	SS	Latitude	Longitude
Srandakan	110	14	46.7	7	56	20.21	-7.9388888888889	110.24611111111
Sanden	110	15	56.34	7	58	4.83	-7.9677777777778	110.26555555556
Kretek	110	19	6.69	7	58	20.21	-7.9722222222222	110.31833333333
Pundong	110	20	37.2	7	57	20.61	-7.9555555555556	110.34361111111
Bambang Lipuro	110	17	47.21	7	56	58.04	-7.9494444444444	110.29638888889

Pandak	110	17	50.46	7	54	35.63	-7.9097222222222	110.29722222222
Imogiri	110	22	52.6	7	55	14.92	-7.9205555555556	110.38111111111
Dlingo	110	27	54.69	7	56	6.9	-7.935	110.465
Banguntapan	110	24	39.72	7	49	42.8	-7.8283333333333	110.41083333333
Pleret	110	24	26.72	7	51	59.73	-7.8663888888889	110.40722222222
Piyungan	110	28	35.23	7	50	19.38	-7.8386111111111	110.47638888889
Sewon	110	21	33.81	7	50	58.21	-7.8494444444444	110.35916666667
Kasihan	110	20	41.26	7	49	42.96	-7.8283333333333	110.34472222222
Sedayu	110	15	30.24	7	48	49.25	-7.8136111111111	110.25833333333

Table 2. GPS and Google Map Coordinates of Villages in Kasihan Sub-district

Village	GPS Coordinates						Google Map Coordinates	
	BT			LS				
	DD	MM	SS	DD	MM	SS	Latitude	Longitude
Kasihan sub-district	110	20	41.26	7	49	42.96	-7.8283333333333	110.34472222222
Bangujiwo	110	18	47.82	7	50	23.16	-7.8397222222222	110.31305555556
Tirtonirmolo	110	20	41.31	7	49	42.96	-7.8283333333333	110.34472222222
Tamantirto	110	19	36.37	7	49	30.60	-7.825	110.32666666667
Ngestiharjo	110	20	46.86	7	48	1.65	-7.8002777777778	110.34611111111

2.2. Research Diagram
Information system development begins with designing a database. The purpose of database design is to fulfil information that contains specific user need. Figure 1 shows flow diagram of the research implementation.

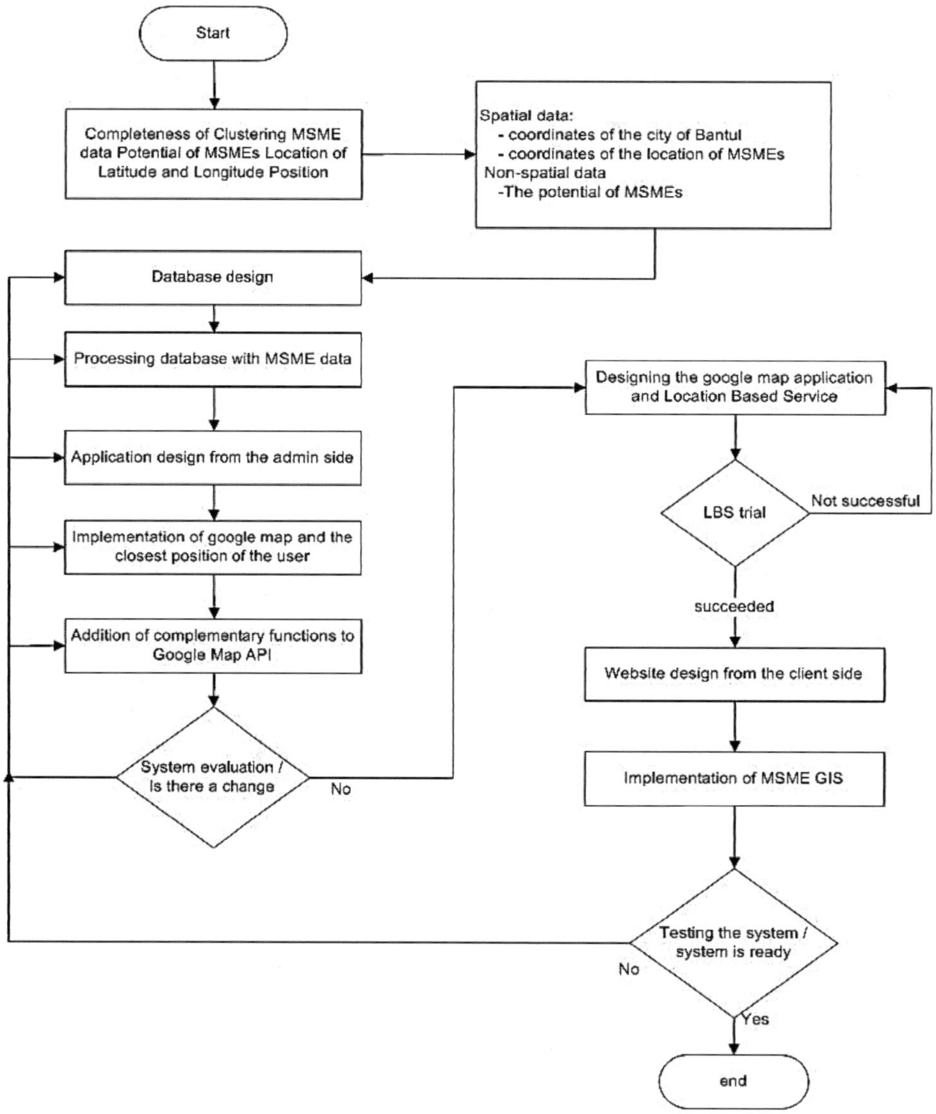

Figure 1. Flow diagram of the research

In summary, the flow diagram of the system development stages are as follows:

a) Identifying the problems by analysing system requirements to be developed
b) Collecting completeness of MSME data, including coordinates location data and potential MSME data
c) Designing the database and carrying out the normalization process to produce a database that is avoided from data duplication
d) Develop applications from the admin side. Application developed to complete all data stored in the database
e) Register on the Google Maps site to get an API Key that contains the code to access Google Maps.

f) Integrating information originating from the database that has been created along with the Google Maps API application into the web that has been designed.
g) In developing applications, the server side is equipped with Google map-based applications, especially in the process of determining coordinates
h) System evaluation from the server application side
i) Developing applications from the client side, the process of creating a web interface is done to design the desired website appearance.
j) After all the scripts have been created and the web interface has been designed, website can be displayed and published with the domain name www.wartacara.com/ukm.

3. Result and Discussion

3.1. Google Map API utilization

This Bantul Regency geographic information system is designed by using Google Map. Google Maps is a Google product that can display a map-based location point. In developing the application (website) that was built, the google map API is used to display the location points of the sub-district, villages and the MSME actors also display routes that can be used to get to a location for MSME actors. The process of displaying location points on Google Map is illustrated in Figure 2. The process of displaying location points is taken from stored tables in the form of latitude and longitude points using the Google map API. This application able to display these data.

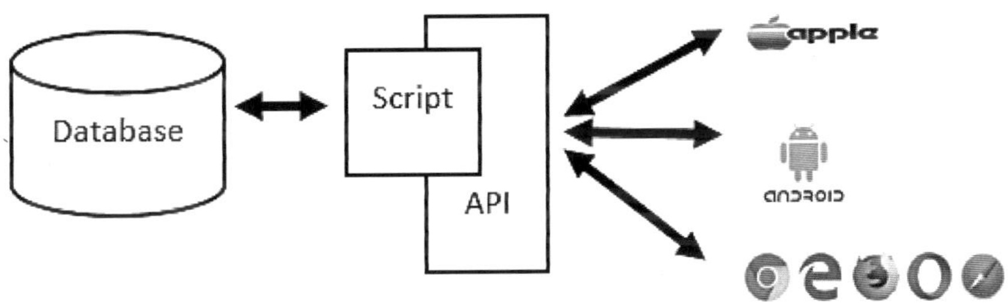

Figure 2. Application developed

In developing this website, the Google Map API functions that are used included in the reference [7]:

- Google Maps API

  This function is used to display maps on the website and set the map mode to be displayed. The implementation of this function is used to display a map of the area on the front page and the location of each MSME.

  Script:

```
<script>
function initMap()
{
var map = new google.maps.Map(document.getElementById('map'), {
zoom: <?=$zoom?>,
center: {lat: -7.801371, lng: 110.364762 }
});
```

```
var secretMessages = [<?php foreach ($sig as $key) {echo "'<div id=content> <div
id=siteNotice></div> <div id=bodyContent> <p><a
href=".base_url('home/detKel/')?><?=$key['kecamatan_id']."#peta-ukm><b>Lihat beberapa
kelurahan di kecamatan</b> <h5 id=firstHeading
class=firstHeading>".$key['kecamatan_nama']."</h5> </a></p> </div> </div> ',";}?>];

<?php $j=0; foreach ($sig as $dt) { ?>
var marker = new google.maps.Marker({
animation: google.maps.Animation.DROP,
title: "<?php echo $dt['kecamatan_nama'];?>",
position: {
lat: <?php echo $dt['kecamatan_latitude'];?>, lng: <?php echo $dt['kecamatan_longitude'];?>
},
map: map
```

- Google Place Auto Complete
  Google Place Auto Complete is used for the process of entering data on address of MSME actors such as street names, city names, regional names and other place names. In the process of entering data, the choice of data is almost the same as the data entered. The process of displaying data is done in the form of auto complete in real-time.

  Script:

```
autocomplete.addListener('place_changed', function() {
 infowindow.close();
 marker.setVisible(false);
 var place = autocomplete.getPlace();
 if (!place.geometry) {
   window.alert("No details available for input: '" + place.name + "'");
   return;
 }
 if (place.geometry.viewport) {
   map.fitBounds(place.geometry.viewport);
 } else {
   map.setCenter(place.geometry.location);
   map.setZoom(17);
 }
 marker.setPosition(place.geometry.location);
 marker.setVisible(true);

 $('#txtLat').val(place.geometry.location.lat());
 $('#txtLng').val(place.geometry.location.lng());
 var address = '';
 if (place.address_components) {
   address = [
   (place.address_components[0] && place.address_components[0].short_name || ''),
   (place.address_components[1] && place.address_components[1].short_name || ''),
   (place.address_components[2] && place.address_components[2].short_name || '')
   ].join(' ');
 }
 infowindowContent.children['place-icon'].src = place.icon;
 infowindowContent.children['place-name'].textContent = place.name;
 infowindowContent.children['place-address'].textContent = address;
 infowindow.open(map, marker);
 });
```

- Google Direction API

By using Google Direction API, this website can display routes that connect between two locations, information in the form of distance, estimated time of arrival to the route that must be passed to get to a location.

Script:

```
<script type="text/javascript">
  var map;
  var options = {
    enableHighAccuracy: true,
    timeout: 5000,
    maximumAge: 0
  };
  function success(pos) {
    var crd = pos.coords;
    dsa(crd.latitude,crd.longitude);
  }
  function error(err) {
    console.warn(`ERROR(${err.code}): ${err.message}`);
    alert ('Terjadi kesalahan saat menghubungkan ke server');
    dsa(-7.875138676242137,110.32486636965064);
  }
  function dsa(mylat,mylon){
    map = new GMaps({
      el: '#map_canvas',
      lat: mylat,
      lng: mylon
    });
    map.renderRoute({
      origin: [mylat,mylon],
      destination: [<?=$ukm->alamat_latitude?>,<?=$ukm->alamat_longitude?>],
      travelMode: 'driving',
      strokeColor: '#131540',
      strokeOpacity: 0.6,
      strokeWeight: 6
    },{
      panel: '#directions',
      draggable: true
    });
  }
  $(document). ready(function(){
    navigator.geolocation.getCurrentPosition(success, error, options);
  });
</script>
```

3.2. Implementation and How the application work from the user side

The main page of website is the first page seen when a user writes an address in *wartacara.com/ukm* on an internet browser. The initial process of this application displays a map of the city of Bantul with marker points in the form of sub-district locations. Figure 3 is the initial appearance of the website that was built.

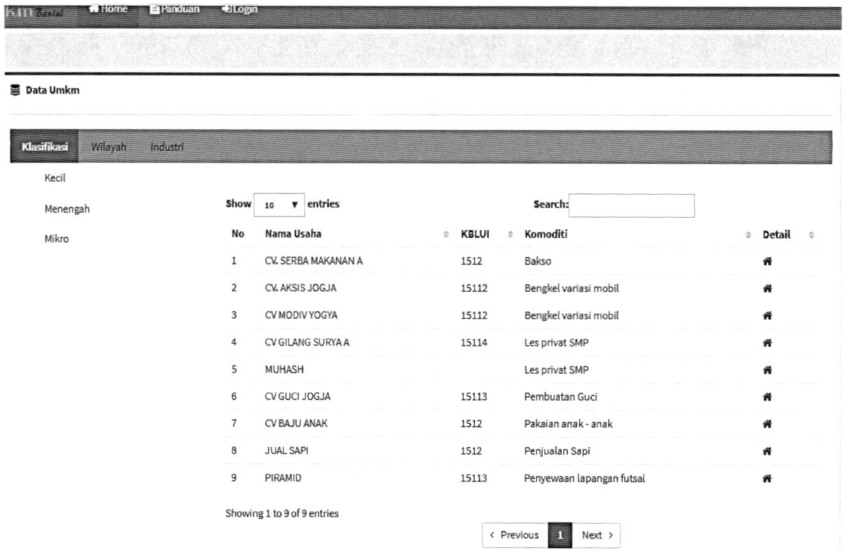

Figure 3. The main page of website

3.3. Potential of MSME

Important information that must be presented on this website is the potential of MSMEs. The main menu that visitors can use to find out the potential of MSMEs is through the regional and industrial classification menu. The main menu of website page display at Figure 4. The display shows a menu that can be selected by website visitors to see the potential of an MSME.

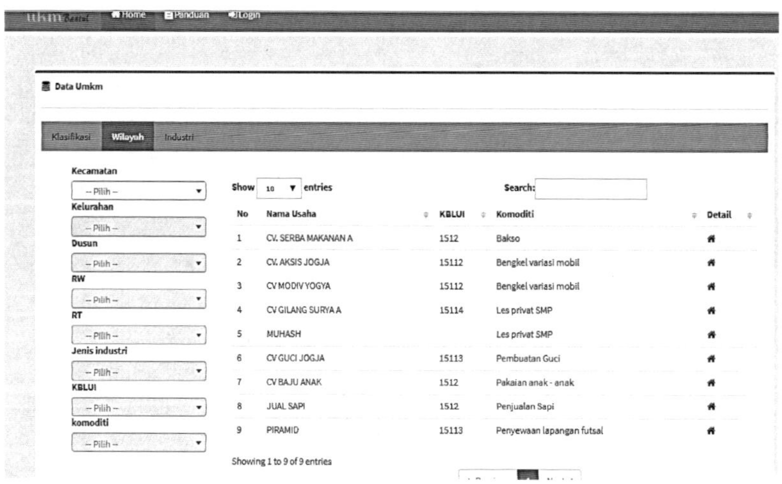

Figure 4. The main menu of website

The website will display detailed information from MSMEs, namely address location information, MSME information in the form of text, images and videos as well as products from MSME players. Figure 5 shows an example of the results of a MSME search.

Figure 5. Detail information of MSME in website

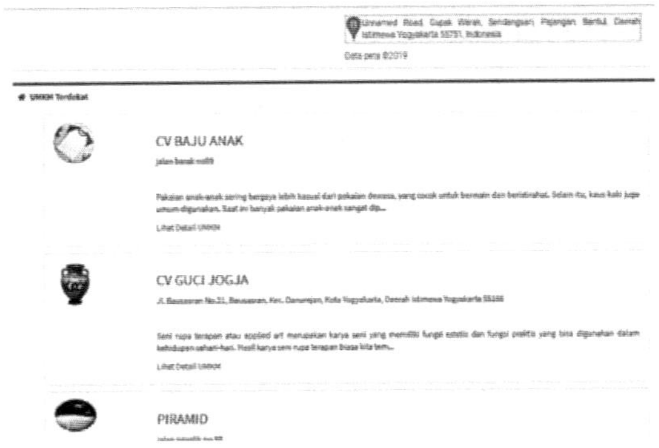

Figure 6. The nearest MSME location

3.4. The nearest MSME Location

Besides informing the location of MSMEs, this application also displays other locations related to MSMEs around the location of the location being searched. Figure 6 displays the other locations closest to the search point.

In addition to informing the nearest location, this application can provide facilities to users to choose and view other locations. With this information, the user obtains other MSME locations adjacent to the first location.

3.5. Display of route pages

This page contains a map display and the route to an MSME location. The results of the page display are found in Figure 8. The route search process begins by entering the starting point and location point searched. This process will display a map to the location as in Figure 7 and the route to the location.

Figure 7. Location point of MSME

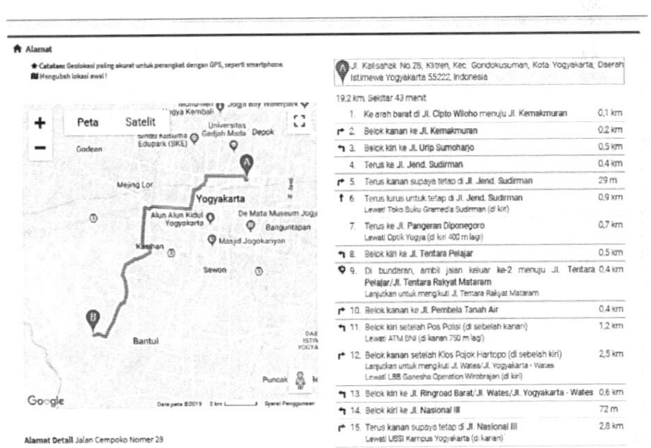

Figure 8. Route to MSME location

This route search program uses an application provided by Google Map. Programs for displaying the results of travel routes use two Google API (Application Programming Interface), namely API service directions and API display directions. Both APIs are useful for knowing the direction or route between two coordinate points or more. By utilizing this API, we can draw these route lines or often known as polyline. Besides the route, we also get information in the form of distance, the estimated time arrives until the step-step path that we must pass.

4. Conclusion

The conclusions of this research are:

a) The design results of this website based on the MSME geographic information system make it easy for users to find information about MSME objects.
b) Information displayed in the form of text information, images of potential areas, and displaying the route to an MSME location.
c) This route application was developed by utilizing the API (Application Programming Interface) from Google Map.
d) By using this website (*https://wartacara.com/ukm*), many MSME actors and government can explore every potential in Bantul Regency

Acknowledgments

The author would like to thank the Directorate of Research and Community Service, the Directorate General of Research and Development, the Ministry of Research, Technology and Higher Education (Kemenristekdikti) for providing funding support for research activities in the scheme of superior applied research in universities in the 2019 and thank to research team.

References

[1] C S Putri 2016 Pengaruh media sosial terhadap keputusan pembelian konsumen cherie melalui minat beli *PERFORMA.Jurnal Manajemen dan Start-Up Bisnis* 1 5 594-603
[2] G Wang, T Wijaya and S Ferianto 2018 Assessing the buyer trust and satisfaction factors in the e-marketplace *J. Theo. App. Elec. Comm. Res.* 13 2 43-57
[3] B D Meilani and Subianto 2019 The application of mobile geographic information system (MGIS) for android-based mapping of micro, small and medium enterprises *Intl. Conf. Adv. Eng. Tech.* 3 4
[4] M E Kusuma and Y Budisusanto 2014 Aplikasi Google Maps api dalam pembuatan web sig guna menunjang kegiatan pariwisata berbasis kerakyatan *Simposium Nasional Geomatika*
[5] S Taryadi, W Binabar and D J Siregar 2019 Geographic information system for mapping the potency of batik industry centre *J. Info. Sys. Eng. Bus. Intel.* 5 1 40-47
[6] A S Susanty, H D P Sari and P Hidayatullah 2016 Using geographic information systems to develop decision support system for supplier selection in batik industry *IOSR. J. Comp. Eng.* 18 4 1-13
[7] A Dincer and B Uraz 2014 *Google Map Javascript API Cookbook* (Birmingham: Packt Publishing)
[8] H N Lengkong and A Lumenta 2015 Perancangan penunjuk rute pada kendaraan pribadi menggunakan aplikasi mobile gis berbasis android yang terintegrasi pada Google Maps *Jurnal Teknik Elektro dan Komputer* 4 2

IOP Conf. Series: Materials Science and Engineering **807** (2020) 012011 doi:10.1088/1757-899X/807/1/012011

Study the Effect of Combination of Shoe Height and Slot Opening width to Reduce Cogging Torque in Permanent Magnet Generator

Marsul Siregar[1*], Liza E. Joe[2] and Tajuddin Nur[3]

[1,3] Department of Electrical Engineering Faculty of Engineering, Atma Jaya Catholic University of Indonesia, Jl. Jend Sudirman no 51 Jakarta 12930, Indonesia.
[2] Department of Electrical Engineering, School of Graduate Engineering. Atma Jaya Catholic University of Indonesia, Jl. Jend Sudirman no 51 Jakarta 12930, Indonesia.

Corresponding e-mail: marsul.siregar@atmajaya.ac.id

Abstract. Cogging torque is one important problem in a permanent magnet generator. With a good understanding of the phenomenon, the cogging torque can be minimized by using the multi-objective technique. This paper presents a combination technique for cogging torque reduction by an appropriate small stator slot opening width, stator shoe height with fractional technique in Permanent Magnet Generator. Cogging torque waveforms, along with other relevant characteristics are investigated in this research and by analysis using the finite element method, the combination width shoe height and magnet edge slotting effect to decline the cogging torque peak value. A significant value of cogging torque, when slot opening width of 1.55 mm and shoe height of 2.3 mm, provide the peak value of cogging torque in the structure is 0.00062 N-m for positive value and about -0.00064 N-m for the negative value. It can be concluded that this type of slot opening width of 1.55 mm and shoe height of 2.3 mm is the best structure of the fractional 8 slot, because it has any significant low of cogging torque and least of unbalance magnetic pull

1. Introduction

Permanent magnet machines have been used in high-performance applications because of their high efficiency, high torque density, and power factor close to unity. However, a drawback of permanent magnet machines is the cogging torque. Cogging torque causes vibration, noise caused and electromagnetic torque ripple which seriously affects the machine performance. The main machine vibration sources are found in ball bearing roughness, cooling air turbulence, electrical commutation and iron core–magnet reluctance forces. Recently, both the electric machine designers and electrical machine manufacturers have developed a better understanding of the cogging phenomenon [1].

Since the last few years, it has been increasing the application of permanent magnet generators (PMGs) for renewable power system. In small renewable power system, the PMG is coupled directly connected to the turbines, leads no any gear box. This kind of electric machine becomes attract by electric machine users since it has some advantages, such as high efficiency, simplicity and reliability of construction. However, the problem of in PMG applications is cogging torque. The presence of CT effects to reduce the PMG performance in electric energy conversion. In low wind speed condition, the blade of wind power is hard to be rotated [1]. Cogging torque in PMSG is the reluctance torque which is affected by the interaction of force from magnet edge in rotor core and stator teeth or slot opening width. The cogging torque is one the most important issue in permanent magnet application, because effect to the PMG performance, especially in low speed condition. Based on the discussion, many cogging torque reduction techniques have been proposed and studied in worldwide. The reports also have been documented [2-7]. Until recently, the research on cogging torque might be classified as shown in Figure 1.

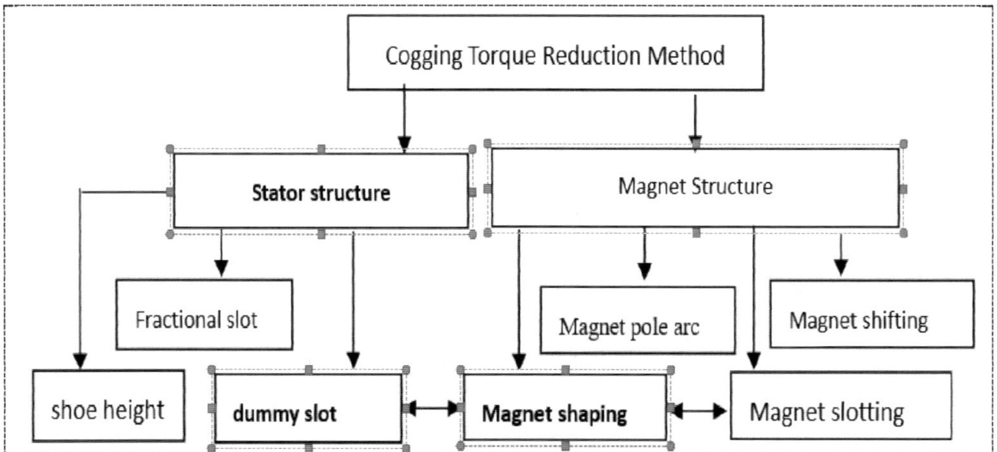

Figure 1. Cogging torque reduction technique

This paper deals with the magnetically generated in electrical machine known as cogging. This is usually large and can cause excessive machine vibrations if not properly controlled**.** Since the last few years, many varieties of techniques for minimizing cogging torque of permanent magnet machine have been reported and proposed [2-8]. In this research, a-8 pole/18 slot (fractional slot) combine with small of stator slot opening and high of stator shoe height for reduction of the peak cogging torque in PMSG is presented.

2. Motivation of the Research

Generally, the electric machine customers require production of a favourable ripple-free torque, good cogging torque waveforms are necessary. However, at the design stage and before the machine prototype is made, it is often difficult to construct any PMSG structure that produces satisfactory cogging torque shape. Moreover, no specific criteria are found to determine the magnet rotor shapes for such cogging torque waveforms. Hence, in this research, a fractional technique with certain of stator slot opening width and stator shoe height for the permanent magnet synchronous generator design is applied to reduce the cogging torque.

In this paper, a proposed permanent magnet synchronous generator (PMSG) design is studied to demonstrate the developed technique to minimize the cogging torque in PMSGs. The magnet rotor used to observe the cogging torque reduction is kept in natural, means that the magnetization is radial and there is no any specific structure in the magnet surface like slotting in the magnet edge, bread loaf or other structure as shown in Fig. 1. In this research, we focus only to find a feasible stator slot opening width and height of shoe in a fractional (8 pole/18 slot) of PMSG to reduce cogging torque and to minimize of unbalance magnetic pull (UMP) effect. The cogging torque of this type of PMSG fractional is compared and presented.

3. Design Development and Proposed of PMSG

In this paper, we consider with the inset magnet rotor type. The initial design is 8 pole / 18 slot of PMSG. The geometry model of shoe height and slot opening width of the PMSG is presented in Figure 1. The result of simulation of magnetix flux distribution illustrated in Figure 2.

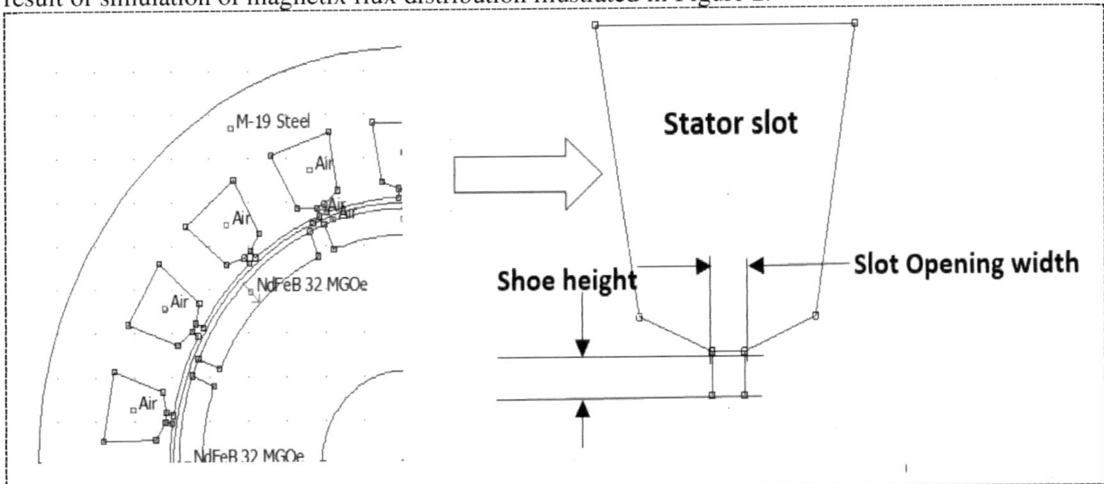

Figure 2. Model of PMSG studied a fractional slot of 8 slot /18 PMSG

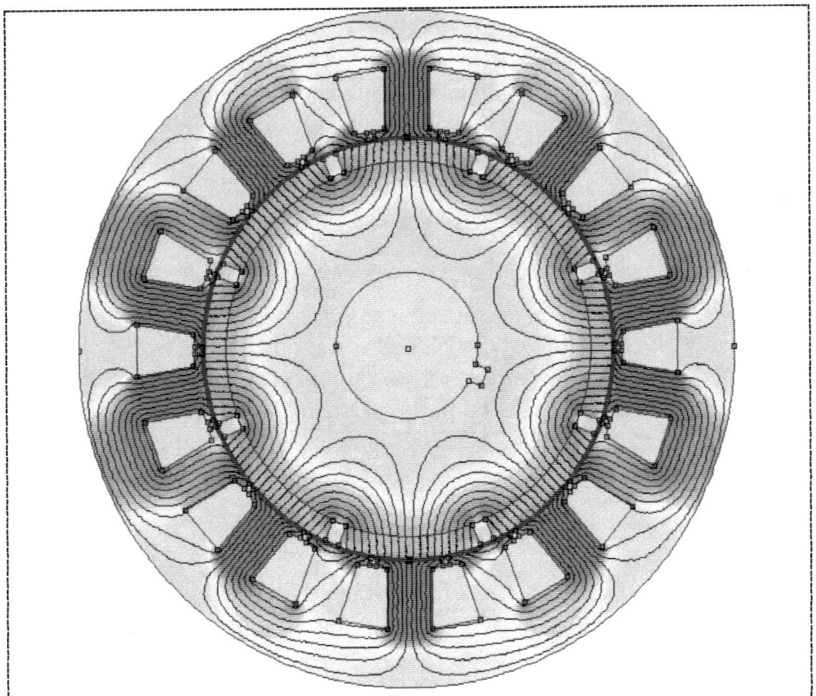

Figure 3. Magnetic Flux Distribution of Model of PMSG studied a fractional slot of 8 slot /18 PMSG

From the Figure.1, the different slot opening and stator shoe height values are applied to a fractional slot of 8 poles/18 slots of PMSG. In this paper, three steps of slot opening width of 1.60 mm (the largest), 1.55 mm and 1.50 mm (the smallest)). For stator shoe, authors investigated six steps of shoe height 2.3

mm (highest), 2.2 mm, 2.1 mm, 2.0 mm 1.9 mm, 1.80 (shortest). The values of slot opening width and shoe height combination are presented in Table 1. The effects of shoe height and slot opening width changing are calculated and simulated using finite element.

Table 1. Combination of SOW and SH in PMSG

Parameters	SOW (16.0 mm)	SOW (15.5 mm)	SOW (15.0 mm)
SH-1 (2.3mm)	2.3	2.3	2.3
SH-2(2.2 mm)	2.2	2.2	2.2
SH-3(2.1 mm)	2.1	2.1	2.1
SH-4(2.0 mm)	2.0	2.0	2.0
SH-5(1.9 mm)	1.9	1.9	1.9
SH-6(1.8 mm)	1.8	1.8	1.8

4. Fractional Techniques in PMSG

In electrical machine design, one of the important issues is cogging torque. Cogging torque can cause noise and vibration. In wind power system, if the cogging torque is too high the blade is not easy to spin, except the blade is forced by strong wind. Considering this fact, in this research we proposed based on the fractional machine with smaller stator slot opening (SOW) width and higher of shoe height (SH) is presented. However, even though fractional slot structure can reduce cogging torque, but some of structures in general create any unbalanced magnetic pull (UPM).

In this paper, we observe some structures of slots opening width and some values of shoe height of a fractional PMSG to find a suitable structure in order to reduce cogging torque and to minimize the effect of unbalanced magnetic pull. One can observe that by using a smaller of opening width and higher shoe can reducing the cogging torque. However, this method can be done so, because the slot opening width (SOW) may be not be constructed smaller than 1.50 mm, unless the winding is going to be difficult to install in the stator slot with conventional way. In the other hand, if the shoe height (SH) is very high, the cross-section area of stator slot will become smaller, means that the number of conductors can occupy the stator slot becomes limited. That is why, the value of SOW and SH is limited. By Using the Combination Fractional Slot Opening Wide and Shoe Height provide the decreasing of CT [9-12].

5. Cogging Torque Waveform and Simulation FEMM Result

In this paper, the 2D FEM (FEMM 4.2) was employed for cogging torque simulation of the PMSG. Graphical presentation and visualization of the FEMM results show the cogging torque of the PMSG. The cogging torque is generated by the tendency of the rotor magnets to align with the stator at positions where the reluctance of the magnetic circuit is minimized. In this study, the cogging torque of the PMSG is solved in the FEMM post processor. For PMSG of a-8 poles / 18 slots, the dependence of the cogging torque on the rotor position are presented in Fig. 4, 5 and 6 respectively

IOP Conf. Series: Materials Science and Engineering **807** (2020) 012011 doi:10.1088/1757-899X/807/1/012011

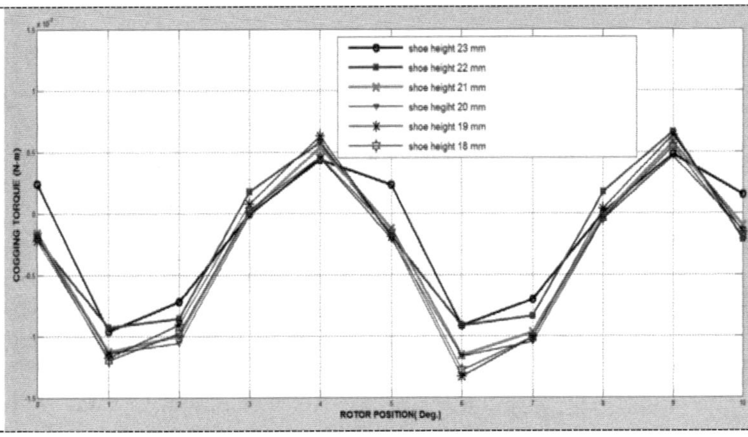

Figure 4. Comparison of Cogging torque of the PMSGs structure (slot opening width of 16.0 mm)

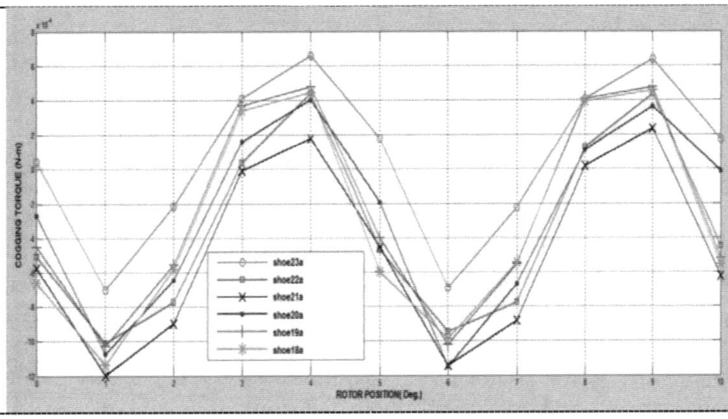

Figure 5. Comparison of Cogging torque of the PMSGs structure (slot opening width of 15.5 mm)

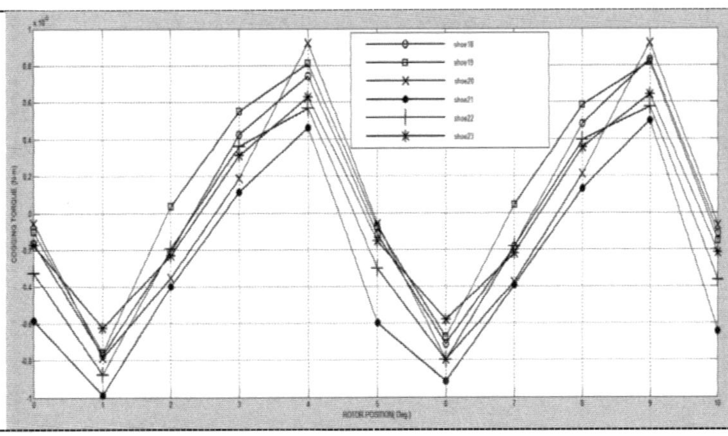

Figure 6. Comparison of Cogging torque of the PMSGs structure (slot opening width of 15.0 mm)

IOP Conf. Series: Materials Science and Engineering **807** (2020) 012011 doi:10.1088/1757-899X/807/1/012011

6. Results and Discussion

The simulation results [9], provide that the slotting design in the magnet edge can effectively reduce the cogging torque of the inset permanent magnet synchronous machine. The peak value of the cogging torque was obtained as much as 92.20% compared with the experimental model.

From the brief analysis of the Fig.3, 4 and 5 a substantial improvement of the peak cogging torque in fractional system combined with smaller slot opening width and higher of stator shoe in the PMSG is evident. For fractional machine (8 pole/18 slot) the cogging torque can be reduced by using any higher stator shoe with smaller slot opening width. However, the wave of cogging torques for all design are unbalance, it means that peak of negative and positive of the cogging torque are different value, except for PMSG structure with slot opening width of 1.55 mm and shoe height of 2.3 mm. From Fig.5 can be seen that cogging torque wave for the PMSG which has 1.55 mm of slot opening width and 2.3 mm of shoe height start from zero or close to zero.

The peak of cogging torque both positive and negative is also almost the same for every cycle. This refers to the effect of unbalance magnetic pull is least. Other advantages of the machine structure are the fact that by using slot opening width of 1.55 mm is still available and easy to install conductor in the slot compared with the slot opening width of 1.50 mm. In practical, slot opening width is limited to 1.50 mm, unless the conductor is difficult to install in the stator slot. From the series (12 models) of PMSG structures are investigated, the PMSG structure with slot opening width of 1.55 mm and shoe height of 2.3 mm it is can be accepted as the best structure.

7. Conclusion

The simulation of fractional structure of 8 pole/18 slot has been presented in this paper. It shows that by using factional structure of PMSG can reduce the cogging torque effectively. However, most of the PMSG structure simulated in this research show that their peak of both negative and positive values of the cogging torque are different. These refer to the effect the static unbalance magnetic pull (UMP). In the simulation result (as shown in Fig.3), the peak value of cogging torque in the structure is 0.00062 N-m for positive value and about -0.00064 N-m for the negative value. And other most important is that structure is not deteriorated with core saturation, the fact that is very important for the smooth of voltage waveform of the generator. It can be concluded that this typical of slot opening width of 1.55 mm and shoe height of 2.3 mm is the best structure of the fractional 8 slot /18 poles of PMSGs, because it has any significant low of cogging torque and least of unbalance magnetic pull affect.

References

[1] L.Petkovska P.Lefley, G.Cvetkovski and S.Ahmed 2013 *Przegląd Elektrotechniczny*, ISSN 0033-2097, R. 89 NR 2b

[2] S.Ahmed and P.Lefley *Proceedings of the 11th Spanish Portuguese Conference on Electrical Engineering* pp. 1–5,

[3] P.Lefley, L.Petkovska, S.Ahmed ,G.Cvetkovski 2010 *Proceedings of the 14th International Power Electronics and Motion Control Conference EPE-PEMC'2010 on CD* pp. T4-96 – T4-101

[4] S,Ahmed and P. Lefley 2009 *Proceedings of the International Conference on Electric Power and Energy Conversion Systems – EPECS.*

[5] N. Levin S.Orlova, V.Pugachov, B. Ose-Zala and E.Jakobsonset 2013 *Elektroka ir elektrotechnika*, ISSN 1392-1215, Vol.**19**, No. 1

[6] L.Dosiek and P.Pillay 2007 *IEEE Transactions on industry applications*, vol.**43**, no.6 pp. 1656-1571

[7] E. Mulyadi, and J.Green 2002 *Proc. Of the 21 American Society of Mechanical Engineers Wind Energy Symposium* pp.1-8

[8] T.Srisiriwanna and M.Konghirun 2012 *ECTI Transactions on Electrical Eng, Electronics and Communication* Vol.**10**, No.2

[9] J.M Ling and T.Nur 2016 *Advanced in Mechanical Engineering* Vol.**8**, No.8, pp. 1-9

[10] D. Meeker 2010 *Finite Element Method Magnetics* Version 4.2 User's Manual

IOP Conf. Series: Materials Science and Engineering **807** (2020) 012011 doi:10.1088/1757-899X/807/1/012011

[11] Nur Tajuddin Mohammed and Tamer Z Fouad 2018 *Advances Science Letter* Vol **24**, No 11 pp 8875-8879
[12] M Arun Noyal Doss, V. Ganapaty, R Brindha, M Saishirinivas Rao, Shivva Rohit and K.Karthik 2016 *International Science Press* pp 6895-6902

IOP Conf. Series: Materials Science and Engineering **807** (2020) 012012 doi:10.1088/1757-899X/807/1/012012

Design of Low Speed Generator 1 Phase Using Permanent Magnetic Type Neodymium

Syafriyudin[1],Mujiman[2],Irawadi Buyung[3]

[1,2]Department of Electrical Engineering Institut Sains & Teknologi AKPRIND Yogyakarta Indonesia

[3] Department of Electrical Engineering Universitas Respati Yogyakarta, Indonesia

Corresponding e-mail: dien@akprind.ac.id

Abstract. The need for electricity in Indonesia and throughout the world is increasing, while most of the energy sources for electricity generation are still using fossil fuels such as oil, natural gas and coal. Fossil energy has a high level of pollution which is the cause of global warming, it is necessary to replace fossil energy that is environmentally friendly, namely renewable energy. The construction of power plants such as micro hydro power plants (PLTMh) which are located in irrigation canals or rivers with relatively small water discharge will result in low wheel rotation. The waterwheel rotation in this PLTMh system has a low rotating speed of around 200 rpm, and has a low torque or no cogging torque effect. Based on the needs in the field, a permanent Neodymium Type (NdFeB) Permanent Magnet Axial Flux Generator is built, which is specifically designed for low speeds of 200 rpm and has no cogging torque or low rotational torque effects. And it is hoped that the design of this generator can be used according to the needs of PLTMH. The results obtained in this study are, the generator can produce a voltage at a rotational speed of 200 Rpm and a frequency of 50 Hz is 100 VAC.

1. Introduction

Energy is one of the main problems faced by almost all countries in the world. This is because energy is one of the main factors for the economic growth of a country. Energy problems are becoming increasingly complex when the increasing need for energy from all countries in the world to sustain economic growth actually makes the supply of conventional energy reserves less and less. Indonesia is one of the countries with the largest growth rate in energy consumption in the world. Based on data from the Directorate of New-Renewable Energy and Energy Conservation, Ministry of Energy and Mineral Resources (ESDM), Indonesia's energy consumption has increased by seven percent per year in recent years. Cumulative global renewable electricity installed capacity grew by 8% in 2014 (from 1,579 GW to 1,712 GW), growth which has remained stable (6.6% per year from 2004-2014) in recent years. The addition of global renewable electricity capacity grew by 24% compared to 2013 (from 108 GW to 134 GW in annual capacity additions), a rate higher than in recent years. Renewable sources accounted for almost 24% of all electricity (5,507 TWh) worldwide in 2014. Global solar PV (Photovoltaic) and CSP (Concentrating Solar Power) increased by 28% and 29% in 2014, each installed capacity. Wind Power Generation installed capacity grew by 16% globally [1].

The total percentage of use of electricity production in the world until the end of 2015, fossil or non-renewable energy is still used for electricity production in the world, namely 76.3% and renewable energy at 23.7%. Hydropower still has the largest capacity in renewable electricity, which is 16.6% [2]. To overcome the threat of future energy deficits, the development of new and renewable energy (renewable energy) in Indonesia is a necessity. Moreover, the potential of the "Thousand Islands" is very abundant. Indonesia's energy consumption pattern also needs to shift from fossil energy to new-renewable energy. This is because, in the midst of declining production levels, national energy consumption is still dominated by non-renewable energy, aka fossil energy. Indonesia plans to increase renewable energy consumption by 25% by 2025 and 31% by 2050 [3]. Figure 1 below shows the share of power plant capacity from NRE and fossil energy.

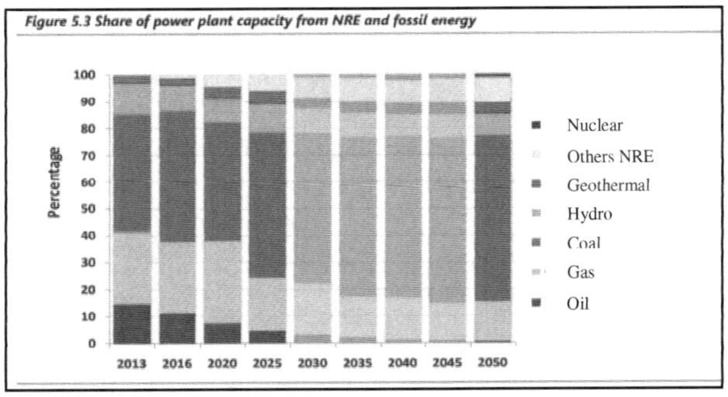

Figure 1. Share of power plant capacity from NRE and Fossil energy.

Electricity needs in Indonesia and the rest of the world are increasing while some of the energy for electricity generation still uses fossil energy such as oil, natural gas and coal. Until now, Indonesia has not been able to meet the electricity needs of the community. For that, a smart solution is needed to overcome these problems. The Microhydro Power Plant (MHP) is an alternative solution to address these problems. Microhydro is a term used for small-scale hydropower plants. The working principle of the MHP itself is basically the same as the hydropower plant; it is only different in capacity or size. MHP in principle utilizes height or slope differences and the number of debits per second in irrigation canals or rivers.

In this study the use of Neodymium (NdFeB) Axial Permanent Magnetic Flux Generator without Core Stator, which is specifically designed for low speeds of 200 rpm, because it uses permanent magnets it is not necessary to direct current reinforcement from the outside and does not have a cogging torque effect or swivel torque low. And it is expected that the design of this generator can be used according to the needs of the MHP.

Excitation is a part of the system of a generator that functions to form / produce flux that changes with time, so that it produces one GGL induction [4].Operating performance in a low speed range and is being studied. A new design method based on a multi-speed design strategy (MSD) is proposed. Using this method with stateless type without core, the efficiency of the total AFPM engine in the HEV operating cycle can be increased. Analysis of the performance of the AFPM in-wheel engine is done using the finite element method (FEM). FEM analysis from a single speed design method (SSD) is also done. Machines designed by MSD and SSD are applied in HEV and simulated using urban cycles and highways. The results obtained show better HEV performance, using machines designed based on MSD in all operating cycles. The experimental results were obtained from practical prototypes of samples, confirming analytical methods [5]. a novel design of axial flux permanent magnet generator using torus construction of stator with plastic material for wind energy generation system. In this Generator the stator is sandwiched between rotor discs where north pole of one rotor disc faces south pole of opposite rotor disc. In this arrangement as the flux does not pass radial in the stator disc, hence does not require magnetic material. This in turn means no iron losses in the designed machine. Designs were prepared for various ratings with same value of design parameters and comparison was made between Axial Flux Permanent Magnet Generator (AFPMG) and Radial Flux Permanent Magnet Generator (RFPM) [6].

Design and manufacturing process for axial flux permanent magnet generators is explained for low-cost rural electrification applications, where local production of small wind turbines is considered. This process is based on existing open source designs and construction manuals, while a systematic approach to designing generators with a nominal power range and for network connections or battery charging schemes is immediately explained. The process of making a 3 kW small wind turbine generator for network connections is used as a case study. Emphasis is placed on using simple tools and techniques to achieve lower costs, while the complex steps in the manufacturing process are explained in more detail. The design of the generator is simulated and the engine built is tested in the laboratory [7]. Design of axial flux permanent magnet generator is using torus construction of stator with plastic material. To maximise the efficiency of the generator, the losses are to be minimized in proposed N-S Arrangement iron loss is absent and only copper losses are present. To minimise the copper loss for a given rating, the per phase stator resistance is to be minimised. The relation between Internal Diameter (ID) to Outer Diameter (OD) Ratio and the per phase resistance has been developed and optimal value is worked out. Also the designs were prepared for various ID to OD ratios to verify the optimal results [8].

Reuse the used stator core and frame from a three-phase induction motor (3Φ) through reconditioning the stator frame and core for the acquisition of a single-phase alternator. The structure of the stator winding is used to fabricate rotor structures with radial flux permanent magnets. The existence of a rotor structure with a radial flux permanent magnet on a single-phase alternator is used as the basis for calculating the theoretical mechanical power required from the starting drive. The structure of the stator winding is formed by six groups of coils (lap) type one lap (lap winding) in the range of 6 (six) channels connected in series. Hollow cylindrical rotor structure with dimensions of 60 mm, 30 mm and 85 mm equipped with 12 permanent magnets. The value of the theoretical electrical power produced by the alternator is 81VA, while the theoretical mechanical power required from the initial drive is 350 watts [9]. The axial flux (disc shape) permanent magnet machine is an attractive alternative to radial flux (cylindrical shape) machines in wind turbine applications. The axial flux configuration is amenable to the low-speed, high-torque operation of a direct drive wind energy system, Direct drive wind energy conversion tends to decrease the system size, weight, and noise, while increasing overall efficiency and reliability [10].

Design of an 8-pole, single-phase permanent magnetic synchronous generator using Neoflux-30 type Neodymium Ferrite Boron (NdFeB) magnet which is operated at a speed of 750 rpm. The test results show that when the generator is not loaded it produces a voltage of 7.91 Volt and when loaded, the voltage generated by the generator is 6.11 Volt with an efficiency of 32.84% [11]. Axial flux permanent magnet synchronous generators (PMSG) are designed as double stators and three rotors and their electromagnetic and structural characteristics are analyzed. The design aimed at the axial flux generator is placed to the single end of the inner rotor of the engine and a permanent magnet is placed into the double end of the middle rotor. One rotor more than the number of stator here is used. Cores are not used in the stator of the machine intended to be designed. The purpose of this study is to provide a reduction in both iron loss and making the engine lighter by reducing the number of rotors to be used. The new axial flux generator is designed to move as a permanent speed of 500 rpm so that the maximum voltage is around 120 V per phase [12].

Microhydro power plants that use irrigation channels or river flow require low-speed generators. At this time in the market it is very difficult to find low-speed ac generators, so that in this study researchers tried to plan low-speed generators, where the stator planning was made from epoxy resin material and the rotor is made of white acrylic material with a thickness of 5mm, with a rotation plan of 200 rpm, with an output voltage of 220 volts at a frequency of 50 Hz.

2. Experimental set up

2.1. Design Generator Axial.

In the research carried out concerning low-speed permanent magnet generator 1 axial type using a permanent magnet type Fe. The faster the rotation the greater the voltage produced. In this study a

IOP Conf. Series: Materials Science and Engineering **807** (2020) 012012 doi:10.1088/1757-899X/807/1/012012

permanent magnet generator was designed, with the design of a generator in the form of an axial flux type 1 phase generator using 15 pairs of permanent magnet NdFeB type and comparison of results and generator performance will be carried out. Figure 2 below shows the design of axial generator permanent flux.

Figure 2. Design axial generator permanent flux

2.2. Magnet neodymium

Neodymium magnets known as NdeFb, NIB or Neo magnets, which are the most widely used types of rare-earth magnets are permanent magnets made of alloys from Neodymium, iron and boron to form Nd2Fe12B tetragonal crystalline structures. The neodymium magnet is shown as in Figure 3.

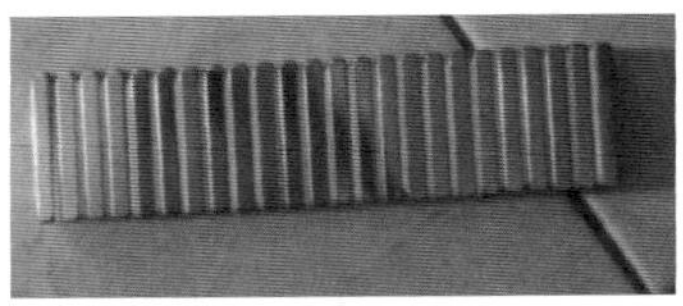

Figure 3. Neodymium magnet

2.3. Axial flux generator stator

Looking from the stator, this axial flux generator can be seen from a variety of variations including: Stator with iron core, stator without iron core, and number of coils. The number of coils in the stator depends on the number of phases and poles that you want to produce and the power produced. An iron-core stator is usually used for low speed and low load torque. Of course this is caused by the absence of an iron core inside. At the stator without an iron core the arrangement of the coils is divided into two types, some are arranged in overlapping and non-overlapping. The design of the stator is shown in Figure 4.

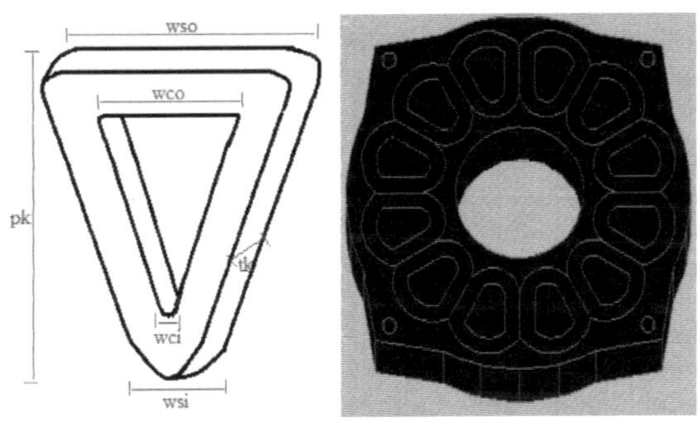

Figure 4. Stator design

2.4. Rotor

The rotor on the axial flux generator is a rotating part, the rotor consists of two parts that flank the stator, the rotor in this generator does not require external reinforcement to produce a magnetic field because it uses neodymium iron boron (NdFeB) permanent magnet. Permanent magnets use these materials which are basically the strongest magnets today compared to other types of magnets. The amount of magnetism in the rotor depends on the number of phases and the poles on the stator; the magnetic arrangement of the rotor is opposite the poles. Figure 5 shows the design of the rotor.

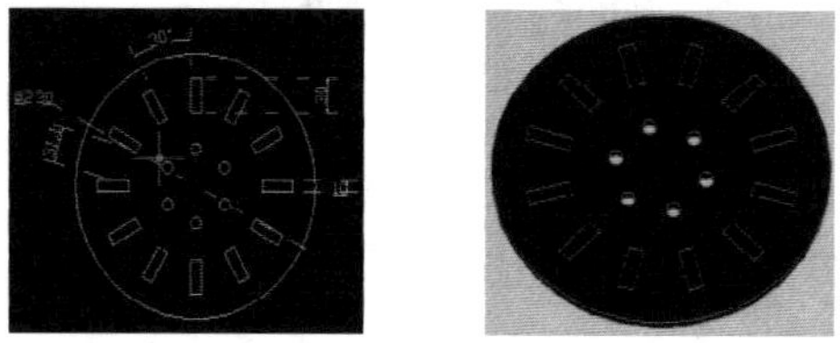

Figure 5. Design of rotor axial flux generator.

In order to find out the maximum flux density that can be produced by a rotor in the air gap equation 1 can be used.

$$B_{max} = B_r \frac{lm}{lm+\delta} \tag{1}$$

where B_{max} is the maximum flux density (tesla), B_r is the magnetic flux density (tesla), lm is the high, and δ is the air gap distance (meter). The maximum magnetic flux value produced can be calculated by using equation 2.

$$\Phi_{max} = A_{magnet} \cdot B_{max} \tag{2}$$

where Φ_{max} is the maximum magnetic flux (Wb), A_{magnet} is the extent of magnetic fields (m^2), and B_{max} is the maximum magnetic field (tesla). Figure 6 shows the measurement in the magnet.

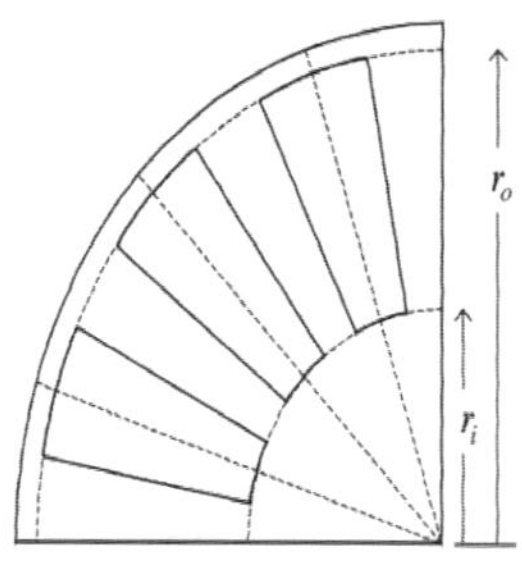

Figure 6. Radius in a magnet (ri), magnetic outer radius (ro).

2.5. Design Stator Generator Flux Axial

The stator of the axial flux generator can be seen from a variety of variations including: Stator with iron core, stator without iron core, and number of coils. The number of coils in the stator depends on the number of phases and poles that you want to produce and the power produced. The shape of the stator on the axial flux permanent magnet generator can be seen in the following figure:

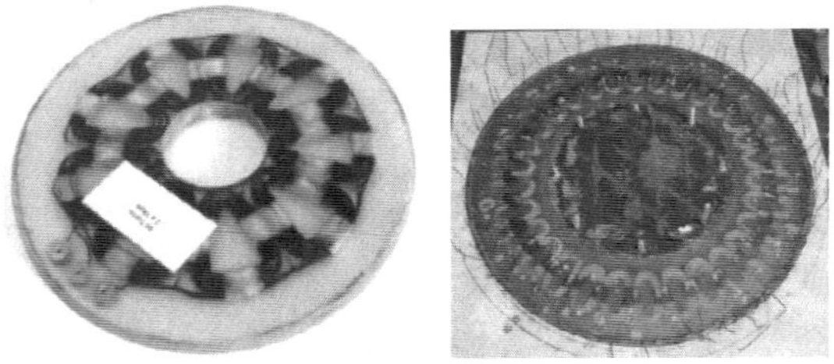

Figure 7. The form of the stator axial flux generator

2.6. Design
The axial flux generator that is to be created has specification as shown in Table 1 below.

Table 1. Generator specification

Specification	Description
Phase:	1
Pole:	30
Frequency	50Hz
Speed	200 rpm
P	74.4 VA
Magnet	Neodymium (NdFeB) grade 50
	Size (L×W×H): 3×1×4 cm

2.7. Result
At the measurement of the generator without load, measurements of frequency, voltage and rotation speed of the generator are carried out. Measurements were taken in eleven measurements, namely the frequency of generators from 10Hz to 110. The measured data is tabulated as shown in Table 2 below.

Table 2. Data comparison of frequency to voltage and rotation

Frequency of generator (Hz)	Voltage of generator (volt)	Rotation speed of generator (Rpm)
10	22,5	40
20	40	80
30	60	120
40	85	160
50	100	200
60	134	240
70	150	280
80	171.6	320
90	193.4	360
100	215.2	400
110	236	440

From the data shown in Table 1, the frequency generator characteristics obtained from generator voltage indicate that the higher the voltage increase in the generator, the frequency will also increase; this is to be able to see the generator frequency characteristics of the voltage at 50 Hz at 100 volts. This can be seen in the following graph in Figure 8 below:

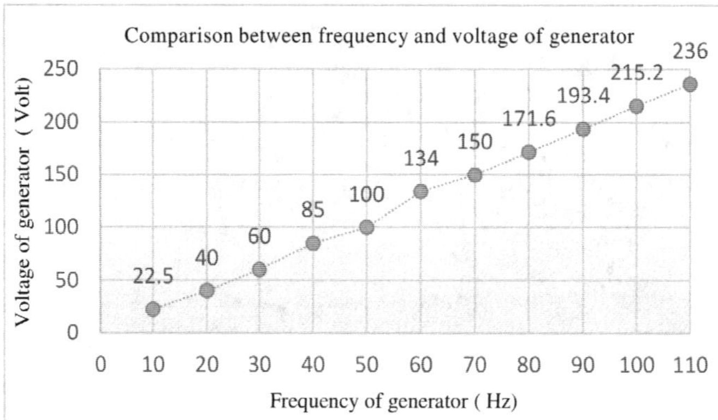

Figure 8. Comparison of Analysis Generator Frequency and Voltage

Also from Table 1, it can be seen that the increase in speed at the generator is linear with an increase in the frequency value, so that the characteristics of the generator at a frequency of 50 Hz are obtained at 200 rpm which is as shown in Figure 9 below.

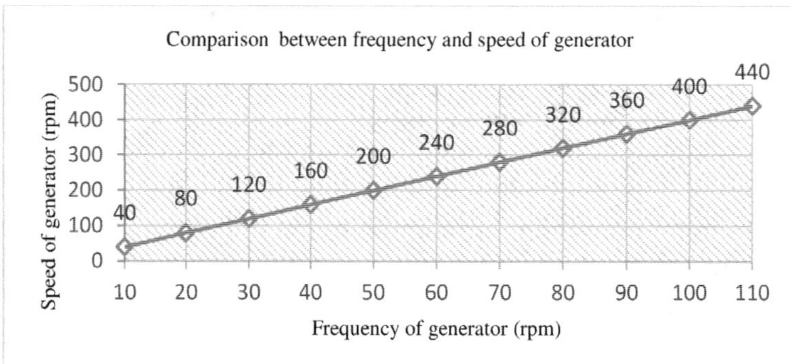

Figure 9. Comparison of speed to generator frequency

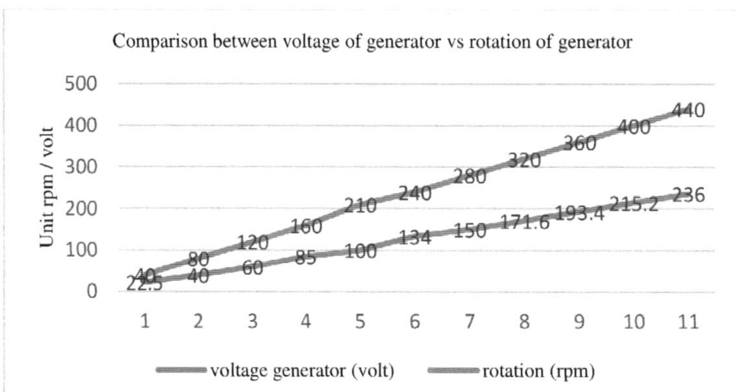

Figure 10. Comparison of generator speed to voltage generator

Figure 10 shows that the rotation of the generator increases as the voltage increases in the generator. This shows that the increase in rotation will increase the magnitude of the magnetic flux that is generated so that it affects the voltage generated by the generator.

On the other hand, the measurement of the loaded generator, generator and current measurements are made. Different generator setting were used. For generator settings at 200Rpm, 50Hz, 100V, the measured data are flow and power. The load used is in the form of 3 led lights arranged in parallel. For setting the generator 420Rpm, 105Hz, 220V, the measured data are flow and power. The load used is in the form of 1 18W LED lamp. Measurement with Generator Settings 200Rpm, 50Hz, 100V is taken 2 measurement parameters including: current and power. Measurement data can be seen in the following Table 3 below.

Table 3. Measurement Data with Generator Settings 200Rpm, 50Hz, 100V

Load	I (ma)	P (w)
1 pcs led 18w	20	2
2 pcs led 18w	51	5.1
3 pcs led 18w	90	9

On the other hand, the measurement with Generator Settings 420Rpm, 105Hz, 220V, there are 2 parameters used in this measurement, namely: current and power. Measurement data can be seen in Table 4 below.

Table 4. Measurement data with Generator Settings 440 Rpm, 105Hz, 220V

Load	I (ma)	P (w)
1 pcs led 18w	120	26.4 w

Comparative analysis is used to determine the characteristics of the current and power that the generator generates against the load can be seen in the graph as shown in Figure 11 below.

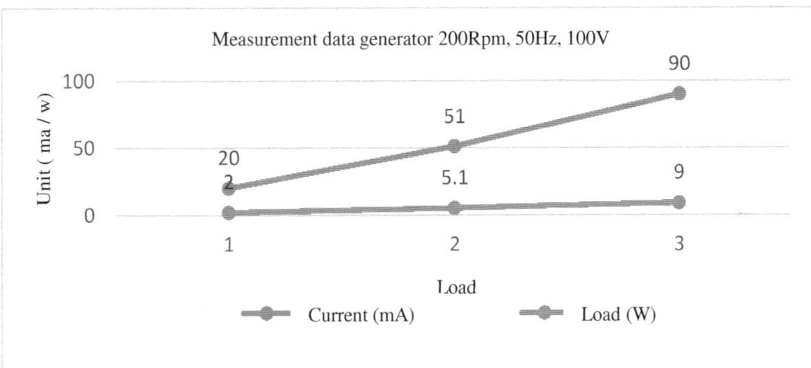

Figure 11. Load characteristics testing (generator settings: 200Rpm, 50Hz, 100V)

The waveform generator is also analyzed using an oscilloscope and viewed the waveform. An oscilloscope (see Figure 12) is used using analog and digital types, how to connect it is a probe on an oscilloscope connected directly to the generator voltage terminal.

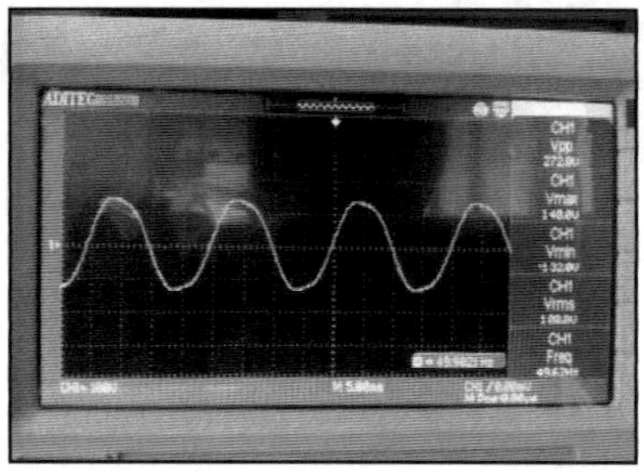

Figure 12. Waveform Generator

3. Conclusion

In planning this axial flux generator can produce 100 volt generator voltage, at a rotation speed of 200 Rpm with a frequency of 50 Hz, the level can rise according to the rotation speed of the generator, as in Figure 10 Rpm and 50 Hz frequency, axial flux generator output voltage is influenced by several parameters, including: rotation speed, magnet type, magnetic volume size, coil shape and size, many turns in one coil and air gap or air gap between the rotor and stator. The output of axial generator flux is influenced by the size of the copper wire email used. The choice of the type of permanent magnet is very influential on the output voltage of the generator. This is very influential, where the volume of the magnet is proportional to the generator voltage.

References

[1] U.S Departement of Energy's National Renewable Energy Labolatory 2015 *2014 Renewable Energy Data Book.*

[2] Renewable Energy – Wind Energy 2016 *Energy in the 21th Century* 198-226 (World Scientific)

[3] Badan Pengkajian dan Penerapan Teknologi 2015 Outlook Energi Indonesia 2015, Pengembangan Energi untuk Mendukung Pembangunan Berkelanjutan (Jakarta)

[4] Jerković V, Miklosevic K, Zeljko, S 2010 Excitation System Models of Synchronous Generator *SIP 2010 28th International Conference Science in Practice, Subotica Tech* 77-82

[5] Price G F, Batzel T D, Comanescu M, Muller B A 2008 Design and Testing of a Permanent Magnet Axial Flux Wind Power Generator *Int. Conf. Eng. & Tech* **190** 202

[6] Saifee A H, Mittal A 2014 Design of novel axialflux permanent magnet generator (AFPMG) for wind energy application *International Journal of Electrical and Electronics Engineering Research* **4(3)** 35-42

[7] Latoufis K C, Messinis G M, Kotsampopoulos P C, Hatziargyriou N D 2012 Axial Flux Permanent Magnet Generator Design for Low Cost Manufacturing of Small Wind Turbines *Wind Engineering* **36(4)** 411-431

[8] Saifee A H, Mittal A 2014 Optimisation of Inner Diameter to Outer Diameter Ratio of Axial Flux Permanent Magnet Generator *IOSR Journal of Electrical and Electronics Engineering* **9(6)** 43-47

[9] Goeritno A 2016 Struktur belitan stator dan rotor bermagnet permanen fluks radial untuk alternator fase tunggal *Proceeding Seminar Nasional Sains dan Teknologi 2016* 2407-1846 (Fakultas Teknik Universitas Muhammadiyah, Jakarta)

[10] Bagher H H M 2013 A New Design Method for low-speed torus type AFPM machine for HEV application *International Journal of Research in Engineering and Technology* **2** 12

[11] Herudin, Prasetyo W D 2016 Rancang Bangun Generator Sinkron 1 Fasa Magnet Permanen Kecepatan Rendah 750 RPM *Jurnal Ilmiah SETRUM* **5** 1

[12] Milnaz M R, Çelebi M 2017 Design and analysis of a new axial flux coreless PMSG with three rotors and double stators *Results in Physics* **7** 183-188

Application of Greenhouse Gas Monitoring System Using General Packet Radio Service on GSM Network

C Iswahyudi[1], M A Novianta[2], Muchlis[3]*and H P Suseno[3]

[1] Department of Informatics Engineering, Institut Sains & Teknologi AKPRIND Yogyakarta, 55222, Indonesia
[2] Department of Electrical Engineering, Institut Sains & Teknologi AKPRIND Yogyakarta, 55222, Indonesia
[3] Department of Environmental Engineering, Institut Sains & Teknologi AKPRIND Yogyakarta, 55222, Indonesia

*Corresponding e-mail: muchlis@akprind.ac.id

Abstract. Emissions of carbon dioxide (CO_2) is the greatest influence on the air quality for the largest contribution to global warming is about 50%. These gases absorb and reflect radiation emitted waves of the earth, so that the heat will be stored on the surface of the earth. The absorption capacity on mangrove plants is one of the efforts to reduce CO_2 emissions. In this work, a gas monitoring system was developed to monitor greenhouse gas utilizing GSM network using GPRS service. Based on experiment result, this system can monitor the absorption of CO_2 emissions and the formation of O_2 gas in mangrove plants as one of the efforts to reduce the impact of global warming. The information can be transmitted using SMS service on GSM network in less packet size and data transmission time, and also can be accessed through web-based application. As conclusion, the use of small and portable gas-sensing air pollution monitoring technologies can obtain more spatially and temporally representative air pollution data.

1. Introduction

A major problem in the 21st century is outdoor air pollution which is contributing to approximately 3.7 million deaths globally [1]. Referring to the WHO-specified limits on air pollutants level, today, approximately 92% of the world's population lives in the regions where air pollution levels are categorized as unsafe [2]. However, air pollution is also the cause for global climate change [3] and environmental problems such as acid rain [4], haze [4-5], ozone depletion [6-7] and damage to crop [8-10]. Thus, efforts are needed to resolve the issues [11]. Carbon dioxide (CO_2) is one of the greenhouse gases that cause global warming, with a contribution of 50% [12]. CO_2 gas absorbs and reflects the radiation emitted by the earth, so that the heat will be stored on the surface of the earth. Traditionally, air pollution is monitored by measuring the concentrations of various pollutants such as carbon monoxide (CO), ozone (O_3), nitrogen dioxide (NO_2), sulphur dioxide (SO_2), and particulate matter (PM) at fixed sites by using accurate and expensive instrumentation [13-17].

Recent advancements in the field of sensors, wireless communication technology, and digital electronics have raised a new paradigm for air pollution monitoring [18-19]. A low-cost sensor for monitoring real-time concentrations of different air pollutants on a ubiquitous network to gather high-resolution spatio-temporal air pollution data implemented by this paradigm. The air pollution data can be then utilised for a variety of air pollution management tasks. Several articles have already addressed this emerging area of sensor-based air quality monitoring. Generally, previous researchers are focus on the needs, challenges, benefits, and future directions of a sensor-based pollution monitoring for widely applications [20-21].

The main advantage of Global System for Mobile Communication (GSM) is the ability for international roaming. With these capabilities, GSM services can reach remote areas [22]. General Packet Radio Service (GPRS) is a packet-switching technology that enables data transfers through the cellular network's global system for mobile communications (GSM). GPRS can transmit the data in the form of multimedia, e-mail, Wireless Application Protocol (WAP), and World Wide Web (WWW) [23-25]. Previous research related to the measurement of hazardous gases, that concerns the process of carbon dioxide (CO_2) uptake in plants by designing a monitoring and controlling system of CO_2 gas concentration have been carried out by [26-31].

The major objective of this research is to develop a system to gather spatio-temporal air pollution data (CO_2, O_2, etc.) by using a ubiquitous network of low-cost sensors for monitoring real-time concentrations of different air pollutants in mangrove plants area, utilizing GSM network using GPRS service. This paper first gives a brief overview of technological advancements of a sensor-based air pollution monitoring system. Next, the architecture of the implemented system is described in Section 2. Section 3 presents measurement results with discussion, while the last section contains some concluding remarks.

2. Experimental apparatus and procedure

This paper discusses the greenhouse gas monitoring system to monitor the absorption of CO_2 gas emissions and O_2 gas concentration in mangrove plant area. The level of emissions and concentration of both gasses monitored continuously, then the data will be stored based on time interval measurements that can be set on part per million (m) units. Periodically the data are sent to the web server as configured period Short Message Service (SMS) which is used in order to configure the monitoring unit asynchronously. The observation system uses a gas sensor of carbon dioxide (CO_2) and oxygen (O_2) placed around the mangrove plant area. Observation station point spread in Muara Opak Mangrove Forest Area located in Baros Hamlet, Tirtohargo, District of Kretek, Bantul Regency. Figure 1 presents the architecture of the system. The monitoring unit is responsible for detecting gas concentration, initialization, and control of the GPRS/GSM modem. Communication between monitoring unit and web server implemented over the wireless link based on GPRS. Then, the data provided by each monitoring unit interpreted by web-based application.

The CO_2 absorption monitoring system is divided into two separate systems that work in an integrated way:

i. Monitoring unit

Monitoring unit consist of electronic components that work to organize and read the sensors installed in the monitoring unit. This system also equipped with gas sensors of carbon dioxide, oxygen gas, temperature and humidity, and storage media. This system serves to send data to the web server using the HTTP protocol over the Internet network. Monitoring unit is implemented using Arduino Ethernet Shield 3G (SIM5216E). As a gas monitoring unit, O_2 sensor (OOM202), CO_2 sensor (MG811), temperature and humidity (DHT22) were used. As a GPRS modem, we have used SIM900A module. The system will detect the presence of CO_2 and O_2 gas in the air, then read by the gas sensor and produce analogue output stage. The data in the form of stage is then processed by the microcontroller, resulting in a series of numbers. The microcontroller contains a program to calculate the variable stage of the sensor. After the electrical signal is processed to obtain data which is then displayed on the LED screen as an indicator of the value of CO_2 and O_2 gas content contained in air. The last process is to send information of gas concentration using GPRS service automatically.

ii. Web-based application

GPRS modem is responsible for communication with the remote web server that runs web-based application, which is accessible by the standard web browser and presents the results to the end user for the analysis. The usage of pull communication model eliminates the need of public IP address for the monitoring system, which reduces the total cost of the system. Data and configuration parameters are exchanged between monitoring unit and web-based application using HTTP-POST method. Web-based monitoring application functions to receive data from microcontroller via HTTP protocol over the internet network, store the data in web server, then display information of monitoring result in real-time and online. The observed information is transmitted using GPRS service on GSM network, which is visualized on the web page.

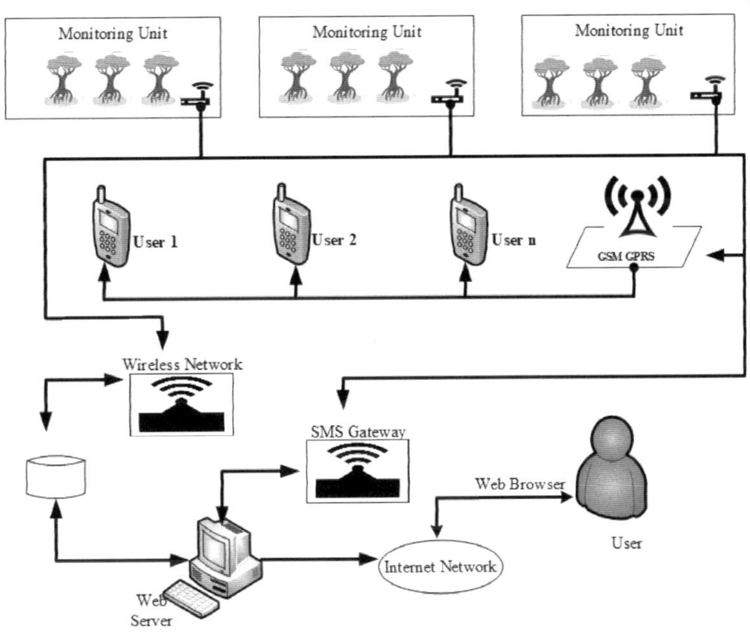

Figure 1. Architecture of the system

3. Results and discussion

In most cases, air pollution is monitored by measuring concentrations of various pollutants such as carbon monoxide (CO), ozone (O_3), nitrogen dioxide (NO_2), sulphur dioxide (SO_2), and particulate matter (PM) at fixed sites by using accurate and expensive instrumentation [16-18]. So far, previous research that concerns to the measurement of hazardous gases by designing a monitoring and controlling system of CO_2 gas concentration have been carried out by [26-31]. Traditional approaches for measuring air quality based on fixed measurements are inadequate for personal exposure monitoring. We need the use of small, portable gas-sensing air pollution monitoring technologies employing portable and mobile methods to obtain more spatially and temporally representative air pollution data [32]. In recent years, researchers have investigated a variety of approaches to sensor-based air quality monitoring that focus on the needs, benefits, challenges, and future directions of a sensor-based pollution monitoring paradigm for different applications [20-21].

The proposed system is capable to display online information containing fluctuations in carbon dioxide, oxygen, temperature, and humidity levels. Information from the monitoring unit is transmitted over GSM networks using the GPRS service to the server. Communication between server and monitoring unit using Raw Socket (TCP/UDP) technology. The information stored in the database is displayed on a web page that has been completed with the marker and value of the monitoring results. Web-based applications allow users to access monitoring units located in multiple locations.

The realized system was installed in the real environment located at Muara Opak Mangrove Forest Area located in Baros Hamlet, Tirtohargo, District of Kretek, Bantul Regency for about 30 days. Data from monitoring unit is recorded every 5 minutes and sent to server every hour. During the testing period, no erratic behaviour was observed in the recording and sending of data in the system.

3.1. Communication between monitoring units and server

Communication between monitoring units and web servers using GPRS data services on GSM networks. The modes contained in the GSM hardware module can work simultaneously between GSM for SMS service for text messaging and GPRS service for data communications. Instructions on the GPRS service to enable internet data communications on the SIMCOM Module which is the user sends an "http on" text message to the GSM number on the monitoring unit, as shown in Figure 2.

```
void kirim_HttpOn(char * NomerHp)
{
DisableTimer();
flagtimer=1;
printf("AT+CMGS=\"%s\"\r\n",NomerHp);
if(wait_karakter('>',5000)==1)Buz_key2=1;
putsf("ID Device: ");kirim_data_eemprom(idAlat);putsf("\nStatus Online");putchar(26);
wait_karakter('+',10000);
wait_Oke();
EnableTimer();
}
```

Figure 2. Program script for communication between monitoring unit and web server

Part of program script to enable GPRS service on GSM module are shown in Figure 3.

```
HTTPINIT:
r=send_USART(commandHTTP[5],"OK",2000);
if(r==0 && ct <3){ct++;goto HTTPINIT;}r=ct=0;

HTTPPARA1:
r=send_USART(commandHTTP[6],"OK",2000);
if(r==0 && ct <3){ct++;goto HTTPPARA1;}r=ct=0;
HTTPPARA2:
printf("AT+HTTPPARA=\"URL\",\"http://monitoringudara.com/data.php?id=%s&co2=%f&o2=-
%f&v=%d&s=%d\"\r\n",id_buf,karbon,oksigen,suhu,lembab);
r=send_USART(NULL,"OK",2000);
if(r==0 && ct <3){ct++;goto HTTPPARA2;} r=ct=0;

HTTPACTION:
r=send_USART(commandHTTP[8],"+HTTPACTION:", 2000);
if(r==0 && ct <3){ct++;goto HTTPACTION;}
if(r==1){
    lcd_clear();lcd_gotoxy(0,1);lcd_puts("Transfer: OK");delay_ms(1000);
    }
if(r==0 && ct ==3){lcd_gotoxy(0,1);lcd_puts("Transfer: ERROR");delay_ms(1000);}r=ct=0;
```

Figure 3. Programme script to activate GPRS service on GSM module

3.2. Display the results of monitoring on the web browser

The data sample obtained from the monitoring unit located at the mangrove plant area shows the concentration of CO_2, O_2, temperature, and humidity during the observation period for 24 hours as shown in Table 1. This measurement aims to determine the level of CO_2 gas pollution during the process of photosynthesis of mangroves plant during the day, as well as changes in CO_2 gas concentrations at night.

Table 1. Measurement results in the mangrove plant area

No.	Time	CO_2 (m)	O_2 (%)	Temp (^{0}C)	Humidity (%)
1	5:30	280.0	23.8	27.0	85.0
2	5:45	290.0	23.7	28.0	84.5
3	8:00	360.0	22.5	29.7	81.5
4	8:15	366.4	22.4	29.9	81.2
5	10:00	403.0	21.3	31.3	78.7
6	10:15	407.0	21.2	31.5	78.4
7	11:30	420.6	20.4	32.4	76.6
8	11:45	423.4	20.3	32.6	76.3

9	12:00	426.2	20.1	32.8	75.9
10	12:15	429.0	20.0	33.0	75.6
11	13:45	453.8	19.1	34.1	73.5
12	14:00	462.3	18.9	34.3	73.1
13	14:15	469.9	19.1	34.5	72.8
14	17:15	561.1	18.1	36.8	68.6
15	17:30	568.7	18.0	37.0	68.2
16	17:45	576.3	17.9	34.0	67.9
17	18:00	583.9	17.8	35.0	67.5
18	18:15	591.5	17.8	33.0	67.2
19	23:30	751.1	18.2	27.0	68.1
20	23:45	758.7	18.3	28.6	69.3
21	0:00	766.3	18.3	29.0	70.5
22	0:15	773.9	18.4	28.0	71.8
23	2:45	849.9	18.8	25.8	84.0
24	3:00	827.1	18.8	25.6	85.2
25	3:15	834.7	18.7	25.4	86.4

HTTP-POST method is used to exchange the configuration and data between monitoring unit and web-based application. Web-based application functions to receive data from microcontroller via HTTP protocol over the internet network. Communication with the web server that runs web-based application is done by GPRS modem, which is accessible by the standard web browser and display the results to the end user for the analysis. Monitoring data is transmitted using a simple and short package format using GPRS services to minimize packet data transmission time. GPRS transmission rate can be improved to 56Kbps or even 114Kbps, so the connection and transmission become more convenient and easier.

Figure 4 shows the data obtained from the monitoring unit, located at 10 locations. Recorded data consist of location, date and time of data transmission, as well as information on temperature, humidity, and concentrations of CO_2 and O_2. Furthermore, the data is processed and displayed in graphical form as shown in Figure 5 to 8.

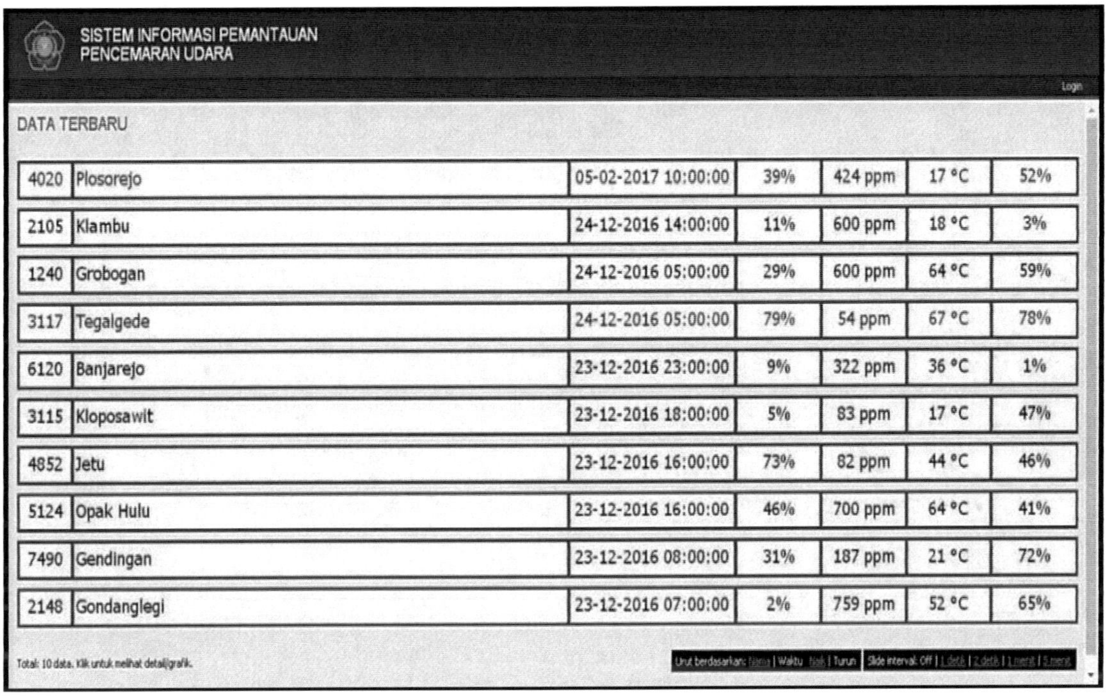

4020	Plosorejo	05-02-2017 10:00:00	39%	424 ppm	17 °C	52%
2105	Klambu	24-12-2016 14:00:00	11%	600 ppm	18 °C	3%
1240	Grobogan	24-12-2016 05:00:00	29%	600 ppm	64 °C	59%
3117	Tegalgede	24-12-2016 05:00:00	79%	54 ppm	67 °C	78%
6120	Banjarejo	23-12-2016 23:00:00	9%	322 ppm	36 °C	1%
3115	Kloposawit	23-12-2016 18:00:00	5%	83 ppm	17 °C	47%
4852	Jetu	23-12-2016 16:00:00	73%	82 ppm	44 °C	46%
5124	Opak Hulu	23-12-2016 16:00:00	46%	700 ppm	64 °C	41%
7490	Gendingan	23-12-2016 08:00:00	31%	187 ppm	21 °C	72%
2148	Gondanglegi	23-12-2016 07:00:00	2%	759 ppm	52 °C	65%

Figure 4. Web page displayed of CO2, O2, temperature, and humidity levels

Figure 5 to 8 show a graphical display of data recorded from one of the monitoring unit located at Opak. Figure 5 shows the fluctuations of CO_2 levels at the observation location over a 24-hour period with random time interval, while Figure 6 shows the O_2 concentration. Figure 7 displays the temperature, while Figure 8 displays humidity.

Based on the display in Figure 5 to Figure 8 it can be stated that the monitoring unit placed at the observation location can send data properly to the server. Data are exchanged between monitoring unit and web-based application by using HTTP-POST method. Web-based monitoring application functions to receive data from microcontroller via HTTP protocol over the internet network. The data is then processed and displayed through a web-based application.

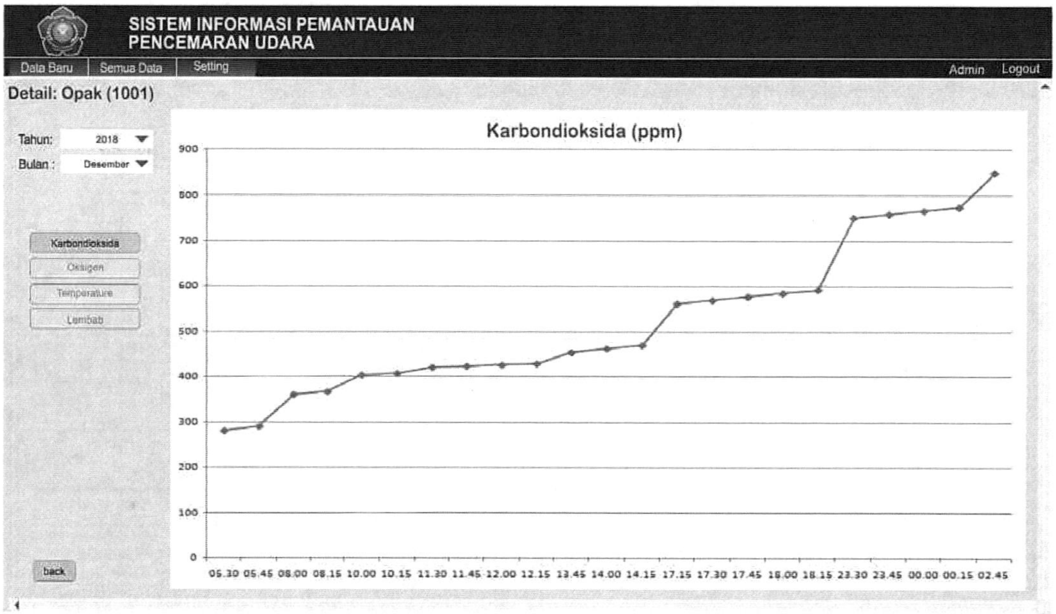

Figure 5. Graphic display of CO2 concentration

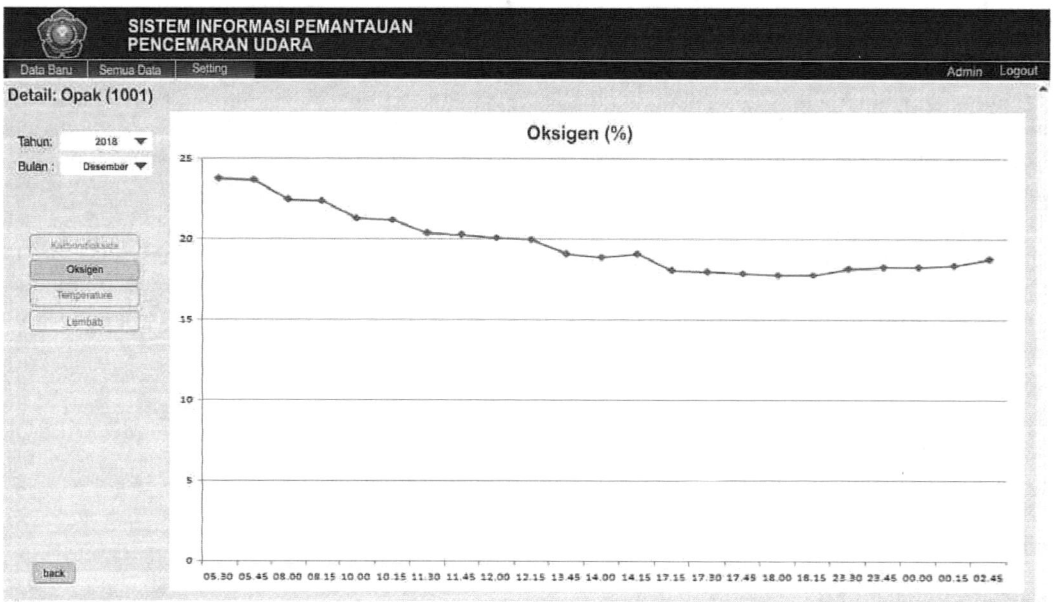

Figure 6. Graphic display of O2 concentration

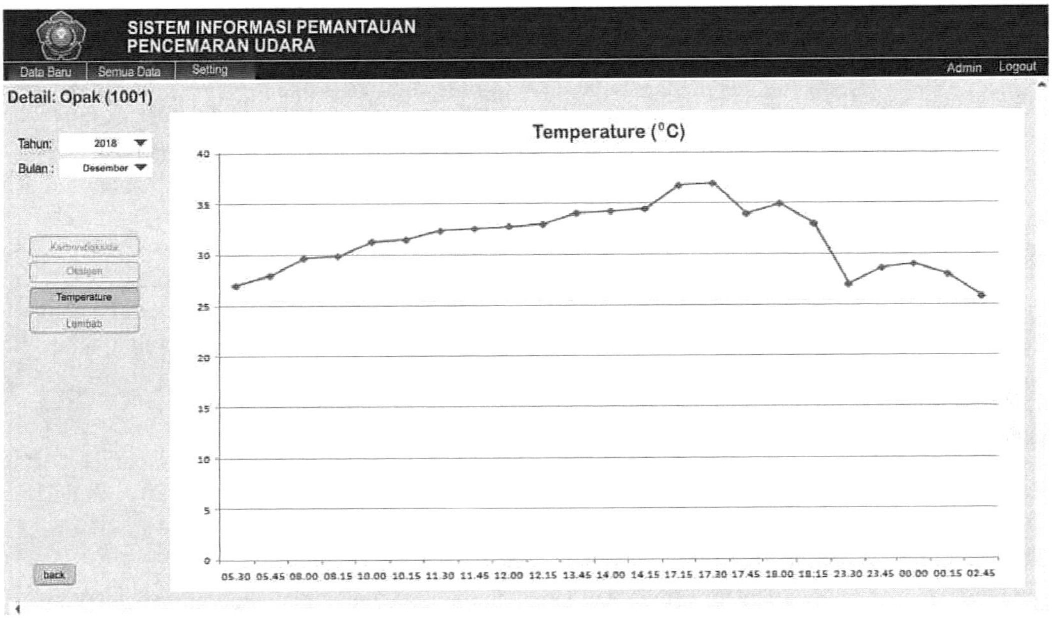

Figure 7. Graphic display of temperature

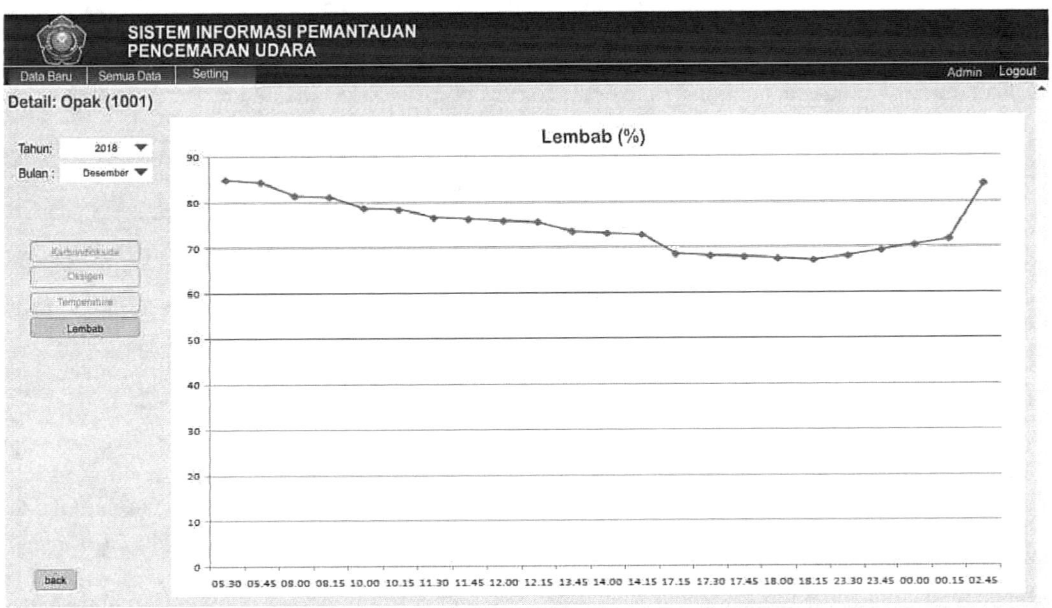

Figure 8. Graphic display of humidity

4. Conclusion

A low-cost architecture of gas monitoring system based on GPRS communication was described and implemented. The system was installed in the real environment at mangrove plant area during the period of 30 days and we have noticed no erratic behaviour during its exploitation. Based on experiment result, this system can monitor the absorption of CO_2 emissions and the formation of O_2 gas in mangrove plants as one of the efforts to reduce the impact of global warming. The information can be transmitted using SMS service on GSM network in less packet size and data transmission time, and also can be accessed through web-based application. As conclusion, the use of small, portable gas-sensing air pollution monitoring technologies employing portable and mobile methods can obtain more spatially and temporally representative air pollution data. In future, the advancement of the system can be improved to provide accurate information about the spatial distribution of pollutants or identify pollution hotspots.

Acknowledgments

The authors would like to greatly thank to the Directorate of Research and Community Service, Directorate General for Research and Development, Ministry of Research, Technology and Higher Education (KemenRistekdikti) who has funded the research activities in accordance with Implementation of Research Grant: DIPA-042.06.1.401516/2017.

References

[1] Ramanathan V and Feng Y 2009 Air pollution, greenhouse gases and climate change: global and regional perspectives *Atmos Environ* **43** 37–50
[2] Menz FC and Seip H M 2004 Acid rain in Europe and the United States: an update *Environ Sci Pol* **7** 253–65
[3] Xu P Chen Y and Ye X 2013 Haze, air pollution, and health in China *Lancet* **382** 2067
[4] Solomon S 1999 Stratospheric ozone depletion: a review of concepts and history *Rev Geophys* **37** 275–316
[5] Solomon S Garcia R R Rowland F S Wuebbles D J 1986 On the depletion of Antarctic ozone J. *Nature* **321** 755–58
[6] Avnery S Mauzerall D L Liu J and Horowitz LW 2011 Global crop yield reductions due to surface ozone exposure: 2. Year 2030 potential crop production losses and economic damage under two scenarios of O_3 pollution *Atmos Environ* **45** 2297–309
[7] Van Dingenen R Dentener F J Raes F Krol M C Emberson L and Cofala J 2009 The global impact of ozone on agricultural crop yields under current and future air quality legislation *J. Atmos Environ* **43** 604–18
[8] Fenger J 2009 Air pollution in the last 50 years–from local to global *J. Atmos Environ* **43** 13–22
[9] Kumar P Morawska L Birmili W Paasonen P Hu M Kulmala M Harrison R M Norford L and Britter R 2014 Ultrafine particles in cities *Environ Int* **66** 1–10
[10] Mouzourides P Kumar P and Neophytou M K A 2015 Assessment of long-term measurements of particulate matter and gaseous pollutants in South-East Mediterranean *Atmos Environ* **107** 148–165
[11] Hidayati R 2012 Masalah Perubahan Iklim di Indonesia Beberapa Contoh Kasus *Makalah Falsafah Sains*
[12] Sharma P Jain S and Kumar P 2013 An integrated statistical aroach for evaluating the exceedence of criteria pollutants in the ambient air of megacity Delhi *Atmos Environ* **70** 7-17

[13] Holmes N S and Morawska L 2006 A review of dispersion modelling and its application to the dispersion of particles: An overview of different dispersion models available.

[14] Sudarsono M A Novianta and Purnawan 2018 Development of SMS Gateway Information System for Detecting Air Quality *IOP Conf. Series: Materials Science and Engineering* **434** 1 1–6

[15] Sudarsono M A Novianta and Purnawan 2018 Performance of AVR Microcontroller-SMS Gateway for Detecting SO2 and NO2 *The 3rd Intl. Conf. of Energy, Environmental and Information System* **73** 1–5

[16] Kumar P, Ketzel M,Vardoulakis S, Pirjola L and Britter R 2011 Dynamics and dispersion modelling of nanoparticles from road traffic in the urban atmospheric environment: A review *J. Aerosol Science* **42** 580–603

[17] Castell N Viana M Minguillón M C Guerreiro C and Querol X 2013 Real-world alication of new sensor technologies for air quality monitoring *ETC/ACM Technical Paper*

[18] Snyder E G, Watkins T H, Solomon P A, Thomas E D, Williams R W, Hagler G S, Shelow D, Hindin D A, Kilaru V J and Preuss P W 2013 The changing paradigm of air pollution monitoring *Environ Sci Technol* **47** 11369–77

[19] White R M, Paprotny I, Doering F, Cascio W E, Solomon P A and Gundel L A 2012 Sensors for community-based atmospheric monitoring *EM: Air and Waste Management Association's Magazine for Environmental Managers* **5** 36–40

[20] Murota K and Hirade K 2001 GMSK Modulation for Digital Mobile Radio Telephony, *IEEE Transactions on Communications*-**29** 7 1044-50.

[21] Mladen K, Zeljko I, and Branko D 2012 GPRS-Based Electrical Energy Monitoring System *Symp. Ind. Elec.* 282-284

[22] Raul I, Gabriel V and Septimiu M 2012 GPRS Based Data Acquisition and Analysis System with Mobile Phone Control *Measurement* **45** 6 1462-70

[23] Susilo Y S, Pranjoto H and Gunadhi A 2014 Sistem Pelacakan Dan Pengamanan Kendaraan Berbasis GPS Menggunakan Komunikasi GPRS *Jurnal Ilmiah Widya Teknik* **13** 1

[24] Zhang H 2012 A self-adaptive greenhouse CO_2 concentration monitoring system based on ZigBee, cloud computing and intelligent systems (CCIS), *IEEE:The 2nd Intl. Conf.* 1137–40

[25] Gour P 2014 Application of fuel cell in power plant to reduce the carbon-dioxide emission *Intl. J. of Core Engineering & Management* **1** 5

[26] Løkke M M, Seefeldt H F, Edwards G and Green O 2011 Novel wireless sensor system for monitoring oxygen, temperature and respiration rate of horticultural crops post-harvest *Sensors* **11** 845-864

[27] Hagler G S, W Solomon P A and Hunt S W 2013 New Technology for Low-cost, Real-time Air Monitoring *EM Magazine Air & Waste Management Association*

[28] Kumar P, Morawska L, Martani C, Biskos G, Neophytou M, Di Sabatino S, Bell M, Norford L and Britter R 2015 The rise of low-cost sensing for managing air pollution in cities. *Environ Int* **75** 199–205

[29] White R M, Paprotny I, Doering F, Cascio W E, Solomon P A and Gundel L A 2012 Sensors for community-based atmospheric monitoring. *EM: Air and Waste Management Association's Magazine for Environmental Managers* **5** 36–40

IOP Conf. Series: Materials Science and Engineering **807** (2020) 012014 doi:10.1088/1757-899X/807/1/012014

Improvement of Teacher's Ability to Make IT-Based Teaching and Evaluation Materials Using Scribe, Camtasia and Quiz Creator (Case Study in SMK Negeri 1 Tembarak, Temanggung, Central Java, Indonesia)

E K Nurnawati[1*], E Sutanta[1] and R D Bekti[2]

[1] Informatics Engineering Department, Institut Sains & Teknologi AKPRIND Yogyakarta, 55222, Indonesia
[2] Statistic Departement, Institut Sains & Teknologi AKPRIND Yogyakarta, 55222, Indonesia

[*]Corresponding e-mail: ernakumala@akprind.ac.id

Abstract. Teachers are an important element in the formal education system, both in primary, secondary, and high education. The ability of teachers to prepare teaching materials is very influential on student learning outcomes. Learning materials that are easy to understand and look interesting will be preferred and easily understood by students. Evaluation is one of the benchmarks for knowing the results achieved by students. Success in delivering subject matter certainly influences the results achieved in student evaluation. This study aims to measure the extent to which training in making teaching materials and exam questions based on information and communication technology (ICT) affects improving the ability of teachers in the learning process and evaluation of students. The method used in this study is a method of collecting data with experimental methods for teachers before and after training. The software used in this training was Scribe 1.0.4 and Camtasia 8.0 video for making IT-based learning material and Quiz Creator 4.5.1 used for making student evaluation materials. Survey data collection was conducted on forty teachers who attended training and sixty students at SMK N 1 Tembarak Temanggung. Wilcoxon Test is used to analyse survey data. The results of this study are that there is a significant correlation with the ability of teachers to develop teaching materials before and after training. The results of a survey of students found that 80% of students preferred teachers to use ICT-based teaching materials and 75% of students preferred to work on ICT-based questions rather than conventional exam questions, 87% teachers found to benefit from the training. The Wilcoxon test proves that the teacher's ability to make teaching materials and exam materials increases after training.

1. Introduction

Along with the development of this globalization era, the world of Information and Communication Technology (ICT) is also growing rapidly. Currently, ICT has become a primary need for many groups. By using ICT, a process and activity can be carried out more quickly, easily, and efficiently. Therefore, mastery of ICT equipment needs to be taught at all levels. From the data from the Ministry of Communication and Informatics (Kominfo), it is known that in January 2019 internet users in Indonesia had reached 152 million people or around 55.92% of the Indonesian population which reached 268.2 million. The population increased from 2018, where at that time, Internet users were around 146.26 million or 54.68% [1]. The use of ICT has penetrated all fields, both education [2-10], business [11-13], advertising, social [14], agriculture, industry, smart city [15] and daily necessities. Many studies have been conducted regarding the use of ICTs and social media [16]. In schools, ICT is used to disseminate information on school activities. Research on the use of ICT in education was also carried out by [17], who stated that improving the ability to write teaching materials and self-regulation through representation approach. If initially educational technology (in a very limited sense) was only acting on the level of the implementation of the curriculum in the classroom, the new concept requires educational technology as input, even from the curriculum planning stage. Thus, since curriculum planning must also be studied and determined the form of educational technology to be applied. The new paradigm of education favoured the selection of quality education by prioritizing the principles and practices of autonomy, accountability, accreditation, and evaluation. Success in the teaching and learning process is determined by the relationship of the teacher and students and how the teacher makes an approach in the classroom [18], the impact of learning experience [19], the impact of beliefs and knowledge of technology among teachers [20], level of students awareness level toward ICT and mobile technology [21],a pedagogical model using ICT [22], students literacy [23], essential ingredients to literacy [24] and the state of arts of the models and the methodologies educational with ICT [25] and [26] also by teacher behaviour [27] and even using social media [11]. ICT professional development and ICT practice in the classroom will be enacted in the teacher's classroom [3]. Using ICT will stimulate teacher to integrate pedagogical practice and ICT itself to improve student's material and evaluations [6] and even embedding virtual reality in the classroom [28].

The application of technology in education in the global era of information is how applied the latest types of ICT in educational practice [29] make review and appraisal of e-learning and ICT systems in the teaching process. The teaching and learning process that applies the latest ICT can be the use of electronic media such as radio, TV, internet, and computer network systems, as well as other forms. The use of ICT in education has been used by all countries, both developed and developing countries. Research has been done by [23] exploring student literacy across 44 countries, Bangladesh [30], India, and Hungary [21], Saharan Africa [31], and Korea [32]. In Indonesia, teacher quality improvement program made using various way, by empowering teachers by psychological approach [33], by using ICT [34-37]. The use of ICT for making computer-based exam questions has also been carried out by [38] by utilizing the Wonder Share Quiz Creator to make Arabic language questions, while [39] uses the same software to create computer-based questions (Computer Based Test) at Harapan 1 Vocational School Medan. The use of ICTs to develop teaching materials has been carried out by [36] who uses information technology to obtain teaching materials from the internet. Whereas [38] use ICT to develop teaching materials at SMAN 1 Bulok Tenggamus Lampung.

The purpose of this study was to determine the effect of the use of ICT in improving the ability of teachers and students in SMK N 1 Tembarak Temanggung by training ICT based Teaching and Evaluation material using Camtasia, Video Scribe, and Quiz Creator.

2. Experimental Setup

2.1. Domain of Study
The object of the study were 40 teachers and 30 students at SMK Negeri 1 Tembarak Temanggung. The training includes training on making ICT-based teaching materials using Camtasia 8.0 and Video Scribe 1.0.4 and making exam questions using Quiz Creator 4.5.1. Before and after the training was given a list of ten questions which the questionnaires are about the effects of the training towards the respondents. The analytical method used is the Wilcoxon test which it used to observe the effects of training on these teachers [40].

2.2. Method of Study
Research methods include observation, training, data collection, and analysis of survey results with the following steps:

i. Observation
   Observations were made on the teachers at SMK N 1 Tembarak and provided a survey of what was used to make teaching materials and exam materials. From the observations, it was found that 80% answered they used word processing applications, and 100% answered that they wanted to be given training in making teaching materials using ICT-based applications that could be managed online. From these observations, training in the making of teaching materials was carried out using Camtasia and Video Scribe, and training in making and circumventing exam questions using Quiz Creator. The training was attended by 40 teachers.

ii. Training
   From the results of observations, training in the making of teaching materials was carried out using Camtasia 8.0 and Video Scribe 1.0.4, and training in making and circumventing exam questions using Quiz Creator 4.5.1. The training was conducted for 6 hours for each application. Training includes the creation of multi-media-based teaching material content and making questions and managing exam questions.

iii. Questionnaire
   After the training, a survey was held about the effects of the training. Ten questions to 30 students and 40 teachers were given and processed using the Wilcoxon test with the help of SPSS software.

iv. d. Analysis of test results
   After testing, the results obtained from the test are analysed and reported in the documentation.

2.3. Wilcoxon Test.
This theory was used to compare between two groups of interconnected data used the Wilcoxon test. This test has more test strength than the sign test. Assumptions for the Wilcoxon test, the data used is at least an ordinal scale.[43]. The equations for statistical calculation are as in equation (1).
The hypothesis used in the Wilcoxon test is as follows.
H0: d = 0 (there is no difference between the two treatments given)

H1: d ≠ 0 (there are differences between the two treatments given) by d shows the difference in value between the two treatments

$$Z = \frac{T - \left[\frac{1}{4N(N+1)}\right]}{\sqrt{\frac{1}{24N(N+1)(2N+1)}}} \quad (1)$$

Explanation:
N = the amount of data that changes after being given a different treatment
T = number of ranks from a negative difference value (if the number of positive differences is more than the number of negative differences)
Z = number of ranks from a positive difference value (if the number of negative differences is more than the number of positive differences)

Critical area: H0 is rejected if the value is $|Z| > Z\alpha / 2$ or P-value $<\alpha$.

3. Results and discussion

Before the training, a survey was conducted on the tools used in the production of teaching materials and evaluation materials. The results are shown in Figure 1, where 75% of teachers make teaching materials with the Word Processing and Presentation Processing Applications, and there are still 20% using conventional methods with blackboards without documentation, and only 5% have used ICT-based tools. Furthermore, the students' examination results were carried out by the teachers, and the average was taken for Bahasa Indonesia subject.

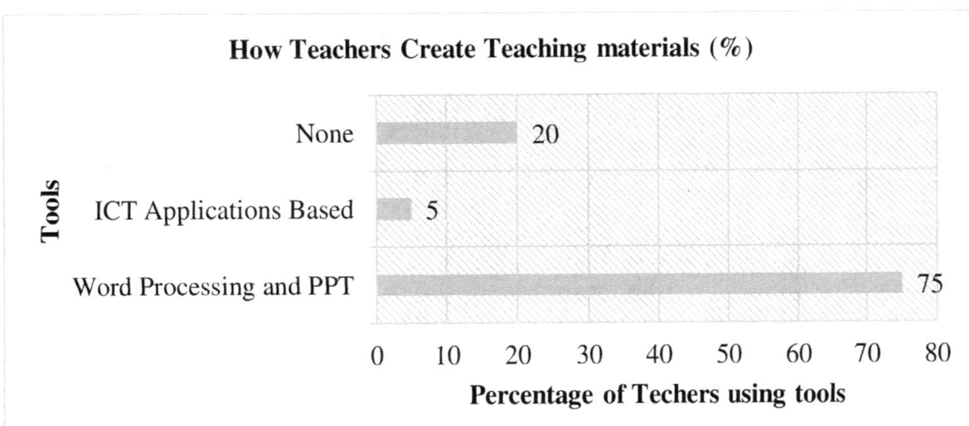

Figure 1. Results of a Preliminary Survey of 40 Teachers in SMK N 1 Tembarak

Then the training on making ICT-based teaching materials and evaluation materials is conducted by using Camtasia 8.0 and Video Scribe 1.0.4. Camtasia Studio is a multi-screen recorder and a video editor used for Windows operating system computer. This application allows one to design, make, and record professional-looking videos easily. It does create an impressive presentation for teachers, learning materials videos, and tutorials for many purposes. The feature in Camtasia including screen recording for Windows operating system computer, webcam recording, simultaneously record video and audio, built-in video editor, transitions and animations, free music and audio library, make interactive quizzes and record and import presentations from other media. Scribe video is a software for creating teaching materials as if you were teaching using video. Video Scribe is often referred to as whiteboard maker. Interesting and easy-to-understand teaching materials will surely be enjoyed by students. With these two applications, teachers can easily create teaching materials as needed and can be given online to students or recorded and can be studied again at home. Making evaluation materials is something that must be done by the teacher in the learning process to measure students' abilities. Regarding with advances in modern technology products, ICT-based learning models are able to help teachers to facilitate the learning process in achieving goals. Therefore, teachers need to be able to use software that can be used to make teaching materials and student evaluation materials. Quiz Creator is a software that is easily run by the teacher. Because, the display of the software is similar to word processing software, and no less important, the results of questions or quizzes made with Quiz Creator can be in flash files formats so that it can be accessed on any gadget as long as Flash Player is provided, also in web pages and even on CD.

After the training, teaching and evaluation materials were tested on Bahasa Indonesia class in three meetings and evaluated. The result is the ability of students increased by 25% with parameter test scores obtained from the material compared with the previous grades before the teacher made ICT-based teaching materials. Figure 2 shows the results of a score of 30 students before and after training.

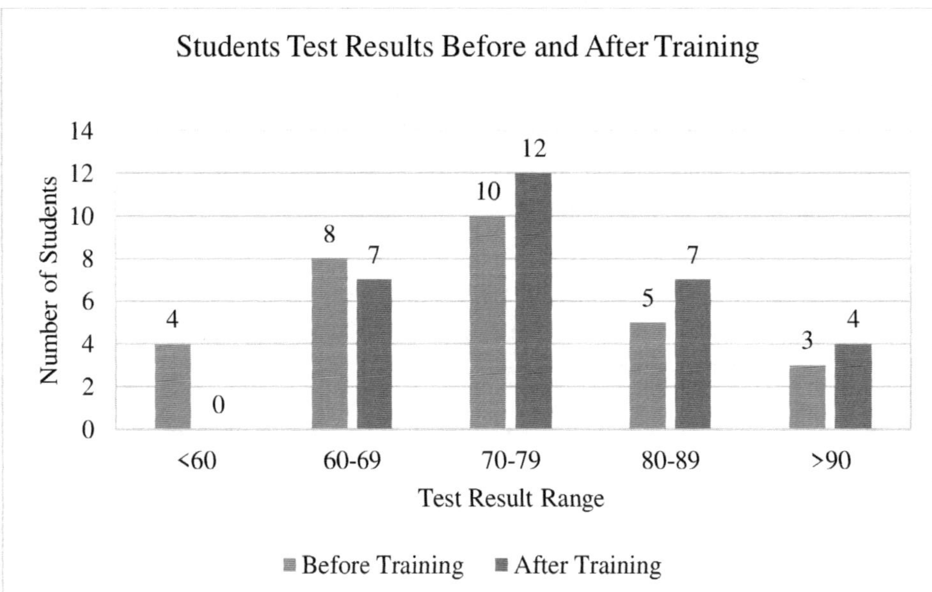

Figure 2. Results of a Preliminary Survey of 40 Teachers in SMK N 1 Tembarak

From Figure 2, the result of the evaluation increase by 25% from averages 69.67 to 72,80. We also asked students if the subject matter was more interesting and easier to understand and the results were that

85% of students answered yes, 10% of students answered not knowing and only 5% of students thought the teaching material was not interesting. Figure 3 shows the result. From the result can be concluded most students prefer the ICT bases student materials and enjoy them more the old one.

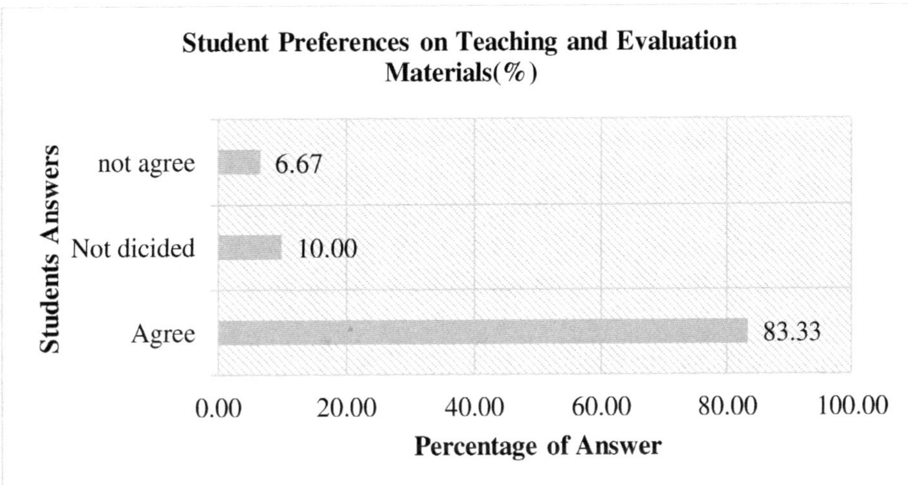

Figure 3. Student Preferences on Teaching and Evaluation Materials.

From the teacher's side, we surveyed whether the training in the making of teaching materials and evaluation materials was useful. From the survey results obtained, 62% of teachers strongly agree, 25% agree, abstain 10 and no one of the teacher answers disagrees or strongly disagree. Figure 4 shows the result.

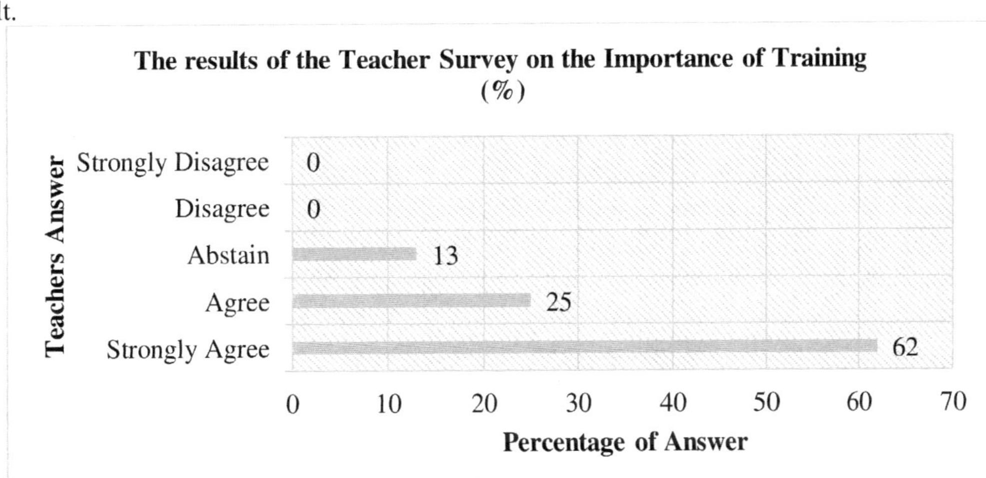

Figure 4. The results of the teacher survey on the importance of training

From these results, the Wilcoxon test was done to find out whether the assumption was that this training was beneficial for the teachers by using another question: It was found that it was easier to manage ICT-based teaching materials and exam questions. By using these applications, managing teaching, and evaluation materials are better. The result of SPPS is shown in Figure 5.

Table 1. The result of the Wilcoxon test for teachers

Test	N	Ranks	
		Mean Ranks	Sum of Ranks
Negative Ranks	0[a]	0.00	0.00
Positive Ranks	40[b]	20.50	820.00
Ties	0[c]		
Total	40[d]		
Test Statistics		Z=-5.617[a]	
Assump. Sig (2-tailed)		0.000	

[a] The values are based on the negative ranks
[b] Wilcoxon Signed Rank Test

Where:
H0: There is no improvement in the ability to manage teaching and evaluation materials after training
H1: There is an increase in the ability to manage teaching and evaluation materials after training.
H0 rejected if Sig (P-value) < α 5%,
Conclusion of result: H0 is rejected; there is an increase in the ability to manage teaching and evaluation materials after training. Thus, training provides a major influence on teachers in managing teaching and evaluation materials.

We also did the Wilcoxon Test using questions before the training on Making Learning Materials, and ICT-based Exam Questions teacher felt that making multimedia-based learning content was difficult, and If they use ICT, the learning content that I create will be more interesting for students. The result is shown in Table 2:

Table 2. The result of the Wilcoxon Test for teachers

Test	N	Ranks	
		Mean Ranks	Sum of Ranks
Negative Ranks	0[a]	0.00	0.00
Positive Ranks	40[b]	20.50	820.00
Ties	0[c]		
Total	40[d]		
Test Statistics		Z=-5.597[a]	
Assump. Sig (2-tailed)		0.000	

[a] The values are based on the negative ranks
[b] Wilcoxon Signed Rank Test

Where,
H0: there is no improvement in the quality of multimedia-based learning content after training
H1: there is an increase in multimedia-based learning content after training
H0 rejected if Sig (P-value) < α 5%,
Conclusion of the result: H0 is rejected; there is an increase in multimedia-based learning content after training. Thus, training provides a major influence on teachers in making learning content.

And last, we conduct Wilcoxon test on how training improves teachers' quality using the questions: by the training on making ICT based learning and evaluation materials were support their duties as a teacher and if teacher use ICT in developing teaching and evaluation materials then it will improve their quality as a teacher. The result of the Wilcoxon test is shown in Table 3.

Table 3. The result of the Wilcoxon Test for teacher

Test	N	Ranks	
		Mean Ranks	Sum of Ranks
Negative Ranks	0[a]	0.00	0.00
Positive Ranks	40[b]	20.50	820.00
Ties	0[c]		
Total	40[d]		
Test Statistics		Z=-5.607[a]	
Assump. Sig (2-tailed)		0.000	

[a] The values are based on the negative ranks
[b] Wilcoxon Signed Rank Test

Where,
H0: there is no improvement in the quality of teachers after training
H1: there is an increase in the quality of teachers after training
H0 is rejected if Sig (P-value) <α 5%,
Conclusion of the result: H0 was rejected; there was an increase in teacher quality after training. Thus, training provides a major influence on teacher quality.

4. Conclusion
The results of the study proved that training in the making of teaching and evaluation materials had increased students interest and ability of students in receiving lessons and evaluations by 25% and preference by 85%, as well as increasing the ability of teachers in making teaching materials and exam materials by 87%. By using the Wilcoxon test it has also been proven that this training has a good impact on teachers with alpha results of α =-5.607 for improving teacher's ability to manage teaching and evaluation materials, α =-5.597 for improving teacher's ability in multimedia-based learning content, and α= -5.607 for improving the quality of teachers after training. The Wilcoxon test proves that the teacher's ability to make teaching materials and exam materials increases after training.

Acknowledgments
The authors would like to thank profusely to Institut Sains & Teknologi AKPRIND that has given funding of the conference activities, and to all member of SMK Negeri 1 Tembarak, Temanggung, Central Java, Indonesia for all of their support during the project.

References
[1] Ming H L, Huang C C and Kuang S L 2017 A study of the effects of digital learning on learning motivation and learning outcome *EURASIA J. Math.* 1-12
[2] Alt D 2018 Science teachers' conceptions of teaching and learning, ICT efficacy, ICT professional development and ICT practices enacted in their classrooms *Teaching and Teacher Education* **73** 141-150
[3] Palomino M C P 2017 teacher training in the use of ict for inclusion: differences between early childhood and primary education *Procedia – Social and Behavioral Sciences* **237** 144-9
[4] Comi S L, Argentin G, Gui M, Origo F and Pagani L 2017 Is it the way they use it? Teachers, ICT and student achievement *Economics of Education Review* **56** 24-39
[5] Kreijns K, Acker F V, Vermeulen M and Buuren H V 2013 What stimulates teachers to integrate ICT in their pedagogical practices? The use of digital learning materials in education *Computers in Human Behavior* **29** 1 217-25
[6] Fontán C M, Señorán M G, Rodríguez A C and González A 2019 Teachers' ICT-related self-efficacy, job resources, and positive emotions: Their structural relations with autonomous motivation and work engagement *J. Computers & Education* **134** 63-77
[7] Moreira F and Rocha A 2018 A special issue on disruption of higher education in the 21st century due to ICTs *Telematics and Informatics* **35** 4 930-2
[8] Tondeur J, Aesaert K, Prestridge S and Consuegra E 2018 A multilevel analysis of what matters in the training of pre-service teacher's ICT competencies *J. Computers & Education* **122** 32-42
[9] Ferraro S 2018 Is information and communication technology satisfying educational needs at school *Computers & Education*.**122** 194-204
[10] Prestridge S 2019 Categorising teachers' use of social media for their professional learning: A self-generating professional learning paradigm *Computers & Education* **129** 143-158
[11] Vitanova V, Pachemska T A, Iliev D and Pachemska S 2015 Factors affecting the development of ICT competencies of teachers in primary schools *Procedia – Social and Behavioral Sciences* **191** 1087-94
[12] Šimandl V and Vaníček J 2017 Influences on ICT teacher's knowledge and routines in a technical e-safety context *Telematics and Informatics* **34** 1488-1502
[13] Copriady J 2015 self-motivation as a mediator for teachers' readiness in applying ICT in teaching and learning *Procedia – Social and Behavioral Sciences* **176** 699-708
[14] Nurnawari E K and Ermawai 2018 Design of integrated database on mobile information system: a study of Yogyakarta smart city apps *IOP Conf. Ser.: Material Science and Engineering* 1-8
[15] Batanero J M F, Rebollo M M R and Rueda M M 2019 Impact of ICT on students with high abilities. bibliographic review (2008–2018) *Computers & Education* **137** 48-58
[16] Sinaga P., Suhandi A and Liliasari 2014 Improving the ability of writing teaching materials and self-regulation of pre-service physics teachers through representational approach *Int. J. Sci: Basic and Applied Research* 80-94

[17] DeLuca C, Coombs A and McEwan D L 2019 Assessment mindset: Exploring the relationship between teacher mindset and approaches to classroom assessment *J. Computers & Education* **61** 159-169
[18] Valtonen T, Kukkonen J, Kontkanen S, Sormunen K and Sointu E 2015 The impact of authentic learning experiences with ICT on pre-service teachers' intentions to use ICT for teaching and learning *J. Computers & Education* **81** 49-58
[19] Taimalu M and Luik P 2019 The impact of beliefs and knowledge on the integration of technology among teacher educators: A path analysis *Teaching and Teacher Education* **79** 101-10
[20] Verma C, Stoffová V and Illés Z 2019 Prediction of students' awareness level towards ICT and mobile technology in Indian and Hungarian University for the real-time: preliminary results *Heliyon* **5** 6 1-10
[21] Hardman J 2019 Towards a pedagogical model of teaching with ICTs for mathematics attainment in primary school: A review of studies 2008–2018 *Heliyon* **5** 5
[22] Hu X, Gong Y, Lai C and Leung F K S 2018 The relationship between ICT and student literacy in mathematics, reading, and science across 44 countries: A multilevel analysis *J. Computers & Education* **125** 1-13
[23] Piper B, Zuilkowski S S, Dubeck M, Jepkemei E and King S J 2018 Identifying the essential ingredients to literacy and numeracy improvement: Teacher professional development and coaching, student textbooks, and structured teachers' guides J. *World Development* **106** 324-36
[24] Halili S J and Sulaiman H 2018 All-learning: The state of the art of the models and the methodologies educational with ICT *Kasetsart Telematics and Informatics* **35** 4 944-53
[25] Sánchez V M G 2017 ICT-based active methodologies *Procedia – Social and Behavioral Sciences* **237** 606-12
[26] Hatlevik I K R and Hatlevik O E 2018 Students' evaluation of digital information: The role teachers play and factors that influence variability in teacher behaviour *Computers in Human Behavior* **83** 56-63
[27] Southgate E, Smith S P, Cividino C, Saxby S and Bergin C 2019 Embedding immersive virtual reality in classrooms: Ethical, organisational and educational lessons in bridging research and practice *Intl J. of Child-Computer Interaction* **19** 19-29
[28] Nikolić V, Petković D, Denić N, Milovančević M and Gavrilović S 2019 Appraisal and review of e-learning and ICT systems in teaching process *Statistical Mechanics and its Applications* **513** 456-64
[29] Jamil M G 2018 Technology-enhanced teacher development in rural Bangladesh: A critical realist evaluation of the context *Evaluation and Program Planning* **69** 1-9
[30] Asongu S A, Roux S L and Biekpe N 2018 Enhancing ICT for environmental sustainability in sub-Saharan Africa *J. Technological Forecasting and Social Change* **127** 209-216
[31] Schopf J C 2019 Room for improvement: Why Korea's leading ICT ODA program has failed to combat corruption J. *Telecommunications Policy* **43** 501-19
[32] Ramdhani N, Ancok D, Swasono Y and Suryanto P 2012 Teacher quality improvement program: empowering teachers to increasing a quality of Indonesian's education *Procedia - Social and Behavioral Sciences* **169** 1836-41
[33] Janum, and Maryono J 2017 Dampak Teknologi dalam Dunia Pendidikan *Jurnal Pendidikan dan Kebudayaan Missio* **10** 1-136

[34] Rosidin, Firdauz and Herpratiwi 2017 Evaluasi pemanfaatan sumber belajar berbasis teknologi informasi dan komunikasi di SMAN 1 Bulok Kabupaten Tanggamus *Laporan Penelitian FKIP Unila Lampung*

.[35] AndiraniT 2015 Sistem pembelajaran berbasis teknologi informasi dan komunikasi, sosial, budaya *Media Komunikasi Ilmu-ilmu Sosial dan Budaya* **12** 1 50-62

[36] Gellerstedt M, Babaheidari S M and Svensson L 2018 A first step towards a model for teachers' adoption of ICT pedagogy in schools *Heliyon* **4** 9 e00786

[37] Budiman 2017 Peran Teknologi Informasi dan Komunikasi dalam Pendidikan Al-Tadzkiyyah *Jurnal Pendidikan Islam* **8** 1 1-10

[38] As'ari DR 2017 Pemanfaatan Wondershare Quiz Creator dalam Pembuatan Soal-soal Berbasis Arab *J. Arabic Studies* **2** 137-48

[39] Dafitri H 2017 Pemanfaatan Wondershare Quiz Creator dalam Tes Berbasis Komputer *Query:Jurnal Sistem Informasi* **1** 1 1-9

[40] Tang L L and Balakrishnan N 2011 A random-sum Wilcoxon statistic and its application to analysis of ROC and LROC data *J. Statistical Planning and Inference* **14** 1 335-44

Observation of the Saibatin Traditional House with Biological – Architecture Approach

Dewi Fadilasari[1]* and Diana Lisa[2]

[1]Faculty of Engineering of Universitas Malahayati, Bandar Lampung, Lampung, Indonesia
[2] Faculty of Engineering of Universitas Lampung, Bandar Lampung, Lampung, Indonesia

*Corresponding e-mail: d.fadilasari@malahayati.ac.id

Abstract. Traditional house was one of the symbol of a previous culture that was inherited from generation to generation of each tribe, with a design adapted from the natural conditions of its time, which were built in such ways as to be used and utilized as needed at that time. The Saibatin traditional house is one of Lampung traditional house type, often called lamban or nuwo, also known as stilt houses, where full of symbol, philosophies, and cosmological content of Lampung culture. The objective of this research is to observe the Saibatin traditional houses located in Pekon Kenali, West Lampung Regency, with the biological–architecture approach. The research was conducted by observation methods and descriptive research to identify the object by comparing between data from literature and the results of documentation, as well as observations from research samples. The data from the object of research would be described and juxtaposed with the theories of biologic-architecture. The observations of The Saibatin Traditional House were executed on the whole part of the house inclusive of roof, wall, floor, columns, beam, stairs, foundation and basement. The results show that most parts of The Saibatin Traditional House are using natural materials which are part of biological architecture. The conclusion of observation research of The Saibatin Traditional House that biological architecture is not only suitable for the ancient building of old time, it also adaptable to nowadays situation, proven their existence can still adjust to the development of the era without leaving local wisdom.

1. Introduction

Spanning from Sabang to Merauke with total area of 1.905 million km^2, Indonesia is one of the largest archipelago countries in the world, with a variety of tribes and cultures. The wealth of ethnicity and culture that Indonesia has includes various local languages, traditional clothing, to traditional buildings (traditional houses).The existence of traditional houses in Indonesia is very diverse and has significance in the perspective of history, heritage, and the progress of society in civilization. Further explored, we could see the diversity of culture, especially traditional buildings with their respective characteristics [1]. Traditional architecture in Indonesia had been through a long journey and received outside influences related to trade and religion relations in the archipelago. One of the tribes in Indonesia is the Lampung tribe, which is in Lampung province, the southernmost part of the island of Sumatra.

IOP Conf. Series: Materials Science and Engineering **807** (2020) 012015 doi:10.1088/1757-899X/807/1/012015

Lampung traditional house is often called a stilt house which is full of symbols, philosophy, and cosmological content of Lampung culture. Most of them are made of wood like most of the houses in the tropical forest areas with sloping roofs that are typical of Asian architecture [2]

Figure 1. Lampung traditional house.

Pekon Kenali is the forerunner of the Saibatin Lampung community and belongs to a traditional or historic area in West Lampung Regency, Lampung Province, Indonesia. Pekon Kenali was feared to be extinct. Some of the buildings that can still be found in Pekon Kenali are inclusive of traditional houses of the local community known as Lamban, surau and walai or hall that function as rice barns. The Saibatin people in this region, in general are from the Lampung ethnicity, and commonly called 'Ulun Lampung' (Lampung people). Traditional – geographically are tribes that occupy the entire Lampung Province and partly in Southern part and Middle part of Southern Sumatera province, such as Martapura, Muaradua in Ogan Komering Ulu (Oku), Kayu Agung, Tanjung Raja, Ogan Komering Ilir, Merpas in southern part of Bengkulu, also Cikoneng in west coast of Banten. The philosophy of Lampung society is Sang Bumi Ruwa Jurai, which mean a household of two bloodlines, each of them delivered a pepadun and saibatin civil society.

Figure 2. Several types of Saibatin traditional house in Pekon Kenali

Figure 3. Several types of traditional houses in Muaradua.

Generally, they use organics building materials from their surroundings, such as wood, bamboo, palm fibres. All these materials were processed in a simple way. These inclusive of the building structure. The structure is the framework of the whole building structure that supports the building to allow the building to stand perfectly. There are many types of building structures in architecture, such as frame structures (poles and beams), bearer wall type structures, large span type structures, combined type structures, shell type structures, tent type structures, and others.

Biological architecture is the driving force of architectural activities. Biological architecture is very compatible with the climate of traditional buildings, where the location of the observations in this study exist. Biological architecture reflects a harmonious, original, rhythmic and dynamic way of life, which is intertwined between human life and the surrounding environment. Biological architecture will use natural technology to penetrate the critical state of nature that has begun to be threatened, to improve the quality of life, namely spirituality, and the quality of buildings with material parts. Building materials used in biological architecture are building materials from nature, such as wood, bamboo, thatch, reeds and palm fibres. Planning of biological architecture always pays attention to construction in accordance with where the building is located. The technology is simple, the shape of the building is also determined by the series of building materials and by functions according to the basic needs of residents by building them.

Figure 4. Biological architecture building.

Biological architecture as a supporting component of architectural buildings in this research area makes a more value in the study area. Where, the community around prioritizes building materials in their natural surroundings. Wood, bamboo and organic roof coatings. Wooden beams are used to support building structures, while processed wood in the form of boards is used as floor coverings and doors and windows. Whereas bamboo, in some buildings, is used as wallcoverings, ceiling coverings as well as supporting the roof construction. palm fibres are used as roof coatings for crop storage.

The purpose of this research is to observe the Saibatin traditional house with biological – architecture approach by collecting data from several samples, identifying the objects, comparing the documentation data on the field with the literature, and observing the samples. The result of observation shows that even though there are some parts of the shape of the buildings have been changed, but the local community still retains the biological architecture in the building where they live.

2. Methodology

This research used direct and intensive field observation methods to collect and document data. Right after that, writers used descriptive research method to analyse data. According to Malgorzata [3], observation is one of the most important research methods, used in a range or research strategies. Then, Kothari [4] revealed that the major purpose of descriptive research is description of as it exists at present.

Data collection is done by observation, documentation and interviews. Observations were made on selected samples which were the object of research, in this case were the building of Saibatin traditional houses found in Pekon Kenali, West Lampung regency. All of elements of the houses were observed and documented by writers. The elements observed were consist of roof, wall, floor, beam, pillars, stairs, foundation and basement. The source of the data was divided into two, the primary data and secondary data sources. The primary data was essential sources coming from Saibatin traditional houses. The secondary data was taken from literatures such as journals, books, that were relevant to Lampung traditional houses and biological architecture.

Based on the results of data collection, documentation and observation, writers analysed the data collected by describing and detailing the building elements of the existing Saibatin traditional houses. Furthermore, writers also comparing the collected data with the reference literatures of biological architecture. The methodology used was summarized at the Figure 5 below.

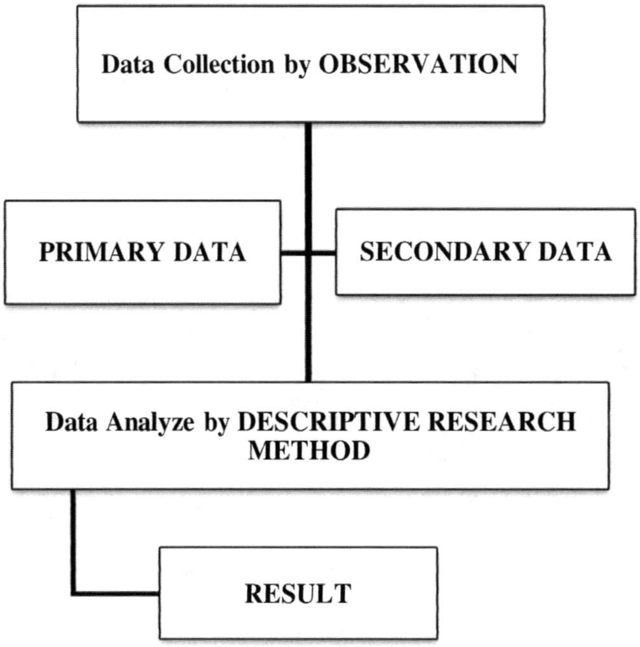

Figure 5. Methodology flow chart

3. Result and Discussion

Based on the results from the field observations, the data obtained is inventoried, identified and analysed, so that from the data and facts of field observations the following discussion results are obtained:

3.1 Saibatin Traditional House

Just like most of traditional house in Indonesia, according to Ibrahim [5], Saibatin traditional house also had a philosophy of head, body and legs in its architectural form. Nevertheless, there were differentiated between Saibatin traditional house and another traditional house. The differentiation was at the column or leg of a Saibatin traditional house. They look like they seem separated from the body and head.

3.1.1 Roof

Generally, the roofs of Saibatin traditional houses have geometrical pyramid shaped. And there also could be found hat certain Saibatin traditional houses (typically 400 years old house) used the organic material such as palm fibres or thatch as the roof covering materials. The roof truss for the houses were constructed entirely from the wood in the form of blocks while some of them did use bamboo. For the ceiling, it was found that the ceiling frames use the wooden blocks or wooden boards as the materials whereas some houses did use bamboo in substituting the wood. From the outside of the house, it could be observed that the canopy width is wide where the wide canopy with an average width of up to one meter or more made of wood and bamboo.

Figure 5. The pyramid shaped roof.

Figure 7. 400 years old Saibatin traditional house with palm fibres roof cover.

Figure 8. Ceiling frames use wooden blocks and bamboo as a substitute.

Figure 9. Wide canopy made of wood and bamboo.

3.1.2 Wall

The wall of the traditional houses was made up from the same materials as the roof which are the wood and bamboo. The wooden and bamboo framed walls in the form of round cylindrical shaped. The frames are then covered with coverings. The wall coverings are made of wooden board while in some houses, the wall frame was covered by bamboo in the form of bamboo stems that have been processed and assembled become the desired part for wall coverings.

There are also openings made at the walls in which it is for the doors and windows. The doorframes and window frames are generally made of wood. Door shutters and window shutters are generally made of wooden boards. The wooden frames on the doors and windows use a simple joint system, known as an angular connection.

Vent, not all houses have ventilation holes. However, there are also those who use vents with the same material as walls in general, namely wooden boards and wooden beams.

IOP Conf. Series: Materials Science and Engineering **807** (2020) 012015 doi:10.1088/1757-899X/807/1/012015

Figure 10. Framed walls, wallcoverings, window frames, window shutters, those are made of wooden board.

3.1.3 Floor and Floor Beam

The framework on the wooden floor was entirely processed timber in the form of blocks while the floor cover using wooden boards with variety sizes. The size of the wooden boards is approximately found 30 cm x 400 cm with a thickness of 3 cm. In some houses, it could be found that bamboo was used as midrib for the floor coating.Most of the Saibatain traditional houses use timber as floor beams. This is because of the wooden natural resources are still abundant. From the beginning until now timber is still used to build houses.

Installation of floor beams done by supporting each other beams and without joint to one another. The connection system is carved into the connection area and with the pen. At the end of the beam pillar there is a distinctive ornamentation of the Lampung tribe called 'paguk'[6], which signifying the philosophy of the boat community.

Figure 11. Processed timber in the form of blocks for framework on the wooden floor.

Figure 12. Framework and 'paguk' of the wooden floor and the floor cover using wooden boards

3.1.4 Building Structures/ Building Pillars

The main frame or the main structure were in the form of a pole, with an approximate size of 20 cm x 20 cm. Specifically, the building support poles in the form of foundations, measuring of 30 cm x 30 cm.

The height of the foundation pile varies. Some of them have 200 cm - 220 cm height whereas some are 80 cm - 100 cm height. As a matter of fact, the height of the foundation pillars is adjustable to the needs. This foundation pillar affects the laying or position of the stairs and the number of steps which amounts to an odd number.

Figure 13. Building support poles in the form of foundations.

Figure 14. The main structure in the form of a pole, with a size of 20 cm x 20 cm.

3.1.5 Stairs

The function of stairs is an intermediary room between the lower floor to the upper floor. The location of the stairs for each house is on the side facing forward and side facing. While in 'lamban adat' building, the stairs are at the back position of the houses. This is in accordance with philosophical and culture of Pekon Kenali. This also is like the philosophy found at OKU [7].

As had been stated before, the height of the pole is adjusted to the height of the walls of the house up to the beam (ring balk) measuring 275 cm to 300 cm without a wooden connection. The ladder material is generally timber, both on the frame and on the footsteps area. Ornamentation were also found used in the area of the stairs, especially at the edge of the fence.

Figure 15. The location of the stairs for each house is on the side facing forward and side facing

Figure 16. Ornamentation used in the area of the stairs

3.1.6 Foundation

Foundation piles of the houses were made from the timber with larger dimension from house poles. The size of the foundation was around 30 cm x 30 cm, and it made from an intact timber. The size and height of the foundation piles are varying, from 80 cm to 200 cm, according to the function of the room [8].

Figure 17. Foundation piles made of an intact timber with larger dimension from house poles.

3.1.7 Basement

The space under the floor (literally the basement) is a compartment or area that is not fully utilized by the occupant of the house. Previously, Saibatin traditional house were built on stilts to anticipate the disturbance of wild creatures. Nowadays, the lower floor space has changed its function into a space that can be used for various activities, such as a crops barn, or to store building materials and many more. In some buildings, the basement area is no longer exposed. It had already covered by wooden wall; therefore, the house looks like a two-storey building.

Figure 18. The existing traditional Saibatin houses still retain the concept of houses on stilts

Figure 19. The Saibatin traditional house with wall covered at the lower ground.

3.2 Biological – Architecture

The term of biological architecture appears to show the close relationship between humans and the environment or the natural surroundings. Biological – Architecture is a science that connect humans and the environment. It also a knowledge about the integral relationship between humans and the environment according to Sopa [9]. A residence can be considered as an organic arrangement, which functions as a protector for humans.

The basic concept of this biological architecture already applied by our ancestor through the buildings of traditional houses. These traditional houses were using materials taken from nature so as not to pollute the environment and consider a design that can withstand all kinds of natural threats, such as wild animals and disasters such as floods, landslides, earthquakes, and many others. Traditional houses in the form of houses on stilts are the examples of the biological architecture of ancient Indonesian society.

4.0 Conclusion

Based on the results of research with field observations, inventory, documentation and data collection of Saibatin Traditional House, we can make conclusion that almost all parts of the Saibatin traditional house in Pekon Kenali including the structure of the building still using local materials that can be found

IOP Conf. Series: Materials Science and Engineering **807** (2020) 012015 doi:10.1088/1757-899X/807/1/012015

in their surrounding such as timber, bamboo, palm fibres and natural stones. These materials are known as organic materials and are included in the concept of biological architecture.

Nowadays, the preservation of houses on stilts concepts can be found in some houses, where the lower ground space remains open. While some others have been covered with walls, but still utilize the organic materials found around them. The use of a wide canopy building still can be found on the Saibatin traditional house in Pekon Kenali. To sum up, the existence of buildings with biological architecture must be maintained and preserved. Their existence proven can still adjust to the changing of times without leaving local wisdom.

References

[1] Prijotomo J 2018 *Arsitektur Nusantara* (Surabaya: Wastu Lanas Grafika)
[2] Nasrul 2018 *Tinjauan Tipologi Rumah Tradisional Lampung Saibatin di Pekon Kenali Kabupaten Lampung Barat* Laporan Seminar Arsitektur, Universitas Bandar Lampung
[3] Ciesielska, Malgorzata, Boström, Katarzyna, Öhlander and Magnus 2018 *Observation Methods*
[4] Kothari C R 2004 *Research Methodology (Methods and Techniques) Second revised Edition* Published by (India: New Age International (P) Ltd.)
[5] Ibrahim and William 2011 Arsitektur tradisional kenali salah satu kearifan lokal daerah lampung *Jurnal Rekayasa* **15** 1
[6] Hanan H 2012 Modernization and cultural transformation: the expansion of traditional Batak Toba house in Huta Siallagan *Procedia: Social and Behavioral Science* **50** 800-11
[7] Pratiwi R A and Gunawan 2017 Study of Lampungnese traditional home garden design *IOP Conf. Series: Earth and Environmental Science* 91012024
[8] Farisa D and Prihatmaji Y P 2017 The Butonese traditional house typology according to the social status in Buton *Proc. UIA 2017 Seoul World Architects Congress* 0749
[9] Sopa M 2018 Local wisdom in the cultural symbol of Indonesian traditional house *Proc. of KnE Social Science* **3** 4 524–31

IOP Conf. Series: Materials Science and Engineering **807** (2020) 012016 doi:10.1088/1757-899X/807/1/012016

Feasibility Study of Micro Hydropower Plants in High-Rise Buildings in Indonesia By Utilizing A Working Scheme for AC Condenser Water Pumps (Ventilation & Air Conditioning System)

Purbantoro Firma[1*], Marsul Siregar[1]

[1]Electrical Engineering, Faculty of Engineering, Atma Jaya Catholik University of Indonesia, JL Jenderal Sudirman Kav No 51, Jakarta 12930, Indonesia

Corresponding e-mail: firma.purbantoro@sampoernaland.com

Abstract. In Indonesia the number of high-rise buildings above 150 m tends to increase, according to CTBUH (Council on tall building and urban habitat) in 2018 there are 85 buildings and those that are under construction are 26 buildings, Super tall buildings with a height above 300 meters are 4 buildings and proposed mega tall building with a height of more than 600 meters is 1 building. All these buildings have very large electrical loads. The percentage of electricity consumption in the building that is the object of research is 30-50% coming from the Ventilation & Air Condition (VAC) system, 20% - 30% lighting systems, 10% computers and the rest from other utilities. Several papers and journals discussing the simulation or feasibility study of micro-hydropower plants in high-rise buildings have not yet been discussed how to utilized kinetic energy from the working scheme of the water condenser pump in the VAC system. in general, the existing papers discuss how to utilized gravity potential energy through the working scheme of the building's clean water tank which is on the roof (rooftop) when distributing water which is not optimal because of the flow of water from the clean water system is not continuous or steady-state. Research Object by taking a case study of a high-rise building in Jakarta with a height of 157 meters. The data was taken from the report on utility use for the 2018 period in the building. From the simulation, the potential of water flow rate (Q) is 0.499 m3 / sec and the electrical power that can be generated (P) is 733.83 kW and this is a Micro-hydro Power Generation category. Economic analysis for building energy savings of Rp. 541,765,714.29 / month and finally roughly Return on Investment (ROI) can be achieved in 3 years and 6 months.

1. Introduction

In Indonesia, the number of high-rise buildings above 150 m continues to increase, according to CTBUH (council on tall building and urban habitat) in 2018 there are 85 buildings and those that are being built are 26 buildings while buildings with a height above 300 meters there are 4 buildings under construction in the super tall category and a building with a height of more than 600 meters that is in the mega tall category has 1 building. In the future, it is estimated that the number of skyscrapers will increase [1-3].

All of these buildings have very large electrical loads. The percentage of electricity consumption in a building is 30%-50% coming from the HVAC system, 20%-30% from the lighting systems, 10% from the computers and the rest from other utilities. Some papers and journals that discuss the simulation of the utilization of micro-hydropower plants in high-rise buildings have not yet discussed the use of condenser water pump work schemes in general, the papers discuss the use of work schemes from building clean water tanks on roofs (rooftop) when distributing water floor below it [4- 9]. This paper focus on a case study of a high rise building in Jakarta with a height of 157 meters which illustrated in Figure 1 and the schematic diagram of the micro hydropower was shown in Figure 2 below.

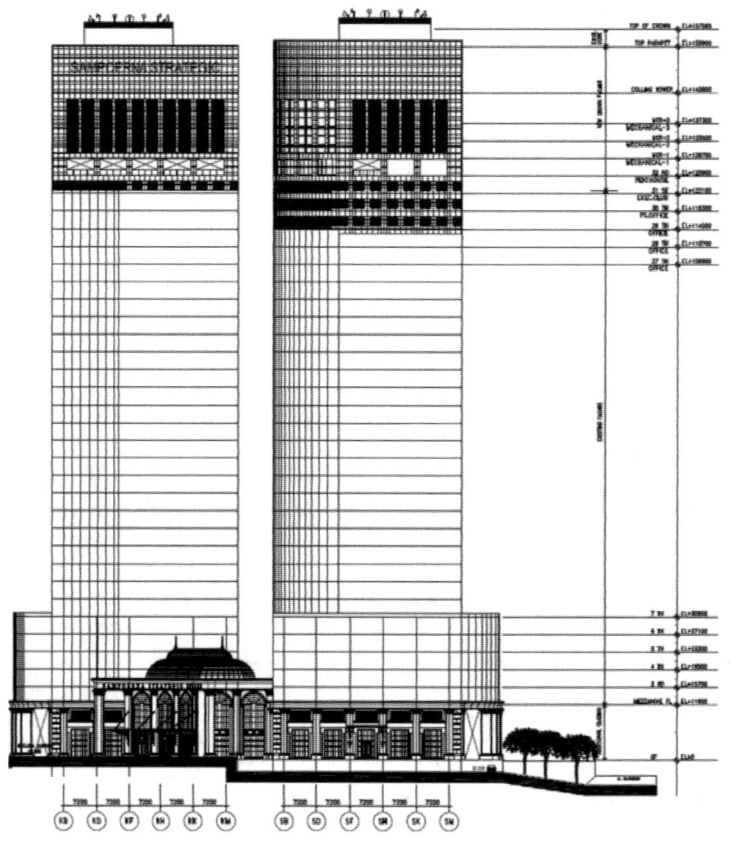

Figure 1. Sampoerna Strategic Square Building

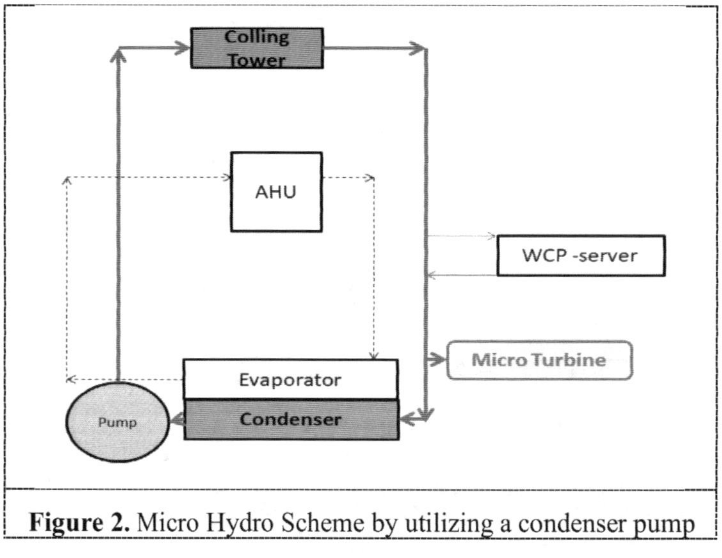

Figure 2. Micro Hydro Scheme by utilizing a condenser pump

IOP Conf. Series: Materials Science and Engineering **807** (2020) 012016 doi:10.1088/1757-899X/807/1/012016

2. Building's Electricity Consumption

2.1. Building's Electricity Consumption in 2018

The average electricity consumption in 2018 was tabulated in Table 1. It can be seen that the average electricity consumption in 2018 was 2,083,714.29 kWH / month or electricity payments per month on average was Rp2,354,609,724.57.

Table 1. Electricity Consumption in 2018 years

Month	PLN		Total (Rp)	Eletricity Bills (Rp)
	LWBP (kWH)	WBP (kWH)		
January	1,932,360.00	283,200.00	2,215,560.00	2,500,101,165.00
February	1,688,400.00	252,840.00	1,941,240.00	2,193,040,256.00
March	1,878,120.00	287,520.00	2,165,640.00	2,449,439,060.00
April	1,887,960.00	286,200.00	2,174,160.00	2,457,775,681.00
May	1,839,600.00	285,480.00	2,125,080.00	2,405,337,702.00
June	1,572,960.00	236,880.00	1,809,840.00	2,045,208,432.00
July	1,878,720.00	275,760.00	2,154,480.00	2,431,365,776.00
August	1,840,080.00	281,160.00	2,121,240.00	2,398,973,869.00
September	1,785,840.00	277,320.00	2,063,160.00	2,335,335,546.00
October	1,948,680.00	278,520.00	2,227,200.00	2,509,959,105.00
November	1,818,480.00	270,960.00	2,089,440.00	2,359,836,300.00
December	1,773,240.00	265,080.00	2,083,320.00	2,302,498,172.00
Total	12,678,120.00	1,907,880.00	14,586,000.00	16,482,268,072.00
Average	1,811,160.00	272,554.29	2,083,714.29	2,354,609,724.57

2.2. Monthly Electric Power / Load Trend

2.2.1. Daily Electricity Load in 1 month

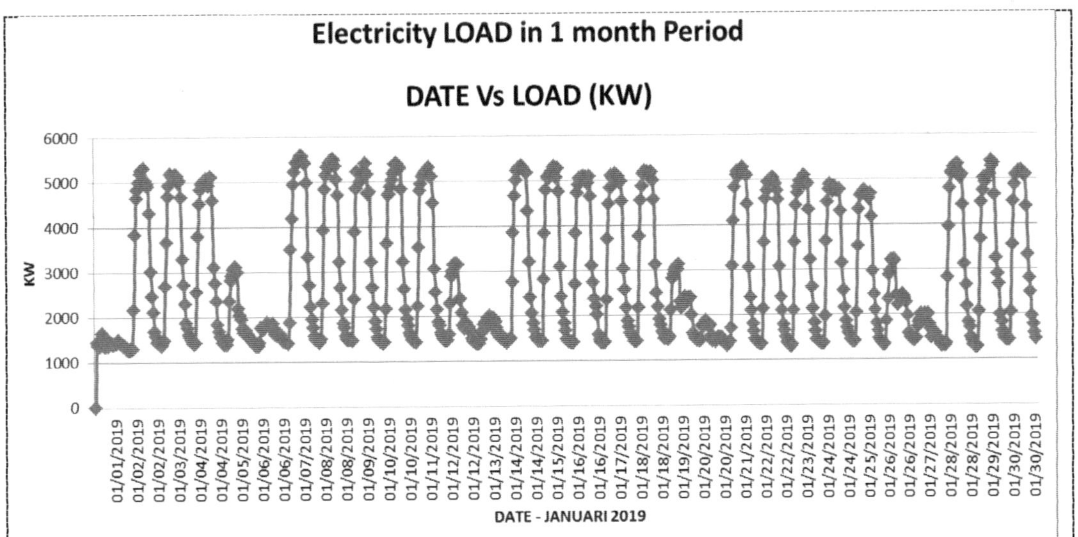

Figure 3. Electricity Load in 1 month Period

2.2.2. Daily Electricity Load on Hourly Bases in 1 Month

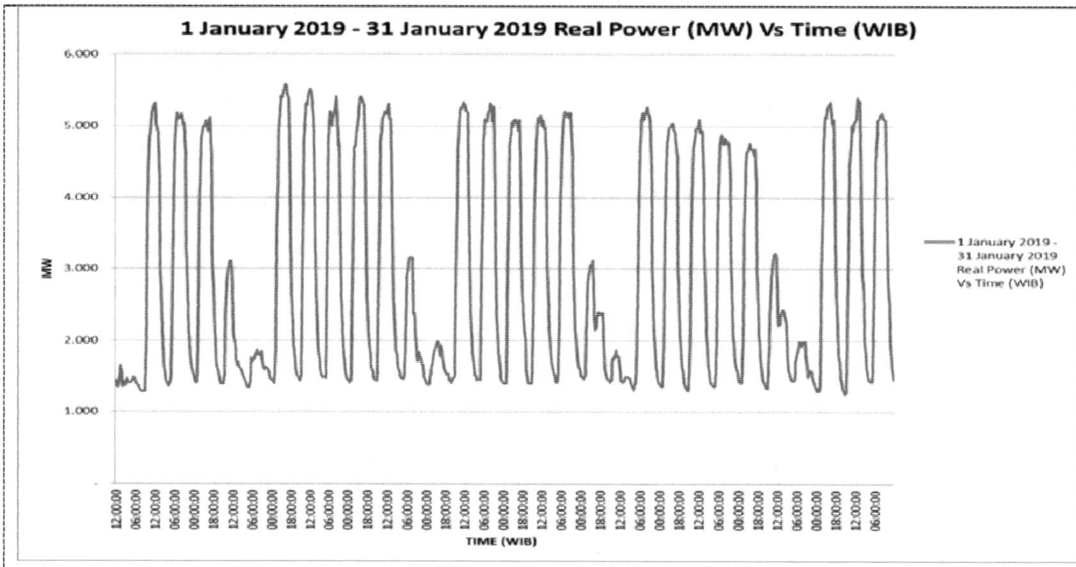

Figure 4. Daily load for 1 month of building electricity

2.3. Daily Electricity Load on Hourly Bases in 24 Hours

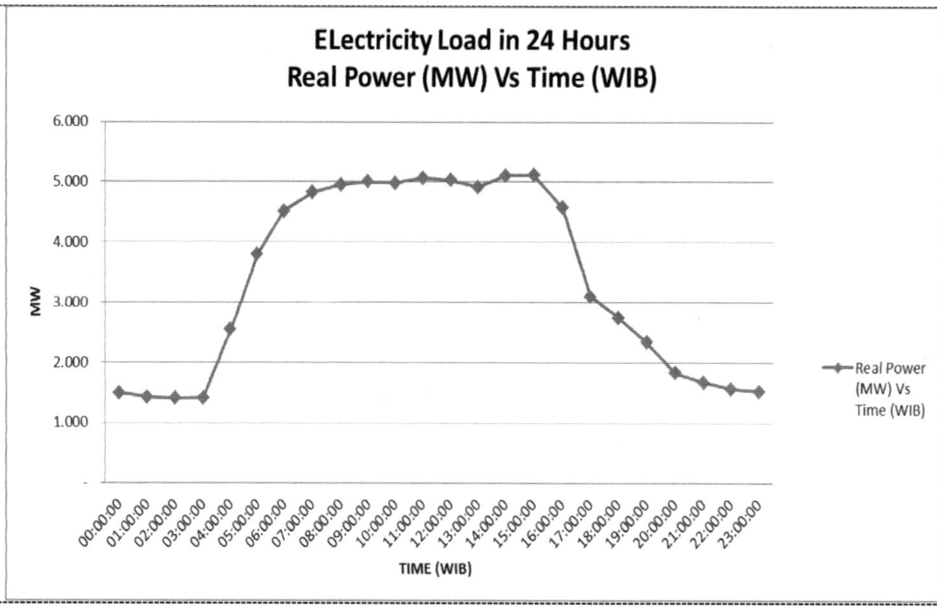

Figure 5. Building Electricity load for 24 hours

IOP Conf. Series: Materials Science and Engineering **807** (2020) 012016 doi:10.1088/1757-899X/807/1/012016

3. Research Methodology

3.1. By extracting data from the Building Automation System (BAS) from the control metasys system by Johnson controls

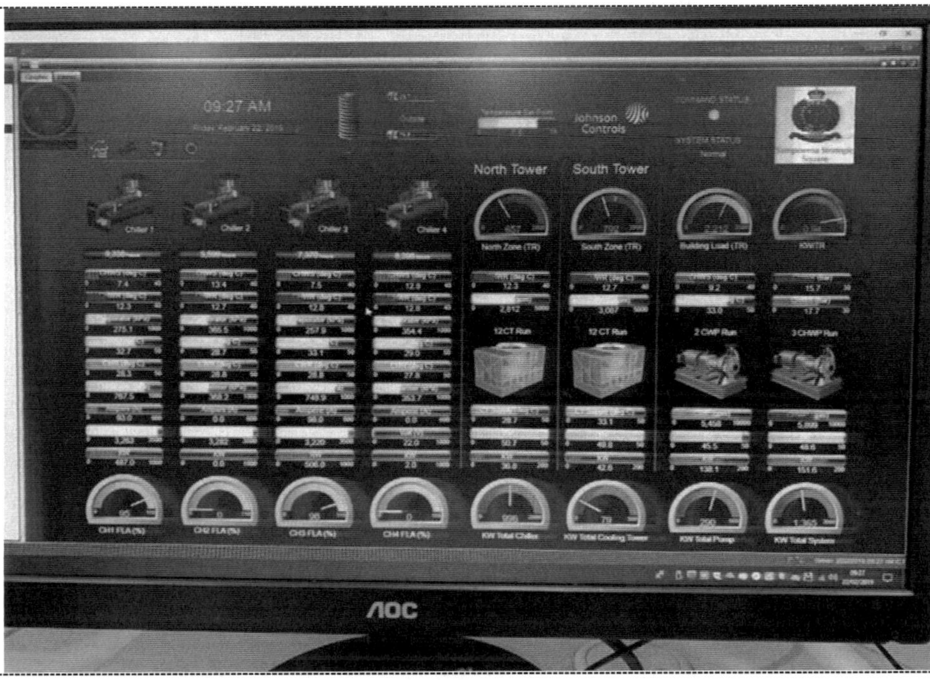

Figure 6. Utility control system, *Metasys* by Johnson control

4. Potential Use of VAC system water flow as a power plant

4.1.1. Potential Utilization

Building data, Average Pressure of the condenser water system pump 7,909 gpm or 0.499 m3 / second, the height of the 150 meters building consists of 33 floors and 3 basements, PLN subscription 8.3 MVA and the highest usage 6.1 MVA

$$P = \rho \times Q \times g \times H \times \eta \tag{1}$$

Where:
ρ = water density (kg / m3)
Q = water discharge (m3 / sec)
H = high waterfall effective (m)
η = overall efficiency of the hydropower plant

$$\text{Power (P)} = \text{Flow Rate (Q)} \times \text{Head (H)} \times \text{Gravity (g)} \times \text{Water Density } (\rho)$$
$$P = 0.4989 \times 150 \times 9.8 \times 1$$
$$= 733{,}383 \text{ kW or } 0.733 \text{ MW}$$

Based on its output capacity, the Classification of Hydroelectric Power Plants (PLTA) shown in Table 2. Potential Generators can be classified as Micro Hydro Power Plants (100 kW to 1 MW).

Table 2. Classification of Hydropower Plants (PLTA) [4]

No.	Classification of PLTA	Capacity
1	PLTA – Big Scale	>50 MW
2	PLTA - Medium	10 – 50 MW
3	PLTM (Mini hydro)	>1 MW – 10 MW
4	PLTMH (Micro hydro)	100 kW – 1 MW
5	Pico Hydro	< 100 kW

The Potential of Electrical Power from Utilizing Condenser Water Pump can be seen in Figure 4, where the highest Power can be generated on 11.00 PM and lowest Power can be generated on 20.00 PM to 03.00 AM.

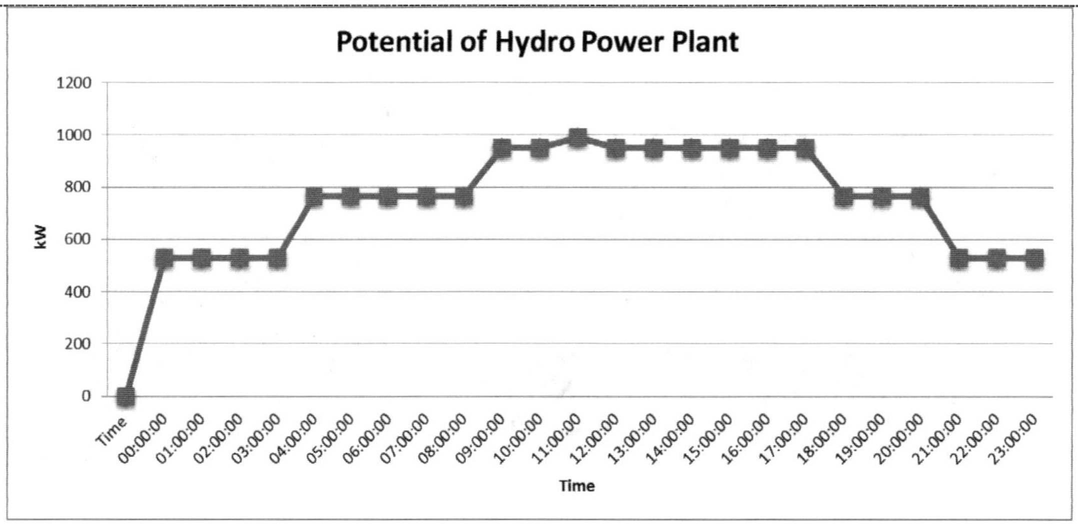

Figure 7. Potential of Power Plant from Utilizing Condenser Water Pump

4.1.2. Potential Power Plants from Utilizing Condenser Water Pump to replace daily electricity consumption

Potential Power Plants from Utilizing Condenser Water Pump to replace daily electricity consumption can be seen on Figure 8 & 9, which can replace the building's power source from 15 %to 40 %.

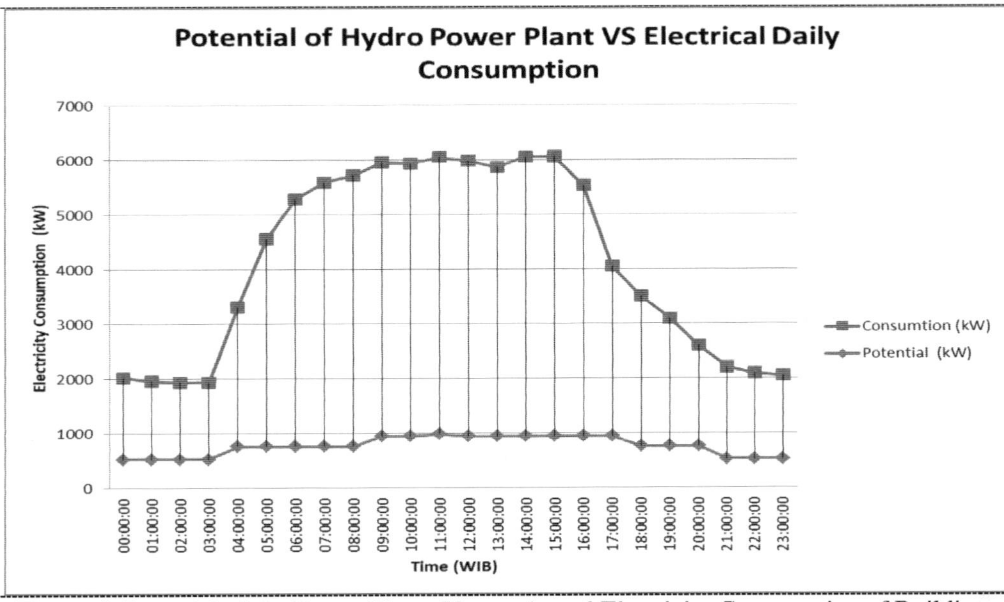

Figure 8. Potential of the Power Plants Versus Actual Electricity Consumption of Buildings

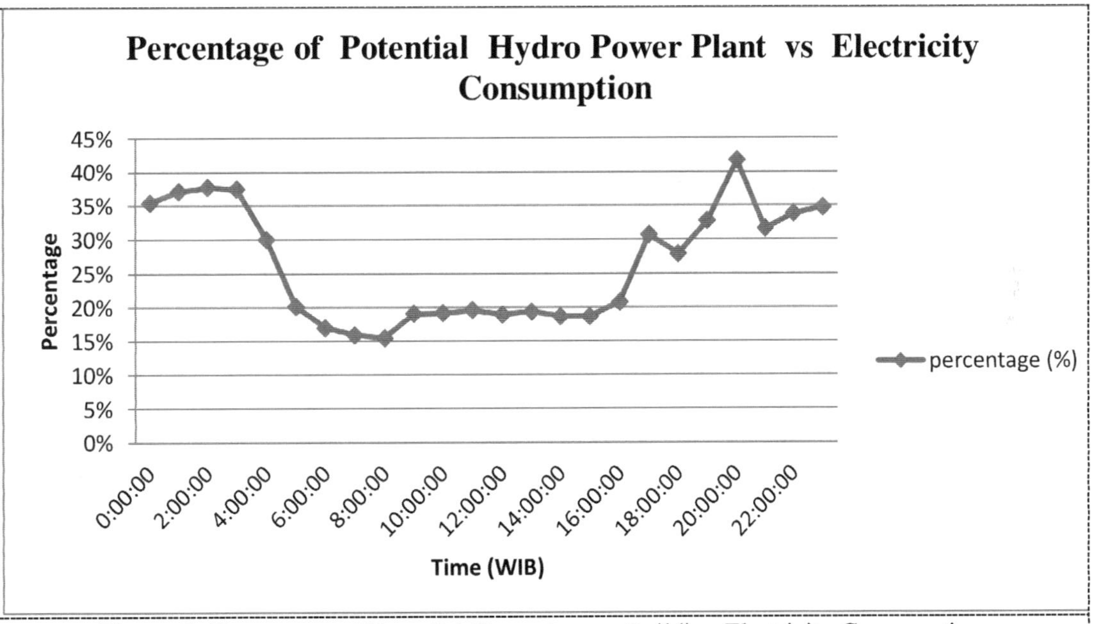

Figure 9. Percentage of Potential Generators vs Building Electricity Consumption

From the figure above, the average potential utilization of the CWP pump system for micro hydro power plants can save electricity consumption by 26% or an average savings per month of Rp541,765,714.29 or savings every year Rp.6,501,188,571.43

4.1.3. Potential Savings

Savings Potential use benchmark data for the cost of making micro-hydro power plants in Indonesia is 1500 USD/kW. At 1000 kW Planning costs around 1,500,000USD or around Rp22,500,000,000 (exchange rate 1USD = Rp15,000,000) So that the savings potential is obtained roughly before considering operational costs, plant efficiency, bank interest or other cost components, then: 26% x Rp2,354,609,724.57 = Rp541,765,714.29 / month

$$\text{Return On Investment} = \frac{22.500.000.000}{541.765.714,29} = 42 \text{ months or 3 years and 5 months}$$

5. Conclusion

It is found that The Potential of Electrical Power from Utilizing Condenser Water Pump can be obtained that the highest Power can be generated on 11.00 PM and lowest Power can be generated on 22.00 PM to 03.00 AM, It is obtained that the average potential utilization of the CWP pump system for micro hydro power plants can save electricity consumption by 26% or an average savings per month of Rp541,765,714.29 or savings every year Rp6,501,188,571.43 and roughly, the ROI or payback period can be achieved in 3 years and 5 months.

6. Further Research

Scope of Further research can be modelled a complete hydro power plan system from a condenser pump (VAC) water flow and simulated by MATLAB [10].

References

[1] Jianmin Zhang and Qianzhi Zhang 2014 *IEEE PES General Meeting | Conference & Exposition*
[2] Jianmin Zhang and Qianzhi Zhang 2013 *IEEE Power & Energy Society General Meeting*
[3] Du J., Yang H., Shen Z., Chen J 2017 *Energy 2017* **137**, 431–440
[4] Badan Standarisasi Nasional 2017.SNI 8396-2017: *Klasifikasi Pembangkit Listrik Tenaga Air*
[5] L. N. Hannett, B Fardanesh 1994 *IEEE Transactions on Power System* Volume **9(4)**
[6] X. Zhang and M. Zhang 2008 *International Conference on Publication* 4139 – 4143(**2002**).
[7] J. Chang and Y. Peng 2009 *Digital Object Identifier*: 10.1109/ICICTA.2009.30 Publication, 87 – 90
[8] D. E. Hesmondhalgh and E. R. Laithwaite Volume **70**, 1987-1993.
[9] P. Drtinaand and M. Sallaberger 1999 *ProcInstnMechEngrs* Volume **213** (C)
[10] G. Singh and D. S. Chauhan 2011 *International Journal of Advanced Engineering Sciences and Technologies* Volume **10 (1)**, 042 – 047

Application of SKATER and Ward's methods in grouping Indonesian provinces based on monthly expenditure per capita of food commodity groups

Yudi Setyawan[1], Rokhana Dwi Bekti[1] and Firman Isarlin[1]

[1]Department of Statistics, Faculty of Applied Sciences, IST AKPRIND Yogyakarta, Jl. Kalisahak 28 Komplek Balapan, Yogyakarta 55222, Indonesia

Corresponding e-mail: setyawan@akprind.ac.id

Abstract. Clustering is a fundamental data mining instrument that intends to find inherent cluster structure in data. Spatial clustering methods are usually used to assess the demographic data characterization. This study aims to classify provinces in Indonesia based on monthly expenditure per capita according to food commodity groups by using Ward's and Spatial 'K'luster analysis by tree edge removal (SKATER) methods and to identify a better classification between the two methods. The variables of this research constitute percentages of expenditure per capita for 14 groups of food commodities of 34 provinces in Indonesia during March 2018. The results of the first analysis (excluding outliers) revealed that SKATER method produced standard deviation rasio of 0.236, better than Ward's method that produced standard deviation rasio of 0.370. However, from the second analysis (including outliers), the outcomes showed that the Ward's method generated standard deviation rasio of 0.170, better than SKATER method that delivered standard deviation rasio of 0.199. Moreover, it can be concluded that the second analysis is better than the first analysis because it produced smaller standard deviation ratios based on the Ward's and SKATER methods contrasted with the first one.

1. Introduction

People's welfare is always an interesting topic to discuss. As in every country, the main goal in development is to improve people's welfare. In Indonesia, people's welfare is one of the state goals stated in the Preamble of the 1945 Indonesian Constitution [1]. One important thing to note in order to identify the society welfare is the expenditure per capita. The expenditure per capita is the cost incurred for the consumption of all household members for a month divided by the number of household members [2]. The proportion of expenditure for food and non-food can reveal the pattern of household consumption. The composition of household expenditure can be used as a measure to assess the level of economic well-being of the population [2-4]. The lower the percentage of expenditure on food on total expenditure, the better the level of welfare. Moreover, household expenditures are differentiated according to food and non-food groups. Changes in one's income will affect the shift in spending patterns. The higher the income, the higher the expenditure on non-food. Thus, expenditure patterns can be used as a tool to measure the level of welfare of the population, where changes in composition are used as an indication of changes in welfare levels [2].

A method is needed to describe the percentage of monthly expenditure per capita according to food commodity groups in each province in Indonesia so that it can be classified in several groups. The

IOP Conf. Series: Materials Science and Engineering **807** (2020) 012017 doi:10.1088/1757-899X/807/1/012017

statistical method that can be used is cluster analysis. Johnson and Wichern in Suhaeni [5], states that, cluster analysis is a multivariate analysis method to group the n objects into m clusters $(m \leq n)$ based on their characteristics. Grouping is based on the nature of similarities or the nature of dissimilarities between objects. Cluster analysis is important for investigating the number of clusters of natural data in several areas. Clustering data is a complex assignment including the choice between a wide range of methods and strategies, parameters and performance measurements, with implications in some true issues [6]. Different clustering methods have been proposed [7-9]. However, it is very hard to choose the method best suited to the type of data [10-14].

Clustering techniques are divided in two classes namely hierarchical and non-hierarchical methods. Hierarchical clustering methods have two different sub-classes, agglomerative and divisive approaches. Non-hierarchical clustering methods are divided in four sub-classes namely partitioning, density-based, grid-based and other approaches [15].

Agglomerative clustering treats every data point as a singleton cluster and then progressively, step by step unions two most similar clusters, called linkages, until all data points have been converged into a single final cluster [16]. Several methods used to choose pairs of clusters to merge such as single linkage, complete linkage, average linkage, centroid, and Ward's methods.

In this research, Ward's method was used because it is the only agglomerative clustering procedures that relies upon the sum of squares criterion, resulting groups minimize within-group dispersion at each binary combination [17]. In Ward's method, the similarity of objects within a cluster is not a single measure of similarity, but the sum of squares across all variables in a cluster [18].

In many cases, spatial clusterings had been applied to classify spatial data such as socio-demographic and geographic data [19-21]. The researchers used a spatial clustering namely the spatial 'K'luster analysis by tree edge removal (SKATER) method. The SKATER method was introduced by Assunção et al. [22] and Bekti [23]. This method uses an algorithm that transforms regional data into partition charts. This method partitions locations that are not neighboring and do not have similar characteristics. The results of Assunçãu works [22] showed that SKATER is an efficient method that produced good quality results and is significantly faster. In this research both methods would be applied to classify 34 provinces in Indonesia based on monthly expenditure per capita according to food commodity groups and the results would be compared.

2. Methodology

2.1. Data Collection Method

The data in this study are percentages of monthly expenditure per capita according to food commodity groups on the total monthly expenditure per capita in 34 provinces in Indonesia during March 2018. Data used in this study are data sourced from the official website of the Indonesian Central Statistics Agency [2].

2.2. Research Variables

The variables in this study constitute 14 groups of food commodities on monthly expenditure per capita of 34 provinces in Indonesia during March 2018. The data units are in percent (%). The variables are shown in Table 1.

Table 1. Research Variables

Variables	
Grains (X_1)	Fruits (X_8)
Tubers (X_2)	Oil and Coconuts (X_9)
Fish/Shrimp/Squid/Shellfish (X_3)	Beverage Ingredients (X_{10})
Meat (X_4)	Spices (X_{11})

Eggs and Milk (X_5)	Other Food Ingredients (X_{12})
Vegetables (X_6)	Processed Food and Beverage (X_{13})
Beans (X_7)	Cigarettes and Tobacco (X_{14})

2.3. Stages of Research

The first stage in this research is to identify spatial patterns and descriptions which is carried out to see the characteristics of the research data used and to look at the description of research data based on 34 provinces in Indonesia. Detecting outliers data is done by looking at the z-score. If there is no outlier or outlier problem can be overcome, then continue to the stages in point c. If there is an outlier in the data that cannot be overcome, then do two types of analyzes namely first and second analyses, then compare them. The first analysis is as in the stages c, but eliminate the outlier data. The second analysis is as in point c using all data, included outliers.

Multi-collinearity test was conducted next. If there is multi-collinearity, then do the main component analysis. Otherwise, do not perform it. After that, cluster analysis was performed with Ward's method by using squared Euclidean distance and Ward's method. The cluster number was determined by using RMSSTD (Root Mean Squared Standard Deviation) criterion. Then, the each formed cluster was interpreted.

The cluster analysis was conducted by using the spatial 'K'luster analysis by tree edge removal (SKATER) method with several stages which are determining weighting factor,where the weight used in this study is binary or queen contiguity, and identifying spatial autocorrelation. The identification of spatial autocorrelation was using Moran's I statistics. Then, the results of clustering obtained from the Ward's and SKATER methods were compared by comparing their standard deviation ratios. Lastly, conclusion was made based on the results.

3. Results and discussions

3.1. Identifying Spatial Patterns and Descriptions

3.1.1. Descriptive analysis

The desription of the data used in this research consisted of the mean, standard deviation, minimum, and maximum values (in percents) of the 14 variables and depicted as in Table 2.

Table 2. Descriptive Analysis of the Data

Variable	Mean	StDev	Minimum	Maximum
X_1	6.600	2.091	2.949	15.072
X_2	0.729	1.333	0.225	8.099
X_3	4.996	1.473	1.664	7.320
X_4	1.919	0.661	0.560	3.409
X_5	2.809	0.348	2.015	3.389
X_6	3.950	0.984	2.438	6.703
X_7	0.836	0.241	0.437	1.548
X_8	2.440	0.527	1.366	4.157
X_9	1.300	0.298	0.688	1.861
X_{10}	1.629	0.355	0.901	2.551
X_{11}	1.021	0.218	0.567	1.430
X_{12}	0.901	0.202	0.549	1.298
X_{13}	15.008	2.674	8.470	19.803
X_{14}	5.967	1.196	2.941	7.774

3.1.2. Spatial patterns

Spatial patterns based on 14 variables are as shown as follow in Figure 1, Figure 2, Figure 3, Figure 4 and Figure 5.

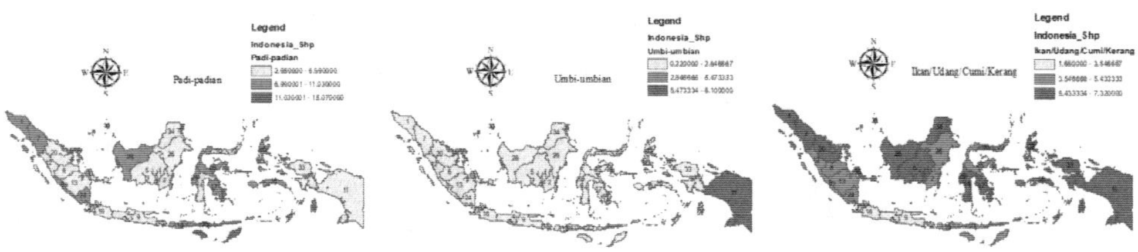

Grains Tubers Fish/shrimp/squid/shellfish

Figure 1. Spatial patterns of grains, tubers, and fish/shrimp/squid/shellfish variables

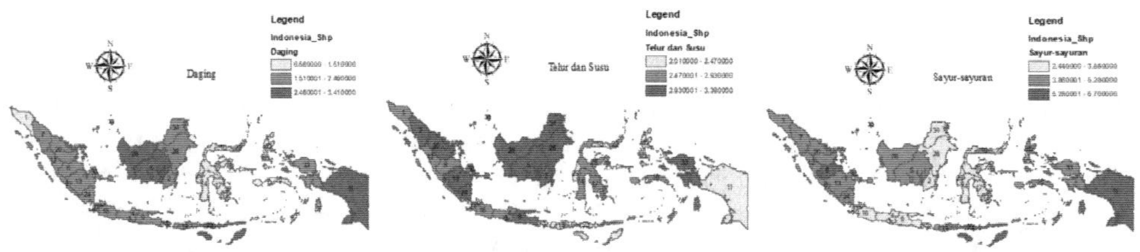

Meat/egg Milk Vegetables

Figure 2. Spatial patterns of meat/egg, milk, and vegetable variables

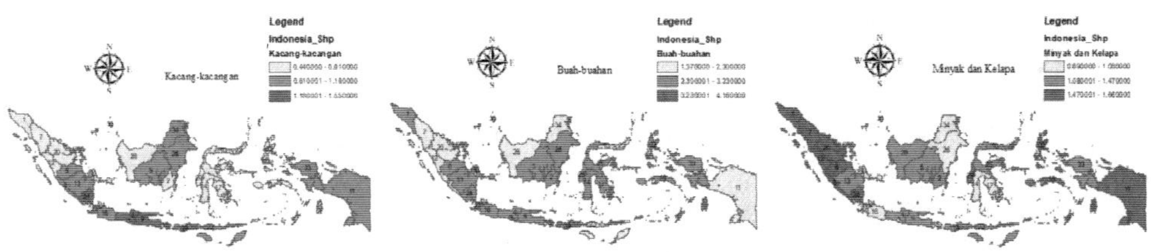

Beans Fruits Oil/coconut

Figure 3. Spatial patterns of bean, fruit, and oil/coconut variables

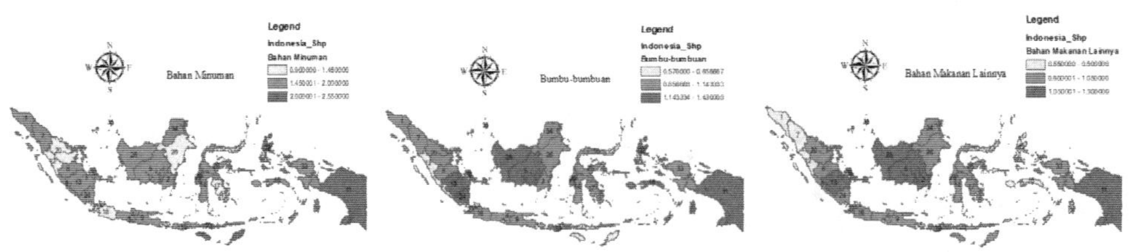

Beverage Spice Other food ingredient

Figure 4. Spatial patterns of beverage, spice, and other food ingredient variables

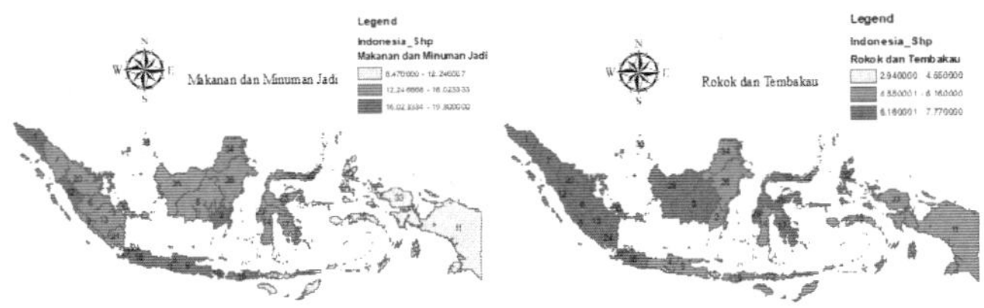

Processed food and beverage   Cigarette and tobacco

Figure 5. Spatial patterns of processed food and beverage, cigarette and tobacco variables

3.2. *Outliers Detection*

In this study, outliers were detected by standardizing or changing each research variable into the z-score. Research data that have z-scores $\leq$ -3.00 or z-scores $\geq$ 3.00 are categorized as outliers. Based on the analysis carried out, it was found that the data categorized as outliers are as in Table 3.

Table 3. Outlier Observation

Variable	Observation number
X_1 (grains)	23 (East Nusa Tenggara)
X_2 (tubers)	11 (Papua)
X_8 (fruits)	19 (West Nusa Tenggara)

To overcome the outlier problem, several data transformations were carried out. The results of the seven data transformations are shown in Table 4.

Table 4. Data Transformation Results

No.	Transform	Result	Variable-Observation number
1	Natural logarithm (ln)	*Outlier*	(X_3-11)
2	Squared root	*Outlier*	(X_1-23) and (X_2-11)
3	Logarithm	*Outlier*	(X_1-23) and (X_2-11)
4	Arcsin	*Outlier*	(X_1-23), and(X_2-11)
5	Inverse	*Outlier*	($X_1 - 21$)
6	Inverse squared	*Outlier*	(X_1-21), (X_2-2), (X_3-3), (X_4-32), (X_8-28), (X_9-21), (X_{10}-21), (X_{13}-11), and (X_{14}-3)
7	Inverse squared root	*Outlier*	(X_1-21), (X_2-11), (X_3-3), (X_4-32), and (X_{14}-3)

Because there are still outliers in the research data even though seven types of transformation have been carried out, the next step to deal with outliers is to eliminate provinces that contain outliers. But in this study all variables were included in the analysis. Therefore, the researchers decided to carry out two types of analyses namely first and second analyses. The first analysis excluded provinces containing outlier values while the second analysis included provinces with outlier values. Then, provinces containing outliers would be identified. Finally, the results would be compared to get the better method for both analyses.

3.3. *First and Second Analyses*

3.3.1 *Multicollinearity test*

In this study, multicollinearity test was identified from the Pearson correlation coefficients. It is said that there is multicollinearity if there is a Pearson correlation coefficient greater than 0.6. In the first analysis, it can be concluded that there is multicollinearity which is characterized by the Pearson correlation coefficients greater than 0.6, namely the correlation coefficients between X_1 and X_{19}, X_1 and X_{10}, and X_9 and X_{14}. In the second analysis, there is multicollinearity in the research data, this is indicated by the Pearson correlation coefficients greater than 0.6, namely the correlation coefficients between X_1 and X_{10}, X_6 and X_9, X_6 and X_{10}, X_9 and X_{10}, and X_{11} and X_{12}. The next step was performing main component analysis in order to eliminate the correlation through the transformation of the origin variable to the new variable that is not correlated.

For the main component analysis, the new main component is formed by searching for the cummulative proportion greater than 80%. In the first analysis, based on the cummulative proportions obtained, the cummulative proportion of W_5 has a value of 0.867 or 86.7% which is greater than 80%, so that components W_1, W_2, W_3, W_4, and W_5 can explain 86.7% of the total proportion variance. Furthermore, those components would be used in the cluster analysis. In the second analysis, based on the cummulative proportion obtained, the cummulative proportion of W_5 has a value of 0.827 or 82.7%, therefore components W_1, W_2, W_3, W_4, and W_5 would be used in the cluster analysis.

For the cluster analysis using Ward's method, the first analysis where the cluster amount was determined, the number of clusters used was determined by finding the smallest RMSSTD value. Based on the calculation, the RMSSTD value of the 2-cluster grouping is 0.233, while the RMSSTD of the 3-cluster grouping is 0.235. Therefore the 2-cluster grouping is a better choice. Based on the dendrogram in Figure 6, the cluster membership for 2-cluster grouping is shown in Table 5.

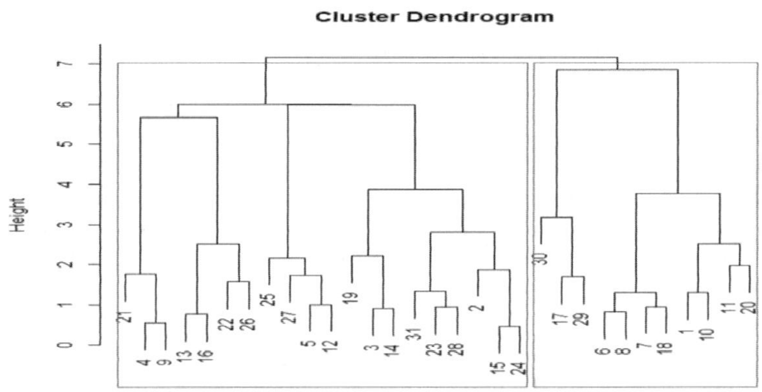

Figure 6. Dendrogram of 2-cluster grouping using Ward's method

Table 5. Cluster membership for first analysis using Ward's method

Cluster	Number of members	Members
1	11	Aceh, Jambi, North Sumatera, Bengkulu, Gorontalo, West Sumatera, Maluku, Riau, North Sulawesi, North Maluku, and West Papua
2	20	South Kalimantan, Special Region of Yogyakarta, Central Java, Central Kalimantan, East Java, South Sumatera, South Sulawesi, Bali, West Java, Southeast Sulawesi, Special Capital Region of Jakarta, Lampung, Central Sulawesi, East Kalimantan, Banten, West Kalimantan, West Sulawesi, Bangka-Belitung Islands, Riau Islands, and North Kalimantan

Profiling or characteristics was then performed. The first cluster consists of 11 provinces. In this cluster, among 14 variables there are 7 commodities having high levels and other 7 commodities having low levels of percentage of expenditure for food commodities. This result is similar to the second cluster consisting of 20 provinces, there are 7 commodities having high levels and other 7 commodities having low levels of percentages. Most provinces in cluster 1 have low levels of welfare and most provinces in cluster 2 have high levels of welfare. In second analysis where cluster amount was determined, based on the calculation of RMSSTD, the results showed that the RMSSTD value of 2 clusters is 0.274, while the RMSSTD 3 cluster is 0.263. Therefore the 3-cluster grouping is more preferable. Based on the dendrogram in Figure 7, the cluster membership for 3-cluster grouping is shown in Table 6.

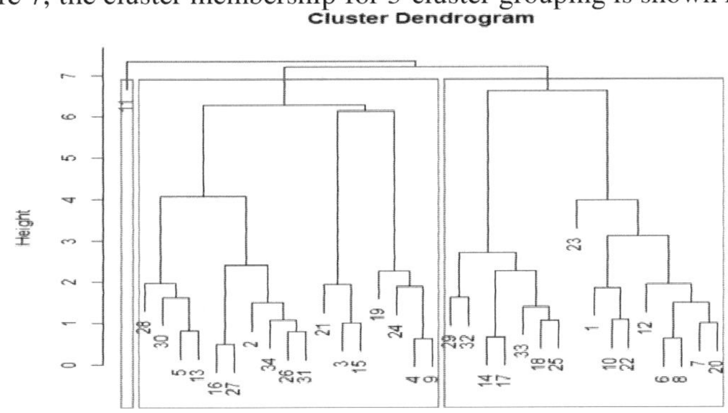

Figure 7. Dendrogram of 3-cluster grouping for second analysis using Ward's method

Table 6. Cluster membership for second analysis using Ward's method

Cluster	Number of members	Members
1	16	Aceh, Jambi, North Sumatera, Bengkulu, Gorontalo, West Sumatera, South Sulawesi, Southeast Sulawesi, Maluku, Riau, North Sulawesi, East Nusa Tenggara, Central Sulawesi, West Sulawesi, North Maluku, and West Papua
2	17	South Kalimantan, Special Region of Yogyakarta, Central Java, Central Kalimantan, East Java, South Sumatera, Bali, West Java, West Nusa Tenggara, Special Capital Region of Jakarta, Lampung, East Kalimantan, Banten, West Kalimantan, Bangka Belitung Islands, Riau Islands, and North Kalimantan
3	1	Papua

The first cluster consists of 16 provinces with medium percentages of expenditure for food commodities. The second cluster consists of 17 provinces with low percentages of expenditure for food commodities. The third cluster consists of one province with a high percentage of expenditure on food commodities. Most provinces (56.25%) in cluster one have low levels of welfare. In contrary, most provinces (58.81%) in cluster two have high levels of welfare.

3.3.2. Spatial autocorrelation test (by Moran's I test) and SKATER method

For the first analysis in Moran's test, based on the results of Moran's I test, it can be concluded that variables W_1, W_2, W_3, W_4 and W_5 have spatial autocorrelations, which is indicated by each of these variables having a p-value of less than 0.05. The results showed that the RMSSTD value for 2-cluster grouping was 0.213, while the RMSSTD value for 3-cluster grouping was 0.211. Therefore the 3-cluster grouping is more preferable. By using SKATER method, the minimum spanning tree partition and the members of each cluster for the first analysis are shown in Figure 8 and Table 7.

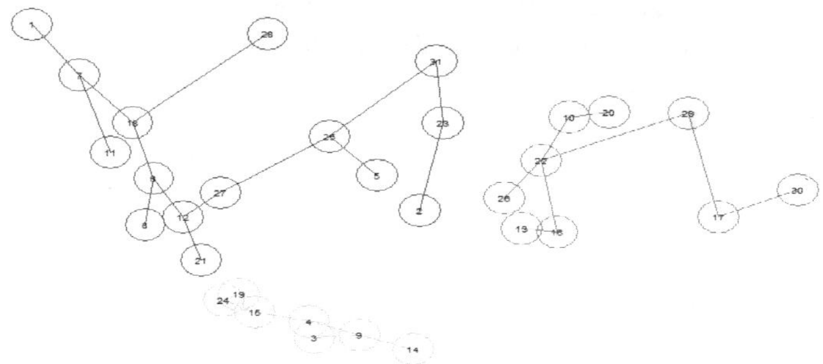

Figure 8. 3-cluster grouping using MST partition for first analysis

Table 7. Cluster membership for first analysis using SKATER method

Cluster	Number of members	Members
1	9	Gorontalo, South Sulawesi, Southeast Sulawesi, Maluku, North Sulawesi, Central Sulawesi, West Sulawesi, North Maluku, and West Papua
2	7	Special Region of Yogyakarta, Central Java, East Java, Bali, West Java, Special Capital Region of Jakarta, and Banten
3	15	Aceh, South Kalimantan, Central Kalimantan, Jambi, North Sumatera, Bengkulu, West Sumatera, South Sumatera, Riau, Lampung, East Kalimantan, West Kalimantan, Bangka Belitung Islands, Riau Islands, and North Kalimantan

The first cluster consists of 9 provinces with low percentages of expenditure for food commodities. The second cluster consists of 7 provinces with high percentages of expenditure for food commodities. The third cluster consists of 15 provinces with moderate percentages of expenditure on food commodities. Most provinces (66.67%) in cluster one have moderate levels of welfare, all provinces in cluster two have high levels of welfare, some provinces (26.67%) in cluster 3 have low levels of welfare.

In the second analysis in the Moran's test, the results are obtained that the variables W_1, W_2, W_4, and W_5 are said to have spatial autocorrelation, which is indicated by each variable having a p-value of less than 0.05. The results obtained are that the RMSSTD value for the 2-cluster grouping is 0.262, while the RMSSTD value for the 3-cluster grouping is 0.259. Therefore the 3-cluster grouping is more preferable.

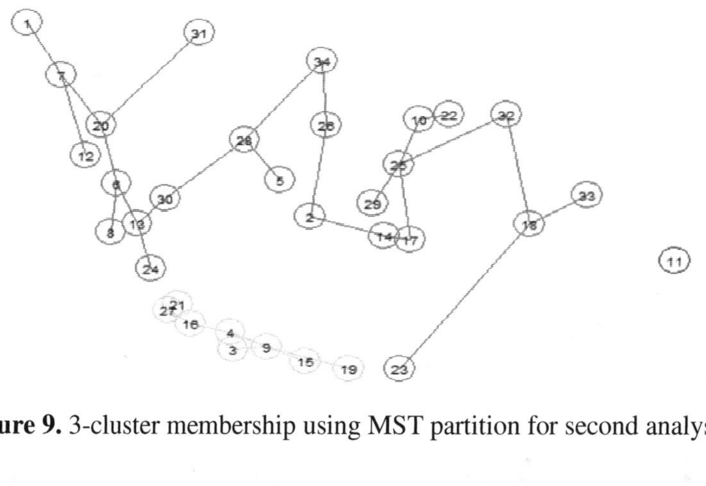

Figure 9. 3-cluster membership using MST partition for second analysis

Table 8. Cluster membership for second analysis using SKATER method

Cluster	Number of members	Members
1	25	Aceh, South Kalimantan, Central Kalimantan, Jambi, North Sumatera, Bengkulu, Gorontalo, West Sumatera, South Sumatera, South Sulawesi, Southeast Sulawesi, Maluku, Riau, North Sulawesi, East Nusa Tenggara, Lampung, Central Sulawesi, East Kalimantan, West Kalimantan, West Sulawesi, Bangka Belitung Islands, Riau Islands, North Maluku, West Papua, and North Kalimantan
2	8	Special Region of Yogyakarta, Central Java, East Java, Bali, West Java, West Nusa Tenggara, Special Capital Region of Jakarta, and Banten
3	1	Papua

The first cluster consisted of 25 provinces with moderate percentages of expenditure for food commodities. The second cluster consists of 8 provinces with high percentages of expenditure for food commodities. The third cluster consists of 1 province with low percentage of expenditure on food commodities. Some provinces (40%) in cluster one have high levels of welfare, most provinces (87.5%) in cluster two have high levels of welfare, one province (100%) in cluster 3 has a low level of welfare (see Figure 9 and Table 8).

3.4. Comparison Between Methods and Analysis

The standard deviation ratio values between the different methods and analysis is shown in Table 9 below.

Table 9. Standard Deviation Ratio Values between Analysis and Method

Method	Standard deviation	
	First Analysis	Second Analysis
Ward	0.37	0.18
SKATER	0.23	0.20

For the first analysis, the SKATER method is much better than Ward's method because it has a smaller standard deviation ratio than Ward's method. However, for the second analysis Ward's method gave a slightly better result than the SKATER method. Furthermore, from the results of the study, it can be concluded that the second analysis is better than the first analysis because it has a smaller standard deviation ratio value using both the Ward's and SKATER methods.

From this research it turned out that the SKATER and the Ward's methods do not always give satisfied result and this case supports the conclusion in [10-14]. To improve the result, modified SKATER method can be applied [24]. Modified SKATER technique is performed by modifying the computation of weights and the partition procedure to oblige the homogeneity within and the heterogeneity between clusters.

4. Conclusion

Based on the results discussed in previous section, it can be concluded that:

1. From the first analysis, the Ward's method produced 2 clusters where the first cluster consisted of 11 provinces with high welfare levels while the second cluster consisted of 20 provinces with low welfare levels. While the SKATER method obtained 3 clusters where the first cluster numbered 9 provinces with moderate welfare levels, the second cluster amounted to 7 provinces with high welfare

levels, and the third cluster amounted to 15 provinces with low welfare levels. In this case the SKATER method is better because it has less standard deviation ratio than that of the Ward's method.
2. In the second analysis the Ward's method produced 3 clusters where the first cluster numbered 16 provinces with moderate welfare levels, the second cluster numbered 17 provinces with high welfare levels, and the third cluster numbered 1 province with a low welfare level. Whereas the SKATER method obtained 3 clusters where the first cluster numbered 25 provinces with moderate welfare levels, the second cluster numbered 8 provinces with high welfare levels, and the third cluster amounted to 1 province with a low welfare level. In this case the Ward's method is slightly better than the SKATER method.
3. Based on the two analyses above, it can be concluded that the second analysis is better than the first analysis because it generates smaller standard deviation ratios for both the SKATER and Ward's methods.

References

[1] Alwi W and Hasrul M 2018 Analisis klaster untuk pengelompokan kabupaten/kota di propinsi Sulawesi Selatan berdasarkan indikator kesejahteraan rakyat *Jurnal MSA* **6**(1) pp 35-42.
[2] Anonimous 2018 *Consumption expenditure of population of Indonesia by province, March 2018.*
[3] Zezzaa A, Carlettoa C, Gennarib P and Jolliffea D 2017 Food counts measuring food consumption and expenditures in household consumption and expenditure surveys (HCES) Introduction to the special issue *Food Policy* **72** pp 1-6.
[4] Sununtnasuk C and Fiedler JL 2017 Can household-based food consumption surveys be used to make inferences about nutrient intakes and inadequacies? a Bangladesh case study *Food Policy* **72** pp 121-31.
[5] Suhaeni 2018 Perbandingan hasil pengelompokan menggunakan analisis cluster berhirarki, K means cluster, dan cluster ensemble (studi kasus data indikator pelayanan kesehatan ibu hamil). *Jurnal Media Infotama* **14**(1) pp 31-8.
[6] Rodriguez M Z, Comin C H, Casanova D, Bruno O M, Amancio D R and Costa LD 2019 Clustering algorithms: a comparative approach. *PLoS ONE* **14**(1).
[7] Jing L, Ng M K and Huang JZ 2007 An entropy weighting K-means algorithm for subspace clustering of high-dimensional sparse data *IEEE Transactions on Knowledge and Data Engineering* **19**(8) pp 1026–41.
[8] Suzuki R and Shimodaira H 2006 Pvclust: an R package for assessing the uncertainty in hierarchical clustering *Bioinformatics* **22**(12) pp 1540–2.
[9] Camastra F and Verri A 2005 A novel kernel method for clustering *IEEE Transactions on Pattern Analysis and Machine Intelligence* **27**(5) 801–5.
[10] Rujasiri P and Chomtee B 2009 Comparison of clustering techniques for cluster analysis *Kasetsart J (Nat Sci)* **43** pp 378-88.
[11] Guha S, Rastogi R and Shim K 2001 Cure: an efficient clustering algorithm for large databases *Information Systems* **26**(1) pp 35-58.
[12] Aggarwal CC and Reddy CK 2013 *Data clustering: algorithms and applications* 1st ed: Chapman & Hall/CRC.
[13] Karypis G, Han E H and Kumar V 1999 Chameleon: Hierarchical clustering using dynamic modeling *Computer* **32**(8) pp 68-75.
[14] Huang J, Sun H, Kang J, Qi J, Deng H and Song Q 2013 ESC: An efficient synchronization-based clustering algorithm *Knowledge-Based Systems* **40** pp 111-22.
[15] Gülağız FK and Şahin S 2017 Comparison of hierarchical and non-hierarchical clustering algorithms *International Journal of Computer Engineering and Information Technology* **9**(1) pp 6-14.
[16] Rafsanjani M K, Varzaneh Z A, and Chukanlo N E 2012 A survey of hierarchical clustering algorithms *The Journal of Mathematics and Computer Science* **5**(3) pp 229-40.

[17] Murtagh F and Legendre P 2014 Ward's hierarchical agglomerative clustering method: which algorithms implement Ward's criterion? *Journal of Classification* **31**.

[18] Hair F H J, Black W C, Babin B C, and Anderson R E 2014 *Multivariate data analysis* 7th ed.

[19] Gelawa Y A, Magalhães RJS, Assefaa Y, and Williams G 2019 Spatial clustering and socio-demographic determinants of HIV infection in Ethiopia, 2015–2017 *International Journal of Infectious Diseases* **82** pp 33-9.

[20] Hughey S M, Kaczynski A T, Porter D E, Hibbert J, and Liu J 2018 Spatial clustering patterns of child weight status in a southeastern US county *Applied Geography* **99** pp 12-21.

[21] Liew H P, Brooks T 2017 Spatial clusters of AIDS in Indonesia *Health Policy and Technology* **6**(2) pp 208-13.

[22] Assunção R M, Corrêa M, Câmara N G, and Freitas C D C 2006 Efficient regionalisation techniques for socio-economic geographical units using minimum spanning trees *International Journal of Geographical Information Science* **20**(7) pp 797-811.

[23] Bekti RD 2015 Metode spasial skater untuk pengelompokan lokasi berdasarkan fasilitas air bersih dan sanitasi *Jurnal Teknologi* **8**(1) pp 53-8.

[24] Rinaldi I K Y, Theresia M B, Efendi A, and Pramoedyo H 2017 Application of modified spatial 'k'luster analysis by tree edge removal (SKATER) method on the level of crime in Way Kanan district Lampung *Cauchy-Jurnal Matematika Murni dan Aplikasi* **4**(4) pp 155-160.

IOP Conf. Series: Materials Science and Engineering **807** (2020) 012018 doi:10.1088/1757-899X/807/1/012018

Finite Element Stress Analysis of Acrylic with Embedded Optical Fibre Sensors

O Kurdi[1,2*], S A Widyanto[1], G D Haryadi[1,2], A Suprihanto[1], I Yulianti[3], A Rahman[1]

[1]Mechanical Engineering Department, Diponegoro Universiti, Semarang, Central Java, Indonesia
[2]National Center for Sustainable Transportation Technology, Indonesia
[3]Physics Department, Universitas Negeri Semarang, Central Java, Indonesia

Corresponding e-mail: ojokurdi@ft.undip.ac.id

Abstract. Finite element (FE) techniques are used to analyse mechanical stress in optical fibre sensors embedded in acrylic. The models represent many features observed in real materials with embedded fibre optic for stress and strain sensor application such as for health monitoring of structures. Three thicknesses, namely 6 mm, 8 mm and 10 mm of acrylic were analyzed with the optical fibre embedded at the mid-plane of acrylic. The simulation was done by implementing Finite Element Method using ABAQUS software. Contact between acrylic and optical fibre was modelled as frictionless and with friction. The friction coefficient was varied for various values which are 0.3, and 0.5. This paper investigates the effect of friction coefficient and the effect of fibre optic's depth embedded in acrylic to the their von Misses stress. Simulation result shows that the higher the friction coefficient the higher von Misses stress in fibre optic and the higher von Misses stress in acrylic. The higher the fibre's depth the higher von Misses stress in fibre and the lower von Misses stress in acrylic. The maximum stress occurred for the coefficient of friction 0.5, with magnitude of 116.4 MPa and 58.72 MPa for acrylic and optical fibre respectively. The most significant effect of fibre's depth to the von Misses stress was obtained for 1 mm depth for acrylic and 5 mm for optical fibre, with magnitude of 72.64 MPa and 57.65 MPa for acrylic and optical fibre respectively.

1. Introduction

Fibre optic as sensor has many advantages namely: small size, light weight and immunity to electromagnetic interference [1]. Fibre optic sensor with these advantages can be embedded into the host material to form an ideally smart structure system. Nowadays fibre optic has been used as sensor in many applications such as for structural health monitoring and aerospace engineering, industrial processing and defence industry [2-8]. The information on numerous important parameters such as micro strain, micro vibration, and stiffness, which are vital to the in situ structural health monitoring as well as the optimizations of the system performances can be obtain from embedded fibre optic sensor [9-18].

Fibre optic has been embedded in some material as a pressure sensor in health monitoring of structures and bridges. Acrylic has been used widely for roof and wall of some buildings. Therefore, to ensure the safety of fibre optic that embedded in acrylic it is necessary to investigate the stress condition inside the fibre optic under pressure load. Some researchers have carried out the stress analysis embedded in composites and other materials [3], [11-12], [14-16], [19-24]. The contact condition between fibre optic and acrylic has also been defined and varied as frictionless and with certain various

IOP Conf. Series: Materials Science and Engineering **807** (2020) 012018 doi:10.1088/1757-899X/807/1/012018

friction coefficient, namely 0.3, 0.5 and 0.8. The depth of fibre optic embedded in acrylic also will be varied by 1mm until 5 mm. The effect of friction coefficient and the depth of fibre optic was embedded in acrylic will be investigated in this research using FE simulation technique.

2. Finite Element Simulation

2.1. Acrylic and fibre optic model

Model of of fibre optic embedded in acrylic has been drawn using Solid works software. The model depicted in Figure 1, for acrylic and fibre optic respectively. The fibre optic was modelled as a solid rod with diameter of 1 mm, and the acrylic has a 50 mm in length and 40 mm in width, with various of thicknesses, 6 mm, 8 mm and 10 mm. The material properties of acrylic and fibre optic were shown in Table 1.

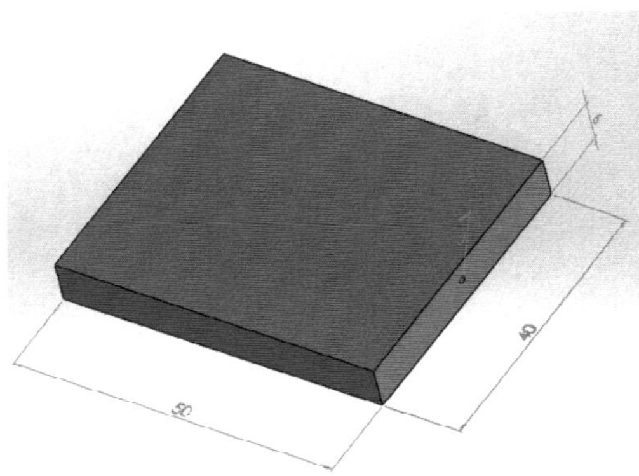

(a)

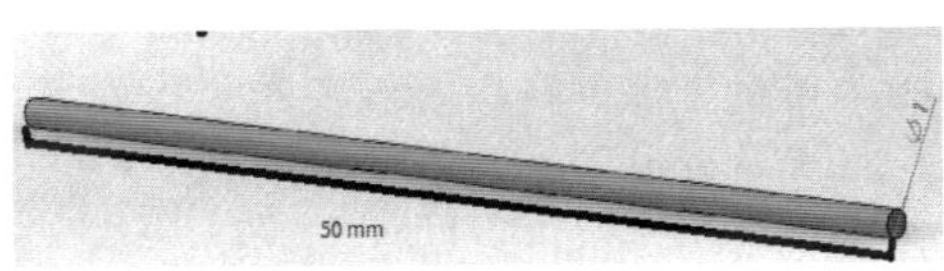

(b)

Figure 1. Model (a) Acrylic; (b) fibre optic

Table 1. Properties of acrylic and fibre

Mechanical Property	Material	
	Fibre Optic	Acrylic
Density (kg/m^3)	7200	7850
Ultimate Tensile Strength (MPa)	450	450
Tensile Yield Strength (MPa)	280	435
Modulus of Elasticity (MPa)	140	190
Poisson's Ratio	0.211	0.27

2.2. Load and boundary condition

Load was given as distributed pressure on top of surface of acrylic with fibre optic has embedded inside. The magnitude of pressure is 80 MPa. That magnitude was considered according to maximum pressure that can be applied to the acrylic with has 6 mm in thickness.

There are two boundary conditions (BCs) on the model. First boundary condition is fixed that was applied in the bottom of the acrylic, this BC restricted the model to translate and rotate in all directions. The second BC was applied on both of two ends of fibre. This kind of BC restricted the fibre translation in axial direction and restricted rotation in all directions. Figure 2 show the load and two boundary conditions of the model.

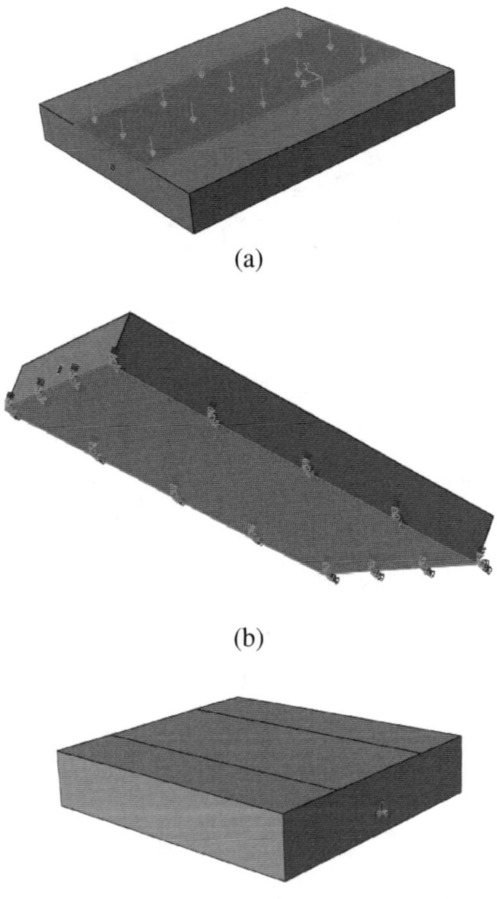

(a)

(b)

(c)

Figure 2. Load and Boundary Condition; (a) Load; (b) BC 1; (c) BC 2

3. Result and Discussion

The simulation result of von Misses stress for fibre with various coefficients of friction were shown in Figure 3 for 4 mm thickness of acrylic and 1 mm of fibre's depth. The maximum von Misses stress of acrylic for frictionless contact condition is 64,04 MPa, while the von Misses stress for coefficient of friction 0.3 is 116.4 MPa. The von Misses stress of acrylic tend to increase by increasing the coefficient of friction.

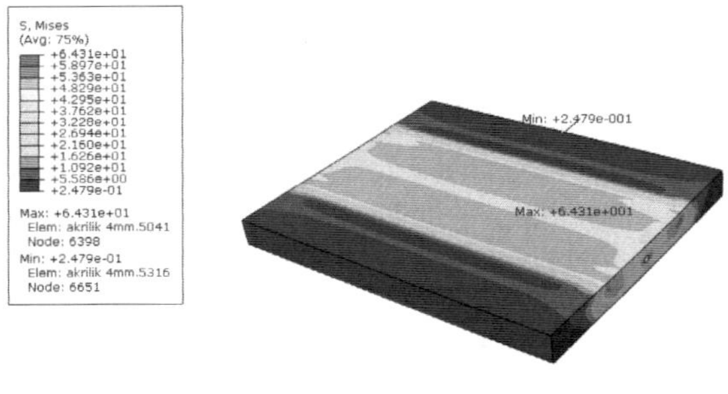

(a)

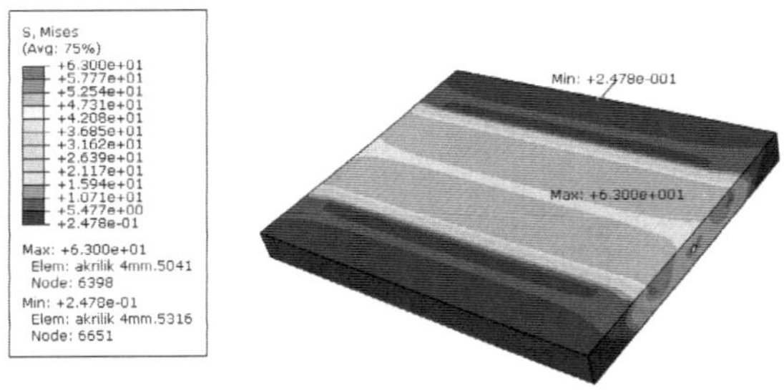

(b)

Figure 3. Von Misses stress in acrylic of 4 mm acrylic thickness; (a). for frictionless condition; (b). for 0.3 coefficient of friction

Von Misses stress simulation result for 4 mm thickness of acrylic and 1 mm of fibre's depth can be seen in Figure 4. The trend of stress in the fibre also same with that of stress in the acrylic. The stress is increase by increasing of coefficient of friction.

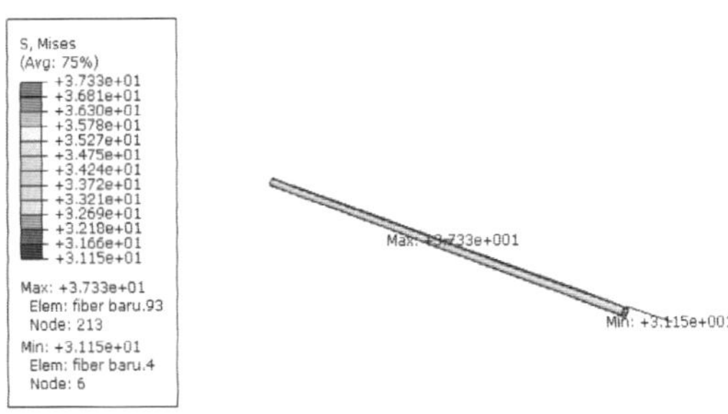

(a)

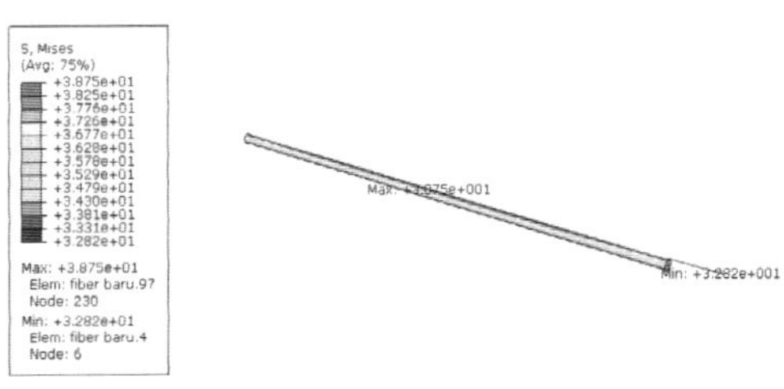

(b)

Figure 4. Von Misses stress in fibre of 4 mm acrylic thickness; (a). for frictionless condition; (b). for 0.3 coefficient of friction.

The relation between von Misses stress and the coefficient of friction was stated in Figure 5. The von Misses stress of both acrylic and fibre were increase by increasing of coefficient of friction.

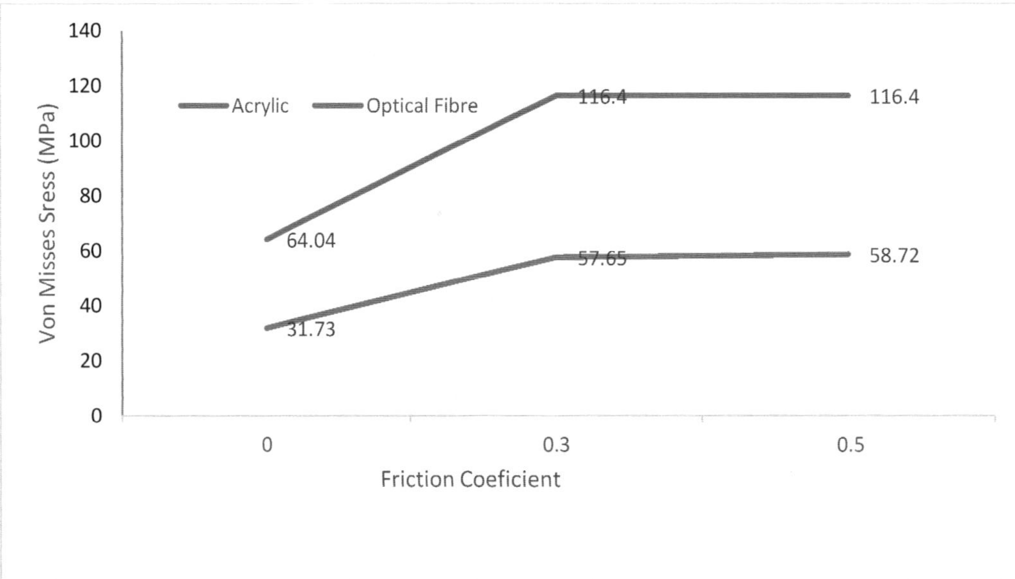

Figure 5. Effect of the coefficient of friction to the stress in fibre optic and acrylic for 1 mm depth.

Figure 6 shows the von Misses stress of acrylic for 8 mm thickness of acrylic and 3 mm of fibre's depth, frictionless condition. It can be seen that the maximum von Misses stress was occurred at near of surface with magnitude of 64.67 MPa, while the acrylic with 1 mm of fibre's depth has the von Misses stress of 72.46 MPa which occurred at the surface of acrylic which contact with fibre.

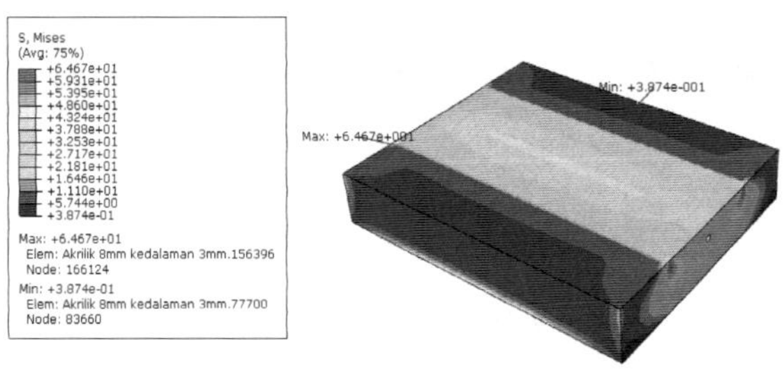

(a)

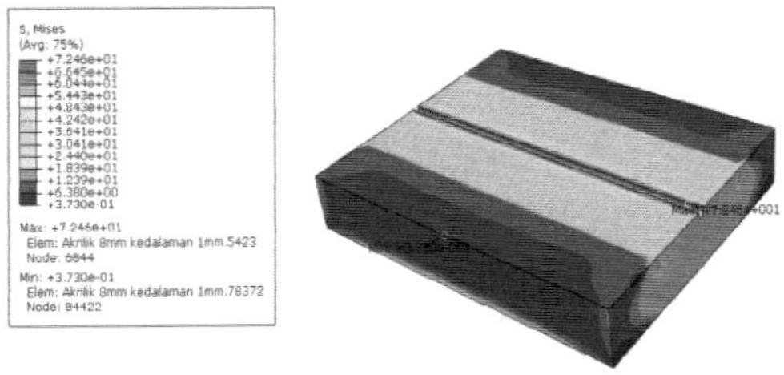

(b)

Figure 6. Von Misses stress in acrylic of 8 mm acrylic thickness (frictionless); (a). for 3 mm of fibre's depth; (b). for 1 mm of fibre's depth

Figure 7 shows the von Misses stress of fibre for 8 mm thickness of acrylic and 3 mm of fibre's depth, frictionless condition. It can be seen that the maximum von Misses stress was occurred at surface near to the right end of fibre with magnitude of 70.32 MPa, while the fibre with 1 mm of fibre's depth has the von Misses stress of 57.65 MPa which occurred at the same region with that of 3 mm of fibre's depth.

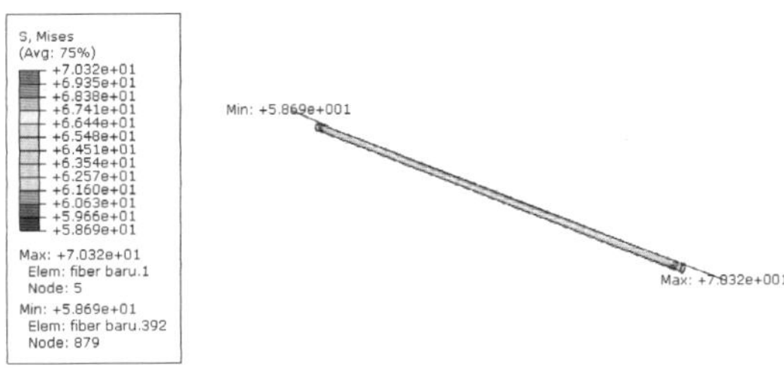

(a)

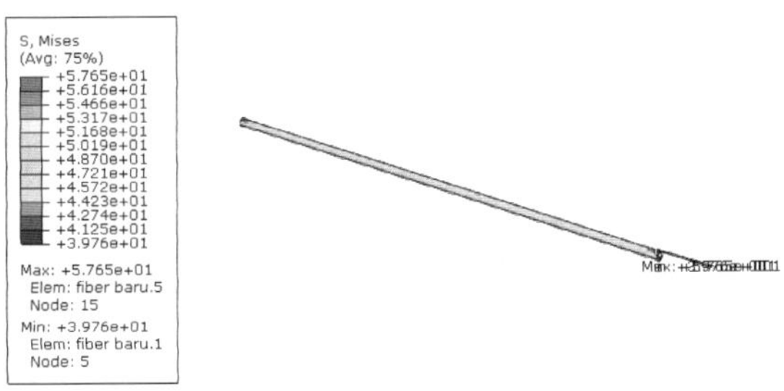

(b)

Figure 7. Von Misses stress in fibre of 8 mm acrylic thickness (frictionless); (a). for 3 mm of fibre's depth; (b). for 1 mm of fibre's depth

The relation between von Misses stress and the fibre's depth was displayed in Figure 8. The von Misses stress of fibre was increase by increasing of fibre's depth, but The von Misses stress of acrylic was decrease by increasing of fibre's depth.

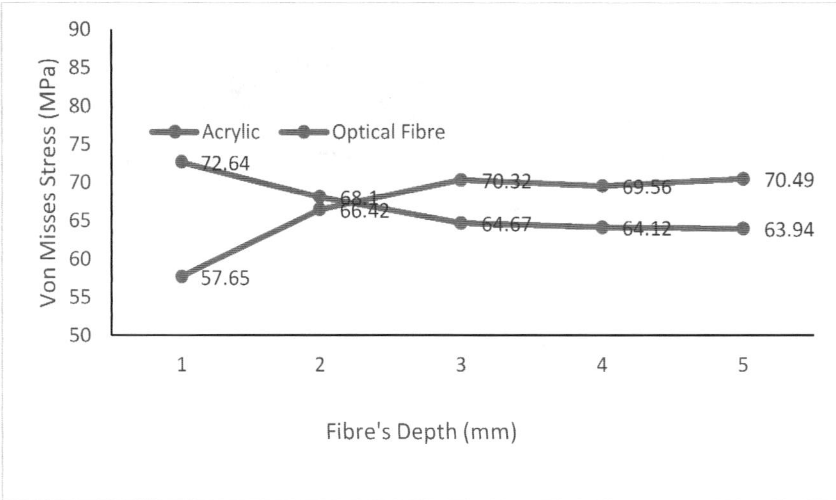

Figure 8. Effect of the fibre's depth to the stress in fibre optic and acrylic for frictionless condition.

4. Conclusion
The stress analysis of fibre optic sensor embedded in acrylic was accomplished by finite element method. Based on the simulation result, it can be concluded that von Misses stress of fibre optic and acrylic were increase by increasing of coefficient of friction. The von Misses stress of fibre was increase by increasing of fibre's depth, but The von Misses stress of acrylic was decrease by increasing of fibre's depth. The maximum von Misses stress of acrylic and optical fibre sensor was occurred for the coefficient of friction of 0.5, for 1 mm depth the magnitude of stress is 116.4 MPa and 58.72 MPa for acrylic and optical fibre respectively. The maximum von Misses stress for fibre's depth variation was occurred at 1 mm for acrylic and 5 mm depth for optical fibre, with magnitude of 72.64 MPa and 57.65 MPa for acrylic and optical fibre respectively.

Acknowledgments
The authors gratefully acknowledge partial funding support from the USAID under the SHERA program.

References
[1] K. T. V. Grattan and T. Sun, "Fibre optic sensor technology: an overview," *Sensors Actuators A Phys.*, vol. 82, no. 1–3, pp. 40–61, May 2000.
[2] T. H. T. Chan *et al.*, "Fibre Bragg grating sensors for structural health monitoring of Tsing Ma bridge: Background and experimental observation," *Eng. Struct.*, vol. 28, no. 5, pp. 648–659, Apr. 2006.
[3] L. P. Canal, R. Sarfaraz, G. Violakis, J. Botsis, V. Michaud, and H. G. Limberger, "Monitoring strain gradients in adhesive composite joints by embedded fibre Bragg grating sensors," *Compos. Struct.*, vol. 112, pp. 241–247, Jun. 2014.
[4] H.-N. Li, D.-S. Li, and G.-B. Song, "Recent applications of fibre optic sensors to health monitoring in civil engineering," *Eng. Struct.*, vol. 26, no. 11, pp. 1647–1657, Sep. 2004.
[5] M. Majumder, T. K. Gangopadhyay, A. K. Chakraborty, K. Dasgupta, and D. K. Bhattacharya, "Fibre Bragg gratings in structural health monitoring—Present status and applications," *Sensors Actuators A Phys.*, vol. 147, no. 1, pp. 150–164, Sep. 2008.
[6] Z. Djinovic, M. Tomic, and C. Gamauf, "Fibre-optic interferometric sensor of magnetic field for structural health monitoring," *Procedia Eng.*, vol. 5, pp. 1103–1106, Jan. 2010.
[7] J. Alvarez-Montoya, A. Carvajal-Castrillón, and J. Sierra-Pérez, "In-flight and wireless damage detection in a UAV composite wing using fibre optic sensors and strain field pattern recognition," *Mech. Syst. Signal Process.*, vol. 136, p. 106526, Feb. 2020.

[8] C. Du, S. Dutta, P. Kurup, T. Yu, and X. Wang, "A review of railway infrastructure monitoring using fibre optic sensors," *Sensors Actuators A Phys.*, p. 111728, Nov. 2019.
[9] L. Rodriguez-Cobo, A. Cobo, and J.-M. Lopez-Higuera, "Embedded compaction pressure sensor based on Fibre Bragg Gratings," *Measurement*, vol. 68, pp. 257–261, May 2015.
[10] A. Y. Fedorov, N. A. Kosheleva, V. P. Matveenko, and G. S. Serovaev, "Strain measurement and stress analysis in the vicinity of a Fibre Bragg grating sensor embedded in a composite material," *Compos. Struct.*, p. 111844, Jan. 2020.
[11] D. C. Lee, J. J. Lee, and S. J. Yun, "The mechanical characteristics of smart composite structures with embedded optical fibre sensors," *Compos. Struct.*, vol. 32, no. 1–4, pp. 39–50, Jan. 1995.
[12] N. Lammens, G. Luyckx, E. Voet, W. Van Paepegem, and J. Degrieck, "Finite element prediction of resin pocket geometry around embedded optical fibre sensors in prepreg composites," *Compos. Struct.*, vol. 132, pp. 825–832, Nov. 2015.
[13] D. Savastru, S. Miclos, R. Savastru, and I. I. Lancranjan, "Study of thermo-mechanical characteristics of polymer composite materials with embedded optical fibre," *Compos. Struct.*, vol. 183, pp. 682–687, Jan. 2018.
[14] P. Zhu, X. Xie, X. Sun, and M. A. Soto, "Distributed modular temperature-strain sensor based on optical fibre embedded in laminated composites," *Compos. Part B Eng.*, vol. 168, pp. 267–273, Jul. 2019.
[15] S. Miclos, D. Savastru, R. Savastru, and I. I. Lancranjan, "Transverse mechanical stress and optical birefringence induced into single-mode optical fibre embedded in a smart polymer composite material," *Compos. Struct.*, vol. 218, pp. 15–26, Jun. 2019.
[16] R. Prussak, D. Stefaniak, E. Kappel, C. Hühne, and M. Sinapius, "Smart cure cycles for fibre metal laminates using embedded fibre Bragg grating sensors," *Compos. Struct.*, vol. 213, pp. 252–260, Apr. 2019.
[17] M. Koerdt *et al.*, "Fabrication and characterization of Bragg gratings in perfluorinated polymer optical fibres and their embedding in composites," *Mechatronics*, vol. 34, pp. 137–146, Mar. 2016.
[18] J. Peng, S. Jia, Y. Jin, S. Xu, and Z. Xu, "Design and investigation of a sensitivity-enhanced fibre Bragg grating sensor for micro-strain measurement," *Sensors Actuators A Phys.*, vol. 285, pp. 437–447, Jan. 2019.
[19] Y. Yu, B. Zhang, Z. Tang, and G. Qi, "Stress transfer analysis of unidirectional composites with randomly distributed fibres using finite element method," *Compos. Part B Eng.*, vol. 69, pp. 278–285, Feb. 2015.
[20] D. Li, Q.-S. Yang, X. Liu, and X.-Q. He, "Experimental and cohesive finite element investigation of interfacial behavior of CNT fibre-reinforced composites," *Compos. Part A Appl. Sci. Manuf.*, vol. 101, pp. 318–325, Oct. 2017.
[21] R. M. Measures, "Fibre optic sensing for composite smart structures," *Compos. Eng.*, vol. 3, no. 7–8, pp. 715–750, Jan. 1993.
[22] M A Pulungan, Sutikno and M S M Sani "Analysis of Bulletproof Vest Made from Fiber Carbon Composite and Hollow Glass Microsphere (HGM) in Absorbing Energy due to Projectile Impact" IOP Conf. Ser. Mater. Sci. Eng. 206 012001, 2019
[23] M S M Sani, N A Z Abdullah, S N Zahari, J P Siregar and M M Rahman "Finite element model updating of natural fibre reinforced composite structure in structural dynamics" MATEC Web of Conferences 83 03007, 2016
[24] R M Yaacob, M A H Hashim and M S M Sani "Finite element modeling and updating of the composite plate structure" IOP Conf. Ser. Mater. Sci. Eng. 100 01201, 2019

Low Cycle Fatigue Properties of Aluminizing Coating on Cold-Drawn AISI 1018 Steel

M Badaruddin[1]*, R D Kurniawan[1], Sugiyanto[1], A Pambudi[2]

[1] Department of Mechanical Engineering, Faculty of Engineering, Universitas Lampung, Bandar Lampung, 35145, Indonesia
[2] Department of Mechanical Engineering, Faculty of Engineering, Universitas Malahayati, Bandar Lampung, 35144, Indonesia

*Corresponding e-mail: mbruddin@eng.unila.ac.id

Abstract. The mechanical and the low cycle fatigue (LCF) properties of cold-drawn AISI 1018 steel (CDS 1018) with and without hot-dip aluminizing (HDA) coating had been experimentally investigated at room temperature. The tensile properties and the LCF properties were determined from tensile test results and cyclic strain-fatigue test results, respectively. The aluminide coating on the CDS 1018 significantly decreases the mechanical properties and the strain-fatigue life of the steel. With increasing in strain amplitude levels, CDS 1018 experienced continuously cyclic softening behaviour after cyclic loading in few cycles. By contrast, the cyclic softening of HDA steel was observed in few cycles and HDA steel continuously exhibits the stably cyclic hardening behaviour until failure. CDS 1018 with aluminide coating results the longer transition fatigue cycles (N_t) than those of CDS 1018 without aluminide coating by a factor of ~3.0. CDS 1018 with HDA coating can be used for material components which the components are safely designed under cyclic loading below the yield strength of the material.

1. Introduction

A low carbon steel (mild steel) price is relatively cheaper than those of high alloy steel grades and stainless-steel grades prices, however mild steel properties are acceptable for many applications in engineering component structures. Therefore, mild steel used for the high temperature material components such as a fire piping of boiler, incinerators and heat exchanger should be protected against high temperature oxidation attack via the hot-dipping aluminizing (HDA) coating. It is well known that the aluminide coating on the steel significantly improve the oxidation resistance at elevated temperatures in air and mixtures of steam environments [1–3] and hot corrosion containing of chloride and sulphur environments [4]. Aluminide layer consisting of the Fe_2Al_5 layer on the steel substrate can supply continuously the aluminium atoms to form the protective aluminium oxide (Al_2O_3) layer. During applications, the material can experience plastic deformation due to strain cyclic loading at starting-up and shut-down periods [5]. Plastic deformation induced the material can cause the component failure in few cycles.

An investigation of an aluminide coating impact on the strength and the fatigue life of steel is very important to determine an initial safety design for engineering components when the aluminized steel products are used for high temperature applications. Aguero et al. reported that the aluminide coating deposited on P91 and P92 steel resulted good ductility of these alloys during creep testing at 650 °C and a small reduction of creep strength was mainly attributable to the presence of voids in the diffusion zone because of formation of Fe-Al intermetallic layer [6]. Furthermore, an aluminizing coating applied on K403 superalloy blades resulted detrimental effect of the fatigue property of the alloy, but the alloy was capable to resist the oxidation attack at 950 °C [7]. Moreover, many voids formed in Fe- and Ni-Al intermetallic layer due to the interdiffusion process lead to rapid degradation of mechanical strength of steel and nickel alloy with aluminide coating [8]. Badaruddin et al [9]. reported that a significant decrease in the mechanical properties of a high strength AISI P20 steel was caused by both the formation of the brittle Fe_2Al_5 layer and voids as results of inter-diffusional process between Fe-atoms and Al atoms in the alloy.

In the present experiments, we focused on an investigation of the low cycle fatigue performance of cold-drawn AISI 1018 steel which was coated by the HDA coating. Few publications are available in literature, reporting the LCF properties of steel with aluminide coating. Therefore, it is important to clarify the impact of aluminide layer on the LCF properties of steel. The CDS 1018 with and without an aluminide coating were fatigued under strain cyclic loading and LCF test results obtained were used to determine the LCF properties using the Basquin and the Coffin-Manson approaches. The empirical equations consisting of combination elastic and plastics terms can be used to predict the fatigue life of the steel for the components in engineering applications where the plastic deformation considerably affects the steel properties.

2. Experimental procedure

2.1. Material and preparation of specimen

A commercial AISI 1018 steel under cold-drawn conditions was used in the present study which its chemical composition was similar with that of the previous research [10]. The tensile and the LCF specimens were provided regarding to ASTM standard E8 [11] and ASTM standard E606 [12], respectively. The tensile and the LCF specimens are shown in Fig. 1a and b, respectively. Details of the procedures for HDA coating process were similar with that of the HDA coating for mild steel in Ref. [4]. The specimen with and without aluminide coating is represented as the HDA steel and the CDS 1018, respectively.

IOP Conf. Series: Materials Science and Engineering **807** (2020) 012019 doi:10.1088/1757-899X/807/1/012019

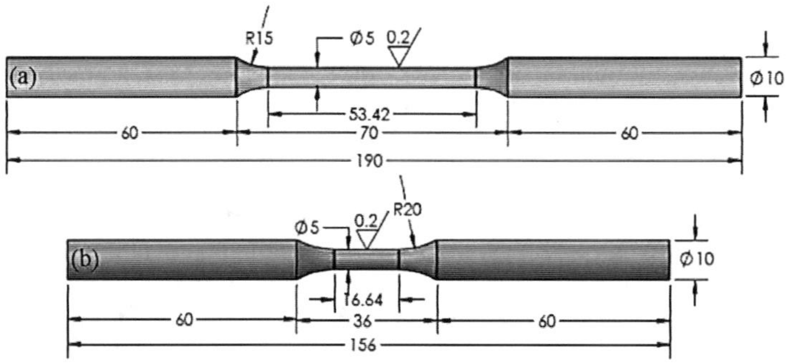

Figure 1. (a) Dimensions of the tensile specimen and (b) the LCF specimen (all in mm)

2.2. Tensile and LCF tests

A servo-hydraulic MTS Landmark test system was used in the present experiments for determining the mechanical properties and the LCF properties of CDS 1018 and HDA steel at room temperature. An MTS model 632.25F-24 axial extensometer with a gauge length of 50 mm was used to measure axial strain during tensile testing at a constant displacement rate of 0.2 mm/min. The mechanical properties for CDS 1018 and HDA steel were determined from the stress-strain curve obtained from the respective three specimens and the typical plot of stress versus axial deformation relationship curve is constructed in Figure 2.

In the present LCF tests, the respective three specimens for CDS 1018 and HDA steel were performed under a fully reversed ramp wave loading controlled by an axial extensometer of MTS 623.12F-20 type under various strain amplitude of 0.003–0.007 mm/mm at constant strain rate of 0.005 s^{-1} with a strain ratio, $R = \varepsilon_{min}/\varepsilon_{max} = -1$, the frequencies were calculated using the equation in Ref. [13], which a 0.417 Hz, 0.25 Hz, 0.179 Hz were for a strain amplitude of 0.003, 0.005 and 0.007 mm/mm, respectively. During LCF testing, the failure cycles of the specimen was automatically determined when a cyclic loading decreased to a 50% of stable loading cycles [12]. LCF test results consist of plastic strain, elastic strain and elastic modulus data at half cycles ($0.5N_f$). Pairs data point consisting of plastic strain and elastic strain with reversals to failure cycles ($2N_f$) at any strain amplitudes were plotted for determining the LCF properties using the Basquin and the Coffin-Manson equations [12].

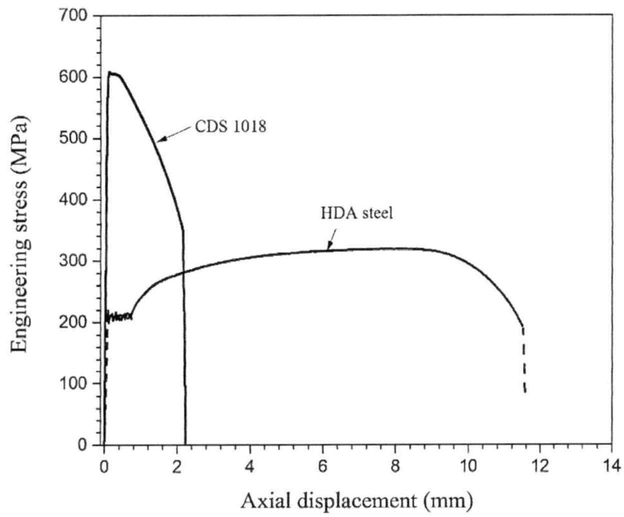

Figure 2. The typical stress-axial displacement relationship curve for CDS 1018 and HDA steel

Table 1. The average of the tensile properties for CDS 1018 and HDA steel

Property	Material	
	CDS 1018	HDA steel
Elastic modulus (GPa)	210.70	209.82
Yield strength, YS – 0.2% proof stress (MPa)	545	229
Ultimate tensile strength, UTS (MPa)	608	329
Percentage of elongation, *e* (%)	8.64	46.12
UTS: YS	1.12	1.44

3. Results and discussion

3.1. Tensile test and LCF test results

Figure 2 displays the stress versus axial displacement curve for CDS 1018 and HDA steel. The tensile properties of the CDS 1018 and the HDA steel are presented in Table 1. Table 2 summarizes the LCF test results at different strain amplitudes. As shown in Figure 2, CDS 1018 does not show well-defined yield strength point, which a typical characteristic of mild steel due to cold-drawing process [10]. In addition, CDS 1018 exhibits a linear plastic behaviour, which can be observed in the region between yield point and ultimate point (Figure 2). Aluminide coating formed on the CDS 1018 substrate resulted significant changes in the steel properties: the tensile strength decreased, and the ductility increased. It could be observed in Figure 2 that CDS 1018 with aluminide coating experienced stress relieving after aluminizing process. Consequently, the yield points for the HDA steel show clearly a similar characteristic of low carbon steel [14] and high strength low alloy AISI 4140 steel under annealing condition [13].

Table 2. LCF test results for CDS 1018 and HDA steel

Specimen	$\Delta\varepsilon_t$ (mm/mm)	ε_a (mm/mm)		E (GPa)	N_f (cycles)
		ε_{ap}	ε_{ae}		
CDS1	0.003	0.00120	0.00180	192.98	7795
CDS2	0.005	0.00305	0.00195	188.90	3980
CDS3	0.007	0.00495	0.00205	191.32	1927
HDA1	0.003	0.00180	0.00120	199.60	6278
HDA2	0.005	0.00341	0.00159	201.57	2085
HDA3	0.007	0.00517	0.00183	191.38	1401

The materials capacity for undergoing strain hardening during monotonic deformation or hardening/softening during fatigue cyclic deformation can be predicted from a value of UTS:YS ratio [15]. The calculated UTS:YS ratio as shown in Table 1 for the CDS 1018 and the HDA steel are an approximately 1.12 and 1.44 respectively. Hypothetically, metallic alloys with UTS:YS ratio lower than 1.2 tend to exhibit cyclic softening behaviour. Conversely, the materials with the UTS:YS ratio > 1.4 generally tend to undergo cyclic hardening behaviour during cyclic loading [13].

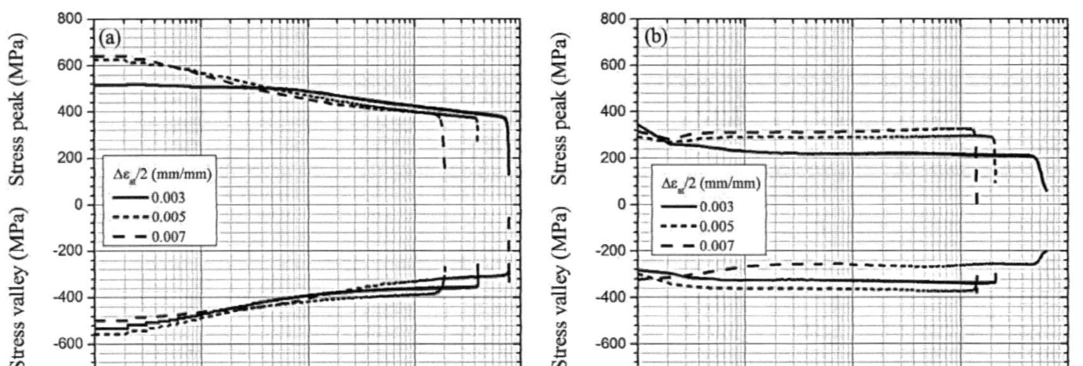

Figure 3. The cyclic stress responses of (a) CDS 1018 and (b) HDA steel with respect to the number of cycles at different strain amplitudes

3.2. Cyclic stress response with respect to fatigue cycles

The cyclic stress responses with the number of cycles at different strain amplitudes for CDS 1018 and HDA steel are displayed in Fig. 3a and b, respectively. The stress peak responses for CDS 1018 steel increase with increasing in strain amplitude levels (Figure 3a). In contrast, HDA steel exhibits a higher peak stress response at lower strain amplitude (Figure 3b). At a strain amplitude of 0.003 mm/mm, the difference of cyclic stress responses in peak stress (tension) and valley stress (compressive) subjected to the HDA1 specimen is about 170 MPa compared to that of the CDS1 specimen. The CDS1 specimen showed slowly cyclic softening in a loading cycle ranging from 3 to 46 cycles and progressively cyclic softening until failure. Furthermore, as shown in Figure 3a when the strain amplitude was increased to 0.005 and 0.007 mm/mm, CDS 1018 exhibited stably cyclic hardening behaviour in few cycles and continuously cyclic softening behaviour until failure. By increasing in fatigue cycles loading at any given strain amplitude, the CDS specimens undergo cyclic softening behaviour, which can be apparently observed via a gradual decrease in a peak stress (Figure 3a). CDS 1018 experienced continuously cyclic softening behaviour, indicating that an AISI 1018 steel under cold-drawn condition is not capable to cyclically plastic deformation, which is a similar cyclic softening behaviours of SAE 1010 steel in cold-rolled condition [16]. High strain hardening induced AISI 1018 steel after cold-drawing processing decreased significantly plastic deformation capacity [11], which lead to continuously cyclic softening behaviours during strain-controlled fatigue. The softening behaviours of CDS 1018 under low cycle fatigue conditions is evidenced in Figure 3a.

Figure 3b shows that the stress peak responses of the HDA steel specimens rapidly decrease at the 2nd loading cycles. Furthermore, at a strain amplitude of 0.003 mm/mm the HDA1 specimen exhibited gradually cyclic softening behaviours in loading cycle ranging from 3 to 46 cycles, and continually underwent stable cyclic hardening for 4527 cycles. The initiation crack started growing at 5292 cycles and cracks propagated rapidly in the HDA1 specimen which could be indicated by a drastic decrease in the stress peak until fail for 6277 cycles. When the strain amplitude was increased to 0.005 and 0.007 mm/mm, the HDA steel underwent stably cyclic hardening starting from the 8th loading cycles until fail. The initiation cracks for the HDA2 and HDA3 specimens respectively grew at 2057 cycles and 1376 cycles. With increasing in strain amplitude levels, the magnitude of compressive stress in the HDA steel observed in Figure 3b increased gradually with respect to fatigue cycles. The higher valley stress in HDA steel compared to those of the peak stress during strain-fatigue loading indicates that strain hardening induced in the material as result of an increase in dislocation density.

Low dislocation interactions possibly result the cyclic softening in few cycles for the HDA steel (Figure 3b) and the dislocations interaction after few loading cycles is responsible to subsequent stable hardening behaviours. Softening has been observed in cyclic deformation of many alloys, which is generally attributed to the rearrangement of dislocations [16,17]. On the other hand, an increase in dislocation density due to multiplication and interaction among dislocation connections resulted by plastic deformation is strongly responsible to the cyclic hardening behaviours of material. In general, an increase in dislocation density and stronger interaction among dislocations results work hardening during cyclic deformation [13]. Work hardening can be observed via the magnitude of valley stress responses (compressive stress) of the HDA steel with increasing of fatigue cycles in Fig. 3b. However, work hardening induced in the HDA steel during strain-cyclic deformation does not give an advantageous increase in the fatigue life of CDS 1018 with aluminide coating.

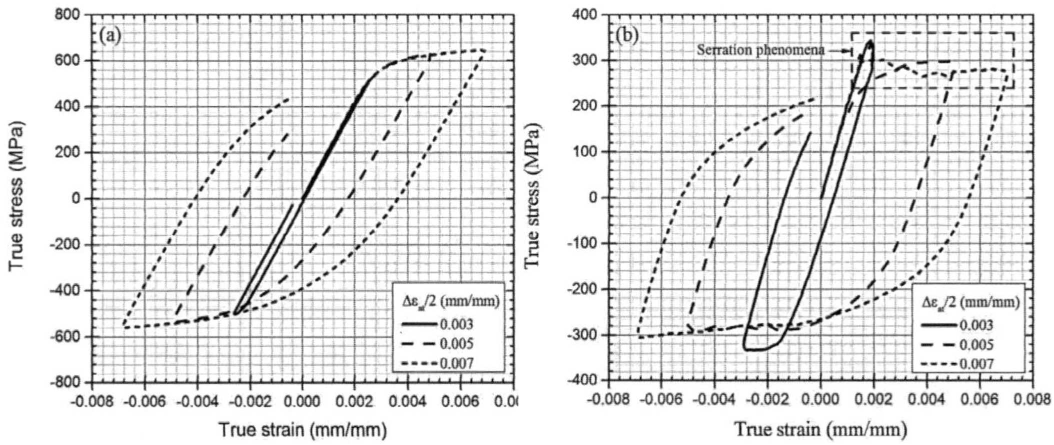

Figure 4. Hysteresis curves at the first cycle for (a) CDS 1018 steel and (b) HDA steel

3.3. Cyclic strain-stress curves analysis

The typical plot of the first hysteresis curves obtained from LCF test with the different strain amplitudes for CDS 1018 and HDA steel are depicted in Fig. 4a and b, respectively. In a strain amplitude ranging from 0.003 to 0.007 mm/mm, CDS 1018 exhibited different characteristic hysteresis loops at the first cycle (Figure 4a). In addition, CDS 1018 exhibits behaviours changes in the peak and the valley stress responses at the first cycle loading, where a valley stress response (compressive stress) is lower value than that of a peak stress response (tensile stress). This behaviours change is known as the Bauschinger effect. The strong Bauschinger effect is clearly observed in Figure 4a at any applied strain amplitudes, which is a similar characteristic with ferrite–bainite microstructures in low carbon steel [18] and an extrusion of aluminium alloy 6061-T6 [19].

Figure 4b shows the influence of the aluminide coating on CDS 1018, which significantly changes the behaviours of the steel during cyclic deformation. The HDA steel evidently shows a different stress response in tension stress at the first cycle for different strain amplitude, which yield point regions exhibited a serration characteristic. The serration phenomenon in Figure 4b is observed in the HDA steel which is a similar characteristic with the monotonic behaviours in a low carbon steel [15,16] and in an AISI 4140 steel under annealing condition [13]. The HDA steel experienced the serration phenomena of yield points at the first loading cycle can be attributed to formation of microplasticity in the Luder band region (Figure 4b). The initial microplasticity formation was as results of low dislocation interactions in the proportional yield points region and the free dislocations in this region were active [20].

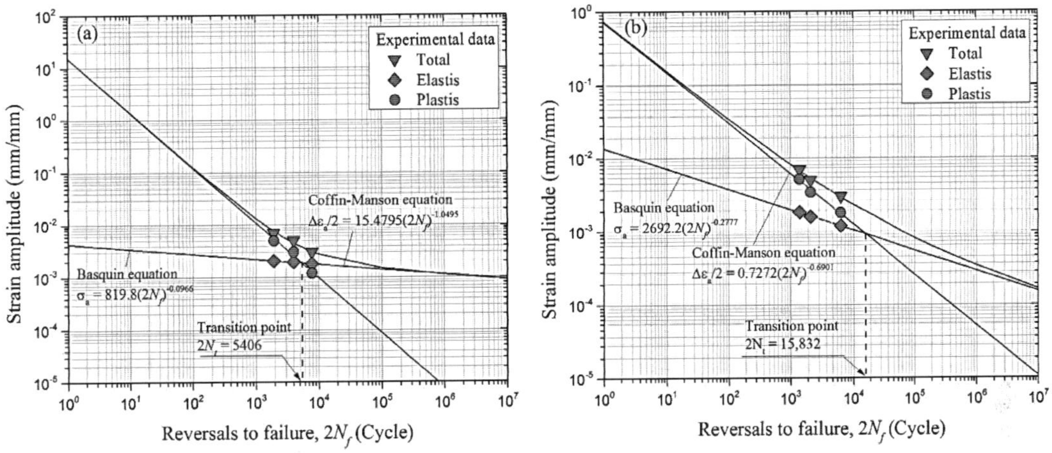

Figure 5. Plot LCF data and calibration curve obtained from the Basquin and Coffin-Manson equations for (a) CDS 1018 and (b) HDA steel

3.4. Effect of aluminide coating on the LCF properties of the steel

The relationships between a total strain amplitude ($\Delta\varepsilon_t/2$) and failure cycles ($2N_f$) are an important factor to investigate the strain-controlled fatigue life of the material due to cyclic loading. In general, the Coffin and the Manson-Basquin methods are used to determine the predictions of fatigue life of the material based on strain controlled LCF test, which is represented by the equation. (1) [12].

$$\Delta\varepsilon_t = \frac{\sigma_f'}{E}\left(2N_f\right)^b + \varepsilon_f'\left(2N_f\right)^c \tag{1}$$

Where E is elastic modulus in LCF condition (for the CDS 1018 is ~191.1 GPa and for the HDA steel is ~197.5 GPa), N_f is the fatigue life or the number of cycles to failure, σ_f' (MPa) is the fatigue strength coefficient, b is the fatigue strength exponent, ε_f' is the fatigue ductility coefficient, and c is the fatigue ductility exponent. Eq. (1) is results of combination of the elastic component and the plastic components obtained from experimental LCF tests as given by equation. (2) and (3), respectively.

$$\frac{\Delta\varepsilon_e}{2} = \frac{\sigma_f'}{E}\left(2N_f\right)^b \tag{2}$$

$$\frac{\Delta\varepsilon_p}{2} = \varepsilon_f'\left(2N_f\right)^c \tag{3}$$

Furthermore, the LCF steel's properties were determined using the curve fitting method which was similar with that of previous research [13]. The total strain versus reversal to failure cycles based on the Basquin and the Coffin-Manson equations (as the calibration curves) obtained is plotted in Fig. 5a and b for the CDS 1018 and the HDA steel, respectively and the experimental data from LCF tests exhibit good correlation with the calibration curves.

The LCF resistance of metals can be represented by a value of fatigue ductility exponent [13,14,16]. In the present investigation, the fatigue ductility exponent of the HDA steel ($c = -0.6901$) is lower than that of the fatigue ductility exponent of the CDS 1018 ($c = -1.0495$), which theoretically gives a significant advantage on the LCF resistance. In addition, the high ductility of the HDA steel ($e = 46.12\%$) is possibly enable to undergo cyclic deformation but experimentally the fatigue life (N_f) of the HDA steel is lower than those the fatigue life of the CDS 1018 at all strain amplitude levels applied. In this case, the low fatigue life of the HDA steel can possibly be caused by formation of the brittle Fe_2Al_5 layer [6,8,9]. The brittle metals generally have the low plastic deformation capacity. Therefore, the Fe_2Al_5 layer was unable to resist strain cyclic deformation, which the initiation cracks were rapidly formed and propagated through this layer. Consequently, the fatigue life of the CDS 1018 with the aluminide coating under strain fatigue condition significantly decreased. The prediction of fatigue life for the CDS 1018 and the HDA steel can be determined by the elastic and plastic components using the Basquin and Coffin-Manson approach, respectively.

For the CDS 1018 steel:

$$\sigma_a = 819.8(2N_f)^{-0.0996} \tag{4}$$

$$\frac{\Delta\varepsilon_p}{2} = 15.4795(2N_f)^{-1.0495} \tag{5}$$

For the CDS 1018 steel with HDA:

$$\sigma_a = 2692.2(2N_f)^{-0.2777} \tag{6}$$

$$\frac{\Delta\varepsilon_p}{2} = 0.7272(2N_f)^{-0.6901} \tag{7}$$

The intersections between the $\Delta\varepsilon_e/2$ and $\Delta\varepsilon_p/2$ versus $2N_f$ relationship curves in Fig. 5a and b result the transition of fatigue life to reversals ($2N_t$). Furthermore, from the Eq. (4) to the Eq. (7) the number of transition cycles ($2N_t$) can be determined by substitution the values of $\varepsilon_{ae} = \varepsilon_{ap}$ and replacing $2N_f$ with $2N_t$. The number of transition cycles ($2N_t$) for the CDS 1018 and the HDA steel are 5406 cycles and 15,832 cycles, respectively. When $2N_f \gg 2N_t$, the elastic strain is more dominant to control fatigue performance of material [13]. On the contrary, when $2N_f \ll 2N_t$, the plastic strain dominantly controls the fatigue performance. In the present LCF test results, the aluminide coating applied on the CDS 1018 significantly results the high fatigue strength coefficient of the HDA steel ($\sigma_f' = 2692.2$ MPa). Therefore, the stress amplitude (σ_a) for the HDA steel can be theoretically calculated by substituting a $2N_t$ value into Eq. (6), which σ_a is an approximate 300 MPa. This recommends that the CDS 1018 with the HDA coating can safely withstand the cyclic stress at a lower stress amplitude conditions when the cyclic fatigue loading is mainly considerable in determination of an initial design for engineering components. A similar characteristic was found in the aluminide coating on K403 superalloy gas turbine blade components, which the alloy exhibits better fatigue performance under the elastic fatigue loading than that of the fatigue performance under the plastic fatigue loading [7].

4. Conclusion

An aluminide coating on the CDS 1018 markedly reduced the mechanical strength and the strain-fatigue life of the steel. The linear-plastic behaviours of the CDS 1018 results the Bauschinger characteristic and the high ductility value of the HDA steel results the serration characteristic in yield point region. The HDA steel under high ductility condition is not capable to resist cyclically plastic deformation. The brittle Fe_2Al_5 layer on the CDS 1018 substrate was considerably believed to lead a rapid formation of initiation cracks in few cycles. A relatively high value of fatigue strength coefficient and fatigue strength exponent of the HDA steel indicate that the HDA coating applied on the CDS 1018 is appropriate for material components where the fatigue loading is the below yield strength of the material.

Acknowledgment

The paper is a part of the basic research grant funded by Kemenristekdikti via the Directorate of Research and Community Service (DRPM) under contract No. 065/SP2H/LT/DRPM/2019.

References

[1] Badaruddin M, Sugiyanto and Asmi D 2019 *Mater. Res. Express* **6** 0865d6
[2] Wang CJ and Badaruddin M 2010 *Surf. Coat. Technol.* **205** 1200
[3] Hakam M, Wang CJ, Wardono H, Asmi D 2018 *AIP Conf. Proc.* **1983** 050002.
[4] Badaruddin M, Sugiyanto and Asmi D 2020 *Bull. Mater. Sci.* **43** 11
[5] Callaghan MD, Humphries SR, Law M, Ho M, Bendeich P, Li H and Yeung WY 2010 *Mater. Sci. Eng.* A **527** 5619
[6] Aguero A, Muelas R, Gutierrez M, Vulpen RV, Osgerby S and Banks JP 2007 *Surf. Coat. Technol.* **201** 6253
[7] Han L, Huang D, Yan X, Zhang Y, Gui M, Tao M, Zhang X and Qi M 2019 *Intl. J. Fat.* **125** 491
[8] Dryepondt S and Pint BA 2010 *Surf. Coat. Technol.* **205** 1195
[9] Badaruddin M, Riza RT and Zulhanif 2018 *AIP Conf. Proc.* **1983** 050004
[10] Badaruddin M, Sugiri A and Wang CJ 2020 J. *Construction and Building Materials* **232** 117193
[11] ASTM E8M Standard test methods for tension testing of metallic materials, ASTM International, West Conshohocken, PA, 2016
[12] ASTM E 606-04 Standard practice for strain-controlled fatigue testing, ASTM International, West Conshohocken, PA, 2012
[13] Badaruddin M, Sugiyanto, Wardono H, Andoko, Wang C J and Rivai A K 2019 *Intl. J. Fatigue* **125** 406.
[14] Yang L, Gao Y, Shi G, Wang X and Bai Y 2018 *Constr. Build. Mater.* **165** 688
[15] Fatoba O and Akid R 2018 *Theo. Appl. Frac. Mech.* **94** 147
[16] Yu MT, DuQuesnay DL and Topper TH 1990 *Intl. J. Fat.* **15** 433
[17] Xie X, Ning D and Sun J 2016 *Mater. Charact.* **120** 195
[18] Smirnov MA, Pyshmintsev IY, Varnak OV, Mal'tseva AN and Goikhenberg YN 2016 *Steel in Translation* **46** 58
[19] Badaruddin M, Zulhanif and Supriadi H 2019 *J. Phys.: Conf. Ser.* **1198** 032002
[20] Tsuchida N, Tomota Y, Nagai K and Fukaura K 2006 *Scrip. Mater.* **54** 57

Analysis of Availability and Water Demand: A Case Study in Sumberejo Village Tanggamus Regency

Anwar[1*], ElzaNovilyanza[1], Kemas Muhammat Abdul Fatah[1]

[1]Faculty of Engineering, University Sang Bumi Ruwa Jurai, Lampung Indonesia

Corresponding email: minakshaka2013@gmail.com

Abstract. Clean water is a basic need for humans so that it is normal if clean water gets priority in handling and fulfilling it. The spring in Sumberejo Village, Sumberejo Subdistrict, Tanggamus District Is needed to facilitate. The purpose of this study is to identify the potential sources of raw water (clean water) in Sumberejo Village. The EPANET 2.0 program was implemented by calculative the number of people to be served as well as the need for clean water, reservoir capacity, and planning for distribution network. The methods used in water availability include measurement of discharge, topographic data, and water quality and for water requirements are projections of population, total water requirements and pipeline network pressure. The results of the analysis and projections of the number of villagers up to 2030 with a total of 2443 people, total water needs of 2842 liters/second, the availability debit in Sumberejo water eyes was 2,979 liters/second, the discharge was still sufficient for water needs until 2030. Hydraulic analysis using EPANET 2.0 obtained pressure at the highest location is 24.75 meters and the lowest is 2.95 meters. The study indicates that availability and water demand is sufficient until 2030

1. Introduction

Water is one of the most important needs for human life and other living things. So far, human needs for water are very large. When viewed in terms of its use, water has never been separated from all aspects of humans. Nature has provided quite a lot of water, but with population growth and increased activity so that the use of water needs increases. Clean water has a very important role in improving environmental health or the community, which has a role in reducing the number of people with diseases, especially those related to water and plays a role in improving the level or quality of life of the community [1-2]. In Indonesia until now the provision of clean water for the community is still faced with a number of quite complex problems and still cannot be fully addressed [3-5].

Sumberejo Village has a water source in the form of springs which have sufficient head potential but are not well distributed. At present the distribution of clean water still uses simple technology, namely by using temporary shelters that are distributed by hoses to the homes of residents. In clean water services to the community, the availability of clean water is a very important part. The main function of the distribution pipeline network is to deliver clean water to all villagers while taking into account the quality, quantity and water pressure factors. The condition desired by the whole community is continuous water availability [6-8].

In this case the author tries to plan a large capacity reservoir to store water so that the distribution can be carried out continuously or continuously, and the water distribution system used will use a gravity system, because the location of Sumberejo Village is at a lower elevation than the raw water source will be used.

With consideration of the above problems, it is necessary to have a study on "Analysis of Availability and Need for Clean Water in Sumberejo Village, Sumberejo District, Tanggamus Regency". The analysis is expected to meet domestic and non-domestic needs that meet the requirements standards.

2. Research Methodology

In carrying out research on the availability of analysis and the need for clean water in Sumberejo Village, Tanggamus Regency, a systematic stage of research planning is needed. This stage of analysis consists of several stages, which include area observation or area identification, evaluation of existing clean water network systems, population size, analysis of water availability, population growth and population needs.

This research was conducted in Sumberejo Village Sumberejo District, Tanggamus Regency. Sumberejo Village is one of 13 villages in the Sumberejo District area. Sumberejo Village is located in a highland area. Geographically the village borders on north side, Pulau Panggung sub-district, east side, Sidomulyo Village, south side, Sumber Mulyo Village and west side, Argopeni Village

Sumberejo Village is located in Sumberejo District, Tanggamus Regency which consists of 4 hamlets. The area of Sumberejo Village is 367.64 km^2, accessibility to Sumberejo Village from the subdistrict capital is 3 km and from the district capital which is about 36 km. The population is 2162 in 2016. In Figure 1, it is a map of village administration.

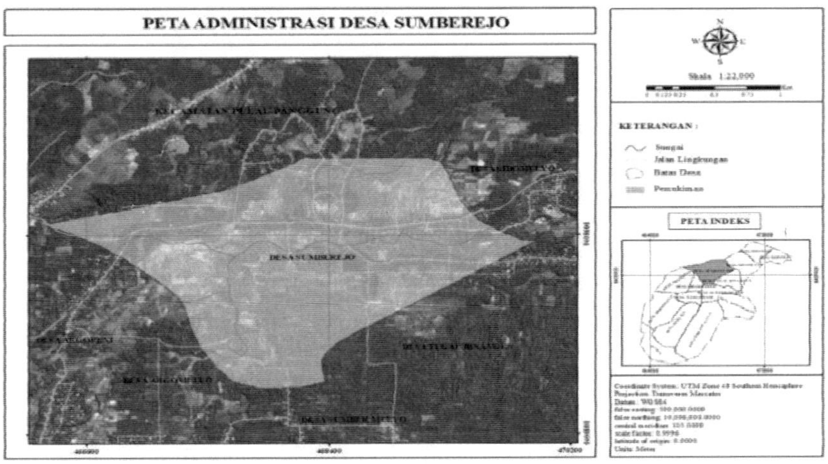

Figure 1. Map of Village Administration.

In this study, the progress of data collection can be grouped into two; primary data, are data obtained from field observations, laboratory tests and interviews. Field observations include potential discharge and potential energy potential (altitude / topography). For laboratory tests carried out is a test of physical and chemical quality of water to be carried out at the Lampung State Polytechnic Laboratory. Secondary data is the data needed to do calculations and analysis. This secondary data is in the form of data including population data and maps of Sumberejo Village, Sumberejo District, Tanggamus Regency.

The analysis was carried out using data that had been obtained, both in the field and in the laboratory, then adjusted to the literature and the results of previous related studies. This water quality analysis was carried out at the Lampung State Polytechnic Laboratory. The method and parameters of the water being analyzed refer to the provisions / requirements for the quality of clean water and drinking water based on RI Minister of Health Regulation No. 416 / MENKES / Per / IX / 1990. Furthermore, water feasibility

IOP Conf. Series: Materials Science and Engineering **807** (2020) 012020 doi:10.1088/1757-899X/807/1/012020

is carried out by comparing the parameters of water samples with the quality standards of clean water and drinking water based on Minister of Health Decree No. 416 of 1990.

Population analysis is carried out to determine the number, density, and population growth of Sumberejo Village, which is geographically possible to be served by springs. The amount of water needs of the rural population refers to the 2001 drinking water SK-SNI, which is 60 liters per person per day. For water needs for public facilities refer to Lampung Province P3P of 20% of domestic water needs. Water losses caused by leakage of piping systems and such are estimated at 20% of total production and referring to Lampung Province P3P data. Fluctuations in water demand for maximum day factor values are 1.5 and the water service system is carried out through public faucets (KU) or public hydrants (HU), so that the population's water needs can be known per year.

In the drainage in the pipe there is friction along the pipe and due to the bend of the pipe so that head loss occurs in the pipe. Energy losses (head losses) due to friction along the pipe (hf) in water distribution are calculated using the formula. Calculation of the amount of energy loss in pipe narrowing (hf1) in water distribution. Jetting energy analysis is carried out to simulate an adequate pipe size / diameter by using EPANET software, so that it is known to what extent the source of the spring can flow its service area.

3. Results and Discussion

3.1. Total population

In 2016 the population of Sumberejo Village Sumberejo Subdistrict was 2,612 people with a ratio of 1,101 men and women of 1,061 people, as shown in Table 1, and Table 2 shown the population and growth rate of Sumbererjo Village.

Table 1 Total Population Based on Gender Ratio in Sumberejo Village [1]

Age	Gender		Total
	Male	Female	
0-4	55	55	112
5-9	69	67	136
10-14	70	68	138
15-19	114	110	224
20-24	163	158	321
25-29	69	67	136
30-34	78	75	153
35-39	130	126	256
40-44	136	132	268
45-49	147	142	289
50-54	78	75	153
55-59	74	71	145
60-64	76	73	149
65+	67	65	132
Total	1328	1284	2612

Table 2. Population and Growth Rate of Sumberejo Village [1]

No	Year	Total Population	Population Growth Rate (%)
1	2010	2043	0.95
2	2011	2078	0.95
3	2012	2095	0.95
4	2013	2112	0.95

5	2014	2129	0.95
6	2015	2146	0.95
7	2016	2163	0.95

3.2. Projected Population Amount

Population increase due to birth, loss due to death, increase and decrease due to migration and increased merging. Each of these elements is influenced by economic and social factors in the community. Other factors such as war, natural disasters, and industrial and commercial activities also influence. These elements, especially industry and commerce can produce sharp, slow growth, stable conditions or population decline.

Calculation of population projections can be done using 3 methods, namely arithmetic methods, geometric methods and least square methods. After knowing the results of the calculation of each method and also determined the value of the standard deviation of each method, determine which method will be used to calculate the projections of water requirements.

3.2.1. Arithmetic Method

Formula of the arithmetic method are:

$$Pn = Po_1 + n(Pr) \tag{1}$$

$$Pr = \frac{Pn_1 - Po_1}{n_1} \tag{2}$$

Where:
Pn : Total population in the nth year projection
Pr : Ratio of population
Po_1 : Total population at the beginning of the data
Pn_1 : Total population at the end of the data
n : Number of years to come
n_1 : Data collection year / final year minus the initial year

$$Pr = \frac{2163 - 2043}{2016 - 2010} = 20$$

$$P_{2030} = Po_1 + n(Pr)$$

$$P_{2030} = 2163 + (14 \times 20)$$

$$P_{2030} = 2443$$

Table 3. Projected Population Growth Using Arithmetic Method

No	Year	Total Population
1	2016	2163
2	2017	2183
3	2018	2203
4	2019	2223
5	2020	2243
6	2021	2263
7	2022	2283
8	2023	2303
9	2024	2323

10	2025	2343
11	2026	2363
12	2027	2383
13	2028	2403
14	2029	2423
15	2030	2443

From the calculation of the standard deviation or standard deviation above in the projected number of populations using the arithmetic method, the standard deviation value is 121.37, as shown in Table 4.

$$s^2 = \frac{(21)x\ (106148594) - (2222933904)}{(21)x(20)}$$

$$s^2 = 14729.928$$

$$s = \sqrt{14729.928}$$

$$s = 121.37$$

Table 4. Standard Deviation in Arithmetic Method

No	Year	Projected Population (xi)	xi^2
1	2010	2043	4173849
2	2011	2078	4318084
3	2012	2095	4389025
4	2013	2112	4460544
5	2014	2129	4532641
6	2015	2146	4605316
7	2016	2163	4678569
8	2017	2183	4765489
9	2018	2203	4853209
10	2019	2223	4941729
11	2020	2243	5031049
12	2021	2263	5121169
13	2022	2283	5212089
14	2023	2303	5303809
15	2024	2323	5396329
16	2025	2343	5489649
17	2026	2363	5583769
18	2027	2383	5678689
19	2028	2403	5774409
20	2029	2423	5870929
21	2030	2443	5968249
$\sum$		47148	106148594
$(\sum xi)^2$		2222933904	

3.2.2. Geometric Method

Formula of the geometric method are:

$$Pn = Po_1(1 + r)^n \tag{3}$$

$$r = \left(\frac{Pn_1}{Po}\right)^{\frac{1}{(n_1 - o_1)}-1} \tag{4}$$

Where:
Pn : Amount in the coming year
Po : Total population at the beginning of the estimate
Pn_1 : Total population at the end of the data
Po_1 : Total population at the beginning of the data
r : Average percentage of population increase
n : Duration of production year from the initial year of calculation
n_1 : Data end year
o_1 : Data start year

$$r = \left(\frac{2163}{2043}\right)^{\frac{1}{(7-1)}-1}$$

$$r = 0.0095$$

$$P_{2030} = 2163(1 + 0.0095)^{14}$$

$$P_{2030} = 2471$$

Table 5. Projected Population Growth Using Geometric Methods

No	Year	Total Population
1	2016	2163
2	2017	2184
3	208	2205
4	2019	2226
5	2020	2247
6	2021	2268
7	2022	2290
8	2023	2312
9	2024	2334
10	2025	2356
11	2026	2379
12	2027	2402
13	2028	2425
14	2029	2448
15	2030	2471

Table 6. Standard Deviation in Geometric Methods

No	Year	Projected Population (xi)	xi^2
1	2010	2043	4173849
2	2011	2078	4318084
3	2012	2095	4389025
4	2013	2112	4460544
5	2014	2129	4532641
6	2015	2146	4605316
7	2016	2163	4678569
8	2017	2184	4768434
9	2018	2205	4860025
10	2019	2226	4953375
11	2020	2247	5048518
12	2021	2268	5145489
13	2022	2290	5244322

14	2023	2312	5345054
15	2024	2334	5447720
16	2025	2356	5552358
17	2026	2379	5659007
18	2027	2402	5767704
19	2028	2425	5878488
20	2029	2448	5991401
21	2030	2471	6106482
	$\sum$	47311	106926403
	$(\sum xi)^2$	2238364345	

From the calculation of the standard deviation or standard deviation above in the projected number of populations using the geometric method, the standard deviation value is 121.37, as shown in Table 4.

$$s^2 = \frac{(21)x\,(87650829) - (1840667409)}{(21)x(20)}$$

$$s^2 = 1688124$$

$$s = \sqrt{1688124}$$

$$s = 1299.28$$

3.2.3. Least Square Method

Formula of the arithmetic method are:

$$y = a + bX \tag{5}$$

$$a = \frac{N\sum y \sum x^2 - \sum x \sum y}{N\sum x^2 - (\sum x)^2} \tag{6}$$

$$b = \frac{\sum x^2 \sum y - \sum x \sum xy}{N\sum x^2 - \sum x \sum x} \tag{7}$$

Where:

y : Total population for the year expected
x : Number of years or years calculated from the base year
a : constant
b : linear regression coefficient

Table 7. Calculations Determine Value A and B Method of Least Square

No	Year	Total Population (y)	x	y^2	x^2	yx
1	2010	2043	-3	4173849	9	-6129
2	2011	2078	-2	4318084	4	-4156
3	2012	2095	-1	4389025	1	-2095
4	2013	2112	0	4460544	0	0
5	2014	2129	1	4532641	1	2129
6	2015	2146	2	4605316	4	4292
7	2016	2163	3	4678569	9	6489
	$\sum$	14766	0	31158028	28	530

$$a = \frac{(7 \text{ x } 530) - (0 \text{ x } 14766)}{(7 \text{ x } 28) - 0}$$

$$a = 18.928$$

$$b = \frac{(28 \text{ x } 14766) - (0 \text{ x } 530)}{(7 \text{ x } 28) - (0 \text{ x } 0)}$$

$$b = 2109.429$$

$$s^2 = \frac{(21)x\,(110800771) - (2317507740)}{(21)x(20)}$$

$$s^2 = 22162.98$$

$$s = \sqrt{22162.98}$$

$$s = 148.872$$

Table 8. Projected Population Growth Using Least Method

No	Year	Total Population
1	2016	2163
2	2017	2261
3	208	2280
4	2019	2299
5	2020	2318
6	2021	2337
7	2022	2356
8	2023	2374
9	2024	2393
10	2025	2412
11	2026	2431
12	2027	2450
13	2028	2469
14	2029	2488
15	2030	2507

Table 9. Standard Deviation in Least Method

No	Year	Projected Population (xi)	xi^2
1	2010	2043	4173849
2	2011	2078	4318084
3	2012	2095	4389025
4	2013	2112	4460544
5	2014	2129	4532641
6	2015	2146	4605316
7	2016	2163	4678569
8	2017	2261	5111475
9	2018	2280	5197423
10	2019	2299	5284087
11	2020	2318	5371468
12	2021	2337	5459566
13	2022	2356	5548380
14	2023	2374	5637911
15	2024	2393	5728158

16	2025	2412	5819122
17	2026	2431	5910803
18	2027	2450	6003200
19	2028	2469	6096314
20	2029	2488	6190144
21	2030	2507	6284691
	∑	48141	110800771
	$(\sum xi)^2$	2317507740	

The results of the 3 (three) population projection methods above and by considering the standard deviation value, it can be concluded that the Arithmetic Method is approaching reality. The reason for choosing this method is because the value of the standard deviation is smaller than the other methods.

3.3. Water Quality

The source of raw water used as a source of raw water in Sumberejo Village is a spring. The Sumberejo Springs are in the coordinates of 5 ° 21 '08.74 "S - 104 ° 42' 06.52" E while for the Springs of the Workspace are at coordinates 5 ° 21 '06.78 "S - 104 ° 42' 44.18" E. The results of the discharge measurement are obtained debit of 3.00 liters / second for Sumberejo Springs and 0.35 liters / second for Sumber Spring works. The results of testing the quality of clean water that has been sampled are as follows.

Table 10.1 Water Quality of Sumberejo Village.

No	Parameter	Maximum Limit Allowed	Test Result	Unit
1	Smell and Taste	No smell/ tasteless	No smell/ tasteless	-
2	Color	Colorless	Colourless	-
3	Temperature	±3°C	27	°C
4	TDS	1	75	mg/ litre
5	Turdibity	5	3,6	FAU
6	Salinity	0	0	-
7	pH	6,5-9,0	7,3	-

Table 11. Water Quality of Sumberkarya

No	Parameter	Maximum Limit Allowed	Test Result	Unit
1	Smell and Taste	No smell/ tasteless	No smell/ tasteless	-
2	Color	Colorless	Colourless	-
3	Temperature	±3°C	26	°C
4	TDS	1	75	mg/ litre
5	Turdibity	5	3,6	FAU
6	Salinity	0	0	-
7	pH	6,5-9,0	7,0	-

Based on the results of measurements and laboratory tests of samples from springs it can be seen that all parameters can be said to meet the requirements to be used as raw water. Clear water flowing with turbidity levels that only ranges from 3.6 FAU because there is no rain, but according to information if the day before the rain occurs it appears that the spring is cloudy, it must be anticipated by using a slow sand filter.

3.4. Topographic Data

Based on the topographical data of Sumberejo Village, the location of the elevation of the spring is located at an elevation of ± 490 m asl and the water service area ranges from ± 488 m asl to ± 380 m asl. So that clean water services do not use pumps and can be done by means of gravitational flow because service areas are in lower elevations than raw water sources. Existing topographic images can be seen in Figure 2, and Figure 3.

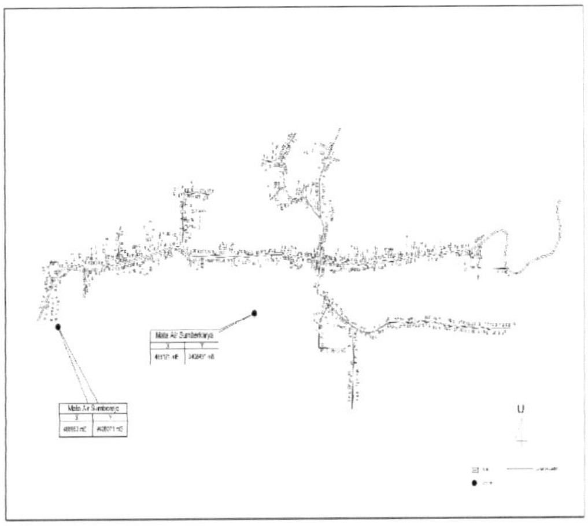

Figure 2. Topographic Sumberejo.

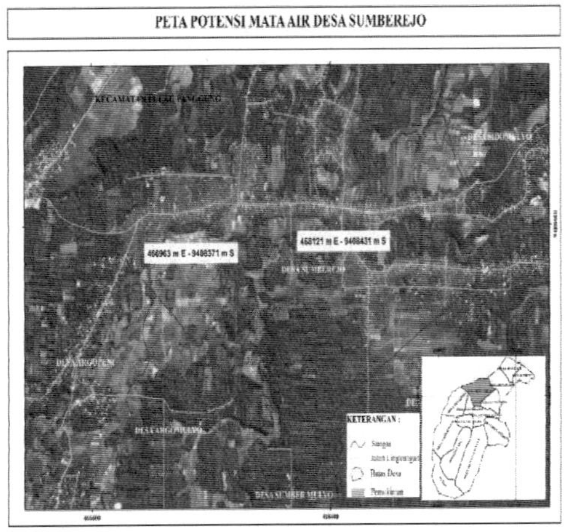

Figure 3. Springs Potential Map.

3.5. Projection of Water Needs

Sumberejo Village already has a clean water supply system, but the supply of clean water still uses simple technology and the clean water obtained now comes from springs.

Table 12. Water Needs in Total Sumberejo Village

No	Year	Total Population	Debit Domestic Water Needs (Qd)	Debit Non- Domestic Water Needs (Qn)	Water Losses (Qa)	Total Water Needs (Qt) in ℓ/s
1	2016	2163	1.590	0.318	0.286	2.194
2	2017	2183	1.592	0.318	0.287	2.197
3	208	2203	1.653	0.331	0.298	2.282
4	2019	2223	1.692	0.338	0.306	2.336
5	2020	2243	1.732	0.346	0.312	2.390
6	2021	2263	1.749	0.350	.0315	2.413
7	2022	2283	1.789	0.358	0.322	2.469
8	2023	2303	1.806	0.361	0.325	2.493
9	2024	2323	1.848	0.370	0.33	2.550
10	2025	2343	1.866	0.373	0.336	2.574
11	2026	2363	1.908	0.382	0.343	2.633
12	2027	2383	1.926	0.385	0.347	2.658
13	2028	2403	1.970	0.394	0.355	2.719
14	2029	2423	2.040	0.408	0.367	2.815
15	2030	2443	2.059	0.412	0.371	2.842

3.6. EPANET 2.0 Piping Network

The clean water distribution pipeline network that will be planned uses pipes with a diameter of 12.5 - 100 mm with the type of planned pipe, namely using PVC pipes. The formula used in calculating Head Loss on a distribution pipeline is to use the Hazen Williams formula.

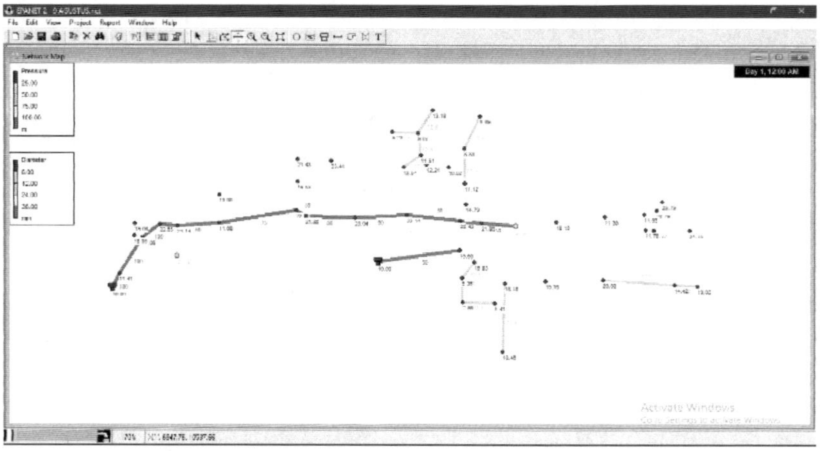

Figure 4. Scheme of Piping Network Using EPANET 2.0 Program.

To analyze the clean water distribution network system using the EPANET 2.0 program, it can be done in 2 ways, namely by Single Period and Extended Period Simulation. In the single period the process of analyzing the distribution of clean water needs is considered the same every hour and the flow is steady, while in the extended period simulation analysis of the distribution of clean water needs varies at each time. In this planning analysis an extended period simulation system is used.

The work steps taken to begin the analysis with the EPANET 2.0 program are as follows:

a. Making a new project.
b. Program settings.
c. Depiction of the scheme of clean water distribution network.
d. Input data on clean water distribution network components.
e. Input water pattern data.
f. Program simulation.
g. Interpretation of simulation results

After the program is run, then see the simulation results. The simulation results can be accessed in the form of tables, graphs, and hourly parameter maps. The flow pressure of each pipe can be seen in the table below.

Table 13. Pressure at Every Point in the EPANET 2.0 Program

Node ID (Junc)	Elevation (m)	Base Demand LPS	Demand LPS	Head (m)	Pressure (m)
2	488	0.229	0.23	499.41	11.41
3	483	0.222	0.22	498.09	15.09
4	480	0.014	0.01	498.05	180.05
5	479	0.007	0.07	498.04	19.04
6	475	0.208	0.21	497.55	22.55
7	472	0.201	0.2	496.14	24.14
8	471	0.007	0.01	496.12	25.12
9	465	0.188	0.19	476.68	11.68
10	463	0.007	0.01	476.66	13.66
11	450	0.174	0.17	472.8	22.8
12	447	0.021	0.02	466.53	19.53
13	445	0.014	0.01	466.43	21.43
14	443	0.007	0.01	466.41	23.41
15	448	0.153	0.15	469.86	21.86
16	437	0.146	0.15	460.04	23.04
17	430	0.139	0.14	452.14	22.14

18	426	0.132	0.13	446.42	20.42
19	423	0.062	0.06	442.6	19.6
20	420	0.056	0.06	436.83	16.83
21	421	0.021	0.02	427.35	6.35
22	421	0.014	0.01	423.95	2.95
23	418	0.007	0.01	423.41	5.41
24	415	0.035	0.03	433.48	18.48
25	415	0.00	0	433.46	18.46
26	412	0.028	0.03	431.76	19.76
27	410	0.021	0.02	430.83	20.83
28	407	0.014	0.01	421.42	14.42
29	402	0.007	0.01	421.02	19.02
30	395	0.007	0.01	419.75	24.75
31	402	0.014	0.01	419.77	17.77
32	399	0.014	0.01	419.79	20.79
33	408	0.021	0.02	419.93	11.93
34	408	0.021	0.02	419.78	11.78
35	396	0.007	0.01	419.79	23.79
36	412	0.049	0.05	423.38	11.38
37	414	0.056	0.06	432.15	18.15
38	418	0.063	0.06	445.25	27.25
39	424	0.069	0.07	445.85	14.79
40	424	0.056	0.06	438.79	14.79

The results of EPANET 2.0 data above the highest elevation point are at 488 m with a pressure of 11.41 m and the lowest elevation is at an altitude of 395 m with a pressure of 24.75 m.

From the results of the below data it can be concluded that the highest head loss is in pipe 18 at 29.13 m / km and pipe 36 at 27.16 m / km and for the lowest head loss in pipe 15 and pipe 53 because it has the same value i.e. 0.04 m / km.

Table 14.2 Speed and Head Loss in Pipes

Link ID (Pipe)	Length (m)	Diameter (mm)	Roughness	Flow LPS	Velocity (m/s)	Unit Head loss (m/km)	Friction Factor
2	299.04	25	100	0.2	0.41	19.3	0.056
3	585.16	25	100	0.1	0.21	5.73	0.062
4	644.96	25	100	0.07	0.14	2.67	0.065
5	893.77	25	100	0.04	0.08	1.04	0.071
6	1121.49	12.5	100	0.02	0.17	8.4	0.072
7	357.32	12.5	100	0.01	0.06	1.1	0.084
8	312.55	25	100	0.04	0.08	1.04	0.071
9	404.94	12.5	100	0.02	0.17	8.4	0.072
10	495.02	12.5	100	0.01	0.06	1.1	0.084
11	1113.51	12.5	100	0	0.01	0.02	0.118
13	690.43	100	100	2.22	0.28	1.91	0.047
14	131.73	25	100	0.02	0.04	0.29	0.078
15	199.62	25	100	0.01	0.01	0.04	0.092
16	350.76	100	100	1.98	0.25	1.54	0.048
17	277.15	75	100	1.77	0.4	5.1	0.047
18	667.98	50	100	1.56	0.8	29.13	0.045
19	483.46	25	100	0.0	0.01	0.04	0.092
20	1229	75	100	1.37	0.31	3.16	0.049
21	176.96	50	100	1.15	0.59	16.58	0.047
22	770.56	50	100	1	0.51	12.74	0.048
23	830.28	50	100	0.85	0.44	9.52	0.049
24	834.55	50	100	0.72	0.36	6.85	0.051

25	327.74	50	100	0.34	0.17	1.73	0.057
26	533.31	50	100	0.27	0.14	1.13	0.059
27	641.51	25	100	0.21	0.42	20.42	0.056
28	763.17	25	100	0.15	0.31	11.49	0.058
29	610.78	25	100	0.1	0.21	5.66	0.062
30	212.49	25	100	0.03	0.06	0.62	0.074
31	165.34	25	100	0.01	0.01	0.04	0.092
32	337.83	25	100	0.01	0.02	0.08	0.086
33	248.09	25	100	0.03	0.06	0.6	0.074
34	132.83	25	100	0.01	0.02	0.8	0.087
35	562.08	25	100	0.01	0.01	0.04	0.092
36	281.04	25	100	0.24	0.5	27.16	0.054
37	337.54	25	100	0.19	0.38	16.8	0.057
38	570.68	12.5	100	0.02	0.17	8.4	0.072
39	578.28	12.5	100	0.01	0.06	1.1	0.084
40	363.8	25	100	0.12	0.24	7.13	0.061

4. Conclusion

Based on the results of the analysis carried out in the previous chapter, conclusions can be taken as follows:

I. Based on the results of the survey at the site there are 2 springs in Sumberejo Village, namely the Sumberejo Springs and Sumber Karya Water Spring. The potential discharge from the two springs is 3.00 liters / second and 0.35 liters / second.

II. From the data and results of the calculation of the projection of the residents of Sumberejo Village by using the arithmetic method for 2030, there are 2443 people.

III. Based on the survey results in the study locations, the population's basic needs were 60 liters / day / capita, so the population was 2443 people for domestic clean water needs of 2.059 liters/ second. As well as non-domestic needs, because there are public facilities around 0.412 liters / day. So that the total water needs of Sumberejo Village are 2.842 liters / second and the available spring discharge is 3.35 liters / second, so that the need for clean water is sufficient until 2030.

IV. Based on the calculation of fluctuations in the need for clean water, the reservoir volume for the Sumberejo Spring is found to be 50 m3 and for the Source Spring 6 m3.

V. From the results of the EPANET 2.0 program simulation, the distribution network design criteria have met until 2030. The planned water distribution network uses PVC pipes with a diameter of 12.5 mm - 100 mm, the total length of pipes needed is 4,916 meters. The results of hydraulic analysis using EPANET obtained the highest pressure (pressure) at 24.75 meters and the lowest 2.95 meters.

References

[1] Badan Pusat Statistik 2017. *Kecamatan Sumberejo Dalam Angka*. Kabupaten Tanggamus
[2] Hidayat A 2017 Peraturan Menteri Kesehatan Mengenai Kualitas Air
[3] Lubis, Zulkifli 2014 Kebutuhan Air Bersih Di Kecamatan Glagah Kabupaten Lamongan
[4] Mashuri 2015 Kajian Ketersediaan Dan Kebutuhan Air Baku Dengan Pemodelan Ihacres di Daerah Aliran Sungai Tapung Kiri
[5] Politeknik Negeri Lampung 2013 *Pengelolaan Air Bersih*. (Politeknik Negeri Lampung, Bandar Lampung)
[6] Posumah, Giovanni David 2015 Perencanaan Sistem Penyediaan Air Bersih Di Desa Paputungan Kecamatan Likupang Barat Minahasa Utara
[7] Suroso, 2018 Metedologi Penelitian Air Bersih
[8] Terunajaya 2016 Analisa Ketersediaan Dan Kebutuhan Air Pada Daerah Aliran Sungai Percut Untuk Memenuhi Kebutuhan Air Bersih Di Kabupaten Deli Serdang

IOP Conf. Series: Materials Science and Engineering **807** (2020) 012021 doi:10.1088/1757-899X/807/1/012021

Analysis Level of Motorcycle Discipline by Using the Use of Left Flow at Jl. Sudirman Metro City, Lampung

Mira Wisman[1] and Trimoyo[2]

[1] Department of Civil Engineering, Engineering Faculty, University of Lampung, Indonesia
[2] Department of Civil Engineering, Universitas of Bandar Lampung, Indonesia

*Corresponding e-mail: mia_wisman@yahoo.com

Abstract. The high number of accidents on Jl. Sudirman Metro city in Lampung is caused by presence of mix traffic, as well as the absence of special treatment for vulnerable motorcycle users, resulting in conflict between motorbikes and other vehicles that have the potential to cause accident. This analysis was carried out to reduce the frequency of high accidents and fatalities due to accident on Jl. Sudirman, an analysis of motorbike lane requirements on Jl. Sudirman and speed management was conducted, one of which was speed limit determination. The analysis samples taken by 7 days survey from 6.00 am – 06.00 pm on Jl Sudirman Metro city Lampung. The result of the surveys showed the criteria for motorcycle lane determination criteria not suitable to implement on Jl.Sudirman segment if it is in accordance with specified indicator because the motorcycle traffic volume is less than 674 smp/hour and 2696 vehicle/hour. This paper indicates that survey analysis on motorcycle driver discipline by using of left flow at Jl. Sudirman Metro city Lampung was not significant. In order to reduce the frequency of high accident we need speed limiting guards with maximum speed 40 km/hours and the signs for motorcycle instruction using the left lane.

1. Introduction

1.1 Background

Jl. Sudirman is one of the Primary Collector roads in Metro City that connects the primary centre to other primary centres, in this case many motorbike users use it. Based on accidents data obtained from Sultanas Metro Police, the number of vehicles involved in accidents that occurred in Metro City in 2017 amounted to 214 accident then while on Jl. Sudirman section amounted to 37 accidents Then the highest number was on motorcycles with as many 23 vehicles. The high number of accidents on the road is caused by the presence of mix traffic, as well as the absence of special treatment for vulnerable motorcycle users, resulting in conflicts between motorbikes and other vehicles that have the potential to cause accidents [1-4].

The number of motorbikes that move a lot on public roads can cause traffic chaos, because many drivers who ride their motorbikes in a hurry, do not comply with traffic signs and regulations, and often do not respect other vehicles [5-9]. When viewed from the aspect of motorcycle safety and security it is very low. First seen from the stability of the movement that the motorbike is only supported by two wheels and the balance of movement depends on the driver's ability to control the vehicle. When driving inadvertently it is easy to slip or crash, that regulates traffic lanes and lanes where motorbikes, vehicles

IOP Conf. Series: Materials Science and Engineering **807** (2020) 012021 doi:10.1088/1757-899X/807/1/012021

with lower speed, trucks and non-motorized vehicles are in the left lane. From these conditions, a study of motorbike compliance in various vcr (volume capacity ratio) is needed, especially on Jl. Sudirman, which is the access road to Metro City.

1.2 Identification of problem

There are many problems that exist today, can be identified as follows:

a. What is the performance of Jl. Sudirman section.
b. Motorcycle movements tend not to follow the same lane
c. Can motorbike lanes be applied on J[. Sudirman
d. Weakness of the existing supervision system, so that there are many violations on the highway that can cause traffic accidents

1.3 Scope and Limitation of Problems

The problem limitation in this thesis is to analyse the design of motorcycle lanes, namely:

a. Analysis without measuring the effectiveness and design of the motorcycle lane.
b. Analysis of traffic volume that occurs during peak hours.
c. Analysis of motorcycle lane needs is carried out from Jalan Jenderal Sudirman
d. Metro City on certain segments

Second is the ability to move at high speed will cause a big impact if the motorcycle experiences a collision. The third is that motorbikes are vehicles that are not designed to protect unprotected riders, not equipped with a cover to protect the riders, so that motorcycle riders must equip themselves with complete safety. With the existence of standard regulations, namely Law 22 of 2009 concerning road transport traffic [10-11].

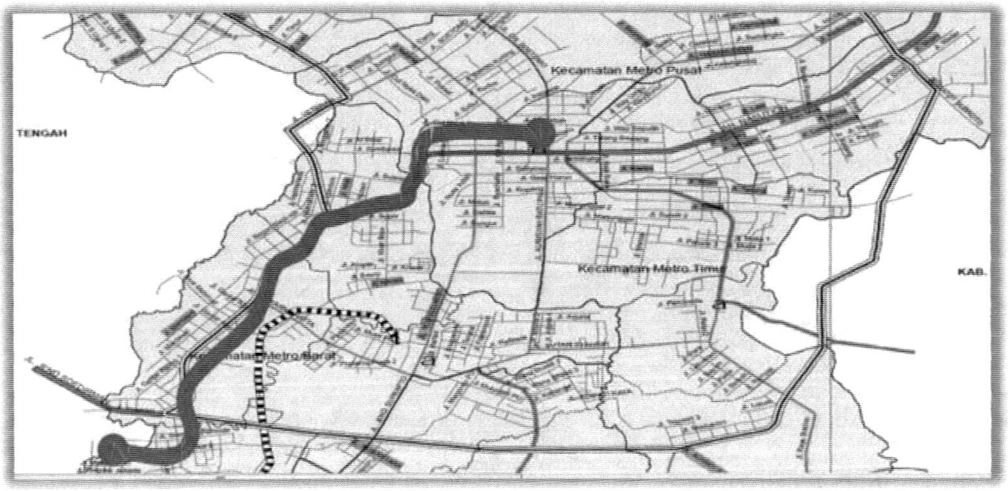

Figure 1 Map of Study Area

2. Methodology

This research will be divided into several stages of activities as follows:

2.1. Location of research

The research conducted at Jl. Sudirman is Metro City, which is a National Road with a road length of 5.53 km which is an access to and out of 5.53 km which is an access to and out of Metro City That Connects Central Lampung Regency with a Population of 162,976 People. And one reason the researchers chose Jl. Sudirman was because of the high number of accidents involving motorbikes.

2.2 Research Data

There are two kinds of research material used in this study, namely primary data and secondary data. Primary data is about the data obtained from surveys in the field, such as traffic.

3. Data Analysis

Data obtained from secondary data on the road data to be studied.

a. Road inventory survey
   Road inventory data is intended to identify the characteristics of road infrastructure, including road length, road width, road conditions, and road equipment facilities visually, considering that these components can affect road capacity.
b. Traffic Volume Survey
   This classified traffic volume survey is intended to determine the level of traffic density on road segments based on classified traffic volumes, the direction of traffic flow, the type of vehicle in a certain time unit carried out by direct observation and enumeration in the field. While the objectives are to find out the periods of peak hours at each survey point
c. Vehicle speed Survey
   Vehicle speed surveys are carried out to determine the travel time, travel speed related to the level of road performance, how the survey was carried out by 2 surveyors using motorbikes along the road. The first surveyor measures the length of the road, the second one to measured volume the vehicle in peak hours.

Processes include:

a. Metro City Statistic Center.
b. The Transportation and Communication Office of the City of Metro, which will be used to find out data on inventory on the road to be studied, and the literature that has been reviewed.
c. Metro City Bappeda to find out the description of the Metro city study area, and Metro City RPJMD 2015-2020.
d. Metro City Public Works Agency, such as road hierarchy data and network maps.
e. Metro City accident data.

3.1 Division of Nodes

The division of Nodes in this study is divided into several nodes namely 002, 003, 004, 011 and 012 can be seen in the table below:

Table 1 Distribution of Node of Jalan Jendral Sudirman

No	Node	Direction
1	002	Bank BRI Intersection
2	003	Maulana Intersection
3	004	Church Intersection
4	011	Soekarno Hatta Intersection
5	012	Nurul Huda Mosque Junction

The division of Node Jalan Jenderal Sudirman where 002, 003 and 004 is a road section with a central location for Central Busines Distric (CBD), and close to the centre of government having non-motorized vehicle traffic, where at node 002 it is not served by public transport traffic is a road close to the center of the activity where there are side barriers. This is because on the right and left side of the road there are shops, as well as the presence of vehicles parked on the roadside.

3.2 Capacity Analysis

In the capacity analysis, the data used ranged from the peak hours of traffic flow that passed the Jl. Sudirman road section at the study site.

Table 2 Result of Capacity Analysis

No	Node		Co	FCw	FCsp	FCsf	Fces	C
	First	End	smp/jam					smp/jam
1	002	003	2900	1.00	0.97	0.81	0.90	2051
2	003	002	2900	1.00	0.97	0.81	0.90	2051
3	003	004	2900	1.00	0.97	0.81	0.90	2051
4	004	003	2900	1.00	0.97	0.81	0.90	2051
5	011	012	2900	1.00	0.97	0.81	0.90	2051
6	012	011	2900	1.00	0.97	0.81	0.90	2051

3.3 Analysis VCR (Volume Capasity Ratio)

From the calculation of the road capacity, the V / C ratio Jl. Sudirman explained in the table below:

Table 3 Result of Analysis of V/C Ratio

No	Node	Peak	Lebar Lajur	Co smp/ja	FCw	FCsp	FCsf	FCcs	C smp/jam	Volume Peak	V/C
1	002-003	Morning	3,50	2900	1,00	0,97	0,81	0,90	2050,7	819	0,40
		Daytime	3,50	2900	1,00	0,97	0,81	0,90	2050,7	681	0,33
		Afternoon	3,50	2900	1,00	0,97	0,81	0,90	2050,7	697	0,34
	003-002	Morning	3,50	2900	1,00	0,97	0,81	0,90	2050,7	839	0,41
		Daytime	3,50	2900	1,00	0,97	0,81	0,90	2050,7	727	0,35
		Afternoon	3,50	2900	1,00	0,97	0,81	0,90	2050,7	727	0,35
2	003-004	Morning	3,50	2900	1,00	0,97	0,81	0,90	2050,7	914	0,45
		Daytime	3,50	2900	1,00	0,97	0,81	0,90	2050,7	737	0,36
		Afternoon	3,50	2900	1,00	0,97	0,81	0,90	2050,7	715	0,35
	004-003	Morning	3,50	2900	1,00	0,97	0,81	0,90	2050,7	935	0,46
		Daytime	3,50	2900	1,00	0,97	0,81	0,90	2050,7	784	0,38
		Afternoon	3,50	2900	1,00	0,97	0,81	0,90	2050,7	828	0,40
3	011-012	Morning	3,50	2900	1,00	0,97	0,86	0,92	2050,7	750	0,34
		Daytime	3,50	2900	1,00	0,97	0,86	0,92	2050,6	590	0,27
		Afternoon	3,50	2900	1,00	0,97	0,86	0,92	2050,6	536	0,24
	012-011	Morning	3,50	2900	1,00	0,97	0,86	0,92	2050,6	738	0,33
		Daytime	3,50	2900	1,00	0,97	0,86	0,92	2050,6	552	0,25
		Afternoon	3,50	2900	1,00	0,97	0,86	0,92	2050,6	648	0,29

3.4 Level of Road Services)

Determination of the level of service desired is an activity of determining the level of service for roads based on service level indicators. There is a level of road services can be seen in the following table:

Table 4 Result of Capacity Analysis

No	Node		Link Length (m)	Volume (Vehice/hours)	V/C	Level Of Service
	Firs	End				
1	2	3	0.239	819.42	0.4	B
2	3	2	0.239	838.87	0.41	B
3	3	4	0.23	913.96	0.45	C
4	4	3	0.23	934.87	0.46	C
5	11	12	0.35	749.84	0.34	B
6	12	11	0.35	648.4	0.33	B

3.5 Proportion of Motorcycle Volume Peak Hours

In analyzing motorcycle lane requirements, it is necessary to analyze the proportion of highest vehicles on Jl. Sudirman in the morning peak hours between 06.45 - 08.00 WIB. The Motorcycle node is (002 - 003) of 59%, node (003 - 002) 60%, node (003 - 004) 60%, node (004 - 003) 61%, node (011 - 012) 53%, node (002 - 003) 64%. Highest vehicles on Jl. Sudirman in the morning peak hours between 06.45 - 08.00 WIB. The Motorcycle node is (002 - 003) of 59%, node (003 - 002) 60%, node (003 - 004) 60%, node (004 - 003) 61%, node (011 - 012) 53%, node (002 - 003) 64%.

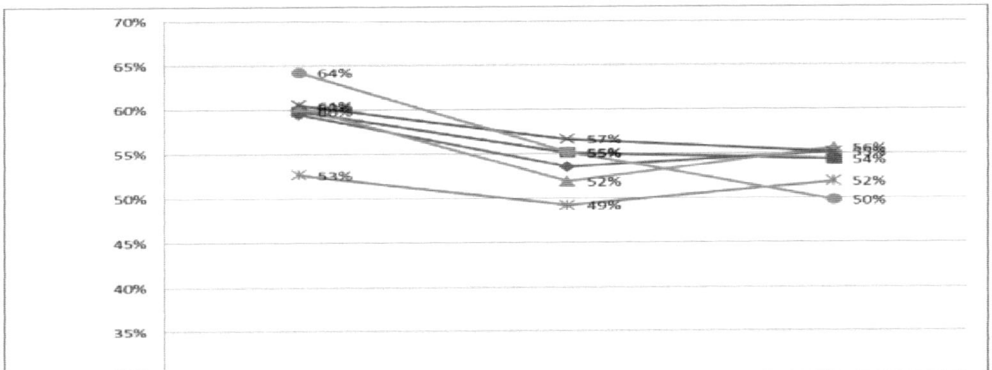

	Morning	Daytime	Afternoon
002 - 003	59%	54%	55%
003 - 002	60%	55%	54%
003 - 004	60%	52%	56%
004 - 005	61%	57%	55%
011 - 012	53%	49%	52%
012 - 011	64%	55%	50%

Figure 2 Proportion of Motorcycle Peak Hour Volume.

3.6 Proportion of MC on PCU/hour

It is necessary to do a proportional analysis of motorcycle passenger cars, below is a graph of the fluctuations in the proportion of motorcycle in various vcr (volume capacity ratio) and the use of the left lane of each node. From the graph above, it can be seen that the highest proportion of motorbikes occurs at the peak, namely 06.45 - 07.45 WIB by 23% with the use of the left lane of 50% and the VC ratio of 0.40. From the fluctuation chart above, it can also be seen the percentage of motorbike riders choosing the left lane in certain traffic conditions represented by the VC ratio where the value of the VC ratio follows the fluctuations in the percentage of left lane motorbikes and fluctuations in the proportion of motorcycles. This means, in heavy traffic conditions, the proportion of motorbikes has an influence on traffic conditions and the percentage of motorbike riders choose the left lane. The percentage level of the distribution of motorbike driver discipline towards the use of the left lane is still evenly distributed between the left and right because the VC ratio is 0.40 where the characteristics are stable currents, but the operating speed is limited by traffic conditions, the riders has enough freedom to choose speed.

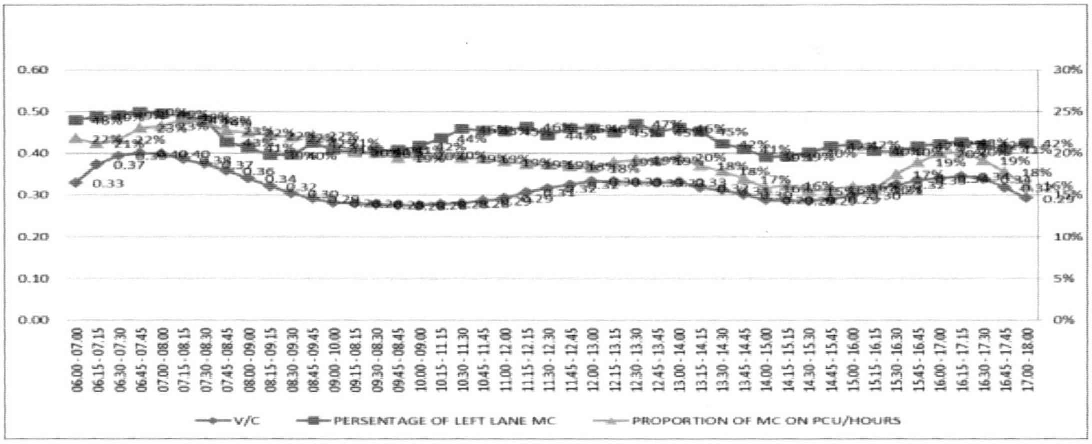

Figure 3 Propotion of MC on Pcu/hour Node (002 – 003).

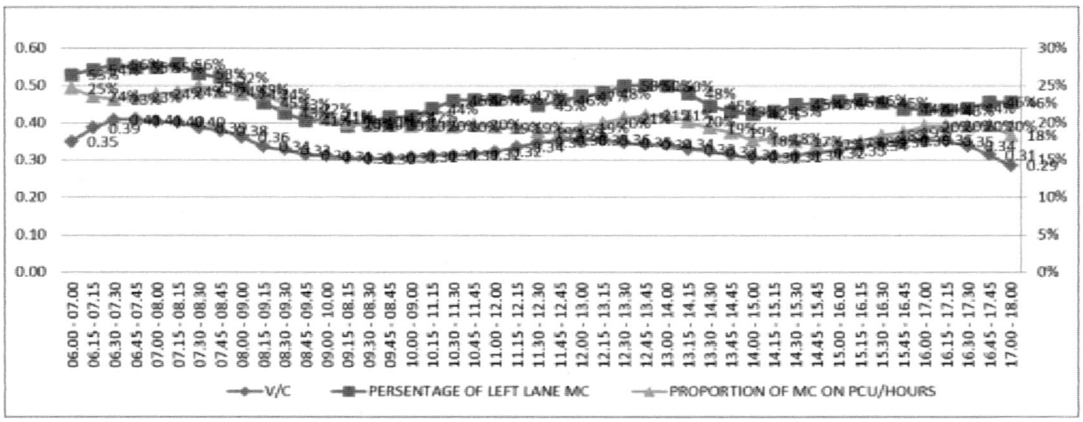

Figure 4 Proportion of MC on Pcu/hour Node (003 – 002).

From the graph above it can be seen that the highest proportion of motorbikes occurs at the peak, namely 06.45 - 07.45 WIB by 23% with the use of the left lane of 55% and the vc ratio of 0.41. From the fluctuation chart above, it can also be seen the percentage of motorbike riders choosing the left lane in certain traffic conditions represented by the VC ratio where the value of the VC ratio follows the fluctuations in the percentage of left lane motorbikes and fluctuations in the proportion of motorcycles. This means, in heavy traffic conditions, the proportion of motorbikes has an influence on traffic conditions and the percentage of motorbike riders choose the left lane. The percentage level of distribution of motorbike driver discipline towards the use of the left lane is still evenly distributed between the left and right because VC ratio is 0.41 where the characteristics are stable currents, but the operating speed starts to be limited by traffic conditions, the riders has enough freedom to choose speed.

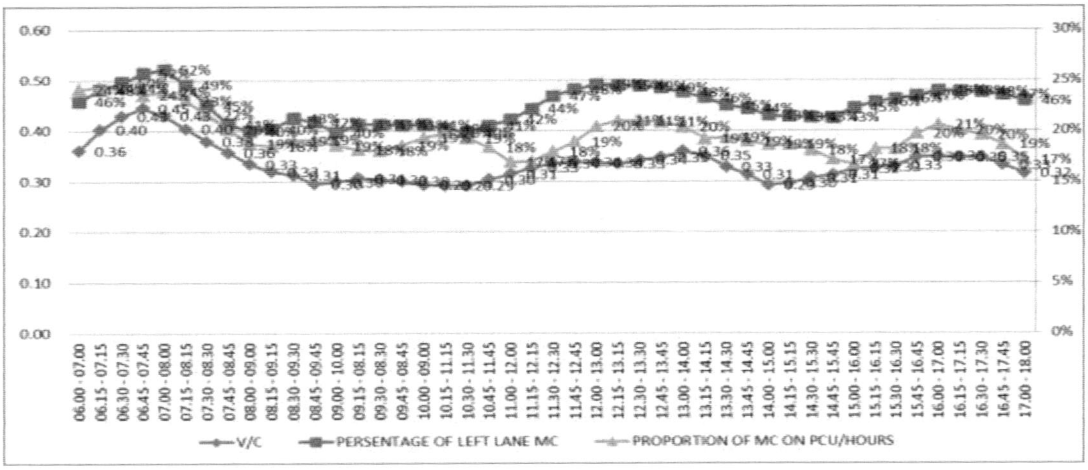

Figure 5 Proportion of MC on Pcu/hour Node (004 – 003).

From the graph above, it can be seen that the highest proportion of motorbikes occurs at the peak hours, namely 06.45 - 07.45 WIB by 24% with the use of the left lane of 52% and the VC ratio of 0.45. From the fluctuation chart above, it can also be seen the percentage of motorbike riders choosing the left lane in certain traffic conditions represented by the VC ratio where the value of the VC ratio follows the fluctuations in the percentage of left lane motorbikes and fluctuations in the proportion of motorcycles. This means, in heavy traffic conditions, the proportion of motorbikes has an influence on traffic conditions and the percentage of motorbike riders choose the left lane. The percentage level of the distribution of motorbike driver discipline towards the use of the left lane is still evenly distributed between the left and right because the VC ratio is 0.45. This ratio has characteristics namely : stable currents, but the speed and motion of the vehicle are controlled, the driver is limited in choosing speed.

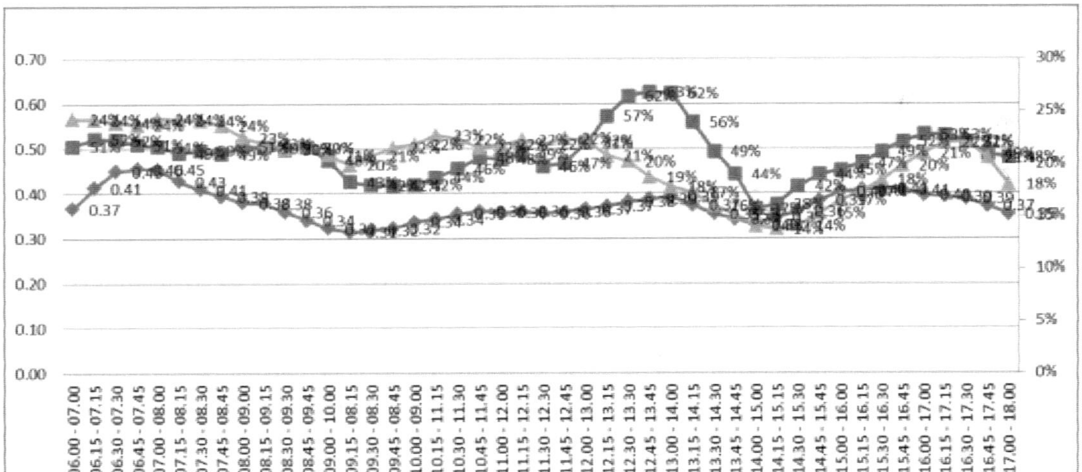

Figure 6 Proportion of MC nn Pcu/hour Node (011 – 012).

From the graph above, it can be seen that the highest proportion of motorbikes occurs at the peak, namely 06.45 - 07.45 WIB by 24% with the use of the left lane of 51% and the VC ratio of 0.46. From the fluctuation chart above, it can also be seen the percentage of motorbike riders choosing the left lane in certain traffic conditions represented by the VC ratio where the value of the VC ratio follows the fluctuations in the percentage of left lane motorcycles and inversely between the fluctuations in the proportion of motorcycles with fluctuations in the percentage of left lane motorbikes, this means that in heavy traffic conditions, the percentage of left lane motorbikes has an influence on traffic conditions and the percentage proportion of motorbikes has no effect on motorbike riders choosing the left lane. The percentage level of the distribution of motorbike driver discipline towards the use of the left lane is still evenly distributed between the left and right because the VC ratio is 0.45 where the characteristics are stable currents, but the speed and motion of the vehicle are controlled, the rider is limited in choosing speed.

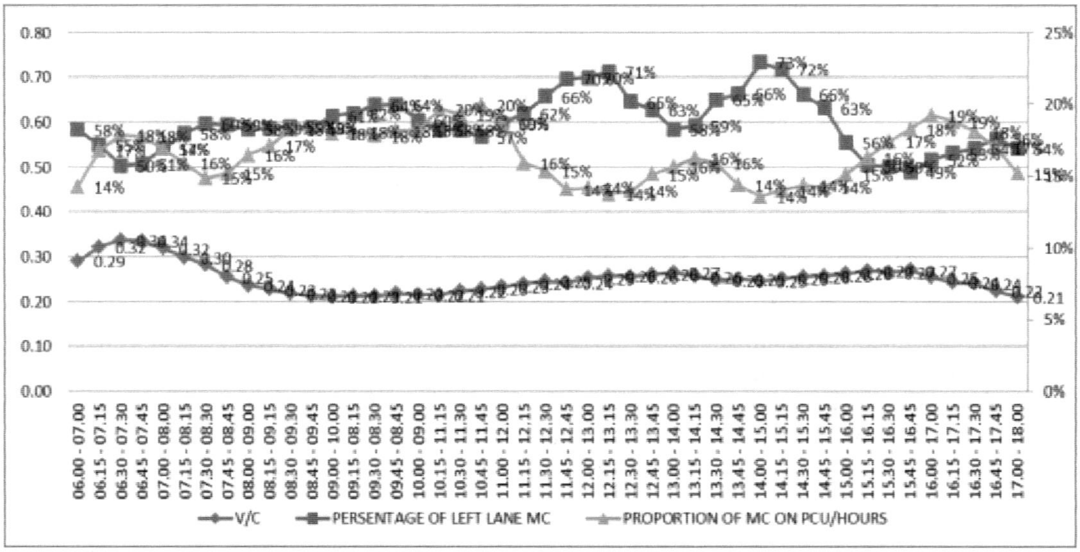

Figure 7 Proportion of MC nn Pcu/hour Node (012 – 011).

From the graph above, it can be seen that the highest proportion of motorbikes occurs at the peak of 06.30 - 07.30 WIB by 18% with the use of the left lane of 50% and the VC ratio 0.34. From the fluctuation chart above, it can also be seen the percentage of motorbike riders choosing the left lane in certain traffic conditions represented by the VC ratio where the VC ratio does not follow the fluctuations in the percentage of left lane motorcycles and inversely between the fluctuations in the proportion of bicycles motorbikes with fluctuations in the percentage of left lane motorbikes. This means, in heavy traffic conditions, the percentage of left lane motorbikes has no effect on traffic conditions and the percentage proportion of motorbikes has no effect on motorbike riders choosing the left lane. The percentage level of distribution of motorbike driver discipline towards the use of the left lane is still evenly distributed between the left and right because VC ratio is 0.34 where the characteristic are Stable currents, but the operating speed is limited by traffic conditions, the riders have enough freedom to choose speed.

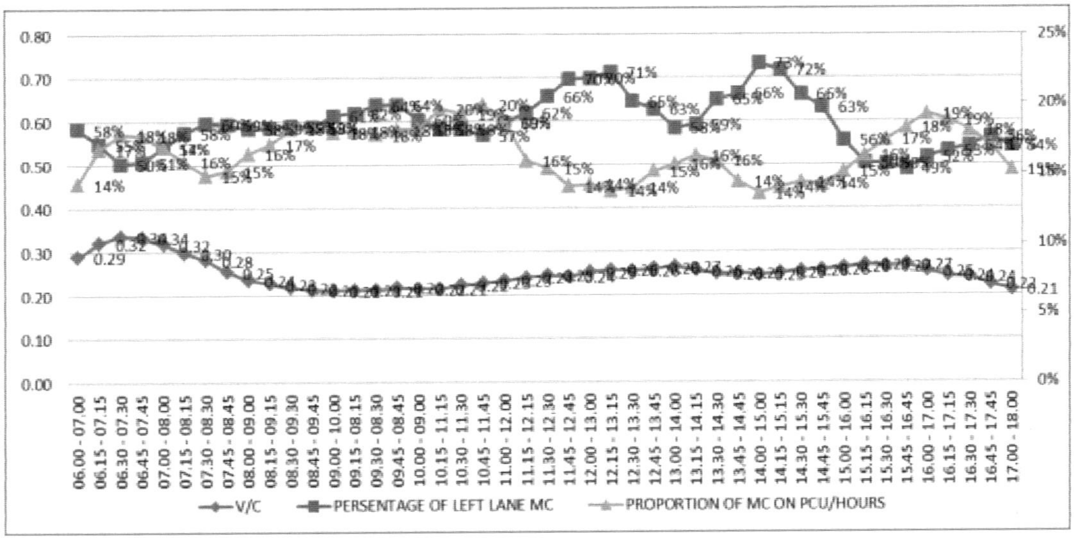

Figure 8 Proportion of MC on Pcu/hour Node (011 – 012)

From the graph above, it can be seen that the highest proportion of motorbikes occurs at the peak, namely 06.45 - 07.45 WIB by 25% with the use of the left lane of 51% and the VC ratio 0.33. From the fluctuation chart above, it can also be seen the percentage of motorbike riders choosing the left lane in certain traffic conditions represented by the VC ratio where the VC ratio does not follow the fluctuations in the percentage of left lane motorcycles and inversely between the fluctuations in the proportion of bicycles motorbikes with fluctuations in the percentage of left lane motorbikes, but fluctuations in VC ratios follow fluctuations in the proportion of motorbikes. This means, in heavy traffic conditions, the percentage of left lane motorbikes has no effect on traffic conditions and the proportion of motorbikes does not give the influence on motorbike riders chooses the left lane. However, the proportion of motorcycles has an influence on traffic conditions. The percentage level of distribution of motorbike driver discipline towards the use of the left lane is still evenly distributed between the left and right because VC ratio is 0.33 where the characteristics are Stable currents, but the operating speed starts to be limited by traffic conditions, the driver has enough freedom to choose speed.

3.7 Proportion of Motorcycle Accidents

To calculate the proportion of motorcycle accidents, use the following formula: Motorcycle accident.

$$\left(\frac{\text{The total motorcycle accident}}{\text{The total accident}}\right) \text{x } 100\% \qquad (1)$$

The number of vehicles involved in the 2017 Jl. Sudirman accident was 37 and the number of motorbikes involved in the accident was 23 units. So, the proportion of motorcycle accidents is as follows

$$\left(\frac{37}{32}\right) \text{x } 100 = 62\%$$

From the results of the calculation above the results of the proportion of motorcycle accidents are 62%.

3.8 Motorcycle Lane Needs
The following is the result of analyzing motorcycle lane requirements on Jl. Sudirman Metro City :

Table 5 Values / Lane Needs Indicators

No	Parameter	Unit	Measurement Result					
			(002 -003)	(003 - 002)	(003 - 004)	(004 - 003)	(011 - 012)	(012 - 011)
1	Operational Speed of MC	km/hour	43	41.17	40.67	40.17	49.67	45.83
2	Operational Speed of LV	km/hour	45.83	42.83	45.83	44	52.17	52.83
3	Operational Speed of HV	km/hour	42.83	40.67	40.5	38.33	45.5	43
4	Proportion OF MC	%	0.59	0.6	0.6	0.61	0.53	0.64
5	Proportion OF MC on pcu	%	0.23	0.23	0.24	0.24	0.18	0.25
6	MC Volume	MC/hour	944	984	1,077	1,109	669	944
7	MC Volume on the pcu	pcu/hour	188.82	196.87	215.36	221.87	133.84	188.89
8	Level of Degree	Q/C	0.4	0.41	0.45	0.46	0.34	0.33
9	Proportion of MC Accidents	%	62	62	62	62	62	62

Based on the table above, it can be concluded that the motorcycle lane is not needed.

3.9 Speed Limit Determination
To reduce the frequency of high accidents and fatalities due to accidents on the general Sudirman road, speed management is carried out. One of them is the determination of speed limits on the Primary Collector road, where on Jalan Jenderal Sudirman Node (002 - 003), Nodes (003 - 002), Nodes (003 - 004), and Node (004 - 003) with the type of land use of the Central Busines Distric (CBD) area determined at a maximum speed of 40 (forty) kilometers per hour. Whereas the road on Jalan Jenderal Sudirman Node (011 - 012), Node (011 - 012) sets the road speed limit with the type of land use housing area (dense settlement) with a maximum speed of 40 (forty) kilometers per hour.

4. Conclusion
a. Based on the analysis of capacity, traffic volume and VC ratio, the level of service road for the JL Sudirman Metro City segment is C because it has the highest VC ratio 0.46 on Jl. Sudirman Node segment (004 - 003) where characteristics are stable currents, but the speed and motion of the vehicle are controlled, the driver is limited in choosing speed has enough freedom to choose speed.
b. Use the Left Lane in peaks hours Jl. Sudirman in Nodes (002 - 003) is 50%, Nodes (003 - 002) is 55%, Nodes (011 - 012) is 50%, and Nodes (012 - 011) is 51% where there is mix traffic or a mixture of traffic flow if the riders of the two-wheeled motor vehicle use a left lane or lane that has mixed traffic

where the distribution of motorbike driver discipline towards the left lane is still evenly distributed between the left and right the characteristics are stable currents, but the operating speed starts to be limited by traffic conditions, the driver has enough freedom to choose speed, while the node (003 - 004) is 52%, node (004 - 003) is 51%, mix traffic or mixing traffic flow if the driver of two- wheeled motorized vehicles uses a left lane or lane that experiences a mixture of traffic flows where the spread of discipline of motorcycle drivers is found The use of the left lane is still evenly distributed between the left and right because its characteristics are stable currents, but the speed and motion of the vehicle are controlled, the driver is limited in choosing speed and has enough freedom to choose speed.
c. Based on the analysis of 9 (nine) parameters (source: Pusat Penelitian Jalan dan Jembatan (Kriteria Desain Lajur Motor, Idris,2007)) the motorcycle lane determination criteria cannot be applied on the Jl. Sudirman road section if it is in accordance with the specified indicators because the motorcycle traffic volume is less than 674 pcu / hour and 2696 vehicles/ hour.

5. Suggestion

a. It is needed to coordinate betweenthe Police and the Department of Transportation in improving regulations and law enforcement which have been stated in Law number 22 of 2009 concerning Road Traffic and Transportation to reduce the level of traffic accidents in Metro City.
b. It is needed to increase the socialization of safety to all the public in Metro City, because of the high behavior of motorcyclists who lack discipline to create disciplined and orderly behavior.
c. The need for infrastructure that supports the left lane such as maintenance of lost road markers and speed limiting signs with a maximum speed of 40 (forty) kilometers per hour, as well as signs for motorbike guidance using the left lane.
d. There is a need for further studies on the level of conflict to be used as an indicator of determining motorcycle lane requirements as an aspect of safety.

References

[1] Bina Marga, 1997. Manual Kapasitas Jalan Indonesia: Jalan Perkotaan, Bina Marga. Bandung
[2] DPU, 1997, Tata Cara Perencanaan Geometri Jalan Antar Kota, Jakarta: Departemen Pekerjaan Umum, Direktorat Jenderal Bina Marga.
[3] Departemen Pekerjaan Umum, Direktorat Jendral Bina Marga, Direktorat Bina Jalan Kota.2007.Manual Kapasitas Jalan Indonesia (MKJI). Jakarta.
[4] Idris, Muhammad, 2007, *Kriteria Lajur Sepeda Motor untuk Ruas Jalan Arteri Sekunder,* Bandung: Puslitbang Jalan dan Jembatan, Balitbang Departemen Pekerjaan Umum.
[5] Idris, Muhammad.2008, Pengembangan Kriteria Kebutuhan Lajur Sepeda Motor Untuk Ruas Jalan Arteri Sekunder 4/2-D dan 6/2-*D*, Makalah Kolokium Laporan Penelitian tahun 2008, Puslitbang Jalan dan Jembatan, Balitbang Departemen PU.
[6] Republik Indonesia. 2009. Undang-undang No. 22 Tahun 2009 tentang Lalu Lintas dan Angkutan Jalan. Sekretariat Negara. Jakarta
[7] Nazir, 2011. Metode Penelitian. Bogor, Penerbit Ghalia Indonesia.
[8] Republik Indonesia.2011. Peraturan Pemerintah Nomor 32 Tahun 2011 Tentang Manajemen dan Rekayasa, Analisis Dampak Serta Manajemen Kebutuhan Lalu Lintas. Sekretariat Negara. Jakarta
[9] Sukor, Nur Sabahlah Abdul dan Fujii, Satoshi.2011. *The Effect of Psychological Factors towards Motorcyclists' Risky Behaviours in Different Type of Motorcycle Lanes.* International Journal of Humanities and Social Science. Vol. 1 No. 9 [Special Issue – July 2011]. Malaysia
[10] Widoyoko, S.E. (2012). Teknik Penyusunan Instrumen Penelitian. Yogyakarta:
[11] Republik Indonesia.2015. Peraturan Menteri Perhubungan Nomor PM 111 tahun 2015 tentang Tata Cara Penetapan Batas Kecepatan

IOP Conf. Series: Materials Science and Engineering **807** (2020) 012022 doi:10.1088/1757-899X/807/1/012022

The analysis of Tsunami evacuation route based on geographic information system: a case study in the coast of Lampung Bay

Rina Febrina[1]*, Evan D K[1], Lusmeilia Afriani[2], Rr M I Retno[2], Susilorini[3], Helmia Adita Fitra[4]

[1] Department of Civil Engineering, Universitas Malahayati, Bandar Lampung 35153, Indonesia
[2] Department of Civil Engineering, Lampung University, Bandar Lampung 65145, Indonesia
[3] Department of Infrastructure and Environmental Engineering, Soegijapranata Catholic University, Semarang 5023, Indonesia
[4] Sumatera Institute of Technology, South Lampung 35365, Indonesia

*Corresponding e-mail: febrinacivil@yahoo.com

Abstract. The most fundamental issues for local people who are potentially at risk to tsunami threat are awareness and preparedness. The aim of this paper is to define the effectiveness of tsunami evacuation routes toward tsunami hazard along the coast of Lampung bay. The parameter applied in this study were elevation and slope created based on Aster GDEM while the coastline distance was made based on vector map and tsunami inundation obtained by Hills and Mader equation. In addition, the road network was derived from Development Planning Agency at Sub-National Level. AHP process was applied in determining the weight of the parameter by using pair-wise comparison via five normalized matrix iteration. Thus, there are five vulnerability classes that were classified. Weighted overlay through spatial analysis in geographic information system (GIS) was used in this study in order to make the final tsunami vulnerability map. The Network Analysis through Closest Facility Analysis in GIS was used to create the effective evacuation route and to point out 34 effective evacuation locations that could be reached by the people before the tsunami strikes. The results presented in this paper could eventually facilitate the improvement in the basic data for city planning related to evacuation process and management strategy during disaster.

1. Background of the study

There is a total of 110 tsunami happened over Indonesia. Among 110 tsunamis occurred, there were 100 tsunamis caused by earthquakes, nine tsunamis were caused by volcanic eruptions and a tsunami by caused by landslide [1]. Tsunamis caused by earthquake is a leading factor for a tsunami to occur in Indonesia. The centre of earthquakes is closely associated with the subduction zones and that caused the path of the earthquakes which occurred often go along the subduction zones [2]. Bandar Lampung, which located in Lampung province, is situated close to the Sunda strait subduction zones. It has very high level of seismicity because of the movement of the plates by the zone. There is a total of five vulnerabilities zones in Lampung, one of it is located Bandar Lampung which is the third zone (Zone

III). Based on the previous studies conducted by BPPT researchers toward the tsunami disaster especially in the Sunda Strait, [3]. Bandar Lampung is considered as one of the tsunami disaster vulnerability zones since the south coast of Bandar Lampung is a bay, which makes it prone to both tsunamis and earthquakes as well as floods and environmental damages on a very dense settlements along the coast.

In order to decrease the impact of the tsunami disaster, a sequence of attempts could be done which including the uses of alternative methods that are more acceptable for the local people. Reducing the risk of casualties and developing plans can be done through the mapping of coastal areas that are safe from the impact of tsunami which will later be used as an evacuation point location. The data then could be used in order to determine the evacuation routes hence minimizing the fatality impact from the disaster of tsunami.

The impact of the Tsunami is not only affecting the human and infrastructures but also the quantity of water absorption and the nature of the soil that would change due to the excessive water infiltration [4]. Consequently, there would be landslides in hill areas around the coast. In addition to that, the nature of soils in loam sandy areas which have excessive water infiltration or strong water flow would most likely change. Therefore, this paper will emphasize to human safety not only due to the tsunamis but other secondary factors.

There are a number of studies on tsunamis and mitigation in Lampung that have been carried out by Alhamidi and Pakpahan [5] that discussed on the coastal resilience of Lampung based along the coastline, Jokowinarno [6] discussed tsunami mitigation, Dewi and Armijon [7] investigated the composing of tsunami hazard map and Nurhasanah [8] is regarding on the tsunami evacuation route simulation. Whereas in this study the researcher conducted several other points of view in obtaining an effective tsunami evacuation route for Bandar Lampung people.

The uses of remote sensing application and geographic information system (GIS) manage to construct the possible evacuation route as well as the distribution of damaged area caused by tsunami disaster and able to evaluate the vulnerability of the zones. GIS with the spatial multi-criteria examination can really help in making a need to identify the basic leadership procedure utilizing the geo-reference information. Additionally, the preferences of the decision maker's concerning with a parameter assessment, analysis of spatial multi-criteria needed all the data from the criterion attribution and the geographical references [9-10]. Some prior works have investigated tsunami hazard using multi-scenario approach [11], the application of GIS technique for tsunami mapping and overlaying by using official land-use map [12] and also analysing the vulnerability by using remote sensing data and integrated analysis which utilizing the GIS towards the physical built-up infrastructure or buildings. There were also studies done in order to identifying the inundation of the surrounding area based on the configuration and the biggest recorded tsunami disaster concerning to the vulnerability of the buildings and human [13-14].

The aims of this study are to draw the tsunami vulnerability areas, to evaluate its impact by designing the possible tsunami inundation area and to make the evacuation routes by utilizing GIS. Elevation, slope, coastline distance and inundation are the main factors that influence the vulnerability. The Analytical Hierarchy Process for estimating weights of the parameters was applied in this study. Evacuation routes were created with network analysis processing through GIS.

2. Research method

2.1. Data

The elevation data applied in this study was collected from The ASTER Global Digital Elevation Model (ASTER-GDEM) version 2. The Ministry of Economy, Trade, and Industry (METI) of Japan and the United States National Aeronautics and Space Administration (NASA) are the ones who developed and made the Advanced Space-borne Thermal Emission and Reflection Radiometer (ASTER) GDEM as a collaborative accessible available inclusively for the public. ASTER-GDEM acquired the data via a space-borne observing optical device. The ASTER-GDEM covers the whole Earth surface at high resolution. In order to enhance global digital elevation resolution and elevation accuracy, the version 2 of the ASTER-GDEM was used. It was also used to advance algorithm and reprocessing a total of 1.5 million scene data including additional 250,000 scenes acquired after the previous release. The data were then producing a 1 arc per second posted grid or approximately 30 m at the equator.

Tsunami's vulnerability, especially for evacuation route, was analysed in this study. In addition, a tsunami heights and inundation information from the 1883 tsunami that was reported in the report book were also used during the research [15]. The steps of analysis started from data collection, proceeded with surface analysis of DEM data, vector data processing, inundation processing, continued with AHP process, raster overlay processing through GIS approach and ended with network analysis processing with road network data as in Figure 3 and by using GIS approach as in Figure 1.

2.2. Study Area
The research area was set up in four districts in Bandar Lampung City, namely Teluk Betung Timur districts, Teluk Betung Selatan districts, Bumi Waras districts and Panjang districts (Figure 2.). The four districts are several districts that have a dense population and were directly affected by the tsunami due to the eruption of Mount Krakatau in 1883. Since all four sub-districts are located not too far away from the Rakata, it makes them prone to tsunami disasters.

2.3. Spatial Analysis

2.3.1. Elevation
One of the primary data set needed for the sample to create vulnerability and tsunami inundation is elevation of data. A digital elevation model was created by using elevation map in order to derive a set of parameters that can show the physical vulnerability using data taken from ASTER GDEM version 2.0. This data was downloaded from ASTER-GDEM website *(https://gdex.cr.usgs.gov/gdex/).*

The elevation is classified into five vulnerability categories based on the height of the surface. It shows that low elevation, measured in meters, of the surface will have high tsunami vulnerability wave, as shown in Table 1. Vulnerability mapping based on elevation data showed that most of the coastal area is in the rather high-class vulnerability as shown in Figure 4.

2.3.2. Slope
Slope is the maximum change rate in z-value from each cell of the image. Slope map was made using the third-order finite difference method and ASTER-GDEM version 2.0. The slope percentages ranges vary from zero to almost infinity. A flat surface is considered as 0%, meanwhile as the surface becomes more vertical the slope percentage becomes increasingly higher. The slope map was categorized into five categories according to the value of tsunami vulnerability as described in Table 1. The vulnerability map on Figure 5 was made by applying the slope classification based on tsunami vulnerability class through reclassify process. Based on the information above, the urban areas are considered as the high vulnerability areas.

2.3.3. Coastline Distance
Dividing areas into five classes of vulnerability was done by using multi-buffering from the calculation of the distance from the coastline to the ground calculation of the distance from coastline to the land by using multi-buffering was done to divide areas into. The possibility range of the tsunami reaching the ground was used to determine the buffering distance. The distance is depending on the historical report of the maximum tsunami height in the area of study and it is expressed by the following equation (1) [11,16].

$$logX_{max} = log100 + \frac{4}{3} log\,(\frac{Y_o}{10}) \quad (1)$$

Based on the equation above, it shows that X_{max} provides the maximum reach of the tsunami over land, and while Y_o is the tsunami height at the coast. The buffering distance was determined based on the range of possible tsunami flows flowing towards inland. The score of the X_{max} in this study was based on the history of tsunami occurred in Lampung which happened on August 27th, 1883. A tsunami occurred in the Sunda Strait with a maximum height of 30 meters above sea level and had the height on the coastlines high as 4 meters on the southern coast of Sumatra where Lampung Province is located (based on the Soloviev and Go catalogue) [15]. We used coastline distance in vector map of Lampung through GIS processing to generate vulnerability map based on coastline distance (Figure 6).

2.3.4. Inundation

Arcgis 10.6 software with the Raster calculator functions was used as the simulation of tsunami. Hills, and Mader [17] had formulated an equation to calculate the tsunami run up as in equation (2):

$$X_{max} = (Hs)^{1.33} n^{-2} k [cos\,S] \quad (2)$$

Based on the equation above, we can know that X_{max} is maximum inundation (m), Hs is the maximum tsunami run-up (m), k is constants which correspond to 0.06 for most tsunami cases, whereas Manning coefficient of land covers is represented by n and the slope of land is represented by S. We applied X_{max} in vector map of Lampung through GIS process in order to generate the vulnerability map based on inundation as in Figure 7.

Table 1. Vulnerability class based on elevation data [18]

Elevation (m)	Slope (%)	Coastline Proximity (m)	Inundation (m)	Vulnerability (class)
<5	0 – 2	0 – 65	0 – 30	High
5 – 10	2 – 6	65 – 164	30 – 75	Slightly high
10 – 15	6 – 13	164 – 282	75 – 128	Medium
15 – 20	13 – 20	282 – 413	128 – 188	Slightly low
>20	>20	>413	>188	Low

IOP Conf. Series: Materials Science and Engineering **807** (2020) 012022 doi:10.1088/1757-899X/807/1/012022

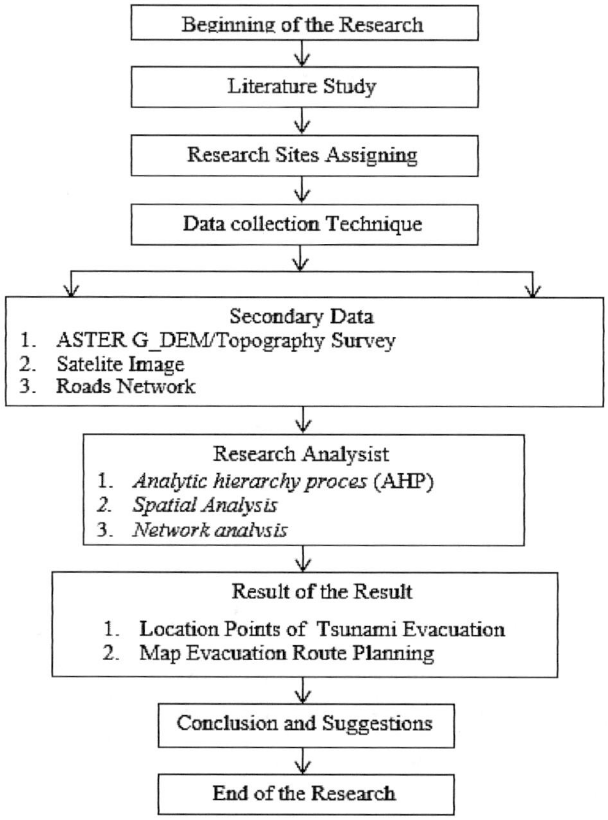

Figure 1. Research flow chart

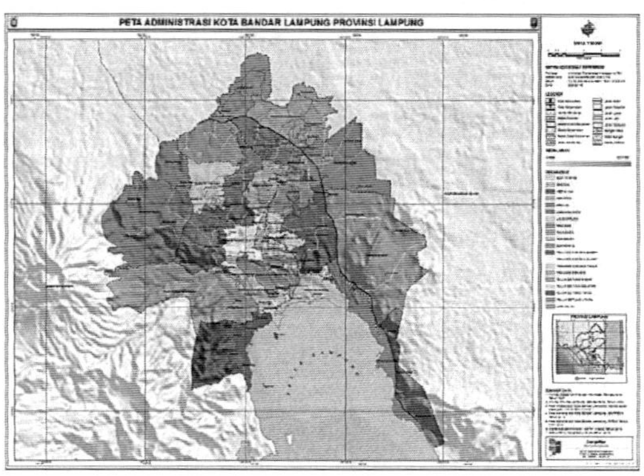

Figure 2. Map of study

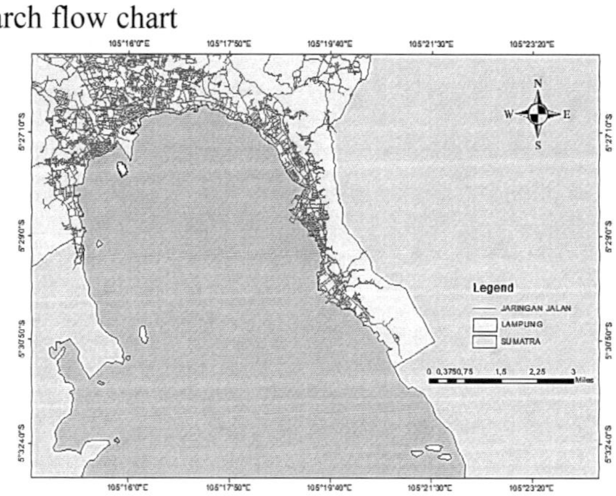

Figure 3. Map of roads network

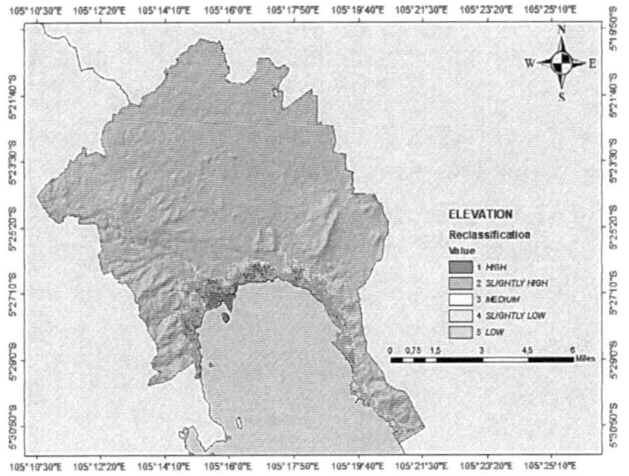

Figure 4. Tsunami vulnerability map based on the parameter of Elevation

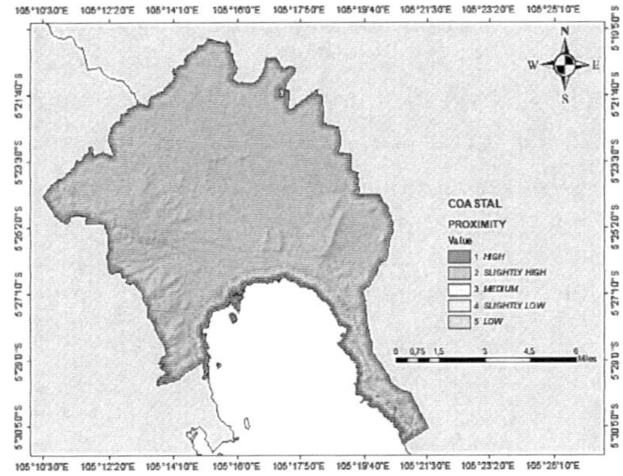

Figure 5. Tsunami vulnerability map based on the parameter of Coastline Proximity

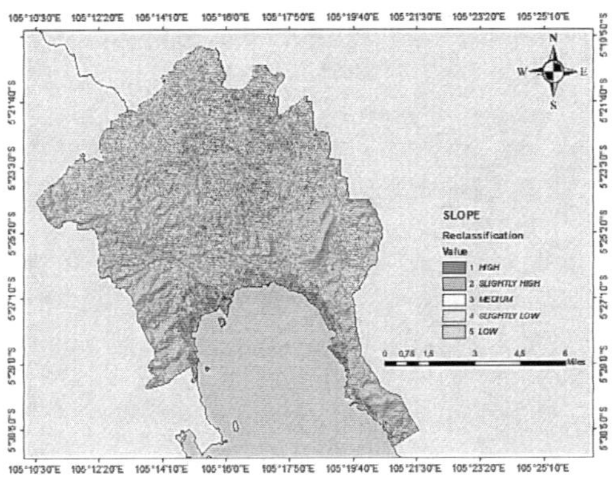

Figure 6. Tsunami vulnerability map based on the parameter of Slope

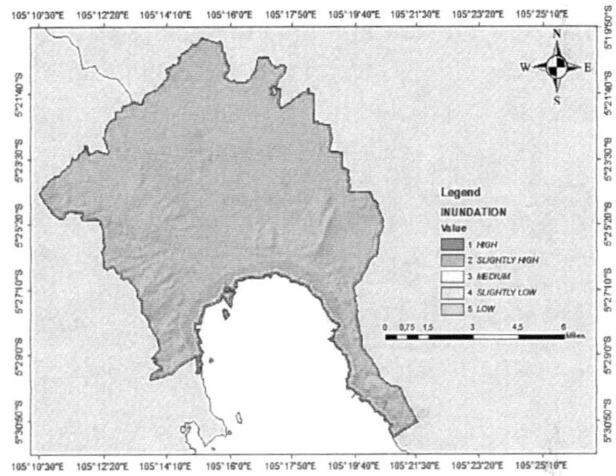

Figure 7. Tsunami vulnerability map based on the parameter of Slope

3. Results and Analysis

3.1. Analytical Hierarchy Process (AHP)

Based on research conducted by [16-18], they provide tsunami vulnerability parameters based on elevation, slope and coastline distance. As for the inundation, parameter determination is based on research conducted by [17]. All parameters in grid cells were then classified into five classes of vulnerability. For the vulnerability categories the following numbers of 1, 2, 3, 4, and 5 were used. They represent low, slightly low, medium, slightly high and high vulnerability. The whole parameter of tsunami vulnerability was overlaid by using weighting and scoring system. Scoring was designed to assess the limiting factor was scored in order to asses on each parameter, while the class of tsunami vulnerability was determined by weighting based on the dominant influence of these parameters.

AHP was applied in assigning weights to each of the parameters. By applying pair-wise comparisons, AHP assisted in making a scaled set of preferences and explaining the significance of each parameter relatively to other parameter [13-15]. In addition, AHP also enable the researchers in evaluating the vulnerability related to the natural hazard. A series of number from 1 until 9 as shown in Table 2 were

used in comparison scale to make pair-wise comparison matrix. These numbers rely on the relative importance of each parameter. Meanwhile the relative weights of each parameter will be calculated through pair-wise comparison as shown in Figure 8.
Prior to the spatial analysis, the consistency level computation was required. Consistency Ratio (CR) is a procedure to determine the index of consistency. It shows that the probability of the matrix findings was randomly calculated [20]. In the other word, AHP allows inconsistency using the consistency ratio computation. The tolerable consistency ratio is less than or equal to 10 percent [21-22]. CR is the ratio between the consistency index (CI) and random consistency index (RI), it can be expressed using equation (3) and equation (4).

$$CR = \frac{CI}{RI} \tag{3}$$

$$CI = \frac{\lambda_{max} - n}{n - 1} \tag{4}$$

Based on the formula above, λ_{max} represents the largest eigenvalue, and N the size of comparison matrix. λ_{max} is computed from the sum of all parameters which later then multiplied by its eigenvector. RI is based on the random consistency index ($RI = 0.09$ for four parameters).

Pair - wise comparison matrix

	Elevation	Slope	Coastial Proximity	Inundation
Elevation	1,00	2,00	3,00	0,33
Slope	0,50	1,00	2,00	0,25
Coastial Proximity	0,33	0,50	1,00	0,20
Inundation	3,00	4,00	5,00	1,00
	4,83	7,50	11,00	1,78

Normalized matrix

Elevation	0,2069	0,2667	0,2727	0,1869
Slope	0,1034	0,1333	0,1818	0,1402
Coastial Proximity	0,0690	0,0667	0,0909	0,1121
Inundation	0,6207	0,5333	0,4545	0,5607

5th iteration	0,2101	0,2101	0,2101	0,2101	21,01%
	0,1355	0,1355	0,1355	0,1355	13,55%
	0,0949	0,0949	0,0949	0,0949	9,49%
	0,5595	0,5595	0,5595	0,5595	**55,95%**

Figure 8. Pair-wise comparison and normalized matrix

Table 2. Nine-point comparison scale [21]

Intensity of importance	Definition
1	Equal importance
3	Weak importance of one over another
5	Essential or strong importance
7	Demonstrated importance
9	Absolute importance
2, 4, 6, 8	Intermediate value between the two adjacent judgments

From the above data, as shown in Figure 8, it indicates that CI was 0.0244, while CR was 2.7% and it also shows that normalized matrix was computed from the pair-wise comparison in five iterations. The pair-wise comparison explains that elevation is the most significant factor and followed by slope, coastline distance, and vegetation density.

3.2. Spatial Analysis for Vulnerability Mapping

The maps parameter was created after applying the weighted overlay using spatial analyst in GIS. The weighted overlay device was used to resolve multi criteria issues [10,23]. It is a procedure in order to apply a common scale of values to diverse input parameters which have different values in establishing an integrated analysis. A series of numbers from 1 (low vulnerability) up to 5 (high vulnerability) were applied to represent the vulnerability categories. The vulnerability categories as displayed in Table 1 was made by using the subtraction from the maximum and minimum values and then divided by the number of the categories. Afterwards, the tsunami's vulnerability in five categories could be mapped (Figure 9).

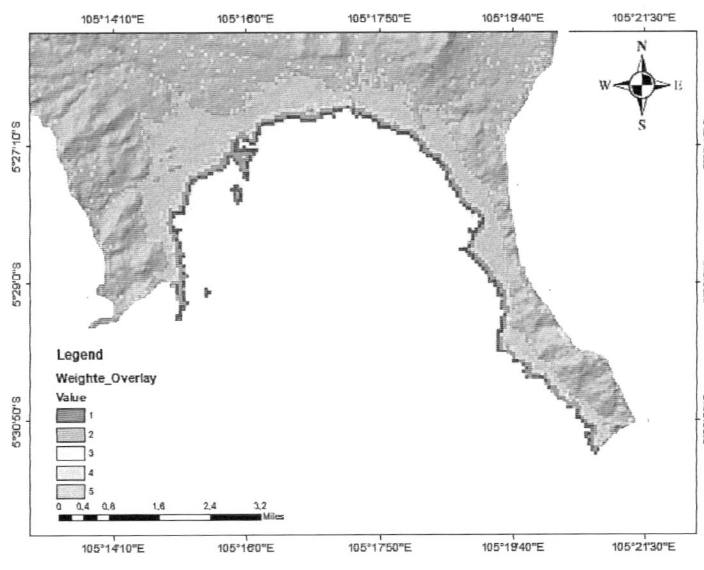

Figure 9. Tsunami vulnerability map

3.3. Network Analysis

Network analysis is carried out using a map of the Bandar Lampung city roads network and using ArcGIS 10.6. Closet Facility Analysis is used as a method for which facility is closer to a point. Like Route Analysis, the determination of facilities can be derived based on distance or travel time. The following are some of the processes and results of the network facility closet analysis in ArcGIS 10.6:

3.3.1. Evacuation Point selection

The analysis was done by considering the distance perpendicular to the coastline where the distance must exceed 516 meters, have a good height and slope, and have a fairly wide location if possible. The determination of these points was done with the help of Google Maps, Google Earth, Peta Tsunami vulnerability and road network maps. The location of evacuation points must also be able to serve all regions (Figure 10).

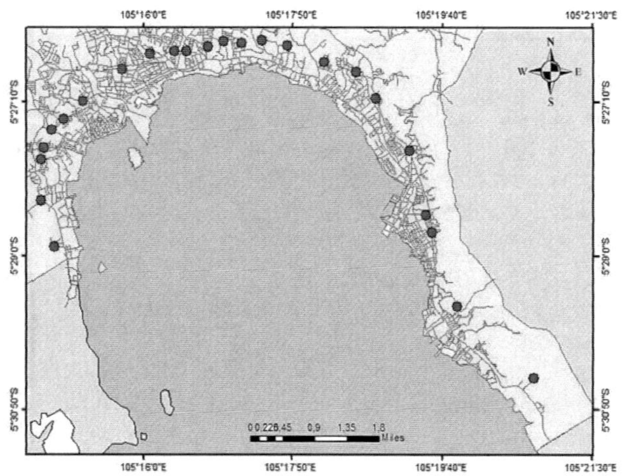

Figure 10. Evacuation point map

3.3.2. The Selection of Incident Points

Incident Points mean the point that is in the tsunami hazard zone and is directly proportional to the level of population density at that point. The position of this point is obtained from the interpretation of satellite images that have been taken previously, and it is a sample point for determining tsunami evacuation routes in Bandar Lampung City (Figure 11). From these points a tsunami evacuation route is assigned to the nearest evacuation point.

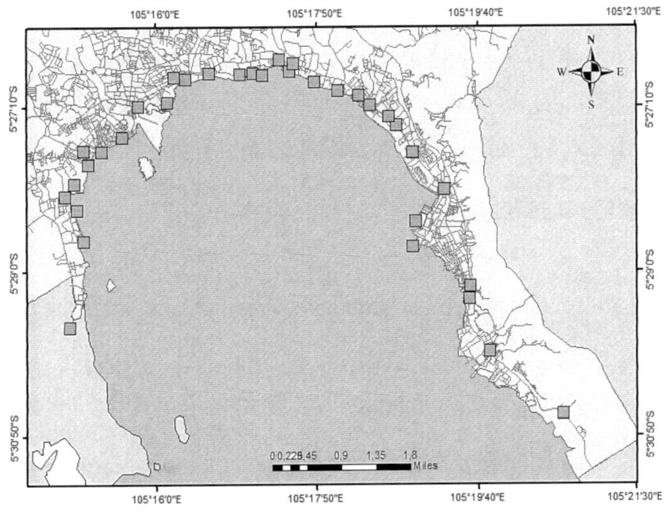

Figure 11. Incident point map

3.3.3. Making Evacuation Routes

In the network analysis, as much as 24 points were assigned to be used as evacuation points and 34 event points were used for the initial reference of movement. After determining these points, there were 34 evacuation routes and 24 evacuation points (Figure 12).

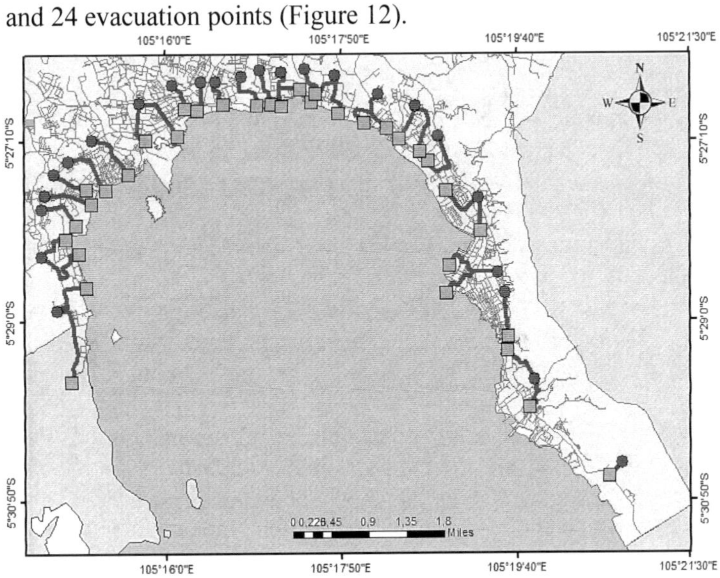

Figure 12. Evacuation Route Map

4.0 Conclusion

Based on the AHP analysis that has been done, the results obtained are as follows: the highest inundation level of influence in this study is scoring 2 for Elevation, scoring 3 for Slope and scoring 4 for Coastline. Spatial Analysis could be made into maps that can be used in determining evacuation points which are assisted with satellite images in the form of Google Earth and Google Maps. In this study as many as 24 evacuation points and 34 evacuation routes were obtained. Each of these routes is the closest route to the evacuation points which is the safest point of the tsunami with a run-up height of 5 meters. The evacuation route also has a height and slope that is suitable for the evacuation.

Acknowledgement
We would like to thank Ministry of Research, Technology, and Higher Education, Republic of Indonesia which funded the research by Applied Research Scheme Grant, Contract No.010/L6/AK/SP2H.1/PENELITIAN/2019.

References

[1] Arief M D 2011 Study of coastal line change using landsat data satellite in Kendal District *J.Remote Sensing* **8** 71-80
[2] Naryanto H S 2008 Analysis of potential earthquake and tsunami in West Coast Region Lampung in relation to mitigation and regional planning *J. Science and Technology of Indonesia* **10** 71-77
[3] Naryanto H S 2003 Mitigation of the south coast region of bandar lampung against the tsunami *J. Alami: J. Water, Land, Environment, and Disaster Mitigation* **8** 24-32
[4] Lusmeilia A 2018 Pengaruh Variasi Waktu Pemeraman dan Perendaman Sampe Tanah Terhadap Nilai Daya Dukungnya *Prosiding Semnas SINTA FT UNILA* **1** 219-14.
[5] Alhamidi, V H Pakpahan and J.E.S Simanjuntak 2018 Analysis of tsunami disaster resilience in Bandar Lampung bay coastal zone *IOP Conference Series: Earth and Environmental Science* **158**
[6] Jokowinarno D 2011 Mitigasi bencan tsunami di wilayah pesisir Lampung *Jurnal Rekayasa* **15** 1
[7] Dewi C, Armijon and Romi F 2014 Analisa Pembuatan Peta Zona Rawan Bencan Tsunami Pada Daerah Pesisir studi lokasi pesisir Kota Bandar Lampung *Proseding Seminar Bisnis dan Teknologi* 2407-6171
[8] Nurhasanah A, Yulfriwini and Susilowati 2016 ACE 3-039 Pola Pemetaan Jalur Evakuasi Tsunami Pesisir Kota Bandar Lampung Prosiding *Seminar ACE Padang* 445-52
[9] Carver S J 1991 Integrating Multi-Criteria Evaluation with Geographical Information Systems *Intl. J. Geo-Information System* **5** 3 321-39
[10] P. Jankowski 1995 Integrating Geographical Information Systems and Multiple Criteria Decision-Making Methods *Intl. J. of Geo-Information System* **9** 3 251-73
[11] Strunz G, Post J, Zosseder K, Wegscheider S, Muck M, Riedlinger T, Mehl H, Dech S, Birkmann J, Gebert N, Harjono H, Anwar H Z, Sumaryono, Khomarudin R M and Muhari A 2011 Tsunami Risk Assessment in Indonesia *Natural Hazards and Earth System Sciences* **11** 1 67-82
[12] Sinaga T P T, Adhi N, Lee Y W and Yongcheol S 2011 gis mapping of tsunami vulnerability; case study of the Jembrana Regency in Bali, Indonesia *KSCE J. Civil Eng* **15** 3 537- 43
[13] Papathoma M and Dominey-Howes D 2003 Tsunami Vulnerability Assessment and Its Implications for Coastal Hazard Analysis and Disaster Management Planning, Gulf of Corinth, Greece *J. Natural Hazards and Earth System Sciences* **3** 6 733-47
[14] Papathoma M, Dominey-Howes D, Zong Y and Smith D 2003 Assessing Tsunami Vulnerability, an Example from Herakleio, Crete *J. Natural Hazards and Earth System Sciences* **3** 5 377-89
[15] Soloviev S L and Go Ch N 1974 A catalogue of tsunamis on the western shore of the Pacific Ocean
[16] Bretschneider C L and Wybro P G 1976 Tsunami Inundation Prediction *Proceedings of the 15th ASCE Conference on Coastal Engineering* 1006-24.
[17] Hills J G and Mader C L 1997 Tsunami Produced by the Impacts of Small Asteroids *Annals of the New York Academy of Sciences* 381-94

[18] Iida K 1963 Magnitude, Energy and Generation Mechanisms of Tsunamis and a Catalogue of Earthquakes Associated with Tsunamis *Proc Tsunami Meeting at the 10th Pacific Science Congress* 7-18
[19] Van Zuidam R A 1983 Guide to Geomorphologic-Aerial Photographic Interpretation and Mapping **6**
[20] Saaty T L 1977 A Scaling Method for Priorities in Hierarchical Structures *J. Math. Psy.* **15** 3 234-81.
[21] Saaty T L 2003 Decision Making with the AHP; Why Is the Principle Eigenvector Necessary *Eu. J. Operational Research* **145** 1 85-91
[22] Saaty T L 1980 The Analytic Hierarchy Process, Planning, Priority Setting, Resource Allocation
[23] Carver S J 1991 Integrating multicriteria evaluation with geographical information systems *Intl. J. Geographical Information Systems*. **5** 3 321-39

IOP Conf. Series: Materials Science and Engineering **807** (2020) 012023 doi:10.1088/1757-899X/807/1/012023

Improvement of Cogging Torque Reduction by Combining the Magnet Edge Shaping and Dummy Slot in Stator Core of Fractional Slot Number in Permanent Magnet Machine

Tajuddin Nur [1]*and Sri Mawar[2]

[1]Department of Electrical Engineering, Graduate School of Engineering, Faculty of Engineering, Atma Jaya Catholic University of Indonesia
[2]Department of Electrical Engineering, Graduate School of Engineering, Faculty of Engineering- Hasanuddin University, Campus Gowa, South Sulawesi, Indonesia

Corresponding e-mail: tajuddinnur559@gmail.com

Abstract. A new technique in reducing the cogging torque by combining the magnet edge shaping and dummy slot in stator core of fractional slot number in permanent magnet machine was studied in this paper. The cogging torque is the most issue in permanent magnet application, since it effects to reduce the machine performance. A permanent magnet machine of fractional slot number with 24 slot /20 pole was presented in the paper. Firstly, the permanent magnet machine with conventional both of magnet structure and stator structure has been investigated as the basic reference of initial structure. Secondly, the magnet edge shaping with conventional stator core was investigated. Thirdly, in proposed structure, the combination of magnet edge shaping with dummy slot in stator was employed. The three different structures of the permanent magnet machine then analysed and compared. By implementation of magnet edge shaping combined with dummy slot in stator core has increased the interaction between magnet and stator core. This leads to increase the cogging torque frequency and reduce the peak value the cogging torque of the machine. Using the FEMM, the cogging torque value of the permanent magnet machine was analysed and computed. It was found that cogging torque reduction of the proposed structure as much as 99.47 % compared with the initial structure.

1. Introduction

Permanent magnet machines (PMMs) has been widely used in many applications such as a motor in pump, electric vehicles and others. In the context of renewable energy systems, the PMMs have been found as generator in wind turbine. For PMMs, the most important parameter in its application is in low speed condition which is the cogging torque (CT). The CT in PMM is complicated. The CT can be divided into two categories which are the stage of design and manufacturing. In the stage of design, the CT is investigated based on the interaction between the flux of magnet (material) in rotor and the slot opening in stator core. The CT is affected by material used in constructing the machine and have existed during the early stage of the machine development. According to researchers [1], it is known as natural CT. In the stage of manufacturing, the CT in permanent magnet do exist as it was caused by the imperfection of material handling when the PMM was fabricated, which is called additional CT. In this paper, authors only focus on the natural CT. As a matter of fact, the CT in PMM is hardly decrease towards zero value since it is influenced by many factors which are related to the PMM structures itself, except for the PMM with slot. On other hand, in renewable energy application such as in wind power system, usually requires a very low cogging torque peak value of the PMM. According to Handershot and Miller [2], a good design of machine is when the CT content of the machine lies between 1 - 2% of the rated torque. Up until now, to achieve a very low CT value in practical use is still a difficult task in electrical machine design. It is in fact that the CT in electrical machine can only be minimized but it cannot be totally removed from the machine, even in the industries a modern machine tools are being utilized for such issues. In the design stage, the CT of any PMM is naturally caused by the interaction of stator slot or slot opening width and the magnet edge of the machine. While in the stage of design, the CT also might be influenced by the imperfection in manufacturing stage. The purpose of this work is to offer and introduce a new technique to reduce the CT of permanent magnet machine by employing the combination of magnet edge shaping and dummy slot in stator core. The shaping of magnet edge and dummy slotting are two different CT reduction technique. Both CTs reduction techniques are combined, in order to achieve the CT reduction of the proposed PMM. In this paper, the CT reduction of the PMMs studied was analysed and focussed only in the design stage instead of manufacture stage. For purpose of study, the analysis the PMM structure, some assumptions have been made:

- The magnetic flux orientation was assumed to be radial.
- End effect of the permanent magnet generator is neglected.
- The magnet poles have the same characteristics.
- Saturation in both stator and rotor core of the machine is neglected.
- No current in stator winding or open circuit.
- The width and length of the stator teeth is not optimized.
- The magnet structure of the magnet has been optimized using response surface method (RSM).

As has been mentioned in the beginning of this paper, a fractional slot number of 24 slots / 20 poles of PMM generic model have been adopted and studied. This type of structures was popular based on reference [2]. The permanent magnets used for the PMM structure proposed have been adopted to be bread loaf type. According Petkovska et al. [3], many commercially available machines drivers own a typical cogging torque ranging from 5 - 15% from the rated torque of the machine. And, due to the imperfection of manufacturing in mass production, the CT of machine may increase more than 25 % even it has been introduced a properly design permanent magnets.

Since the last two decades, many CT techniques have been proposed by researchers. The results of the research on CT reduction methods have been reported and documented. Some of the CT reduction techniques have been proposed such as dummy slot in stator teeth [4-6], slotting in magnet edge [7-10], and magnet shaping [11-14]. However, the most effective way to minimise the CT in permanent magnet machine is by combining the magnet shaping and dummy slot in stator core. The advantages of this technique are the fact that the distance between the points at magnet edge to the slot opening might be adjusted to a proper value. This technique aims to improve the proper contact between magnet edge and slot opening in stator core. Moreover, the CT frequency would increase and hence decrease the peak of CT. In this paper, the PMM model of a-24 slots / 20 poles structure have been selected and studied [2]. In the beginning, the initial magnet structure of the PMM was analysed. It was expected the CT of the PMM can be reduced as low as possible and could be realized in real structure or practical. The CT of the PMM have been analysed with considering the possibilities to the real machine structure such as the bolts in the rotor and stator core.

In this paper, some axial channels in rotor core of the PMMs models also have been introduced. The effect of axial channels in rotor of the PMM do not have a significance effect on the value of CT of the PPMs studied. The effect of axial channels in rotor core on the PMM performance may rarely discussed in research. However, the presence of some axial channels in the rotor core may be influence the magnetic flux distribution in rotor core of the PMM. Thus, to investigation further the effect of axial channels in the rotor core on the CT peak value, one should pay attention on the type and size of the axial channels in the rotor core of the PMM. In this paper, 5 axial channels have been employed. In this study, the effect of combining the magnet edge shaping and dummy slot in stator core on the CT reduction in the proposed PMM was presented in this paper. In order to increase the accuracy of the CT measurement, some dummy line in the air gap of the PMMs studied have been introduced. Using finite element based on the FEMM, authors have found that by combining the magnet edge shaping and dummy slot in stator core, the CT of the proposed PMM could be reduced dramatically to around 99.47 % compared with the initial structure of the PMM.

2. Permanent Magnet Machine Structure

2.1. Initial Structure
The PMM structure proposed in this paper refers to the cogging torque reduction in the stage of design. In this paper, the cogging torque of the PMM of fractional slot number with 24 slot / 20 pole structure was selected to minimise the cogging torque. The cogging torque reduction developed in this paper was by combining two of existing techniques which is the shaping in magnet edge and dummy slotting in stator core. Combining the magnet edge shaping with dummy slotting in stator core is one of new technique in achieving the cogging torque reduction in fractional slot number type of machine. The effectiveness of proposed structure was validated using the finite element analysis. In this study, a 24 slot / 20 pole with conventional magnet structure as the initial structure has been considered. In general, the cogging technique reduction can be summarized as in Figure 1. In the stage of design, it shows that the cogging torque reduction in PMM can be achieved by optimizing the structure of rotor, stator or combined of both of stator and rotor structure. The initial structure of the PMM study in this paper as shown in Figure 1.

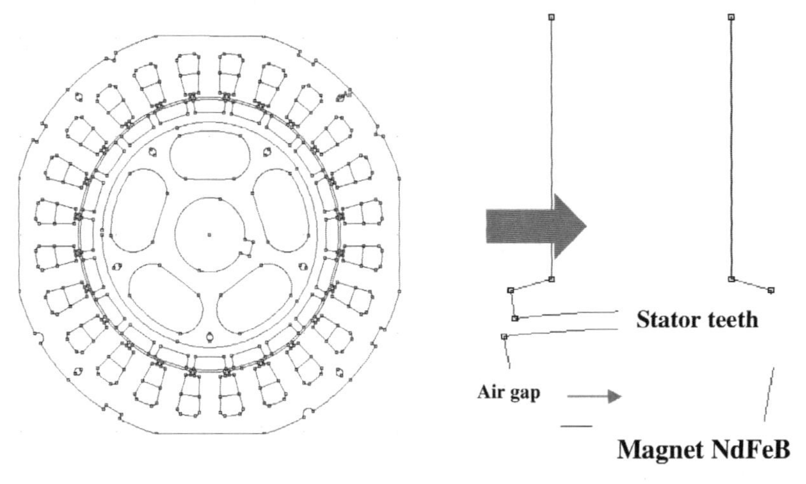

Figure 1. Model of PMSG studied a fractional slot of 8 slot /18 PMSG

3. Cogging Torque Reduction Classification

As have been mentioned in previous of this paper, many CT reduction techniques have been developed and proposed among researchers. The CT reduction technique can be summarized in Figure 2. It can be observed that the CT technique could be done by one or combination of two cogging torque techniques. Every cogging torque technique has advantages and disadvantages hence it could be stated that there is no exact best technique for the CT. In our proposed structure, a combination of dummy slot in stator core and shaping in magnet edge was developed and investigated.

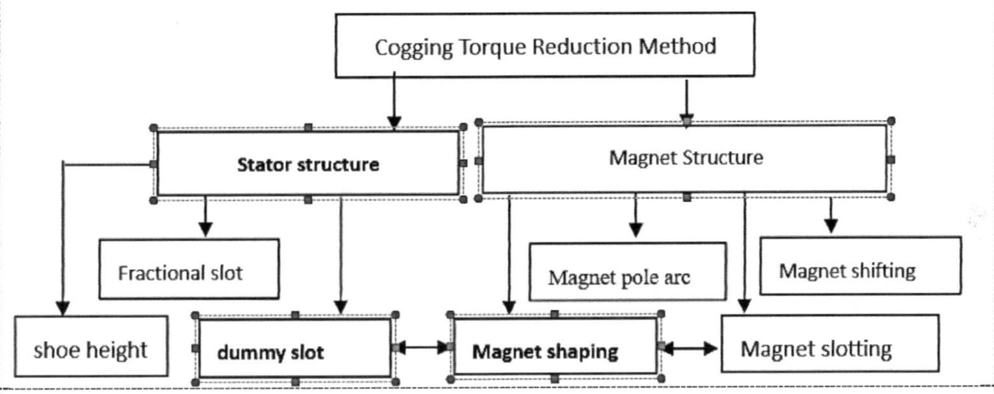

Figure 2. Cogging torque reduction technique classification [13]

3.1. Proposed Structure

As could be observed from Figure.2, the CT in any PMM can be reduced by modification the stator or magnet rotor. In this paper, a combination of dummy slot in stator core and shaping in magnet edge of rotor has been offered as one CT reduction solution in PMM. The advantage of this method may be lied in the fact that the source of CT in the edge of the magnet can be cut directly. It leads to minimize directly the source of the CT in magnet structure [15].

The stator teeth of proposed structure were much more similar with the conventional CT, the only main difference is in the proposed structure of PMM that has a dummy slot in every stator tooth. The slot opening in the stator core was set to be 2mm, which may be the most common in conventional stator structure. With the value of slot opening width of 2mm, it might be quite large and easy to insert the conductor into the slot of the PMGs when they are being manufactured. It is also easy for rewinding the machine. The other advantages of using a larger slot opening width, is the fact that by increasing the slot opening, the magnetic loading effect of the machine can be minimized [16]. The PMM structure investigated in this paper was presented in Figure 2

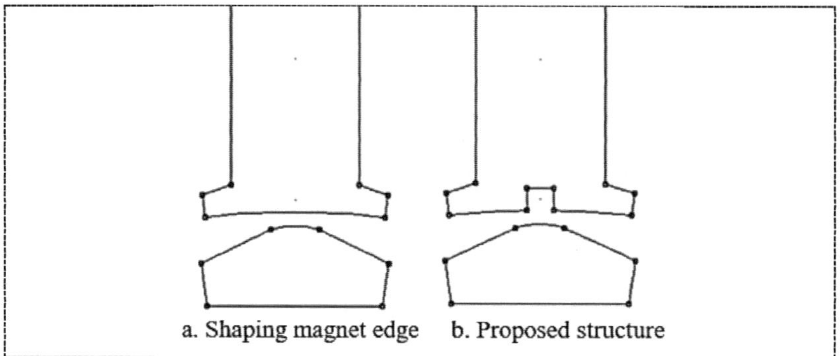

Figure 3. Cogging torque reduction technique classification [13]

From Figure 3 one can observe and compare that proposed structure of PMM with magnet edge shaping. The presence of magnet shaping increases the air gap area of the machine. In Figure 4, the stator core could be observed that the magnet shaping was paired with dummy slot.

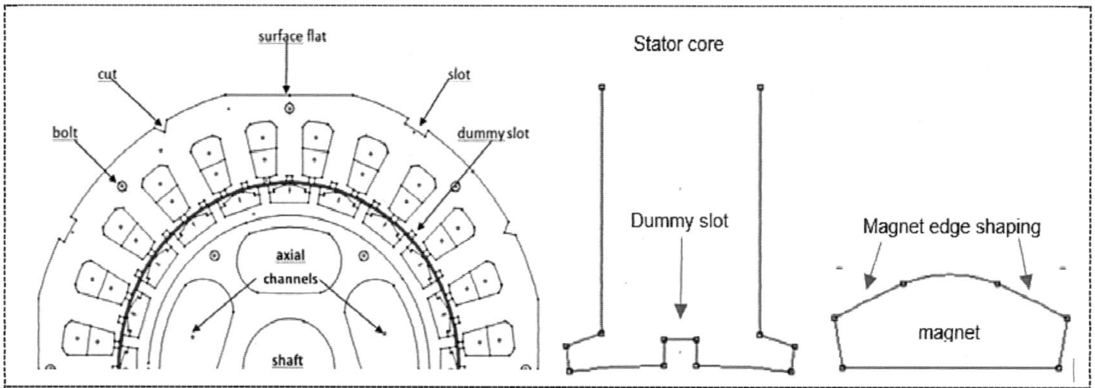

Figure 4. Combining magnet edge slotting and dummy slot in stator core

As has been stated in the beginning of this paper, one of effective technique to minimize the CT in PMM is by combining the shaping in magnet edge with dummy slot. The structure of PMM which employs the dummy slot in magnet edge and dummy slot in stator core. This type of CT technique has been studied and proposed by the previous authors as in [12]. However, the shaping magnet edge may be slightly reduced on the CT reduction of the machine. This research observes the CT reduction of the PMM achieved by employing the shaping in magnet edge paired with dummy slot in stator core. This technique effectively reduces the residual magnet flux in magnet edge which can minimise the CT of the PMM, as has been reported in [7],[10]. By combining the magnet edge shaping and dummy slotting in stator core.

4. Material for Structure Proposed Permanent Magnet Machine

In this paper, as in the initial structure and other structure, material used for the PMM proposed is M-19 steel, both for stator and rotor core. The advantage of using this material is the fact this kind of very material are common be used and very popular in PMM design nowadays. Also, the material could support the high magnet flux density in both stator and rotor core of the PMM. The material M-18, is expected to operate at the flux density not more than 1.5 Tesla, unless the PMM core losses could be increased rapidly. For the magnet in rotor core, the permanent magnet of Nd-Fe-B has been used in this study.

5. Analysis of Magnet and Stator Structure

The CT of the PMM is reduced by minimization of residual magnet flux in the magnet edge or in the magnet surface studied in references [7-10]. In order to achieve the CT reduction of the permanent magnet machine, the combination of shaping in magnet edge and dummy slot in stator teeth were introduced. As can be seen in the Figure 1, by employing the dummy slot in the magnet and dummy slot in stator could reduce the CT of the PMM more effectively as compared to the initial magnet structure and magnet shaping without no dummy slot. In the study, the pole arc of the shaping magnet has been optimized using design of experiment. However, the procedure of design of experiment was not studied detail in this paper. The machine structures then analysed using finite element method magnetics (FEMM).

6. Cogging Torque Calculation

In this paper, the magnets in PMM study are assumed to be unity and there is no current in the stator slot. The magnetic flux density in air gap was generated by the permanent magnet in rotor core. The magnet flux flows into the air gap of the machine radially. The CT reduction is calculated based on equation (1),[4],[9-10].

$$T_c = -\frac{1}{2}\emptyset^2{}_{\varepsilon}\frac{dR_{\varepsilon}}{d\theta} \qquad (1)$$

where $\emptyset^2{}_{\varepsilon}$ is the magnet flux in the air-gap, R_{ε} is the passed air-gap reluctance, and θ is the rotor position of the machine. The calculation procedure based on the past researchers [7],[9-10]. By using the combination of shaping in magnet edge with dummy slot in stator core could eventually reduce both magnetic flux in air gap and air gap reluctance. Based on equation 1, this technique would yield two advantages over the other techniques which are reduce the total flux flowing into air gap and reduce the air gap reluctance as well. The air gap reluctance become reduced since the magnet occupy the air gap the machine become smaller, compared with the initial structure. However, to achieve the cogging torque reduction, the magnet shaping degree must be assigned with a certain degree. In this paper, a 4^0 of pole arc of the magnet edge shaping has been assigned in order to achieve the CT reduction of the propose PMM [17]. The CT reduction of the three PMMs were presented and compared in this paper. The value of the CT reduction of the three different structures was presented and compared in Figure 5.

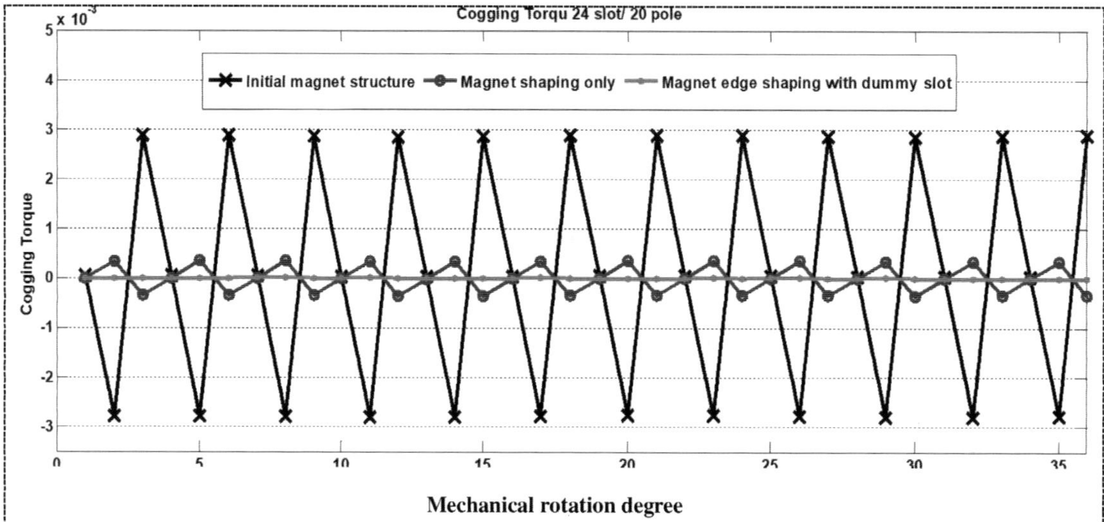

Figure 5. Combining magnet edge slotting and dummy slot in stator core

7. Result and Discussion

Figure 5 represents the comparison of CT reduction of the PMMs studied and investigated in this paper. As can be seen from the Figure 5, the CT value of the initial structure has the highest peak of CT among them of the three different PMM structures. Using the finite element analysis, all the PMMs studied were analysed and the CT reduction were compared. Initially, it has been found that the CT peak of the initial structure of PMM was approximately 0.0028554 N.m. In order to study the effect of the proposed CT reduction technique, the magnet edge shaping was employed in the PMM. The CT peak of the shaping in magnet edge was decreased to 0.0003471 N.m (blue line). The CT reduction is 84.44% compared with the initial structure. In the proposed PMM structure, the combining the magnet edge shaping and dummy slotting in stator core was employed. It was found that the CT peak of the proposed dropped to 0.0000165 N.m (red line). It can be concluded the CT peak reduced significantly, compared with initial structure of the PMM. This can be concluded the CT reduction of the proposed PMM was 99.47% compared with the initial structure.

8. Conclusion

Improvement of CT reduction by combining the shaping in magnet edge and dummy slot in stator core of the proposed PMM has been presented in this paper. The combination of magnet edge shaping with dummy slot in stator core has been developed to reduce the CT of the PMM. The comparison of CT value of each PMM structures was compared as shown in Figure 4. The CT of the proposed PMM reduced significantly, when the shaping in magnet edge was combined with dummy slot in stator core. Using finite element of FEMM 4.2 tools, the CT reduction of the PMM were calculated and compared. It was found that the maximum CT peak value of proposed structure as much as 0.0000156 N.m, while the initial CT peak value of the permanent magnet machine structure is around 0.0028554 N.m. It can be concluded that the CT reduction of the proposed declined as much as 99.47% compared with the initial magnet structure. The combining of magnet edge shaping with dummy slot in stator teeth was the effective way to improve the CT reduction in fractional slot number of 24 slot / 20 pole.

References

[1] Cernogoj A, Gasparin L and Fiser R 2010 Natural and additional cogging torque component of PM synchronous motors-evaluation and reduction *Automatika*.**5**.2.157-65
[2] Handershot J R and Miller T J 1994 Design of brushless permanent magnet motors *Magna Physic Div*
[3] Petkovska L, Cvetkovski G and Lefley P 2015 Analysis of the stator topology impact on cogging torque for surface permanent magnet motor *COMPEL: Intl J. Math. in Electrical and Electronic Engineering* **34**.2457-74
[4] Bianchi N and Bolognani S 2002 Design techniques for reducing the cogging torque in surface mounted PM motors *IEEE Trans. on Ind. App.* **38** 5 1259-65
[5] Kim T and Chang J 2012 Influence of cogging torque reduction methods on torque ripple in a surface-mounted permanent magnet synchronous motor *J.Magnetics* **17** 2 109-14
[6] Ishikawa T and Slemon G R 1993 A method of reducing ripple in permanent magnet motor without skewing *IEEE Trans. on Magnetics* **29** 2 2028-31
[7] Ling J M and Nur T 2016 Influence of edge slotting of magnet pole with fix slot opening width on the cogging torque in inset-permanent magnet synchronous machine *Advanced Mechanical Engineering*.**8** 8
[8] Chen N, Ho S L, Fu W N 2010 Optimization of permanent magnet surface shape of electric motors for minimization of cogging torque using finite element method *IEEE Trans. on Magnetic* **46** 6 2478-81
[9] Scuiller F 2014 Magnet shape optimization to reduce pulsating torque for a five-phase permanent-magnet low-speed machine *IEEE Trans. on Magnetics* **50** 41–9
[10] Chabchoub M, Salah B and Krebas G 2012 Permanent magnet synchronous machine cogging torque reduction: comparison between different shape of magnet *Intl. Conf. Renewable Energies and Vehicular Technology* **26** 28 206-11
[11] Chung D W and You Y M 2015 Cogging torque reduction in permanent magnet brushless generator for small wind turbines *J. Magnetics* **29** 1 176-85
[12] Zhou Y, Li H, Meng G and Cao Q 2015 Analytical calculation of magnetic field and cogging torque in surface-mounted permanent magnet machines accounting for any eccentric rotor shape, *IEEE Trans. N Industry Electronics* **62** 6 3438-47
[13] Zhou Y, Li H, Wang W, Cao Q and Zhou S 2015 Improved method for calculating magnetic field of surface mounted permanent magnet machines accounting for slot and eccentric magnet pole *J. Electr. Eng Technol* **1** 10 30-40
[14] Ruangsinchaiwanich S, Zhu Z Q and Howe D 2005 Influence of magnet shape on cogging torque and back-emf waveform in permanent magnet machines *Intl Conf. Electrical Machines and Systems* **1** 284-9
[15] Doss M A N, Brindha R, Mohanraj K, Dash S S and Kavya K M A 2018 novel method for cogging torque reduction in permanent magnet brushless dc motor using t-shaped bifurcation in stator teeth *Progress in Electromagnetics Research* **66** 99–107
[16] Hwang M H, Lee H S and Cha H R 2018 Analysis of torque ripple and cogging torque reduction in electric vehicle traction platform applying rotor notched design *J. Energies* **11** 3053
[17] Sarac V 2019 Performance optimization of permanent magnet synchronous motor by cogging torque reduction *J. Elect. Eng.***70** 3 218–26

IOP Conf. Series: Materials Science and Engineering **807** (2020) 012024 doi:10.1088/1757-899X/807/1/012024

Study the Influence of Edge Points of Magnet Surface and Stator Core on the Cogging Torque Reduction in Permanent Magnet Machine

Sri Mawar[1] and Tajuddin Nur[2]

[1]Department of Electrical Engineering, Graduate School of Engineering, Faculty of Engineering, Hasanuddin University Campus Gowa, South Sulawesi, Indonesia
[2]Department of Electrical Engineering, Graduate School of Engineering, Faculty of Engineering- Atma Jaya Catholic University of Indonesia , Jl. Jend. Sudirman No.51 Jakarta 12930 Indonesia

Corresponding e-mail: nasirtan415@gmail.com

Abstract. Cogging torque in permanent magnet machine application effects to some undesirable vibration and noises. To minimize the cogging torque in permanent magnet machine's the most important issue recently. Influence the number of edge points in the magnet of one and two steps of magnet edge slotting was studied in this paper'. For purpose of study, the structure of permanent magnet machine with 18 slot / 8 pole has been chosen, and the magnetic flux of all magnets was radial. The finite element of 2D based on FEMM 4.2 software was implemented to analysis the magnet flux distribution in air gap. Simulation results showed that by improvement of edge point at magnet edge of the two steps of slotting in magnet edge could reduce the cogging torque of permanent magnet significantly when compared with the initial structure. Employing the numerical simulation, both of the machine structures were analysed. It has been found that, the cogging torque reduction of edge points with 8 points has the highest reduction around 98.84 % compared with initial structure of the permanent magnet machine.

1. Introduction

Permanent magnet machines (PMM) have been widely used in many applications because of its compact feature, simple and low size mechanical construction, easy maintenance, and good reliability, high torque density and efficiency [1-3]. Nowadays the PMMs have been applied in many systems in our daily life such as in robotics, pumps, electric bike, wind power, water power and many more. However, the most important issue in PMM application is the cogging torque. The cogging torque is produced by the magnetic interaction between permanent magnet of the rotor and the stator teeth or stator core. It refers that by modify the magnet structure or optimise the stator core can minimize the cogging torque. To minimise the cogging torque in permanent magnet machine is one of difficult task in electric machine design and manufacture. Since the last few years, many cogging torque reduction techniques have been developed and proposed in worldwide. The cogging torque reduction techniques have been developed and proposed in world scientific, research and education. All of the cogging techniques mentioned above have advantages or may be drawback. The cogging torque techniques also can be employed with combination of two or more of the cogging torque techniques. In paper, the effect of edge points of magnet edge on the cogging torque reduction in permanent magnet machine was proposed. In the beginning, to create the edge points of the magnet edge, the edge points were created through the aims of slotting in magnet structure. The slotting in magnet structure can be employed in any part of the magnet structure or in magnet surface, and it has been developed and proposed in past as in [1],[2],[4]. However, the most effective way to slotting the magnet structure is by placing the slotting is in the magnet edge [3],[5]. In this study, the slotting was focussed only in the magnet edge. The advantages of slotting in magnet edge to create the edge points was laid in the fact that magnet part cutting is smaller compared with slotting in magnet surface. In general, reducing magnet volume in permanent magnet machine by cutting through the magnet surface can reduce the cogging torque [7]. However, this technique of cogging torque reduction may be not sufficient for special cogging torque reduction value requirement. Thus, using slotting in any part in magnet structure refers to reduce the magnet volume and effects to the cogging torque reduction. It is more efficient compared with reducing the magnet volume through cutting in the all magnet surface. But, even though by reducing more part of the magnet structure can effect to reduce the cogging torque, but the effect of cogging torque reduction may be more efficient by placing the slotting in magnet edge, instead in the centre of magnet. The reason is that, the magnet edge produces both normal and tangential flux density. When the slotting was employed in the edge of the magnet, the tangential flux density usually drops significantly compared with the magnet structure without any slotting, or the slotting magnet centre. As it has been understood, that tangential flux density is one of parameter which gives contribute to increase the cogging torque. Reducing the tangential flux density is more efficient by reducing the magnet edge. Effect of edge points of magnet structure on the cogging torque in permanent magnet machine was developed and proposed in our study. In this paper, the edge points in magnet surface were performed through the combining shaping and slotting in surface or magnet edge. A-two steps of slotting in magnet edge have been used and it was found that the number of edge points of the magnet increase ten times, compared with initial structure, while for the one step of slotting in magnet edge, the edge points were six times compared with the initial magnet structure. The presence of edge points or the number of points in magnet edge, leads to increase the number of contact points between magnet in rotor core and slot opening width in stator core. In addition, the interaction between magnet in rotor and stator more frequent and increase cogging torque frequency and reduce the peak of cogging torque. The presence of edge points also influences the magnetic flux distribution magnet edge, leads the total magnet flux flowing into the air gap become declined. The technique proposed in this study have been validated using numeric analysis based on finite element analysis. For comparison, the cogging torque reduction of different number of magnet edge points with 18 slot / 8 pole of permanent magnet machine have been studied and presented.

2. Motivation of the Research

Nowadays, the users of electric machine have been increased rapidly as result of modern life. Most of the electric machine user require a production of a favourable ripple-free torque, low cogging torque peak are necessary. However, at the design stage and before the prototype of any permanent magnet machines are made, it is often difficult to construct any machine which produces satisfactory cogging torque reduction. In addition, there is not any specific criteria are found to determine the magnet rotor structure for such cogging torque reduction. Hence, in this study, a fractional slot number of 18 slot / 8 pole with certain of stator slot opening of 2 mm and magnet structure with more edge points was design to reduce the cogging torque in permanent magnet machine. In this study, a proposed structure of permanent magnet machine design was investigated to developed technique to minimize the cogging torque in permanent magnet machine. In order to increase the edge points of the magnet structure was performed with the aim of slotting in magnet edge. In this paper, authors focus to find a feasible number of edge points in magnet structure of fractional slot number of 18 slot / 8 pole to reduce the cogging torque. The cogging torque values of the permanent magnet machines with fractional were compared and presented.

3. Theories of the cogging torque in PMM

The cogging torque is caused by the interaction between stator core or slot opening width and permanent magnet in rotor core. Since the permanent magnet machine is constructed from various material such steel, iron, magnet and other, the cogging torque also is influenced by the material of the machine. It refers that the cogging torque in permanent magnet machine depend on the structure and material of the machine. In fact, the part of the magnet which contributes the most cogging torque in permanent magnet machine may be in magnet edge. For normal stator structure, where there is no any dummy slot, dummy teeth, shifting magnet or other, the cogging torque can be declined by reducing the magnet edge. The cogging torque reduction technique by reducing the magnet edge usually called magnet shaping. Another technique to reduce the amplitude of the cogging torque value is by reducing the volume of magnet [7]. By increasing the distance between magnet surface and slot opening width in stator core is another way to reduce the cogging torque [4]. The cogging torque in permanent magnet machine can be calculated as shown in equation (1). In this study, the magnet pole structure of the machines are assumed to be radial magnetization [2],[3],[7].

$$T_c = -\frac{1}{2}\phi_g^2 \frac{dR}{d\theta} \tag{1}$$

where ϕ_g^2 is the magnet flux in the air-gap, $\boldsymbol{R_g}$ is the passed air-gap reluctance, and $\boldsymbol{\theta}$ is the rotor position of the machine. The cogging torque also can be formulated as a Fourier series shown in equation where m is the least common multiple of the number of stator slot (N_s) and the number of pole (N_p), k is an integer and T_{mk} is a Fourier coefficient as shown in equation (2).

$$T_c = \sum_{k=1}^{\propto} T_{mk} \sin(\theta\, mk) \tag{2}$$

For PMM with integral slot number, each pole of the machine has a whole number multiple stator teeth, so that the cogging effects of each magnet are in phase and added, leads the cogging torque tend to be higher compared with the fractional slot number. In contrast, for the permanent magnet machine with fractional slot number as proposed in this study, only some of the poles number multiple stator teeth, leads only some of the magnets in rotor core are in phase and added. This leads the permanent magnet machine with fractional slot number tends to have a lower cogging torque compared with the integral slot number. Another way to calculate the cogging torque can be done as shown in equation (3).

$$T_c = \frac{L_{stk}}{\mu_0} \int_0^{2\pi} r\, B_n B_t d\theta \tag{3}$$

where Lstk = stack length of the machine, Bn = normal flux density, Bt = tangential flux density, θ = mechanical degree of rotor rotation, r = dummy radius in air gap where the cogging torque values are measured. From Equation 3, refers that cogging torque is a function of the normal magnet flux density and tangential flux density. In order to increase the accurate of cogging torque computation in this paper, 4 dummy lines has been employed in the air gap of the machine [10]. The magnetic flux density of the three different of magnet structures was presented and compared.

4. The Proposed Magnet Structure for Permanent Magnet Machine

To study the effect of the edge point in magnet surface on the cogging torque reduction in permanent magnet machine of fractional slot number with 18 slot / 8 pole was proposed. In our study, the slot opening width of all machines studied were 2 mm. In addition, some axial channels in rotor of the machines study also have been introduced. However, the effect of axial channels on the cogging torque reduction was not studied detail in this paper. In the beginning, in order to achieve the number of edge points in magnet surface, the slotting in magnet edge has been done firstly. After assigning the slotting in each of magnet edge, the position of the edge can be observed in Figure 1a, 1b and 1c respectively. As can be observed, the presence of one step of slotting in magnet edge, automatically creates 6 six edges in magnet and does a – 80 in magnet for the two steps of slotting (proposed structure). The presence of edge points also effects to distance between magnet surface and slot opening width or stator core. The magnetic flux distribution also become changed in the magnet edge. The magnet cross section or magnet volume become declined. It can be predicted that total flux in to flow into air gap through the edge of magnet may be reduced significantly. At the same time, the air gap cross section of the machine become increase, since the volume of the magnets become decreased. This leads the reluctance air gap become higher and the air gap reluctance become smaller compared with the initial structure (see Fig.1a). According to the Equation 1, this effect to decreasing of the cogging torque of the machine. The cogging torque reduction then was computed using finite element analysis based on FEMM 4.2. The cogging torque reduction for three different of magnet structures were presented and compared in this paper.

In Figure 1, one can observe the stator core for all PMMs are same, the only different is the slotting in the magnet structures as shown in Figure 1a and 1b respectively. The magnet for initial structure has no slot (see Figure 1a) and only has 2 edge points. In Figure 1b and 1c, the presence of magnet edge slotting generates 6 edge points. For two step of slotting, the edge points are 8 edge points. In addition, the points contact also effects to variation of the distance between magnet in rotor and stator core. This leads to increase the interaction between the magnet edges and the stator slot. As a result, the cogging torque frequency of the machine more frequent and the peak value of cogging torque can be declined.

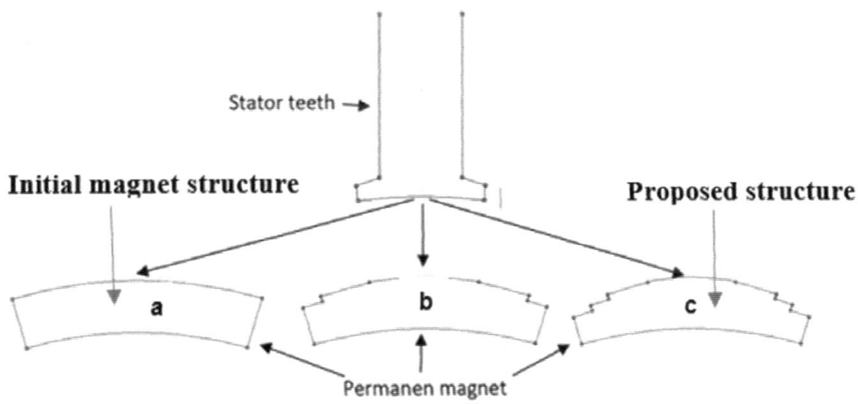

Figure 1. Three different magnet structures

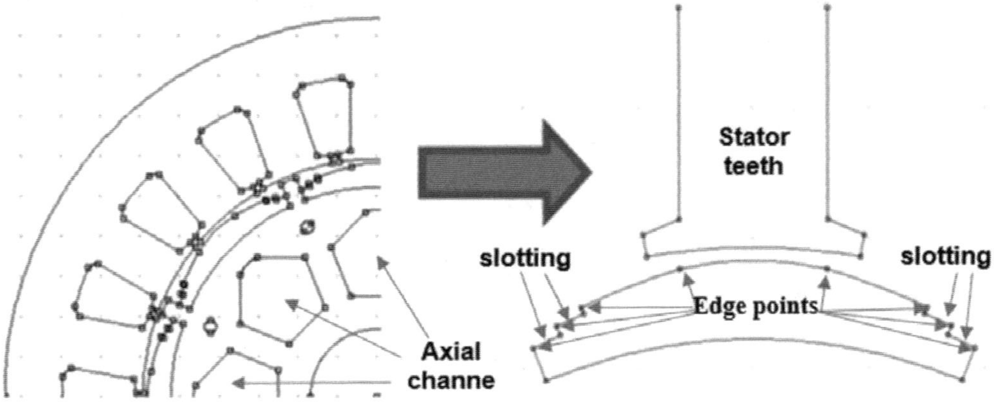

a. Half of pmm structure **b. detail of magnet structure**

Figure 2. Structure of Permanent Magnet Machine Proposed

From Figure 2, one can observe that the edge points were exist as the slotting was employed. The number of slotting effect to number of edge points. For a-two step of slotting in magnet edge, resulting some of 8 edge points in magnet structure. The structures of PMM with 18 slots/8 pole with three different magnet structures were studied and compared. As a comparison, the initial magnet structure is popular magnet structure of permanent magnet without any slotting in magnet. It also shows that the magnet has only has 2 edge point which located in magnet edge (see Figure 1a). This conventional type of magnet structure has the highest peak of cogging torque compared with the 6 edge points and 8 edge points.

5. Effect of edge point on the magnet volume

The presence of more edge points in magnet surface effects to increase the contact point between magnet rotor and stator slot. It can be expected, that the more the number of edge points in magnet the more frequent the interaction between magnet rotor and stator core, leads to increase the frequency of cogging torque. In other hand, it also effects to reducing the magnet volume of the machine. But, from cost effective of magnet material view, it can be stated that it is one of advantage. The drawback of slotting the magnet edge may be, the magnet volume become reduced and effect to declining the power output of the machine.

6. Computing the Magnetic Flux density and Cogging Torque

6.1 Analysis the magnetic flux in air gap

In the beginning, authors computed and presented the normal flux density and tangential flux density in air gap of the initial and proposed structure, as shown in Figure 3a, 3b respectively.

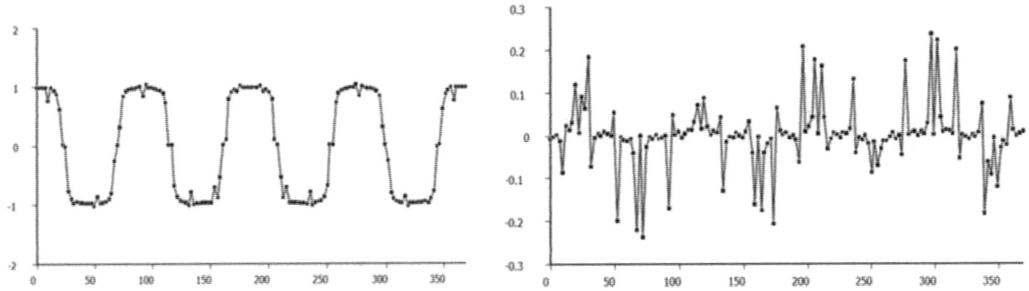

a. Normal Flux Density b. Tangential Flux Density

Figure 3. Magnetic Flux Density of Initial Magnet Structure

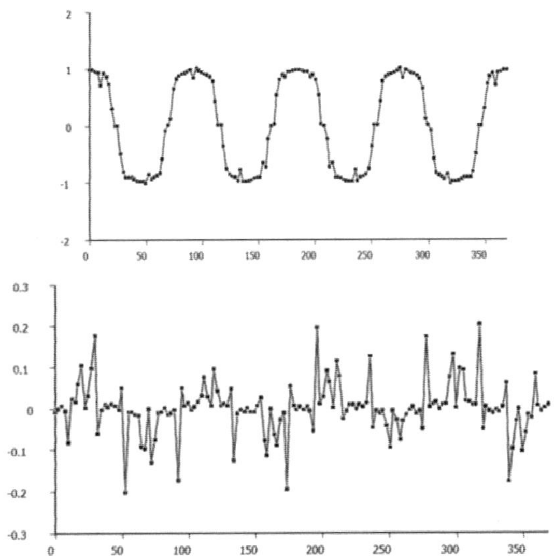

Figure 4. Magnetic Flux Density of Magnet Structure with 6 edge points

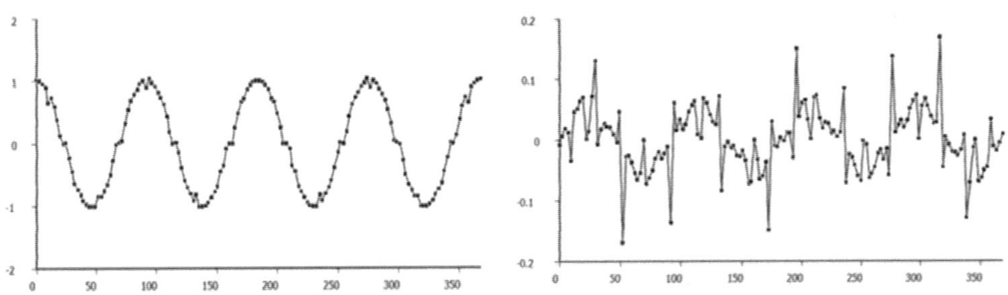

a. Normal Flux Density b. Tangential Flux Density

Figure 5. Magnetic Flux Density of Proposed Structure (8 edge points)

From Figure 3 and Figure 4, shows the magnetic flux density in air gap of the machine. Figure 3a and 3b, represents the normal and tangential magnetic flux in air gap of initial magnet structure. It can be observed that normal magnetic flux density distribution is square, while the tangential flux magnet density in Figure 3b is dense. In Figure 4a, the normal magnetic flux density peak is remaining stable is around 1.0 Tesla and the same as the initial magnet structure does. The only different the shape of the normal flux density wave is more sine, compared with the initial magnet structure. The tangential flux density in Figure 4b tends to decrease compared with the tangential of the tangential flux density of initial magnet structure. The peak tangential flux density also declined to as much as 0.18 Tesla.

6.2. Computation the Cogging Torque using finite element

In this study, the FEMM 4.2 was used to compute the cogging torque values of the permanent magnet machines studied. The procedure of cogging torque computation was based on past researchers [2],[3],[5],[7],[8]. The comparison of cogging torque reduction for three different magnet structures of 18 slot / 8 pole was presented and compared in Figure 6.

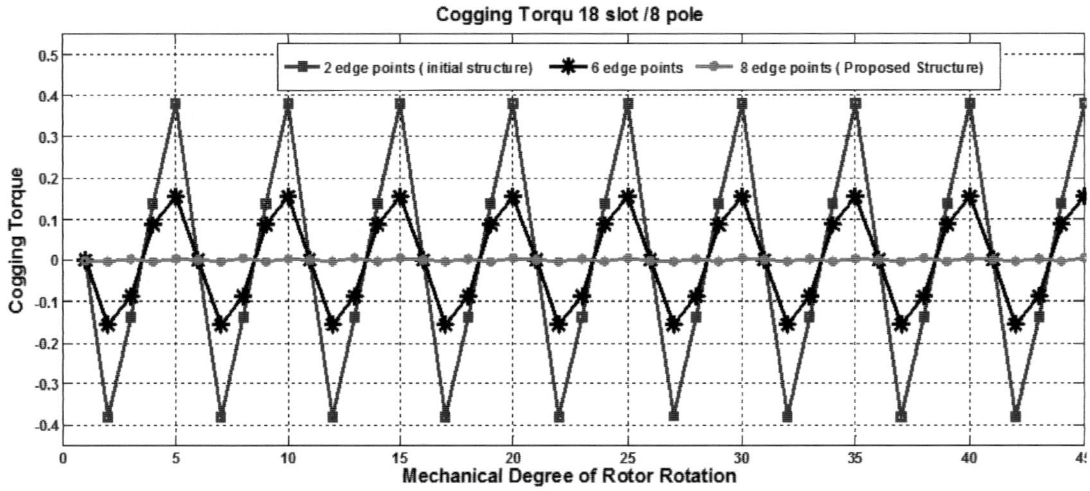

Figure 6. Comparison of Cogging Torque for different magnet structure of 18 slot / 8 pole

7. Discussion

Figure 6 represents the comparison of different magnet structure based on the edge points number of 18 slot / 8 pole. From the Figure 5, one can observe that the cogging torque for initial structure which the number of edge pints 2 has the highest of cogging torque peak value. It has been analysed and it been found as much as 0.3793488 Tesla (as shown in blue line). After performing the slotting and improving the number of edge points in magnet edge, the cogging torque tends to be reduced (as shown in black line). This is caused by the interaction between magnet edge and stator core become more frequent leads to increase the cogging torque frequency. The presence of more slotting in magnet edge, automatically increases the number of edge point of the magnet structure, compared with initial magnet structure. As a result, the peak value of the cogging torque become decreased sharply. The cogging torque of the proposed magnet structure could be reduced to 0.0044001 Tesla. The cogging torque reduction of permanent magnet has been found in this paper was significant. Also, the cogging torque reduction of the proposed machine is a little bit higher compared with the previous research [1]-[5]. It means the cogging torque content of the proposed permanent magnet machine is lower, and it can be accepted to implement for any specific requirement.

8. Conclusion

Influences of edge points in magnet surface and stator core structure in permanent magnet machines of fractional slot have been investigated in this paper The relation between the edge point number in magnet edge or magnet surface on the cogging torque for three different number of edge points were presented and compared. The edge points of permanent magnet machine with 8 edge point has the highest cogging torque reduction, compared with initial structure of 2 edge points and 6 edge points, respectively. Using the finite element analysis, the cogging torque has been computed and it was found that peak of cogging torque as much as 0.0044001 N.m. The cogging torque reduction of magnet structure with 8 points in magnet surface was found as much 98.84 %.

9. References

[1] Chen N., Ho, S.L. and Fu, W.N,2010, Optimization of Permanent Magnet Surface Shape of Electric Motors for Minimization of Cogging Torque using FEM, IEEE Trans. on Magnetics, Vol.46, No.6, pp. 2478-2481

[2] Scuiller F,2014,Magnet Shape Optimization to Reduce Pulsating Torque for a Five-Phase Permanent-Magnet Low-Speed Machine, IEEE Trans. on Magnetics, 2014, Vol. 50, No.4, pp.1-9

[3] Chabchoub M, Salah B. I, Krebs G, et al.2012, PMSM Cogging Torque Reduction: Comparison Between different shapes of Magnet, In: First International Conference on Renewable Energies and Vehicular Technology, Hammamet, 26-28 March; 206-211

[4] Wu L, Jin W, Ni J, Ying J. 2007, A cogging torque reduction method for surface mounted Permanent magnet motor, In: Internal conference on Electrical Machine and Systems (ICEMS), Seoul Korea, 8-11 October, pp.769-773

[5] Ling,J.M Nur T.Nur,T,2016, Influence of Edge Slotting of Magnet Pole with fixed slot opening Width on the Cogging Torque in Inset Permanent Magnet Synchronous Machine, Advanced in Mechanical Engineering, 2016, Vol.8, No.8, pp.1-9

[6] Rahideh, A., Korakianitis, Ruiz P., Keeble T., Rothman M.T.,2010, Optimal Brushless DC Motor Design using genetic algorithms, Journal of Magnetism and Magnetic Materials, Vol.322, pp.3680-3687

[7] Srisiriwanna, T, Mongkol K., 2010, A Study of Cogging Torque Reduction Methods in Brushless DC Motor. ECTI Trans. on Electrical Engineering, Electronics and Communications, Vol. 2012; 10: 146-152.

[8] Meeker, D., Finite Element Method Magnetics, 2018, Version 4.2 User's Manual, October

[9] Nur, T., and Mulyadi, M.,2018,Improve of Cogging Torque Reduction in Inset Permanent Magnet Machine, In IEEE ICASI Conference, Chiba Tokyo Japan, April 2018

[10] Bianchi, N, Electrical Machine Analysis using Finite Elements, 2005,Taylor and Francis.

IOP Conf. Series: Materials Science and Engineering **807** (2020) 012025 doi:10.1088/1757-899X/807/1/012025

Study the Effect of Dummy Slot in Stator and Rotor on the Cogging Torque Reduction in Permanent Magnet Machine

Liza Evelyn Joe[1] and Tajuddin Nur[2]

[1] Department of Electrical Engineering, School of Graduate Engineering. Atma Jaya Catholic University of Indonesia, Jl. Jend Sudirman no 51 Jakarta 12930, Indonesia.
[2] Department of Electrical Engineering Faculty of Engineering, Atma Jaya Catholic University of Indonesia, Jl. Jend Sudirman no 51 Jakarta 12930, Indonesia.

Corresponding e-mail: liza_e_joe@yahoo.co.id

Abstract. In this study, the effect of dummy slot in stator core and magnet in rotor core greatly affects the value of the cogging torque reduction in permanent magnet machine. The structure of permanent magnet machine was analysed employed finite element based on the finite element method magnetic (2D). In this study, 4 type of permanent magnet machine structure has been analysed. Permanent magnet machine structure in rotor have 4 type design magnet, permanent magnet machine with initial structure magnet, bread loaf, magnet edge slotting, and magnet edge slotting-magnet slot of surface. The cogging torque reduction of the proposed permanent magnet machine were computed and compared with the initial structure. Using the finite element, authors have computed and found that the cogging torque reduction of type 3 and type 1. The cogging torque of proposed permanent magnet machine could be reduced to 97.17% by employing the dummy slot in stator core.

1. Introduction

In renewable energy systems, the permanent magnet generators have been applied widely. The reason is that this kind of machines has some advantages over the other electric machine type. However, the most drawback of the permanent magnet generator is the cogging torque. The presence of cogging torque effect to reduce the permanent magnet generator and makes some undesirable performance such as noise, vibration and may be more. Based on the discussion, in this paper authors have offered a kind of cogging torque reduction technique. The effectiveness of the proposed technique was validated using finite element method magnetics. A novelty of this research is the fact that cogging torque can be reduced effectively as much as 93.19% compared with the initial structure. While if the structure was compared with the bread loaf structure, the cogging torque reduction is 97.17 %.

In this paper, the influence of combination of the fractional number of slot per pole of 24 slot / 14 pole has been studied. The Construction of Stator and Rotor in Permanent Magnet Machines are shown in Figure 1 [1-2]. For the purpose of study, shape of slot has been simplified. The structure of the machine was analysed employed finite element based on the FEMM.

The purpose of this study is to come up with and analysed the effect of incorporating the magnetic shaping method with dummy slot to reduce the cogging torque. For the cogging torque prediction, we adopt the finite element method magnetic (2D) as a tool [1].

In the study, the Maxwell Stress Tensor Method for the calculation of the cogging torque has been used. In the simulation, one must rotate the rotor of Permanent Magnet Machine by small steps and at each position has to be calculated the change of cogging torque in air gap in machine. The cogging torque that has been developed in the machine can be written as the partial change in cogging torque with respect to the virtual displacement of the rotor. The cogging torque in permanent magnet machine can be computed using the following equation [2-3] :

$$T_c = \frac{L_{st}}{\mu_0}\int_0^{2\pi} r^2 . B_n . B_t . d_\theta \qquad (1)$$

where, Tc = Cogging Torque, Lst = Stack Length of the Machine, r = Radius where the cogging torque is measured, B_n = Normal Flux Density, B_t = Tangential Flux Density, dθ = Rotor Displacement
From the equation (1), the most effecting on the cogging torque in permanent magnet machine are Normal Flux Density and Tangential Flux Density.

2. Permanent Magnet Machine Structure

In Figure 1 represents the permanent magnet machine structures studied in this paper. Analyse 4 types of permanent magnet machine. Permanent magnet machine with dummy slot and without dummy slot in stator. Permanent magnet machine with initial structure magnet, bread loaf, magnet edge slotting, and magnet edge slotting-magnet slot of surface in rotor.

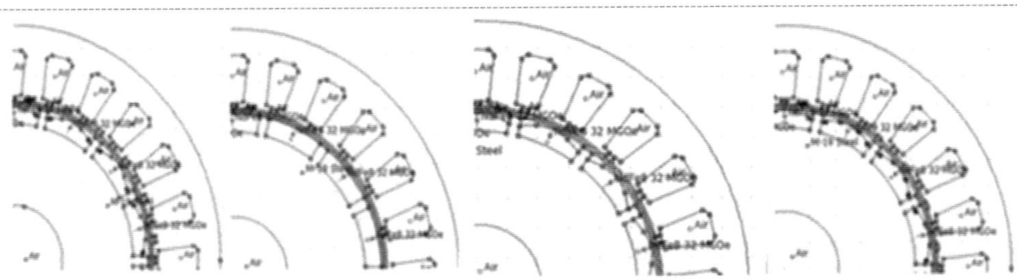

Figure 1. Construction Stator and Rotor in Permanent Magnet Machine.

Based on figure 1. This paper studied 4 types permanent magnet machine structure. This permanent magnet machine has 4 types of permanent magnet in rotor dan 2 type of dummy slot in stator. In the following of this paper, authors have summarized the specification of the 4 types of the permanent magnet machine.

2.1.Type 1

In this type, refers to the permanent magnet structure of 24 slot/14 pole A. As can be observed in the Figure 2, the permanent magnet machine with dummy slot in stator and magnet edge slotting in rotor.

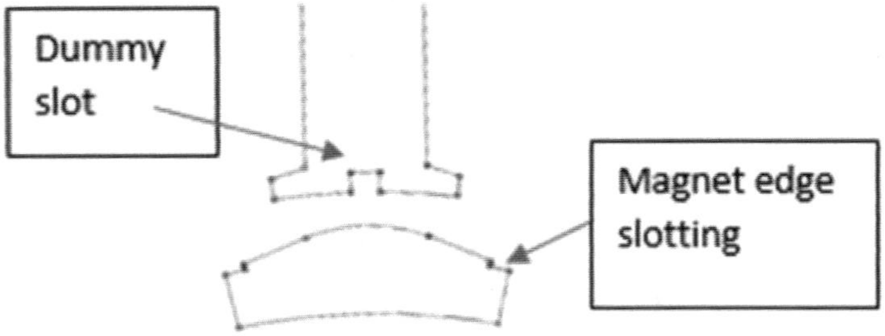

Figure 2. Permanent Magnet Structure 24/14 A

2.2. Type 2

As can be seen in Figure 3, the structure of the permanent magnet machine refers to original permanent magnet structure (24/14 B). In this type, the structure has no slotting both of stator and in rotor core.

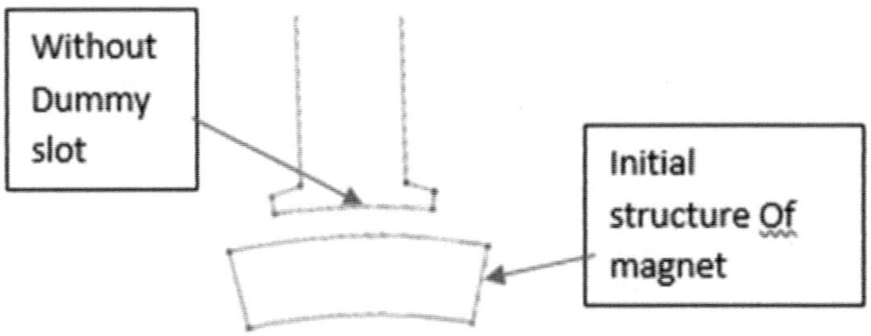

Figure 3. Original Permanent Magnet Structure 24/14 B

2.3. Type 3

For the type 3, is the permanent magnet structure of 24/14 C. In this type of machine, without dummy slot in stator and bread loaf of magnet in rotor core.

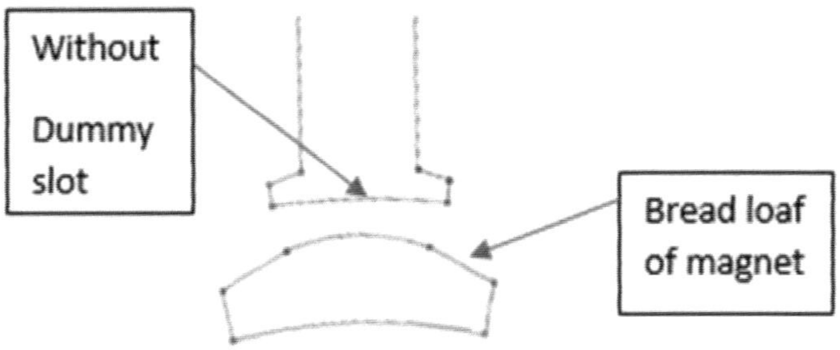

Figure 4. Permanent Magnet Structure (bread loaf) 24/14 C

2.4.Type 4

Type 4 is permanent magnet structure 24pole/14slot D. Permanent magnet machine with dummy slot in stator and used slotted in magnet surface – magnet edge slotting in rotor.

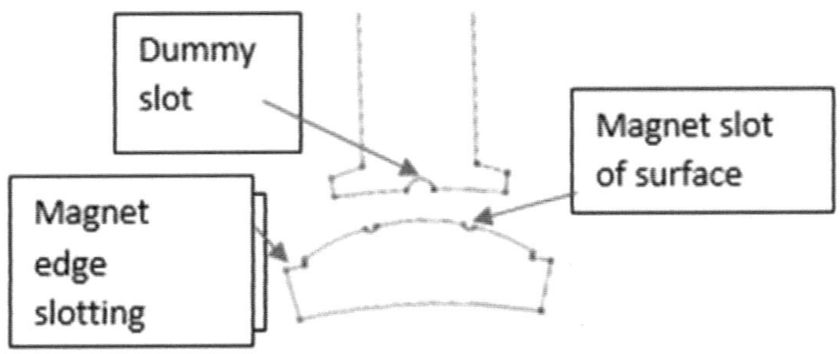

Figure 5. Permanent Magnet Structure 24/14 D

3. Normal flux density and tangential flux density

The slotting in stator and magnet rotor core of the machine effects to the normal and tangential flux density. The normal and tangential flux density are two of important parameters which give the most contribution in the cogging torque value. In the following, authors present the normal and tangential flux density of 4 types of permanent magnet machine and depicted in Table 1.

Table 1. Bn and Bt

	2414 A (Tesla)	2414 B (Tesla)	2414 C (Tesla)	2414 D (Tesla)
Bn	0.8	0.8	0.9	0.9
Bt	0.15	0.21	0.19	0.16

In Table 1 represents the value of normal flux density (B_n) and tangential flux density (B_t) flowing in Stator and Rotor core of the permanent magnet machine form 4 types permanent magnet machine structure.

The profile of normal flux density and tangential flux density can be observed in Figure 6 and Figure 7, respectively. The profile of both normal flux density and tangential flux density was analysed using finite element based on FEMM. Figure 6 is normal flux density (B_n), the dot (black color) is representation FEMM computation, while the line (red color) refers to the analytic of type 1 (24/14 A). It can be seen that wave from is sine. It can be concluded that the normal flux density is not destroyed by the changing of the machine structure, thus, this can be accepted. In Figure 7, we can see the value of tangential flux density (Bt) gained by analytic and finite element computation of 24slot/14pole A (type 1) permanent magnet machine. The peak of tangential flux density tends to be reduced effect of the dummy slot in stator core and edge slotting in magnet of the machine. The reduction of peak tangential flux density has a contribution to decline the tangential force in air gap. In addition, with the lower peak of tangential flux density, the cogging torque of the machine was expected to be reduced significantly.

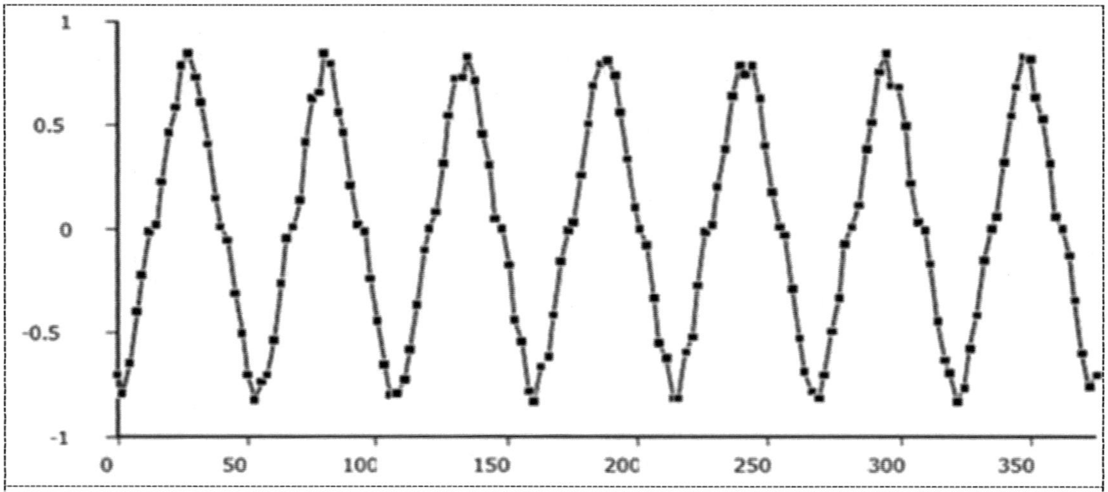

Figure 6. Normal Flux density (Bn) of the propose structure of 24slot /14pole A

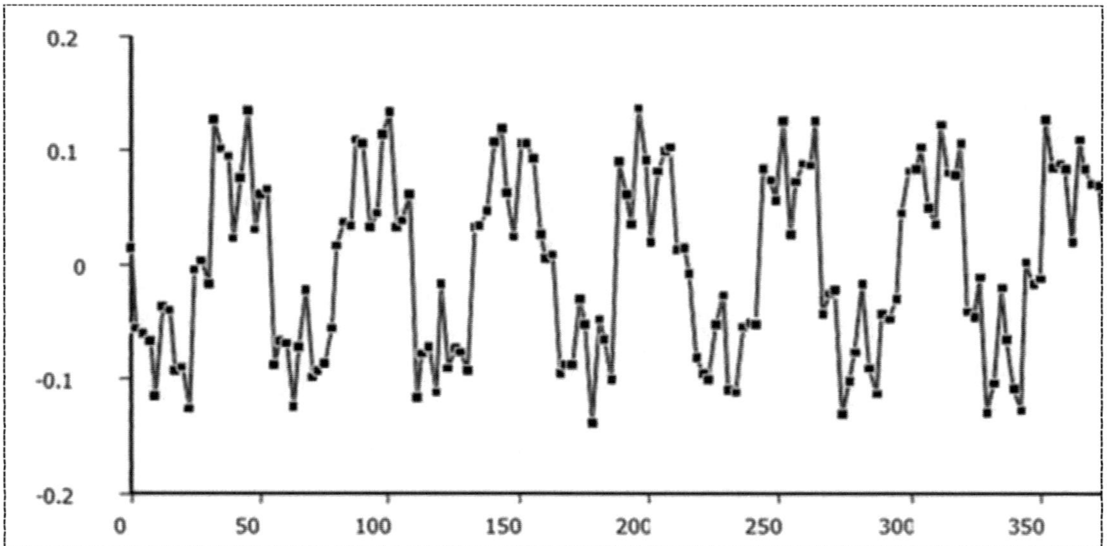

Figure 7. tangential flux density (B_t) of the propose structure of 24slot /14pole A

4. Magnetic flux distribution

The magnetic flux distribution of the permanent magnet machines studied was presented in this paper. One can observed that the effect of slotting in rotor and stator effect to magnetic flux distribution and magnetic flux density in the machine core. For example, the magnetic flux density in 24/14 A, the flux density in core is 1.029 Tesla, while the structure of 24/14 B, the flux density in core is 1.198 Tesla. The structure of 24/14 C, the flux density in the core of the machine is 1.018 Tesla, while the structure of 24/14 D, the flux density in core is 1.069 Tesla. The lowest flux density is 24/14 C structure. In general, the value of flux density in core of machine structure represent the losses in the machine core. In other word, the higher the flux density in the core, the higher the losses in the core. However, for material M-19, the flux density can be allowed.

IOP Conf. Series: Materials Science and Engineering **807** (2020) 012025 doi:10.1088/1757-899X/807/1/012025

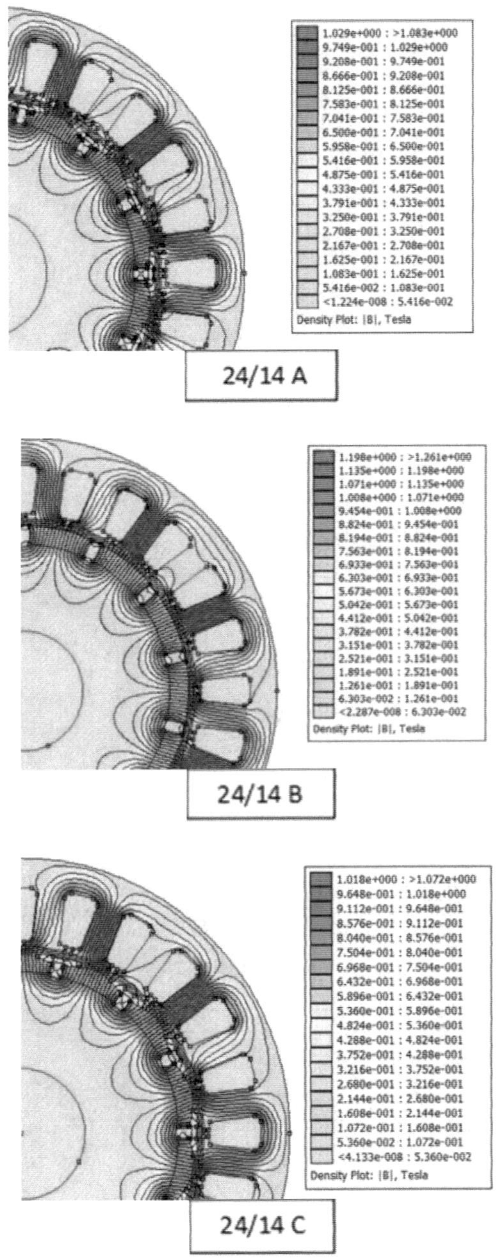

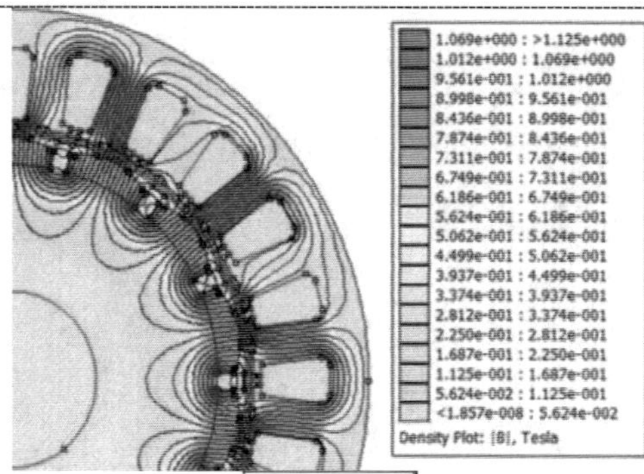

24/14 D

Figure 8. Magnetic flux distribution

5. The cogging torque simulation

In this simulation, the LUA 4.0 scrip has been used to perform the permanent magnet machine analysis [1]. The advantage of using LUA is the fact that simulation process can be done parallel. It save the time of computation of the simulation. The illustration of cogging torque reduction of the propose structure of 24 slot / 14 pole from 4 type of machine structure was given and compared. The peak of cogging torque of 24/14 C (type 3) has been found as much as 0.0008954 N.m and the peak of cogging torque of 24/14 A (type 1) is 0.0000253 N.m. In this study, authors have found that cogging torque reduction technique proposed can be reduced effectively as much as 97.17%. The peak of cogging torque of 24/14 B (type 2) is 0.0003716 N.m, while the peak of cogging torque of 24/14 A (type 1) is 0.0000253 N.m. Another comparison has done in this study. It was found the cogging torque of type 1 can be reduced effectively as much as 93.19% compared with the initial structure (24/14 B) type 2 .

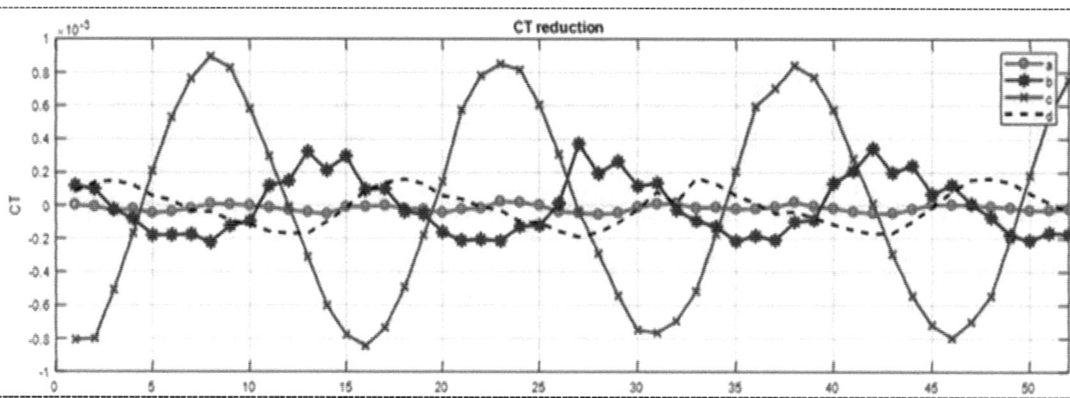

Figure 9. Cogging Torque of the propose structure of 24slot /14pole

6. Conclusion

In this study, the effect of combining of dummy slot in both of stator and rotor on cogging torque reduction in permanent magnet machines have been presented and compared. The greatest cogging torque reduction among the permanent magnet machine studied was the type 1 as much as 97.17 % compared with type 3. When compared with type 2 to type 1, the cogging torque reduction about 93.19 %. It can be concluded that the permanent magnet machine type 1 offers permission to be the best cogging torque reduction among them.

References

[1] Tajuddin Nur and Herlina 2018 Enhancement of Cogging Torque Reduction on Inset Permanent Magnet Generator by Using Magnet Edge Shaping Method (Bangka: International Conference on Electrical Engineering and Computer Science (ICECOS))
[2] Tajuddin Nur and Melisa Mulyadi 2018 Improve Cogging Torque Method in Inset-Permanent Magnet Synchronous Machine (Proceedings of IEEE international Conference on Applied System Innovation)
[3] J.M Ling and T.Nur 2016 Influence of edge slotting of magnet pole with fixed slot opening width on the cogging torque in Inset-Permanent Magnet Synchronous Machine (Advance Mechanical Engineering Vol.8, No.8) pp.1-9
[4] David Meeker 2018 Manual Book FEMM 4.2
[5] Hongbo Qiu, Kaiqiang Hu, Wenfei Yu and Cunxiang Yang 2018 Influence of the Magnetic Pole Shape on The Cogging Torque of Permanent Magnet Synchronous Motor (Australian: Journal of Electrical and Electronics Engineering)
[6] A. J. P. Ortega, S. Paul, R. Islam, and L. Xu 2016 Analytical Model for Predicting Effects of Manufacturing Variations on Cogging Torque in Surface-Mounted Permanent Magnet Motors (IEEE Trans. Ind. Appl, vol. 52, no. 4) pp. 3050–3061
[7] K.-H. Shin, H.-I. Park, H.-W. Cho, and J.-Y. Choi 2017 Analytical calculation and experimental verification of cogging torque and optimal point in permanent magnet synchronous motors (IEEE Trans. Magn vol. 53, no. 6) pp. 1–4
[8] T Nur and TZF Mohamed 2018 Combined Methods to Overcome Unbalanced Magnetic Pull in Fractional Slot Electric Machines, Journal of Computational and Theoretical Nanoscience vol.24, no.11, pp.8875-8879

IOP Conf. Series: Materials Science and Engineering **807** (2020) 012026 doi:10.1088/1757-899X/807/1/012026

Study of The Effect of Height and Length of Slotting in Magnet Edge on the Cogging Torque Reduction of Fractional Slot Number in Permanent Magnet Machine

Tajuddin Nur[1*] and Liza Evelyn Joe[2]

[1] Department of Electrical Engineering, Graduate School of Engineering,
[2] Faculty of Engineering, Atma Jaya Catholic University of Indonesia, Jakarta, Jl. Jend. Sudirman No.51 Jakarta 12930 Indonesia

Corresponding e-mail: tajuddinnur559@gmail.com

Abstract. Cogging torque generated in permanent magnet machine effects to some undesirable vibration and noises. In order to minimize the cogging torque in permanent magnet is the most important issues recently. This paper investigated the influence of the value of height and length of slotting in magnet edge on the cogging torque reduction. For purpose of study the structure of 24 slot/10 pole of permanent magnet machines have been chosen. The 2-D finite element based on FEMM 4.2 software was implemented to compute the permanent magnet machine performance. Simulation results showed that by optimising the height and length of slotting in magnet edge can reduce the cogging torque of permanent magnet machine significantly. The most effective way by employing a-two step slotting with 2.5 mm of height and 0.7 mm of length of the slotting in magnet in the first step of slotting. In the second step of in magnet edge, the length of slotting has been assigned as much as 2.3 mm and the height of slotting was 0.588635 mm. By employing the FEMM, three of different magnet structures of permanent magnet machine were analysed. It was found that the cogging torque of permanent magnet machine proposed magnet structure as much as 99.7 % compared with the initial structure of the permanent magnet machine.

1. Introduction

In renewable energy systems such as wind power, water power, wave power and other related to the special purpose of electrical machine, the utilization of permanent magnet machine (PMM) is one of the most important part in order to harvest the electric energy. The reason is that the type of electrical machine has some advantages over the other type of electrical machines such as high efficiency, lower maintenance, and high-power density. However, the most issues related to the PMM application is the cogging torque (CT). The cogging torque in PMM produces attractive force between magnet in rotor and stator core. In low speed wind situation, the blades are hard to be rotated and rotor experiences to be braked. The CT in PMM is produced by magnet flux in rotor core, then generates magnetic interaction or force between the stator core and permanent magnet in the rotor. Based on the above discussion, it can be concluded that the CT value in PMM depends on the structure and the material used in the PMM manufacturing. Hence, in order to minimise the CT value in PMM, an appropriated design of magnet structure in rotor core and stator core structured must be provided. In addition, the influence of the material used for the machine must be considered. Since the last few years, there have been increasing the research on CT reduction techniques which have been proposed and reported in worldwide where the authors have paid attention at the slotting in magnet structure [1- 5]. In [1- 2], developed a cogging torque reduction technique by employing the slotting in the magnet surfaces of PMM. While in [3] and [5], proposed the slotting in the magnet edge to minimise the CT. and authors [5] investigated the effect of the length and height of slotting in magnet edge on the CT reduction in an integral slot number of 24 slot / 8 pole. Employing the magnet edge slotting technique, the CT of the machine could be reduced as much as 95 % compared with initial structure. This means that the CT content in the PMM was 5 %. In the study, the material used for magnet was NdFeB and for both of stator and rotor core, the M-19 steel have been used. Another scholar [4] employed one step of slotting in magnet edge to reduce the CT and it could be reduced significantly. However, for renewable energy application the CT in the CT contains in the permanent magnet machine must be as low as possible. Past researchers suggested the content of CT for special purpose, the CT content in the range 1 to 2 % of the rated torque.

In this study, the effect of length and height of slotting in magnet edge of a fractional slot number was presented. For the purpose of this study, a- 24 slot / 10 poles of PMM structure have been chosen. Three of different structures of magnet were analysed and compared. All the stator cores of the machines studied have the same structure, the different only in the magnet structures. In the beginning, the conventional of magnet structure or initial structure was analysed, which was no any slotting in magnet. Using the finite element analysis, the value of cogging torque was calculated and recorded. Further, for the second structure, a one step of slotting was employed in the magnet edge. As have been predicted, the cogging torque can be reduced. However, the cogging torque reduction was not significantly for renewable energy system requirement. For wind power application, the cogging torque reduction at least 99 % from the rated torque. Thus, in order to achieve the CT reduction of proposed structure, a 2 step of slotting in magnet edge has employed in this study. The advantage of using a 2-step slotting in magnet edge is the fact that more length and height of the slotting can be obtained. In this paper, the optimum of length and height of the slotting in the proposed structure has been found as much as 2.5 mm and 0.7 mm respectively, then in the second step of slotting the length and height have been determined as much as 2.3 mm and 0.5886 mm, respectively.

2. Theories of the cogging torque in PMM

The CT in any PMM is caused by the interaction between stator core or slot opening width and permanent magnet in rotor core. In fact, the part of the magnet rotor which contributes the most of CT content in air gap may be in the magnet edge. So, in order to minimise the CT in air gap is by reducing the residual magnet flux in magnet edge. Reducing the residual can be done by shaping or slotting or compounding shaping and slotting in magnet structure. The CT can be calculated as shown in equation (1). In paper study, the magnet pole structure of machines was assumed to be radial magnetization [2],[3],[7].

$$T_c = -\frac{1}{2}\emptyset_g^2 \frac{dR}{d\theta} \tag{1}$$

where the $\emptyset_g^2$ is the flux in the air-gap, $\boldsymbol{R_g}$ is the passed air-gap reluctance, and $\boldsymbol{\theta}$ is the rotor position of the machine. The magnet flux in air gap in Eq.1, can be formulated as shown in Equation (2)

$$\emptyset_g = B_g A_g \tag{2}$$

where B_g=magnetic flux density in air gap and A_g = air gap cross section of the machine. The magnetic flux density is depended on the magnet flux remanence and air gap cross section is influenced by the distance between magnet in rotor core and the stator core. Another CT formulation for any PMM is

$$T_c = \frac{L_{stk} r}{\mu_0} \int_0^{2\pi} B_n \, B_t \, d\theta \tag{3}$$

The cogging torque also can be formulated as a Fourier series shown in equation where m is the least common multiple of the number of stator slot (N_s) and the number of pole (N_p), k is an integer and T_{mk} is a Fourier coefficient. For the PMM with integral slot number, each pole of the machine has a whole number of multiple stator teeth, so that the cogging effects of each magnet are in phase and added, leads the cogging torque tend to be increased. In contrast, for the PMM with fractional slot number as in the 24 slot / 10 pole structure, only some of the poles number multiple stator teeth. This leads that only some of the magnets in rotor core are in phase and added. That is one of reason why the fractional slot number usually considered as CT reduction techniques. However, in order to fulfil requirement of low CT for renewable energy application, the fractional slot number technique should be combined together with other CT reduction technique to get a low CT according to the renewable energy system requirement. In this study, it has been found that value of length and height of slotting has affected on the CT reduction of the machine.

3. The proposed of PMM Structure

To study the effect of the height and length of slotting in the magnet edges on the CT reduction in a fractional slot number, a-24 slot/10 pole structure has been studied and proposed in this paper. The stator core structures are like the real machine, such as slot opening width, stator slot number, stator teeth height and stator teeth width were the same and have the same characteristics. The stator core for all permanent magnet machine studied were same, the only different is the magnet structures. As a comparison, in the beginning of this study, the magnet structure was initial t structure of magnet as shown in Figure 1. In Figure 2, a-one step of slotting in magnet with certain height and length have been employed to the magnet structure. While in Figure 3, the two steps of slotting were employed in magnet edge. The presence of slotting in magnet edge effect to the changing the distance between magnet edge and stator core. As a consequence, the magnet flux distribution in magnet edge also become changed.

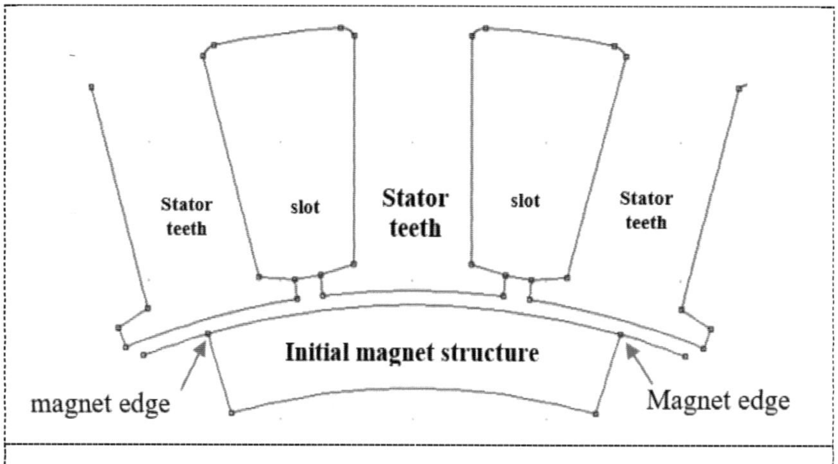

Figure 1. Initial Magnet structure

As can be seen in Figure 1, the structure of magnets is smooth and the distance between magnet rotor and stator teeth/stator core is homogenous. In Figure 1, the distance between magnet and stator core was 1 mm and the magnet flux were radially. Since there is no any slotting in magnet surface, in the initial magnet structure, only has two edges. The magnet edge of initial structure located in the right and left side of the magnet.

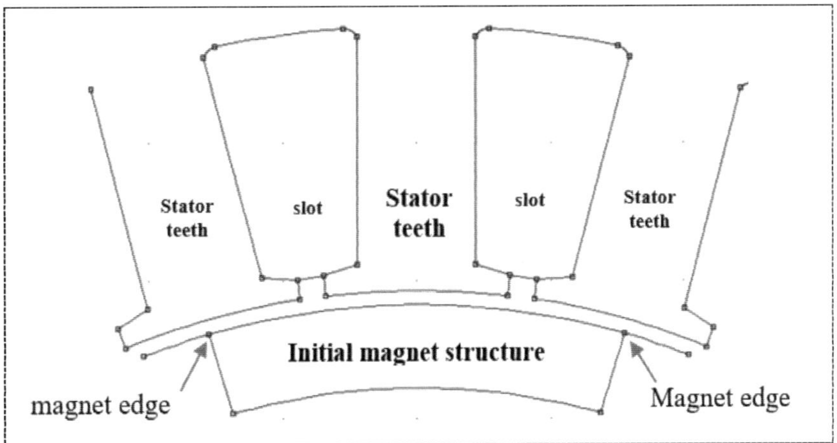

Figure 2. One step of slotting in magnet edge

In Figure 2, every side of the magnet has 2 edges so that there are 4 edges in magnet structure. The length and height of the slotting does not affect to the number of magnet edge, but it can change edge point of the magnet. It means that, the distance between magnet edge to the stator core depends on the length and height of the magnet edge slotting. The cross-section area or magnet volume also become changing caused by the length and height of the magnet slotting. As a consequence, it will affect to the total magnet flux flowing into the air gap and will reduce the cogging torque of the machine [7].

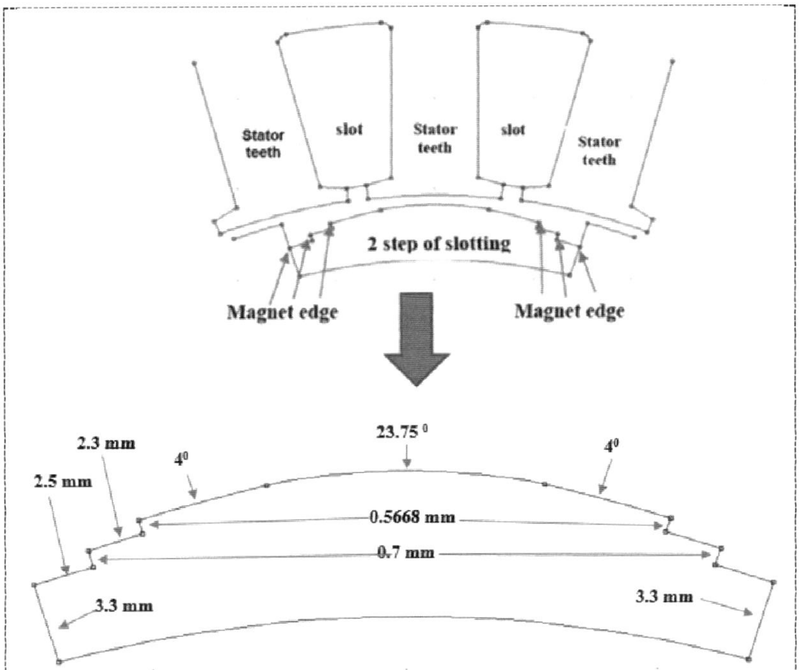

Figure 3. Two steps of slotting in magnet edge

In Figure 3, the presence of 2 steps of slotting in magnet edge with certain value of length and height effect to change magnet structure. As can be observed in Figure 3, there are 3 edges in each side of magnet. Compared with the initial structure, the magnet cross section area is smaller. As a result, the total flux flowing into air gap become decreased sharply. Also, the presence the 2 steps of slotting in magnet edge, effect to the increasing of the air gap cross section since the magnet cross section occupying the rotor core was reduced. This leads to reduce the air gap reluctance of the machine. Thus, the effect of employing 2 steps of slotting with first length of 2.5 mm, first height of 0.7 mm, second length of 2.3 mm, and 0.588635 mm, respectively, have given the best contribution in achieving the CT reduction of the PMM. In this study, the most effective value of length and height of slotting in magnet edge on the cogging torque reduction of the machine, as presented in Table 1.

Table 1. Height and Length of magnet edge slotting

No	1st length of (mm)	1st height of (mm)	2nd length of (mm)	2nd height of (mm)	T_{cmax} (N.m)
Initial structure	No magnet slotting	No magnet slotting	No magnet slotting	No magnet slotting	0.2352206
1 step of slotting	2.5	0.6	No magnet slotting	No magnet slotting	0.0675994
2 steps of slotting	2.5	0.7	2.3	0.588635	0.0006774

The presence of slotting in magnet edges effect to increase the number of edges in magnet structures. The magnet cross section or magnet volume also become decrease. This effect to magnetic flux distribution in magnet edge into air gap of the machine decline which leads to reduce the total flux magnet in air gap. For magnet structure in Figure 3, the magnet edge has been achieved to be 6 edges. The CT was calculated and compared with the initial structure. By employing the one step slotting in magnet edge became reduce. In Figure 3 (proposed structure), the two steps of slotting in the magnet edges was employed to increase the CT reduction. The different between one step of slotting in magnet edge and the two steps of slotting is the fact that the distance between magnet become variable. Also, the air gap cross section of the machine become larger compared with the one step of slotting. As a result, the air gap reluctance become decreases and declines the peak of cogging torque of the machine. The one step of slotting was illustrated in Figure 4 below and the effect of the slotting in magnet edge of the three structure of machine was calculated.

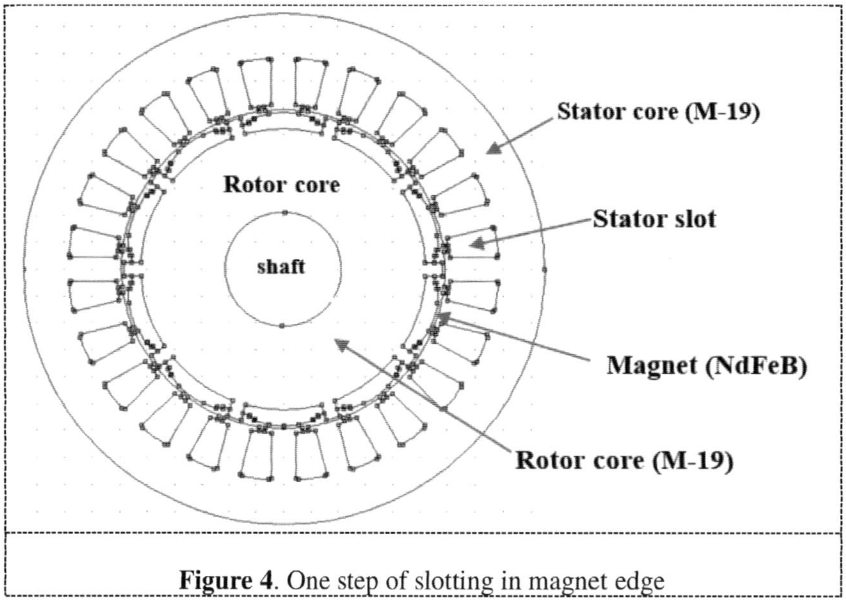

Figure 4. One step of slotting in magnet edge

4. Simulation of Magnetic Flux Distribution and Cogging Torque

4.1. Magnetic Flux Distribution in machine core

In order to predict the performance of the machines effected the CT reduction technique, the flux distribution in the core of the machine has been presented and compared, in this paper. The flux distribution in the part of the machine was performed by FEMM. The FEMM is one of powerful software which can calculate all the parts of the machine. The magnetic flux distribution of the all machines studied as shown in the Figure 4a, 4b and 4c, respectively.

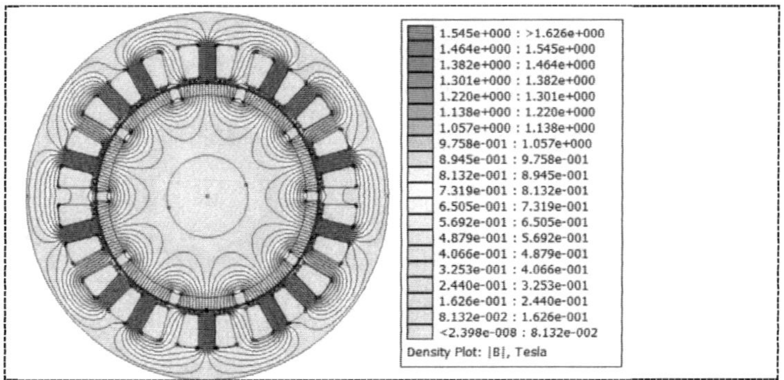

Figure 4a. Initial Magnet Structure

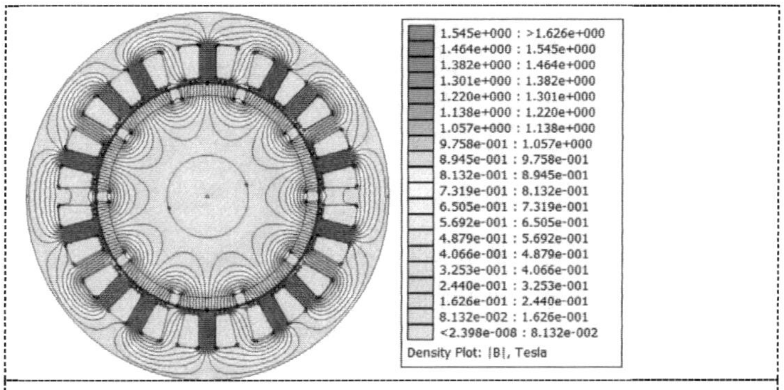

Figure 4b. One step of slotting in magnet edge

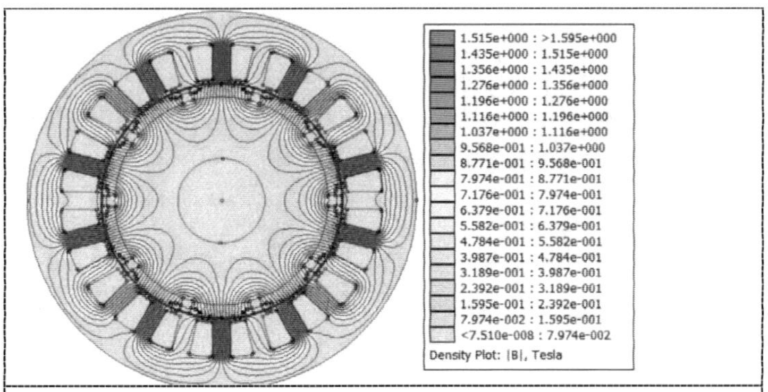

Figure 4c. Two steps of slotting in magnet edge

In Figure 4a,4b,4c, we can observe that the effect length and height of slotting only give a small contribution on the changing the magnetic flux density in the machine core. For proposed structure, the magnetic flux density in the core have been investigated using finite element and it has been found as much as 1.515 Tesla, while for the 1 step of slotting the magnetic flux density was 1. 535 Tesla and 1.545 Tesla for the magnetic flux density of the initial structure.

4.2. Calculation the Cogging Torque

In this paper, the FEMM 4.2 was used to investigate the PMM characteristics based on finite element analysis [2],[3],[5],[7],[8],[9]. The FEMM 4.2 is one of a-2D finite element analysis which has been widely used for analysis for any electrical machine performance since the last few years. The procedure of FEMM 4.2 operation based on three steps [8]. In order to achieve the accurate of CT computing in air gap, 5 dummy lines have been employed in the air gap the PMMs studied. Every structure of the machine was analysed and the CT of the PMM was recorded. Through the analysis of FEMM 4.2, CT of the PMMs studied were presented and compared, as shown in Figure 5.

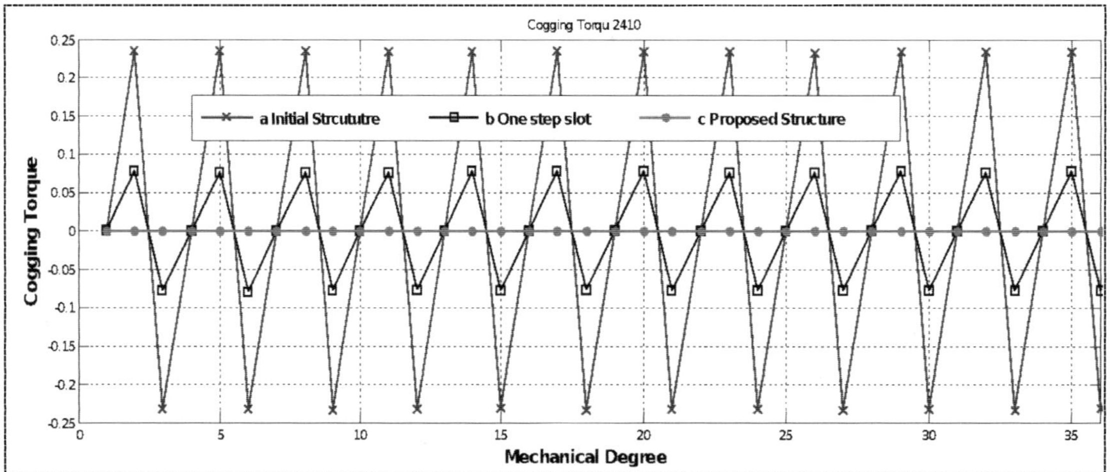

Figure 5. Comparison of Cogging Torque for different magnet structure (24 slot / 10 pole)

In Figure 5, one can observe that the CT for initial structure of PMM 24 slot / 10 pole is the highest. The CT peak value has been analysed and it was found around 0.2250818 Tesla. After employing the appropriate length and height of slotting in magnet edge, the magnet structure of proposed structure has the most edge points among the three different magnet structure. As a result, the magnet flux interaction between magnet surface and stator slot opening become frequent. This leads the achieving the CT frequency and reduce the peak of CT of the machine. The CT reduction was validated with finite element analysis and it has been found as much as to 0.0016483 Tesla. Based on the initial structure, the CT reduction of the machine was calculated as 100 % - (T_{cmax} of proposed structure/T_{cmax} of initial structure)100 % =100 % - (0.0006774 / 0.2352206) % =100 % - 0.29 % = 99.7 %. It could be concluded that the CT reduction for the proposed structure was around 99. 7 %.

5. Conclusion

In the present work, the effects of height and length of slotting in the magnet edges on the CT reduction in fractional slot number of 24 slot /10 pole has been presented. The CT reduction of the three different magnets of slotting have been compared. It was found that the two steps of slotting with 2.5 mm of length and 0.7 mm of height, while in the second step, the length and height of slotting was 2.3 mm and 0.588635 mm, respectively. The CT reduction for the proposed structure was computed using FEMM and it was found as much as 99.7 % compared with the initial structure. It can be concluded that the length and height of slotting in magnet edge of fractional slot number of 24 slot/ 10 pole effects significantly on the CT reduction of the proposed PMM.

References

[1] Chen N, Ho SL and Fu WN 2010 *IEEE Trans. on Magnetics* Vol.**46,** No.6, pp. 2478-2481

[2] Scuiller F 2014 *IEEE Trans. on Magnetics* Vol. **50,** No.4, pp.1-9

[3] Chabchoub M, Salah B. I, Krebs G 2012 *First International Conference on Renewable Energies and Vehicular Technology*, Hammamet pp. 206-211

[4] Wu L, Jin W, Ni J, Ying J 2007 *Internal conference on Electrical Machine and Systems (ICEMS)* pp.769-773

[5] Ling,J.M and Nur,T 2016 *Advanced in Mechanical Engineering*, Vol.**8**, No.8, pp.1-9

[6] Rahideh A, Korakianitis, Ruiz P, Keeble T, Rothman MT 2010 Vol.**322**, pp.3680-3687

[7] Srisiriwanna T, Mongkol K 2012 *ECTI Trans. on Electrical Engineering, Electronics and Communications* Vol. **2012**; 10: 146-152.

[8] Meeker D.2018 *Finite Element Method Magnetics* Version 4.2 User's Manual

[9] Nur,T Mulyadi,M 2018 *IEEE ICASI conference.*

IOP Conf. Series: Materials Science and Engineering **807** (2020) 012027 doi:10.1088/1757-899X/807/1/012027

Modelling Uranium in Vicinity of Groundwater Population by Neural Networks of Multilayers Perceptron

Iing Lukman[1], Noor Akma Ibrahim[2], Natalina[3]

[1]Dept. of Accounting, Faculty of Economics, Universitas Malahayati, Bandar Lampung, Indonesia
[2]Dept. of Mathematics, Faculty of Science, Universiti Putra Malaysia, Serdang Selangor D.E. Malaysia
[3]Dept. of Environmental Engineering, Faculty of Engineering, Universitas Malahayati, Bandarlampung, Indonesia

Corresponding e-mail: iing@malahayati.ac.id; deiingofthelukman@gmail.com

Abstract. The existent of Uranium in vicinity of groundwater population can give a threat to the water supplier for human consumption. The objective of the research was to find the most important variables to the existence of the Uranium. This paper shows some modelling process for above matters by applying Neural Networks of Multilayers Perceptron. Data taken from US Department of Energy. Neural Networks used in this study were learning the representation of the model inside the data, and how best it relation with the output variable that we obtained from prediction. The results showed that the training samples was 87 out of 127, and the testing samples was 40 out of 127. The results were not giving indication that a mathematical model obtained. The conclusion was Conductivity becoming the most important variable to the existence of Uranium, which followed by the second importance that was Arsenic, the third importance was Selenium, the fourth important was Total Alkalinity.

1.0 Introduction

The Radionuclide and the plausible human carcinogens were emit by Uranium [1]. Groundwater sources were relied by rural areas residents as their primary drinking water sources, for example in South Carolina around 40% of its residents regularly use it [2], and in the 10 Southeast Asian and Pacific nations [10]. Colorectal, breath, kidney problems, and total cancer incidences may be increase in region with frequent groundwater use and elevated groundwater Uranium [1]. Uranium is one of the dangerous heavy metals due to its toxic and radioactive, and it is required from its distribution in the environmental region [3]. In fact, many types of mathematical model required the modeller to know things about the system that are unfortunately generally impossible to find [4]. The objective of the research was to find the most important variables to the existence of the Uranium.

2.0 Methodology

The methodology of the Multi-layer Perceptron applied for data analysis. The multilayer perceptron is the most known and most frequently used type of neural network [11]. One perceptron is a linear classifier and an algorithm that classifies input by separating two categories with a straight line. Input is typically a feature vector x multiplied by weights w and added to a bias b so that y = w * x + b. One perceptron produces a single output based on several real-valued inputs by forming a linear combination using its input weights (and sometimes passing the output through a nonlinear activation function). The equation is as follows:

$$y = \varphi(\sum_{i=1}^{n} \omega_i + b) = \varphi(w^T x + b) \qquad (1)$$

Where w denotes the vector of weights, x is the vector of inputs, b is the bias and phi is the non-linear activation function. Rosenblatt [6] built a single-layer perceptron. That is, his hardware-algorithm did not include multiple layers, which allow neural networks to model a feature hierarchy.

3.0 Multilayer Perceptron (MLP)

Subsequent work with multilayer perceptron has shown that they are capable of approximating an XOR operator as well as many other non-linear functions. Just as Rosenblatt [6] based on the perceptron on a McCulloch-Pitts neuron, conceived in 1943, so that, perceptron themselves are building blocks that only prove to be useful in such larger functions as multilayer perceptron.

A multilayer perceptron (MLP) is a deep, artificial neural network [7]. It is composed of more than one perceptron. They are composed of an input layer to receive the signal, an output layer that makes a decision or prediction about the input, and in between those two, an arbitrary number of hidden layers that are the true computational engine of the MLP. MLPs with one hidden layer are capable of approximating any continuous function [7].

Multilayer perceptron often applied to supervised learning problems [8]: they train on a set of input-output pairs and learn to model the correlation (or dependencies) between those inputs and outputs. Training involves adjusting the parameters, or the weights and biases, of the model in order to minimize error. The adjusting parameters is like competing risk in the survival analysis, where the predominantly parameter will have it position in the final output [8]. Applying Backpropagation to make those weigh and bias adjustments relative to the error, and its error be measured in a variety of ways, including by root mean squared error (RMSE).

The data source for this research was 127 groundwater samples (cases or row) and 12 measurements (variables) taken from [9] which was collected by Department of Energy of United States of America. Samples data consist of horizon producing codes (P_GC) by 5 types of P_GC namely TPO (Ogallala Formation), TRD (Dockum Formation), POQ (Quartermaster Group), PGWC (Whitehorse and Cloud Chief Group), PGEB (El Reno Group and Blaine Formation). Variables are consist of Uranium, (U), Arsenic (AS), Boron (B), Barium (BA), Molybdenum (MO), Selenium (SE), Vanadium (V), Sulphate (SO4), Total Alkalinity (T_AK), Bicarbonate (BC), Conductivity (CT), and pH. Table 1 show the layout of groundwater data.

Table 1. Layout of Groundwater data.

Sample Number	Latitude	Longitude	P_GC	U	AS	B	BA	MO	SE	V	SO4	T_AK	BC	CT	pH
2201	33.21	101.445	TPO	7.99	17.6	300	150	-4	0.4	100	35	278	157	640	7.60
....	....	....	....	....	....	....	....	....	....	....	....	....	....	....	....
2210	33.164	100.608	PGWC	3.1	4.0	625	750	-4	-0.2	25	-30	175	101	540	7.56

4.0 Results and Discussion

Data analysed by SPSS Neural Network or Multiple Perceptron give as follows: it seems that the real value and its predicted value had big different or high deviation. The different between predicted to the real value was bigger, for example, as seen on the Figure 1, that if the real value was closed to 80, then the predicted value was closed to 30 or the deviation was around 50. The outlier value of the real data between 120 and 140 did not give predicted value at all, thus this Multiple Perceptron was not that fully suited for this data. Training a multilayer perceptron is often quite slow, requiring thousands or tens of thousands of epochs for complex problems [11].

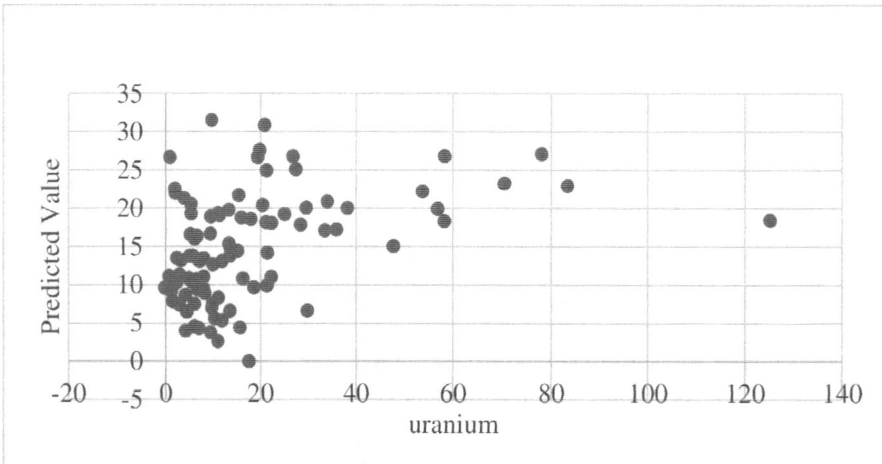

Figure 1. Uranium real and its predicted value

Figure 2 described that the residual of predicted value of Uranium from normalised value of 0 to 15 were mostly around 0 which was good for the prediction. However, some of the residual value of normalised value of 17 to 30 were generally far from zero. Therefore, generally the residual of the predicted value not closed to zero, which then giving problem to the precision of the model.

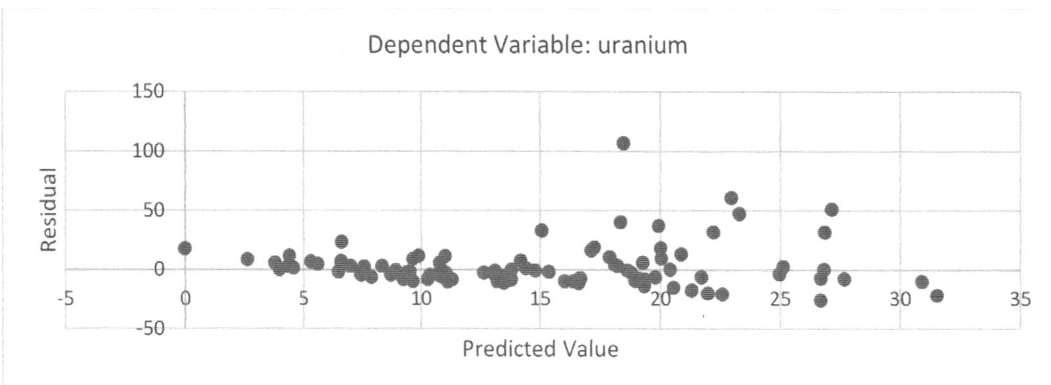

Figure 2. Uranium: it predicted and residual value

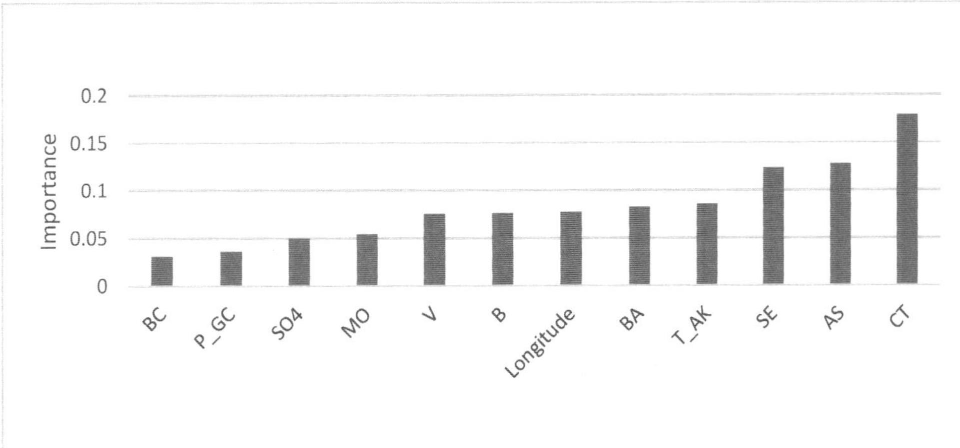

Figure 3. Uranium was a dependent variable against the rest of the variables in the observation

Figure 3 showed that for the dependent variable Uranium, the most independent variable was CT (0.179) followed by AS (0.128) , SE (0.124), T_AK (0.085), BA (0.082), Longitude (0.078), Boron (0.76), MO (0.055), SO4 (.050), horizon (.036). Table 2 showed the importance of the variables and its normalized scale.

Table 2. Independent Variable Importance

Importance	Normalized Importance	
horizon	.036	20.3%
Longitude	.078	43.3%
arsenic	.128	71.5%
boron	.076	42.6%
BA	.082	46.0%
MO	.055	30.4%
SE	.124	69.3%
V	.076	42.3%
SO4	.050	27.9%
T_AK	.085	47.5%
BC	.031	17.3%
CT	.179	100.0%

As seen on Table 2 that CT (conductivity) was the most important variable to the existence of the Uranium followed by Arsenic, Selenium, Total Alkalinity etc. This was in accordance with Hin [12].Then the least important is Bicarbonate. The computation of percentage of importance is as follows: the scale of CT is the biggest one then label it as a hundred percent importance. Afterwards, the normalized scale lower than that then be compared to normalized scale of CT which then give the results as seen on Table 2.

Table 3. Case Processing Summary

		N	Percentage (%)
Sample	Training	87	68
	Testing	40	31
Valid		127	100
Excluded		0	
Total		127	

Table 3 showed that from the total samples of 127, 87 taken for training, and 40 taken for testing. All the 127 samples are valid. Table 4 showed that the horizon producing codes P_GC as a factor, and the rest of the variables as covariates except longitude and latitude.

Table 4. Network Information

Input Layer	Factors	1	P_GC
	Covariates	1	AS
		2	B
		3	BA
		4	MO
		5	SE
		6	V
		7	SO4
		8	T_AK
		9	BC
		10	CT
		11	PH
	Number of Units[a]		16
	Rescaling Method for Covariates		Standardized
Hidden Layer(s)	Number of Hidden Layers		1
	Number of Units in Hidden Layer 1[a]		4
	Activation Function		Hyperbolic tangent
Output Layer	Dependent Variables	1	U
	Number of Units		1
	Rescaling Method for Scale Dependents		Standardized
	Activation Function		Identity
	Error Function		Sum of Squares

a. Excluding the bias unit

Table 5. Model Summary

Training	Sum of Squares Error	33.435
	Relative Error	.778
	Stopping Rule Used	1 consecutive step(s) with no decrease in error[a]
	Training Time	0:00:00,07
Testing	Sum of Squares Error	27.543
	Relative Error	.513

Dependent Variable: U

a. Error computations were based on the testing sample.

Table 5 showed the sum squares error of the training and the testing samples, which suggested that the testing sample has less sum square error than that of training one, and it relative error was smaller than that of training one.

5.0 Conclusion

CT or conductivity was the most important variable to the existence of the Uranium followed by Arsenic, Selenium, Total Alkalinity. The least important was Bicarbonate.

Acknowledgments
Authors wishing to acknowledge assistance or encouragement from colleagues to name a few Rina Febrina, Devi Oktarina, Dewi Fadila Sari, Emy Khikmawati,Weka Indra Darmawan, Marcellya Wijaya, Heri Wibowo Kristianto, and Rani from Engineering Faculty of Universitas Malahayati, Bandarlampung.

References

[1] ATSDR 1999 *Toxicological profile for uranium. US Department of Health and Human Srvices, Public Health Service; Atlanta: Agency for Toxic Substances, Disease Registry*
[2] SCGWA 2009 South Carolina Groundwater Association http://scgwa.org/pages/industrynews/sconomy.html. Accessed on 20 July 2018
[3] Nichols, C.E., Kane V.E., Browning, M.T. and Cagle, G. W. (1976). Northwest Texas Pilot Geochemical Survey, Union Carbide, Nuclear Division Technical Report.
[4] Farlow, S. J. 1984 GMDH Algorithm *Self-Organizing Method in Modeling: GMDH Type Algorithm Stanley J. Farlow, Editor pp 1-24*
[6] Rosenblatt, F 1958 *The Perceptron: A Probabilistic Model for information Storage and Organization in the Brain*
[7] Minsky, M and Papert, S 1988 *Perceptrons: An Introduction to Computational Geometry The MIT Press Cambridge*
[8] Lukman, I, Noor Akma Ibrahim, Isa Bin Daud, Fauziah Maarof, and Mohd Nasir Hassan 2002 Advantage of the modified Lunn-Mcneil technique over Kalbfleisch- Prentics technique in competing risks *Proceedings of SPIE Vol.4730 (2002) Data Mining and Knowledge Discovery: Theory, Tools, and Technology IV, Belur V. Dasarathy, Editor. Pp 203-211*
[9] Andrews, D.F., and Herzberg, A.M. 1985 *Data: A Collection of Problems from Many Fields for the Student and Research Worker. Springer-Verlag.*
[10] Carrard, N., Foster, T., Willetts, J. 2019 Groundwater as a Source of Drinking Water in Southeast Asian and the Pacific: A Multi-Country Review of Current Reliance and Resource Concerns *Water 2019, 11(8), 1605; https:doi.org/10.3390/W11081605, Accessed 15th January 2020*
[11] Popescu, M., Balas, V. E., Popescu, L. P., Mastorakis, N. 2009 Multilayer Perceptron and Neural Networks *WSEAS Transactions on Circuits and Systems*
[12] Hin, Celine 2018 Thermal Conductivity of Metallic Uranium. DOI: 10.2172/1433931

The Ratio Resin Polyester Composite Basal The To Testing Pull

Tumpal Ojahan R[1], Anang Ansyori[1], Zulfikar Alimuddin[1]

[1]Study of the techniqueprogram, machine engineering faculty, MalahayatiUniversity.

Corresponding e-mail: tumpal_ojahan@yahoo.com

***Abstract*.** Dropsy of the pollen and polyester resin constitutes the materials that can make material composite after the mixing of both , namely basal dust as filler and as a matrix. Polyester resin this research do the ratio of basal 40, mesh size of pollen 60, mesh and 150 mesh against 15 %, polyester resin 20 %, and 25 %.To get matrial, composite printed according to standard ASTM-D638 type V as for the purpose of this research is to obtain information regarding the ratio of polyester resin with basal powder against tensile strength and scored a maximum and minimum as well as determine the outcome of tensile strength of the ratio of basal powder 40, mesh 60, mesh150 mesh with resin polyester 15 % , 20 % , 25 % and a mixture of a catalyst 1/10, 1/20, 1/30 of the results of the analysis uses the method taguchi kompsisi resin polyester optimal there are on 25 % and the 60 mesh and catalyst 1/30 gr .The results of minimum found in composite basal the 150 mesh with force 60 Kgf with fariasi resin 15 % and catalyst 1/10 , while a maximum score found in composite basal the 60 mesh with fariasi resin 25 % and catalyst 1/30 with force of 135 Kgf. Of the result analysis using a method of taguchi kompsisi polyester resin levels found in 25%, while basal 60 mesh, powder and a catalyst 1/30 gr.

1. Introduction

Lampung provincial is one of producing basal rocks mfis in indonesia.Basal rocks is in the Lampung dark-colored, has characteristics: fine grained, and composed of mineral plagiokas and pirokesn.Basal rocks due to of the existence of the lava inundated on the surface or who settles hundreds of years on earth often also called to extrusive rock. Rocks basalt this can be processed to be used as a mix to get material composite [1]. Hyndman says that most basal formed as "bedrock" on earth. Basal more formed as an a .Most of the rocks basalt is in the surface of the sea and the earth as illustrated in Figure 1 (a) & (b)

Figure 1 (a). Rocks basalt.

Figure 1 (b). Rocks basalt path in the sea.

The division of basalt rocks in the broad can be divided into two categories which the first one categorized as Rocks Basalt Alkaline where a kind of owning amount minimize the action of an alkali

Na_2O and K_2O which greater than rocks basalt tholeitik. It is also containing titanium augit, fenokris olivine, an oxide of iron, nephelin and petrograph and underaturated right beside the road. The second category is the Rock Basalt Tholeiitic where it contains less Na_2O and K_2O compared to the first one. It is oversaturated and owning amount minimize pigeonit, augit subklasik, and interstitial glass. This type of rocks can be found as lava or very large extraction of magma. Recently, the mineral rocks basalt can be processed to the as a mix to material composite material.

According to Matthews and Rawing [2], composite is a material one which formed from acombination of two or more material one anddid not equally worthless and yet presuming through of a mixture that has are not homogenous, where is of the nature of the mechanics from each material one and did not because they were different from and yet presuming to strive of a mixture of the loan would also be produced material one and did not a composite having the nature of mechanical and of different characteristics of material one and did not equally worthless and yet presuming.

Material composite has been much used as material alternative a substitute for metal diberbagai industry, because matrial metal is already very difficult obtained and the value of selling very high.With the these then done research to get matrial a substitute for namely material composite.

Polyester resin is resin thermoset shaped, liquid with relatively low, viscosity containing monomers stiren so that the temperature thermal deformation lower than in resin termoset and have reached 110-140°C. Heat resistance In an arrangement of resin composite fibers responsible for protecting and tie to work well.The most polyester resin used especially for the application of lightweight construction, has the characteristic of low to be tinged typical, transparent, can be made rigid and flexible, waterproof, weatherproof and chemicals.Polyester resin can be used at a temperature of lower than last year reached 79°C network or higher depends particles resin and its need [3]. The characteristics of polyester resin are shown in Table 1. Hence, this paper aims to obtain information regarding the ratio of polyester resin with powder basalt and catalyst of tensile strength to get a maximum and minimum as well as to determine the tensile strength of the ratio of basal 40, mesh 60, mesh 150 mesh with 15, percent polyester resin 20, percent percent and a catalyst 25 1/10, 1/20, 1/30 taguchi by using the method

Table 1. Characteristic of polyester resin

Item	Units	Tipical Score	Note
The specific gravity of need to resort to violence	gr/cm3	1.4	25°C
The temperature of distortion heat	°C	70	
The uptake of water (room temperature)	%	0.188	24 hours
	%	0.446	7 day
Power fleksural	Kg/mm2	9.4	
Modulo fleksural	Kg/mm2	300	
tensile strength	Kg/mm2	5.8	
modulus of elasticity	Kg/mm2	300	
Elongation	%	2.4	

2. Research Methodology

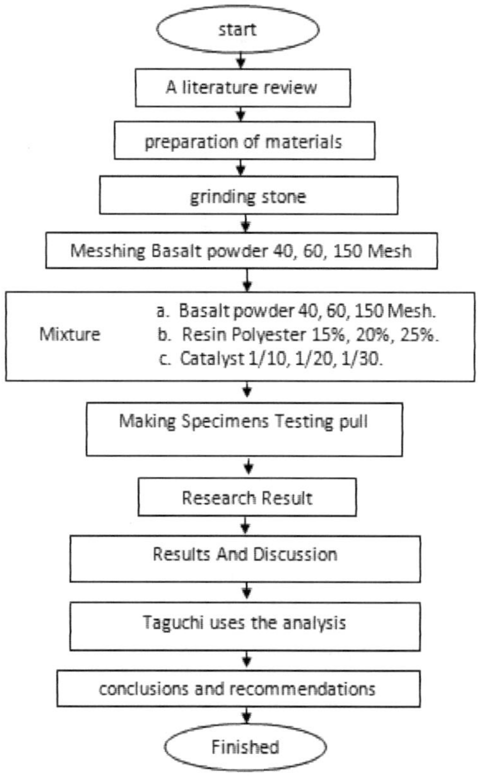

Figure 2. A diagram researches.

Figure 2 above indicates the work flow in order to complete this study. From the rocks basalt ± that has size of 20 cm x 30 cm undergoing a shredding process using a Jaw Crushier with size of 3x3 cm before it go to pegging process using Ballmill until it gained ground ballmill particles of basalt [4]. Figure 3 and 4 below illustrates the equipment used.

Figure 3. Jaw Crushier.

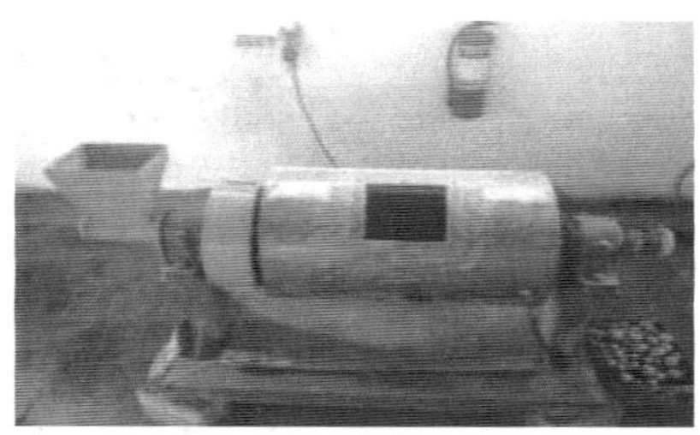

Figure 4. Ballmill.

The basalt was meshed using a smooth ayakan of 40, 60, and 150 mesh. The results were depicted in Figure 5 (a) to (c). The basalt then mixed with resin polyester and catalyst to get material composite with the weight of 40,60,150 mesh, 15%, 20%, 25% and 1/10, 1/20, 1/30 gr. The specimens that have stirred evenly were printed out using printings palaron ½ inches and it is a need to ensure that the specimen test pull conforming to standards ASTM D638 type V [5,9-12] using methods hand lay up. All the specimens were listed in coding system to know the variation of composite that has been done. The composite codes were tabulated in Table 2.

Figure 5 (a). 40 mesh

Figure 5 (b). 60 mesh

Figure 5 (c). 150 mesh

Table 2. Composite codes

No.	Size of Particles (*mesh*)			Resin (%)
	40	60	150	
1.	TA.A.2.1	TA.B.2.1	TA.C.2.1	15%
2.	TA.A.2.2	TA.B.2.2	TA.C.2.2	15%
3.	TA.A.2.3	TA.B.2.3	TA.C.2.3	15%
4.	TA.A.3.1	TA.B.3.1	TA.C.3.1	20%
5.	TA.A.3.2	TA.B.3.2	TA.C.3.2	20%
6.	TA.A.3.3	TA.B.3.3	TA.C.3.3	20%
7.	TA.A.4.1	TA.B.4.1	TA.C.4.1	25%
8.	TA.A.4.2	TA.B.4.2	TA.C.4.2	25%
9	TA.A.4.3	TA.B.4.3	TA.C.4.3	25%

After all the specimens were ready, every specimen was pull tested using type HT-2402 Computer Universal Testing Machines which shown in Figure 6 and the data were processed using Taguchi analysis.

Figure 6. Pull Machine.

3. Results And Discussion

3.1. Research Results

Table 3. The pull the resin polyester and basal of 40 mesh

No.	Basalt 40 mesh				
	Kode	*p*	*t*	*l*	*Max.F*
1	TA.A.2.1	62.90	11.10	14.10	60
2	TA.A.2.2	63.00	10.50	14.00	57
3	TA.A.2.3	63.15	9.91	13.91	58
4	TA.A.3.1	64.11	11.37	15.43	91
5	TA.A.3.2	63.02	11.7	13.98	70
6	TA.A.3.3	62.25	10.43	14.47	69
7	TA.A.4.1	63.11	10.23	14.37	66
8	TA.A.4.2	62.80	11.15	14.56	87
9	TA.A.4.3	64.12	9.74	14.63	96

Table 4. The pull the resin polyester and basal of 60 mesh

No.	Basalt 60 mesh				
	Kode	*p*	*t*	*l*	*Max.F*
1	TA.B.2.1	62.56	11.01	14.50	95
2	TA.B.2.2	62.70	10.81	14.76	98
3	TA.B.2.3	63.67	11.76	15,03	118
4	TA.B.3.1	63.50	10.76	15.68	126
5	TA.B.3.2	62.63	10.17	15.82	124
6	TA.B.3.3	63.13	10.9	14.84	123
7	TA.B.4.1	63.9	11.1	15.62	135
8	TA.B.4.2	62.94	11.49	15.59	139
9	TA.B.4.3	63.85	12.02	15.41	137

Table 5. The pull the resin polyester and basal of 150 mesh

No.	Basalt 150 mesh				
	Kode	*p*	*t*	*l*	*Max.F*
1	TA.C.2.1	62.57	10.97	14.35	53
2	TA.C.2.2	63.12	9.98	15.16	72
3	TA.C.2.3	63.10	9.89	15.02	70
4	TA.C.3.1	62.78	9.52	14.21	80
5	TA.C.3.2	63.08	9.17	14.74	87
6	TA.C.3.3	63.15	9.97	14.2	89
7	TA.C.4.1	63.16	8.78	13.99	92
8	TA.C.4.2	62.12	8.79	14.32	95
9	TA.C.4.3	63.31	9.41	14.7	89

Description:
p (mm) = Long on specimens
t (mm) = High on the specimen
l (mm) = Wide on specimens
Max.F (Kgf) = The power of a specimen strong pull test

3.2. A discussion of test results on pull the results of the power of test pull polyester resin

The results of the power of test pull polyester resin and basal powder 40 mesh the value of the lowest is 57 Kgf while the top is 96 Kgf. On the other hand, the results of the power of test pull polyester resin and basal powder 60 mesh the value of the lowest is 95 Kgf while the top is 139 Kgf and the results of the power of test pull polyester resin and basal powder 150 mesh the value of the lowest is 53 Kgf while the top is 95 Kgf.

On the specimen basal 40 powder mesh showing the value of tensile strength 96 Kgf, while the specimen basal powder 60 mesh showing the value of tensile strength 139 Kgf, that is greater than basal 40 mesh. Powder it was because the dust particles basal 60 mesh that serves as a booster (filler) has the tie (matrix) better than the basal 40 mesh. Powder the ratio of as many as 25 %, polyester resin better resources highchairs against dropsy powder 60 mesh than the ratio of polyester resin 15 % and 20 %.

In specimens basal the 150 mesh are showing value tensile strength 95 Kgf , while specimens basal powder the 60 mesh are showing value tensile strength 139Kgf, larger compared basal the 150 mesh.It was because particles basal the 60 mesh that serves as amplifier (filler) has its headband the (matrix) better than in basal the 150 mesh.The ratio resin polyester 25 %, better power strengh against dropsy 60 mesh than the ratio resin polyester 15 % and 20 %.

In a specimen of basal powder 150 mesh showing the tensile strength 95 Kgf, while the specimen basal of 40 mesh showing the tensile strength 96 Kgf, larger compared by basal of 150 mesh it was because the particles of basal 40 mesh that serves your (filler) has a connective (matrix) its better than this as a basal of 150 mesh the ratio of as many as percent, 25 polyester resin better resources ikatnya against dropsy of 40 mesh ratio than polyester resin 15 % and 20 %.

The results of the analysis are presented in Figure 7 of the curve of the strong pull testing to be the case .

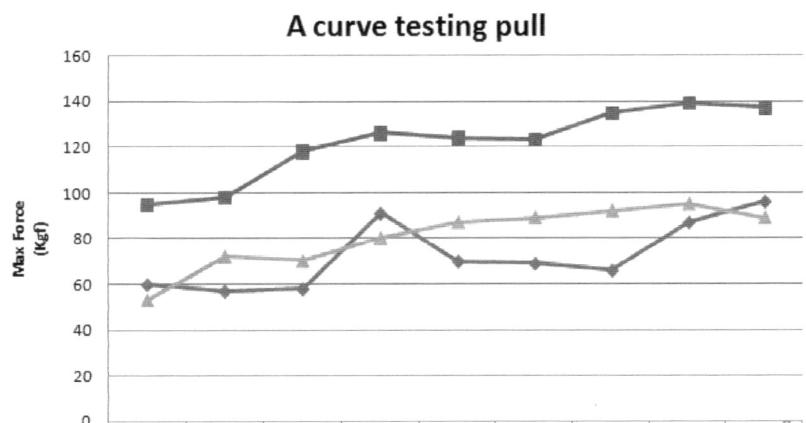

	15% (1/10)	15% (1/20)	15% (1/30)	20% (1/10)	20% (1/20)	20% (1/30)	25% (1/10)	25% (1/20)	25% (1/30)
basalt powder 40 mesh	60	57	58	91	70	69	66	87	96
Basalt powder 60 mesh	95	98	118	126	124	123	135	139	137
Basalt powder 150 mesh	53	72	70	80	87	89	92	95	89

Figure 7. The Pull Curve.

4. Analysis Support For New Uses The Method Taguchi

Through assistance with minitab 16 software able to obtain 9 specimens composite variations selected to analyze a specimen of 27, composite 9 spesiemen variations this is a composite 1 from every representative was chosen from masing-masing variations, basal powder , polyester resin and the catalyst.

The selection of 9 of 27 specimen is referred to as gust with Orthogonal Array (OA) , that is one of the a splinter group from guinea pigs all of whom only using a part of the condition of the total , which is hard for this part only menggambil men learned in the mosaic , a quarter or shall be an eighth of an experiment of factorial is full of .OA advantage is its ability to evaluate a number of factors to the total number of a test by which the a minimum [6-8]. Variation composite is close they can be seen in Table 6

Table 6. A specimen of a composite select with OA Taguchi.

Code Specimen	Factor			
	A	B	C	Max.F (Kgf)
TA.A.2.1	1	1	1	60
TA.A.3.2	1	2	2	70
TA.A.4.2	1	3	3	96
TA.B.2.1	2	1	2	98
TA.B.3.3	2	2	3	123
TA.B.4.1	2	3	1	135
TA.C.2.3	3	1	3	70
TA.C.3.1	3	2	1	80
TA.C.4.2	3	3	2	95

Description:
Factor A = Basalt Powder
Factor B = *Resin Polyester*
Factor C = catalyst

A factor with numbers 1 is 40 mesh, basalt powder the numbers 2 is 60 mesh, basalt powder and figures 3 is 150 mesh. Basalt powder to the factors b with numbers 1 is 15 %, polyester resin the numbers 2 is 20 %, polyester resin and figures 3 is 25%. polyester resin. While the c with numbers 1 is 1/10, catalyst the numbers 2 is 1/20 a catalyst and figures 3 is 1 / 30 catalyst. Figure 8 and Table 7 below shows the effect of each factor and an optimum result for each factor.

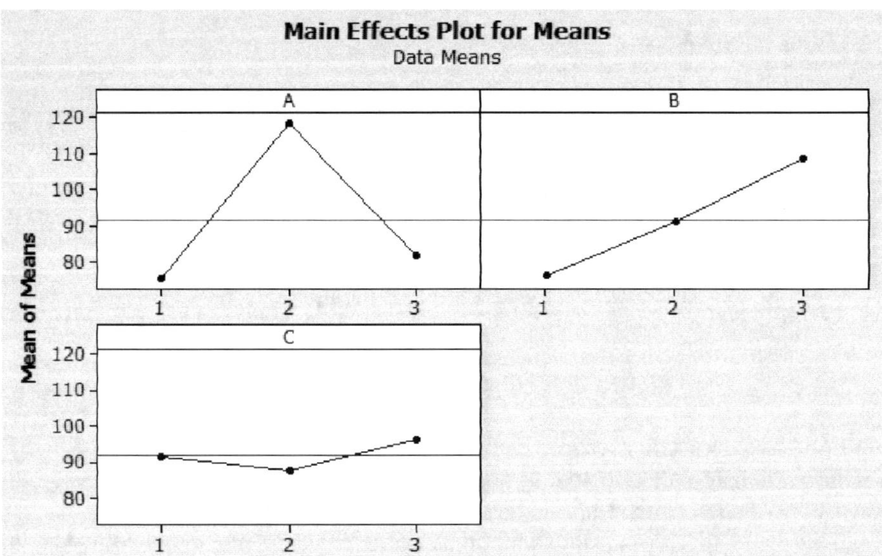

Description:
X-axis: Code Specimen
Y-axis: Max F (Kgf)

Figure 8. A plot the effect of each factor.

Table 7. The results of an optimum every factor

Level	A	B	C
1	37.37	37.43	37.74
2	41.41	38.92	38.76
3	38.17	40.60	39.45
Delta	4.04	3.17	0.70
Rank	1	2	3

5. Conclusion

As a conclusion, it was revealed from the result analysis using a of Taguchi's method, the composition polyester resin levels found in 25%, while basal 60 mesh, powder and a catalyst 1/30 gr. Other than that, the result of minimum was found in composite basal the 150 mesh with force 60 Kgf with fariasi resin 15% and catalyst 1/10, while a maximum score was found in composite basal the 60 mesh with fariasi resin 25 % and catalyst 1/30 with force of 135 Kgf.

References

[1] Hyndman Donald 1985 *Petrology of Igneous And Metamorphic Rocks* (Mc Graw Hill, New York)
[2] Matthews F.L, and Rawing RD 1993 *Composite Engineering and science, Imperal College of Science, Tecnology and Medicine* (London,UK)
[3] M. Schwartz 1984 *Composite Materials Handbook* (McGraw-Hill Book Company, New York)
[4] Balai Penelitian Teknologi Mineral 2019 (Tanjung Bintang, Lampung Selatan)
[5] American Society Of Testing And Matrals *Tensile Propeties of plastics.* Annual Book of ASTM Standards. (American Society for Testing and Matrials Philadelphia).
[6] Soejanto I. 2009 "Desain Eksperimen dengan Metode Taguchi", *Graha Ilmu* (Yogyakarta)
[7] Nandi 2010 Batuan, Mineral dan Batu Bara. *Handouts Geologi Lingkungan* (GG405) (Bandung)
[8] Surdia Tata 1995 *Pengetahuan Bahan Teknik* Edisi **2**
[9] American Society Of Testing And Matrals (ASTM),D20*Tensile Propeties of plastics*
[10] American Soctiety for Testing and Matrials (ASTM) D638 Annual Book of ASTM Standards Philadelphia.
[11] Ferri Rendika,Ferri K. 2011 Definisi Batuan (Universitas Gadjah Mada)
[12] Matthews F.L,, Rawing RD. 1993 *Composite Engineering and science, Imperal College of Science, Tecnology and Medicine,* (London, UK)

IOP Conf. Series: Materials Science and Engineering **807** (2020) 012029 doi:10.1088/1757-899X/807/1/012029

Simulation of Mach-Zehnder Interferometer plastic optical fibre for temperature and refractive index measurement

Ian Yulianti[1], N. M. Dharma Putra[1], Fianti[1], S. Maimanah[1], H. Rumiana[1], M. Y. M. Noor[2], A. S. M. Supa'at[2], O. Kurdi[3]

[1]Physics Department, Faculty of Mathematics and Natural Sciences, Universitas Negeri Semarang, Semarang, Indonesia
[2]School of Electrical Engineering, Faculty of Engineering, Universiti Teknologi Malaysia, Skudai, Johor Bahru, Malaysia
[3]Faculty of Mechanical Engineering, Universitas Diponegoro, Semarang, Indonesia

Corresponding e-mail: ianyulianti@mail.unnes.ac.id

Abstract. This work presents simulation of light propagation inside tapered structure- plastic optical fibre (POF) subjected to various surrounding temperatures and refractive index (RI). The tapered structure served as Mach-Zehnder interferometer (MZI) which split light into core modes and clad modes that recombined at the end of the taper. Simulation was done by using Beam Propagation Method (BPM) solved using Finite Difference Method (FDM). The light launched in to the MZI was varied from 611nm to 655nm. In order to give temperature effect, the temperature was varied from 30°C to 80°C with increment of 10°C. The results showed that the free spectral range (FSR) of the MZI are shifted as temperature increased. The sensitivity for the peak and dip of the FSR are -0.0158 nm/°C and -0.0153 nm/°C, respectively. For RI sensitivity investigation, the surrounding RI was varied from 1.333 to 1.349. The sensitivity of peak FSR and dip FSR to RI change are -98.638 nm/RIU and 105.4 nm/RIU. Therefore, the MZI can be used to measure RI and temperature simultaneously.

1. Introduction

Temperature and refractive index (RI) are important parameters in many applications such as biomedicine, biochemistry, food industry and environment. In chemical or food industry, temperature and RI measurement are used in quality control. Meanwhile, in environmental application, temperature and RI serve as indicators in assessing the level of pollution [1][2]. Therefore, researches have been conducted to develop temperature and RI sensor using various techniques. Among available techniques, optical fibre sensor is interesting since it is free from electromagnetic interference and also suitable for hazardous environment.

Temperature and RI measurement could be done simultaneously by modifying optical fibre structure and exploiting the measurement techniques. Zhang et al. [3] proposed optical fibre sensor using fibre Bragg grating (FBG) and Fabry-Perot interferometer (FPI) using single-mode silica optical fibre (SMF). Beside SMF, photonic crystal fibre (PCF) has also been used by modifying it in to D-shape and coating it by silver film so that it excite surface plasmon resonance (SPR) [4]. Moreover, the PCF was also filled by silver nanowire. The above-mentioned sensors rely on wavelength modulation technique. Other wavelength-based sensor for temperature and RI measurement was developed using microfibre

IOP Conf. Series: Materials Science and Engineering **807** (2020) 012029 doi:10.1088/1757-899X/807/1/012029

resonator and FBGs [5]. Wavelength-based sensor offers higher accuracy compared to intensity-based sensor since its measurement is independent from intensity fluctuation and loss. However, complex fabrication technique limits its advantages.

Wavelength-based sensor could also realize using Mach-zehnder interferometer (MZI). MZI has several advantages such as good linearity [6], low cost [7], compact size and easy fabrication process [8]. In-line MZI using polymer optical fibre (POF) has been proposed as RI and strain sensor by using simple fabrication technique [9]. Graded-index POF (GI-POF) was used to form the MZI by forming taper using heat-and-pull technique. It was shown that the sensor has sensitivity in the order of 10^{-1}pm and 10^{1} nm for strain and RI, respectively. Due to the thermo-optic nature, MZI structure could also be employed for temperature measurement.

In this work, we studied the response of in-line MZI using step-index POF (SI-POF) to temperature and RI theoretically using numerical method. SI-POF was used since it has larger diameter and therefore sturdier than GI-POF. Larger diameter also makes SI-POF is easy to handle. The simulations were done by using Beam Propagation Method (BPM), which was solved numerically using Finite Difference Method (FDM) employing BeamProp software.

2. Sensor Working Principle

The in-line MZI structure consists of two tapers separated by an interferometry region with length of L, as shown in Figure 1. Light launched in to the MZI propagates in the core mode and when it passes through the first taper, some of the light refracted in to the cladding and propagates along the interferometry region. At the second taper, the cladding modes then recombine with the core modes. Due to the difference between RI of core and clad, the light travelled in both paths have a phase difference which further results in interference when the light recombines.

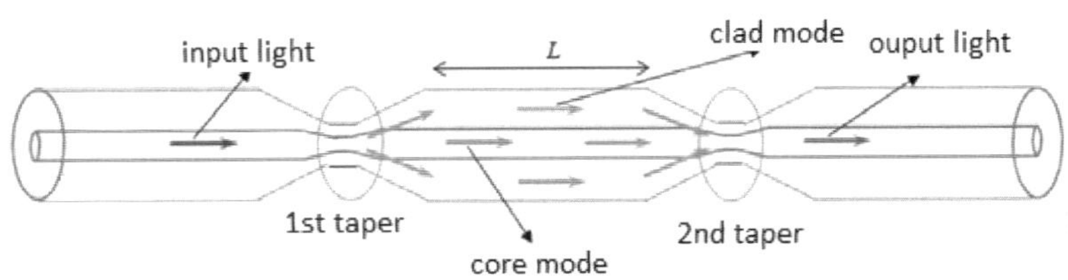

Figure 1. In-line MZI structure and light splitting inside it.

The intensity of the interference light is defined by [10]

$$I = I_1 + I_2 + 2\sqrt{I_1 I_2}\cos\Delta\phi \tag{1}$$

where I_1 and I_2 are the intensity of core modes and intensity of clad modes, respectively. $\Delta\phi$ is the phase difference between core modes and clad modes which is define by:

$$\Delta\phi = \phi_{core} - \phi_{clad} = \frac{2\pi\left(n_{eff}^{core} - n_{eff}^{clad}\right)L}{\lambda} = \frac{2\pi\Delta n_{eff}L}{\lambda} \tag{2}$$

where λ, n_{eff}^{core} and n_{eff}^{clad} are the light wavelength, the effective refractive index (ERI) of core and clad, respectively. Δn_{eff} is the difference between ERI of core and clad. When the phase difference satisfy $\Delta\phi = (2m+1)\pi$, m= 0,1,2,..., the interference light intensity reach a maximum value. The wavelength at which the maximum interference occurs is defined by [6]:

$$\lambda_m = \frac{\Delta n_{eff} L}{2m+1} \tag{3}$$

The free spectral range (FSR) between two maximum intensity is defined by

$$\text{FSR} = \lambda_m - \lambda_{m-1} \approx \frac{\lambda^2}{\Delta n_{eff} L} \tag{4}$$

where λ_m and λ_{m-1} are the mth and $(m-1)^{th}$ order of the peak wavelengths. From (4), it is clear that the FSR is depend on the ERI of both core and clad. The ERI of clad at tapered region is affected by the RI of the surrounding [6] and also by the temperature through the thermo-optic effect. Therefore, by observing the change of the FSR of the output spectrum, temperature and RI can be determined. However, since both temperature and RI contribute to the FSR, then we proposed to measure FSR of peak and dip. Therefore, using sensitivity of FSR of peak and FSR of dip to temperature and RI, simultaneous measurement of temperature and RI could be done.

3. Method

Prior to simulation, the in-line MZI was first fabricated using heat-pull method [9] using SI-POF with n_{core} of 1.49 and n_{clad} of 1.41. The POF core and clad diameter are 980 µm and 1mm, respectively. The thermo-optic coefficient (TOC) of the PMMA is -1.2×10^{-4}°C^{-1} [11]. The interference region (L) was set to be 2 cm. The waist diameter of the first and the second taper are 872 µm and 790 µm, respectively. Complete dimension specification is shown in Figure 2. To obtain the output spectrum, the light propagation inside the MZI was simulated with wavelength in the range of 611nm to 655nm. To investigate the response to temperature change, the surrounding temperature was varied from 30°C to 80°C with increment of 10°C, while the surrounding RI was maintained at air RI which is 1.0. Meanwhile, to obtain sensor sensitivity to RI change, the RI was varied from 1.333 to 1.349 which are the RI of glucose with various concentration [12].

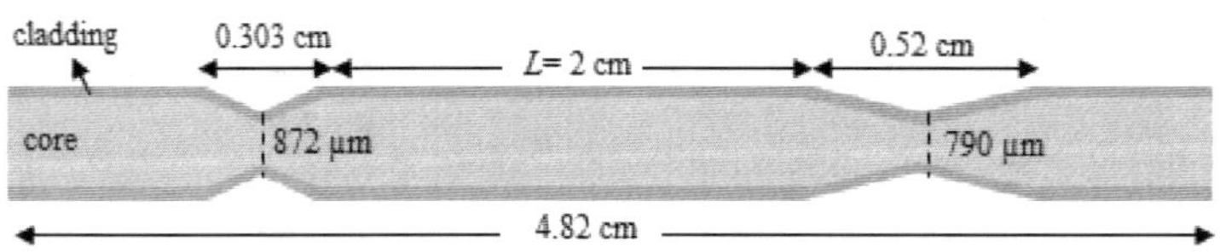

Figure 2. Dimension specification of MZI used in the simulation.

4. Results and Discussions

Simulation was done to obtain output spectrum of the MZI subjected to temperature and RI change. From the output spectrum, the FSR change was then obtained. By plotting the FSR of peak and FSR of dip to temperature and RI, the MZI sensitivity to temperature and RI were determined.

4.1. Temperature sensitivity

Light propagation inside MZI at temperature of 30°C and wavelength of 650nm is shown in Figure 3(a). It is shown that as light propagates through the first taper, some of the light penetrates through the clad as shown by red colour that spread out to the core-clad interface at z of ~1.2 cm, as expected. The steep taper of the first taper caused refraction with high intensity at the core-clad interface. At z of around 3 cm at where the second taper located, the light interference occurs. The smooth structure of the second taper minimize refraction and contributes to the higher intensity of interference of core modes and clad modes [12]. The output spectrum at temperature of 30°C is shown in Figure 3 (b).

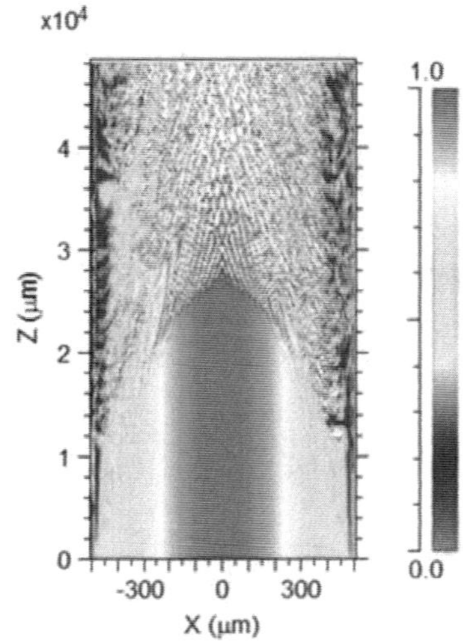

Figure 3 (a). Light propagation inside MZI at temperature of 30°C and wavelength of 650nm.

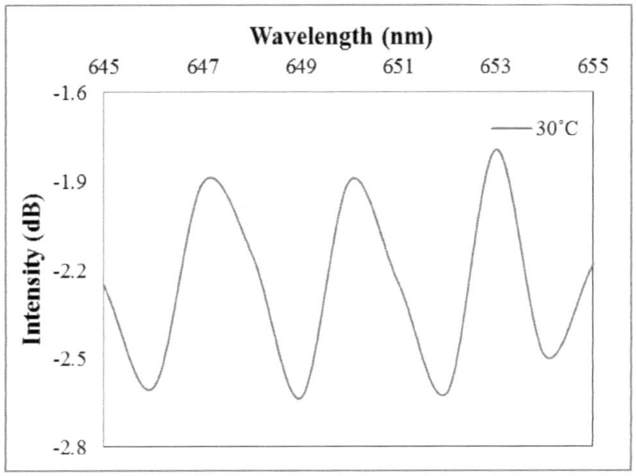

Figure 3 (b). The output spectrum at temperature of 30°C.

The change of FSR of peak and dip were plotted against temperature value as shown in Figure 4. It is shown that as temperature increases, the FSR of both peak and dip decrease. The sensitivity of FSR of peak to temperature is -0.0158 nm/°C with correlation coefficient of 87.83%. Meanwhile, for FSR of dip, the sensitivity is – 0.0153 nm/°C with correlation coefficient of 94.9%.

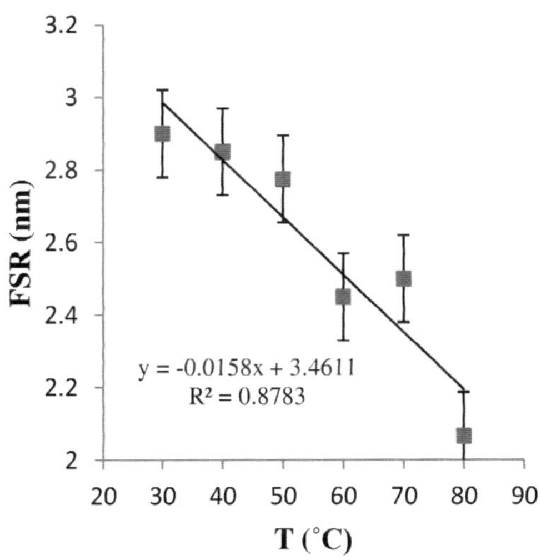

Figure 4 (a). The change of FSR of peak.

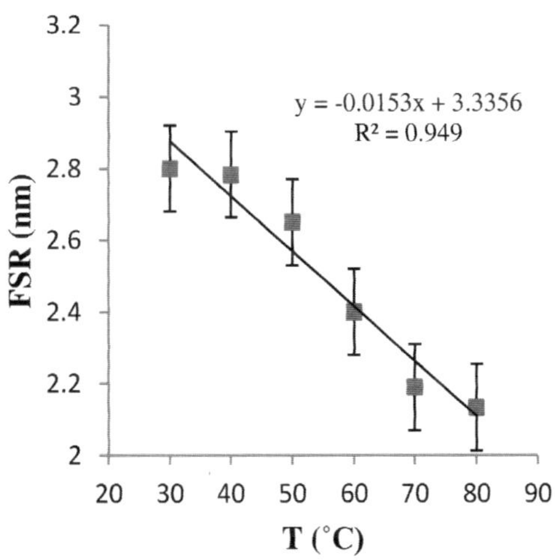

Figure 4 (b) The change of FSR of dip.

*The error bars represent the standard error of the data

4.2. RI sensitivity

For RI sensitivity investigation, the temperature was maintained at 28°C. The change of FSR due to RI change of both peak and dip are shown in Figure 5. As response to temperature, it is also observed that the FSR decreases as RI increases. It was found that the sensitivity of peak FSR to RI change is -98.638

nm/RIU with correlation coefficient of 96.12 %. Meanwhile, higher sensitivity which -105.4 nm/RIU was obtained for dip FSR. However, linearity of dip FSR was lower compared to that of peak FSR as indicated by lower correlation coefficient which is 90.1 %. The sensitivity is higher than that of GI-POF based MZI [9].

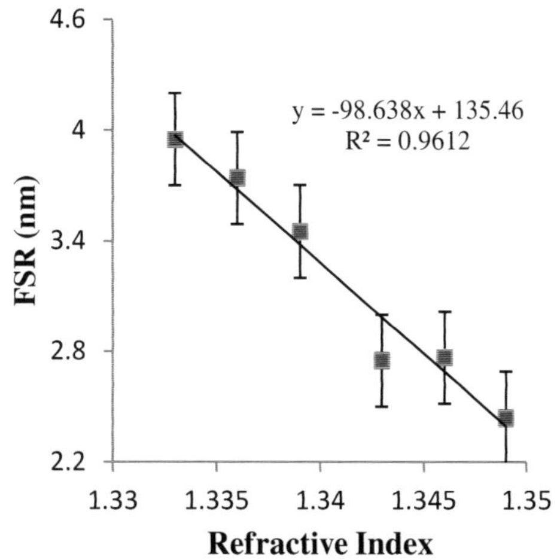

Figure 5 (a). The change of FSR of peak.

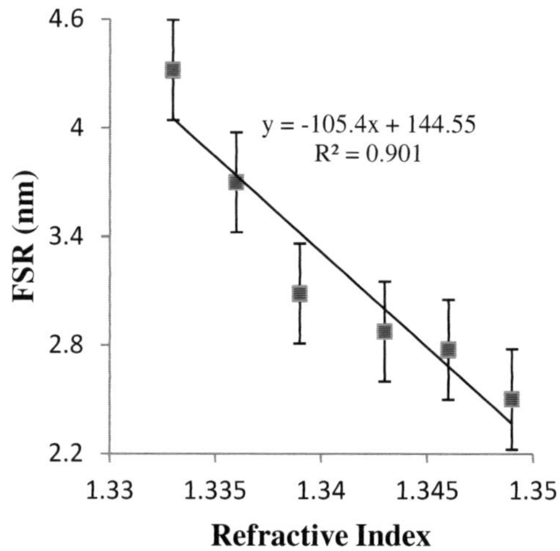

Figure 5 (b) The change of FSR of dip.

* The error bars represent the standard error of the data.

4.3. Temperature and RI determination

Simultaneous determination of temperature and RI can be done by using sensitivity of FSR of peak and dip to temperature and RI. Since both FSR of peak and dip depend on temperature and RI, then the change of both FSRs can be defined by

$$\Delta FSR_p = \frac{\partial FSR_p}{\partial T}\Delta T + \frac{\partial FSR_p}{\partial n}\Delta n \tag{5}$$

$$\Delta FSR_d = \frac{\partial FSR_d}{\partial T}\Delta T + \frac{\partial FSR_d}{\partial n}\Delta n \tag{6}$$

where ΔFSR_p and ΔFSR_d are the change of peak FSR and dip FSR, respectively. $\frac{\partial FSR_p}{\partial T}$ and $\frac{\partial FSR_p}{\partial n}$, are sensitivity of peak FSR to temperature and RI, respectively. Meanwhile, $\frac{\partial FSR_d}{\partial T}$ and $\frac{\partial FSR_d}{\partial n}$ are sensitivity of dip FSR to temperature and RI, respectively. By using sensitivity obtained from the characterization, then equation (5) and (6) can be written as

$$\Delta FSR_p = -0.0158\ \Delta T + (-98{,}638)\Delta n \tag{7}$$

$$\Delta FSR_d = -0.0153\ \Delta T + (-105{,}4)\Delta n \tag{8}$$

IOP Conf. Series: Materials Science and Engineering **807** (2020) 012029 doi:10.1088/1757-899X/807/1/012029

Equation (7) and (8) can be written in matrix form as

$$\begin{bmatrix} -0.0158 & -98.638 \\ -0.0153 & -105.4 \end{bmatrix} \begin{bmatrix} \Delta T \\ \Delta n \end{bmatrix} = \begin{bmatrix} \Delta \mathrm{FSR}_p \\ \Delta \mathrm{FSR}_d \end{bmatrix} \tag{9}$$

Therefore, temperature and RI change can determine simultaneously by solving the matrix equation, as defined by

$$\begin{bmatrix} \Delta T \\ \Delta n \end{bmatrix} = \frac{1}{D} \begin{bmatrix} -105.4 & 0.0153 \\ 98.638 & -0.0158 \end{bmatrix} \begin{bmatrix} \Delta \mathrm{FSR}_p \\ \Delta \mathrm{FSR}_d \end{bmatrix} \tag{10}$$

where D is the matrix determinant.

5. Conclusions

Investigation of the sensitivity of in-line MZI based on POF to temperature and RI have been done numerically. The results showed that surrounding temperature and RI change the output spectrum of the MZI. In terms of FSR, the MZI has temperature sensitivity of -0.0158 nm/°C and -0.0153 nm/°C for peak FSR and dip FSR. It is also observed that the MZI has higher sensitivity to RI change. The sensitivity of peak FSR and dip FSR to RI change are -98.638 nm/RIU and 105.4 nm/RIU. By observing the peak FSR and dip FSR, simultaneous measurement of temperature and RI can be accomplished.

Acknowledgements

We would like to thank to Ministry of Research, Technology and Higher Education, Indonesia for funding the research through grant no of 66.18.3/UN37/PPK.3.1/2019. Our gratitude also goes to Photonic Research Laboratory, School of Electrical Engineering Faculty of Engineering, Universiti Teknologi Malaysia for providing the BeamProp software and to the members of Physics Department, Universitas Negeri Semarang for their helpful discussion throughout the completion of this work.

References

[1] Harris J, Lu P, Larocque H, Chen L and Bao X 2015 *Sens. and Act. B: Chemical* **206** 246-251

[2] Yao Q, Meng H, Wang W, Xue H, Xiong R, Huang B, Tan C, and Huang X 2014 *Sens. and Act. A: Phys* **209** 73-77

[3] Zhang T, Liu Y, Yang D, Wang Y, Fu H, Jia Z, and Gao H 2019 *Opt. Fibre Technol* **49** 64–70

[4] Zhao L, Han H, Lian Y, Luan N, and Liu J 2019 *Opt. Fibre Technol* **50** 165–171

[5] Xiao Y Y, Cai X P, and Chen H 2019 *Optik (Stuttg)* **191** 116–120

[6] Wang Q, Wang B T, Kong L X, and Zhao Y 2017 *IEEE Trans. on Instr. and Meas.* **9** 2483–2489

[7] Ma Y, Qiao X, Guo T, Wang R, Zhang J, Weng Y, Rong Q, Hu M, and Feng Z 2012 *IEEE Sens J.***12** 2081–2085

[8] Zhao N, Lin Q, Jing W, Jiang Z, Wu Z, Yao K, Tian B, Zhang Z, and Shi P 2017 *Sens. and Act.***267** 491-495

[9] Jasim A A, Hayashi N, Harun S W, Ahmad H, Penny R, and Mizuno Y 2014 *Sens. Act. A: Phys* **219** 94–99

[10] Wang Q, Wei W, Guo M, and Zhao Y 2016 *Sens. and Act. B: Chemical* **222**159-165

[11] Luo Y, Yan B, Zhang Q, Peng G, Wen J, and Zhang J 2017 *Sensors* **17** 511

[12] Bhardwaj V and Singh V K 2016 *Sens. and Act. A: Phys* **244** 30-34

IOP Conf. Series: Materials Science and Engineering **807** (2020) 012030 doi:10.1088/1757-899X/807/1/012030

Quality Service Analysis at Pusat Pelayanan Pendidikan Terpadu (P3T) Universitas Malahayati Using Fuzzy and Service Quality Method

M W Wardana[1*], A Sidiq[1], H Wibowo[1], Y Pangestu[1]

[1]Industrial Engineering Department, Universitas Malahayati
Jl. Pramuka No.27 Kemiling, Bandar Lampung 35153

Corresponding e-mail: marcelly@malahayati.ac.id

Abstract. Each institution must try to provide the best service for costumers, but the level of customer satisfaction with a service is different. Every customer always expects to get the optimal service they want. This study aims to determine the level of service quality (Servqual) in the Quality Analysis of Pusat Pelayanan Pendidikan Terpadu (P3T) with Fuzzy Methods and Quality Services at Universitas Malahayati based on five dimensions of Tangible, Reliability, Responsiveness, Assurance and Empathy. This research was conducted by distributing questionnaires to P3T service users and then testing the validity and reliability. The method used is the fuzzy method followed by the calculation of the gap. Respondents in this study were 100 mixed students of various study programs at Universitas Malahayati with an error rate of 10%. The results of the calculation showed that the service performance provided by P3T was almost all the attributes whose performance must be improved. By improving the attributes that have high inequality values, it is expected to improve the quality of the P3T service so that it can increase the trust of service providers to customers.

1. Introduction

Pusat Pelayanan Pendidikan Terpadu (P3T) is one of the facilities engaged in the services provided for Universitas Malahayati students, in order to meet the information service needs of Universitas Malahayati students. Examples of service providers or other similar institutions in the Bandar Lampung area, for example, Universitas Teknokrat and IBI Darmajaya have provided online payment services (to banks) whose results will go directly to the institution and at the same time have been automatically validated, so students do not need to queue taking too long for payment or validation. Therefore, the P3T of Universitas Malahayati in Bandar Lampung needs to improve the quality of service. Its production is intangible and is not linked to a physical product so that service is the behavior of producers in order to meet the needs and desires of consumers for the achievement of satisfaction to consumers themselves [1]. This study aims to identify the perceptions, expectations of students and the level of gap P3T service quality at Universitas Malahayati with the Fuzzy and Service Quality methods. Fuzzy Method and Service Quality can translate students' wants and needs to service quality. This research are related to Fuzzy and Servqual in analyzing service quality as conducted in the hospital inpatient installation services [2], and the public transportation [3].

2. Methods

The data in this study is data obtained from student responses using the questionnaire method. This questionnaire method is used because of the importance of communication between service providers and users to improve service quality [4],[5]. Student response data obtained through questionnaires, observations, direct recording in the field. The data used in the form of a questionnaire relating to the quality of services dimensions tested from Tangible, Reliability, Responsiveness, Assurance and Empathy [2],[3]. They are called servqual that can define the perceptions and expectations of customers with serve received [3],[6]. Then test the adequacy of the data followed by testing the validity of using SPSS on the data collected if the data obtained invalid data then the data needs to be discarded or removed until there are all valid data and continued with reliability testing if the data is not reliable then it returns to the collector data if the data is reliable all the data processing can be done questionnaire. The data is processed using the fuzzy servqual method, namely by calculating fuzzification with TFN each attribute on the questionnaire [3],[7]. Then do the mean calculation again for each attribute. After that the results can be interpreted into defuzzification and then obtained servqual results per attribute and per dimension as a whole, after which the gap calculation is done to determine the quality of the service. Then conclusions can be made towards the end.

3. Results and Discussion

Table 1. Service Quality Attribute

No	Attributes
1	The room is spacious, clean and neat
2	There are enough seats for visitors
3	Availability of complete facilities (air conditioning, chairs, internet)
4	Arrangement of counters according to their function
5	Payment system, KRS input etc. based online
6	Start to work on time
7	Having the ability to process a job
8	Provide completion and handling services quickly
9	The ability to present information correctly
10	The attitude of employees gives trust to visitors
11	The queue time is too long
12	The officer is swift in serving
13	Officers have free time to help
14	All transactions are served or carried out quickly and precisely
15	Ease of getting and clarity of information
16	Officers always smiled and greeted the visitors
17	Polite the officers of treating visitors
18	Friendliness of the officers when serving
19	Honesty of the officers when providing information
20	Officers have the knowledge and ability to provide information
21	Give special attention to visitors
22	Able to help the visitors who are confused
23	Having seriousness in responding to visitor requests
24	Officers do not look at social status and treat visitors with respect and courtesy
25	Officers greets at the beginning and end of the service

Table 2. Results of Attribute Validity of Code P1 (Perception) and H1 Code (Expectations) With SPSS 16.00

	Item_P1	Item_total	Item_H1	Item_total	Explanation
Pearson Correlation	1	0.445**	1	0.745**	VALID
Sig. (2-tailed)		0.000		0.000	
N	100	100	100	100	
Pearson Correlation	0.445**	1	0.745**	1	
Sig. (2-tailed)	0.000		0.000		
N	100	100	100	100	

Table 3. Results of Attribute Reliability of Code P1 (Perception) and H1 Code (Expectations) With SPSS 16.00

	Item_P1	Item_total	Item_H1	Item_total	Explanation
Pearson Correlation	1	0.757**	1	0.943**	RELIABLE
Sig. (2-tailed)		0.000		0.000	
N	100	100	100	100	
Pearson Correlation	0.757**	1	0.943**	1	
Sig. (2-tailed)	0.000		0.000		
N	100	100	100	100	

After recapitulating the perception and expectation questionnaire data, fuzzification is performed by forming Triangular Fuzzy Numbers (TFN). The TFN provisions are where when the minimum lower limit value is not reduced by 1 and when the maximum upper limit is not added 1, in the following way [8],[9]:

For the first respondent data that assesses the attribute P1 = 4 then;
$\text{Lower limit} \quad = 4 - 1 = 3$
$\text{Middle Value} \ = 4$
$\text{Upper limit} \quad = 4 + 1 = 5$

The calculation results for the mean attribute code P1 are (3.08;4.08;4.81). The next step is defuzzification to change TFN to:

Mean value of TFN = (3.08;4.08;4.81)

$$\text{Defuzzification} = \frac{(\text{Lower} + \text{Middle} + \text{Upper})}{3} = \frac{3.08 + 4.08 + 4.81}{3} = \frac{11.97}{3} = 3.99$$

Table 4. Value of Fuzzy Mean (TFN) of Respondents' Perception

No	Attribute Code	Triangular Fuzzy Numbers			Defuzzification	No	Attribute Code	Triangular Fuzzy Numbers			Defuzzification
		Low	Mid	Upp				Low	Mid	Upp	
1	P1	3.08	4.08	4.81	3.99	1	H1	3.64	4.64	4.98	4.42
2	P2	2.30	3.30	4.23	3.28	2	H2	3.68	4.68	4.99	4.45
3	P3	2.56	3.55	4.34	3.48	3	H3	3.72	4.72	4.99	4.48
4	P4	2.67	3.67	4.52	3.62	4	H4	4.24	4.24	4.86	4.45
5	P5	2.51	3.51	4.28	3.43	5	H5	3.81	4.81	4.98	4.53
6	P6	1.91	2.90	3.89	2.90	6	H6	3.78	4.78	4.98	4.51

7	P7	1.72	2.69	3.66	2.69	7	H7	3.75	4.75	4.97	4.49
8	P8	1.68	2.67	3.66	2.67	8	H8	3.75	4.75	4.98	4.49
9	P9	2.15	3.10	4.08	3.11	9	H9	3.81	4.81	4.97	4.53
10	P10	1.90	2.87	3.87	2.88	10	H10	3.69	4.69	4.97	4.45
11	P11	1.28	2.01	3.01	2.10	11	H11	3.87	4.87	5.00	4.58
12	P12	1.71	2.66	3.66	2.68	12	H12	3.73	4.73	5.00	4.49
13	P13	1.76	2.72	3.72	2.73	13	H13	3.69	4.69	4.99	4.46
14	P14	1.42	2.40	3.40	2.41	14	H14	3.81	4.81	4.98	4.53
15	P15	1.95	2.90	3.89	2.91	15	H15	3.81	4.81	5.00	4.54
16	P16	1.55	2.41	3.41	2.46	16	H16	3.79	4.79	4.98	4.52
17	P17	1.76	2.73	3.75	2.75	17	H17	3.81	4.81	4.99	4.54
18	P18	1.30	2.18	3.19	2.22	18	H18	3.82	4.82	4.99	4.54
19	P19	2.33	3.32	4.27	3.31	19	H19	3.80	4.80	4.99	4.53
20	P20	2.24	3.22	4.16	3.21	20	H20	3.77	4.77	4.98	4.51
21	P21	1.87	2.84	3.83	2.85	21	H21	3.71	4.71	4.98	4.47
22	P22	2.07	3.06	4.06	3.06	22	H22	3.69	4.69	5.00	4.46
23	P23	1.85	2.84	3.84	2.84	23	H23	3.71	4.71	4.98	4.47
24	P24	1.98	2.98	3.96	2.97	23	H24	3.81	4.81	4.98	4.53
25	P25	1.61	2.36	3.34	2.44	25	H25	3.80	4.80	4.99	4.53
		Total			72.99						

From the calculation of the perception defuzzification value obtained, an average search is 72.99/25 = 2.92. Value 2.92 is the value of service quality. This value if represented by the value of satisfaction (in the questionnaire column) is not satisfied.

After obtaining the perception value and the expectation value of respondents in the form of defuzzification, then the servqual value is calculated using the following formula to find the gap value [9],[10]:

GAP Value = Mean Perception - Mean Expectation
For the 1st attribute = 3.99 - 4.42 = - 0.43

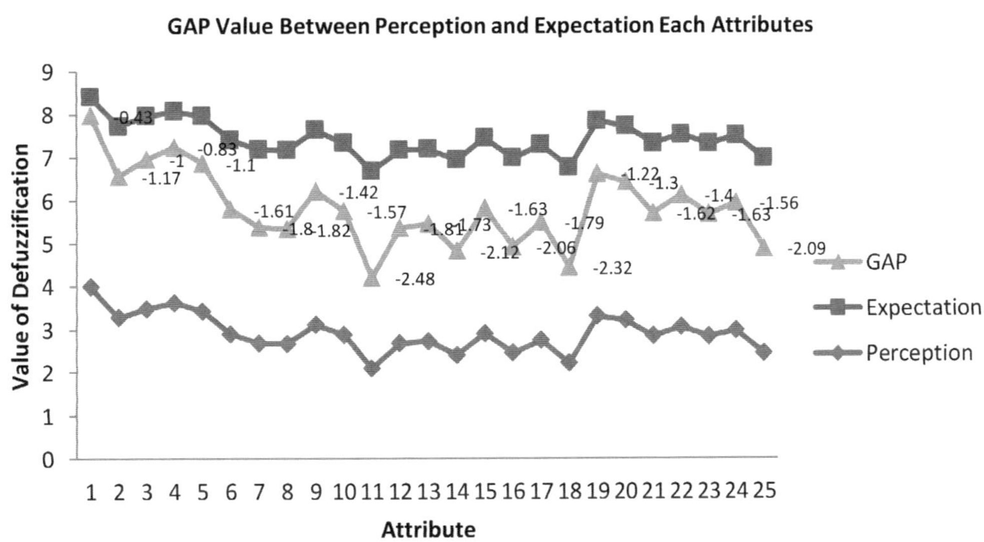

Figure 1. Gap Between Perception and Expectation

The largest gap value indicates that the attribute has a low value of customer satisfaction while the attribute that has the smallest gap value indicates that the quality of services provided is almost in line with the expectations of the respondent. It turned out that respondents considered that the queue time was too long, something that needed to be considered and had to be minimized by P3T. Especially for financial counters by students who want to make payments at financial counters, there is often a queue build-up due to too long service time and too few employees at financial counters. Service providers or other similar institutions in the Bandar Lampung region, for example, Universitas Teknokrat and IBI Darmajaya have provided online payment services (to banks), the results of which will go directly to institutions and at the same time have been automatically validated, so students do not need to queue too for payment or validation. Electronic payments via banks or others are automatically validated at the Universitas Teknokrat and IBI Darmajaya can be used as a good example for the future in financial services at the Universitas Malahayati.

The criteria that have almost met the respondents' expectations are the spacious, clean and tidy room, the majority of service users feel quite comfortable because every time there is a janitor who always controls the cleanliness of the P3T room.

The main priorities in improving the quality service of P3T are some of the attributes of the questionnaire that are too long to queue, the friendliness of the officers when serving, the officers greeting at the beginning and end of service, all transactions are served or carried out quickly and accurately, the officer always smiles when greet customers, have the ability to process a job. After knowing which attributes are influential, companies need to immediately fix related to the queue time that is too long for students of Universitas Malahayati, at least there is a change in the queuing system or accelerate transactions to students so that the quality of services provided is able to meet customer expectations. Settling gradually from the largest gap to the smallest value in the attributes mentioned in the questionnaire.

Table 5. Highest and Lowest Gap Value

No.	Attribute	Expectations	Explain
1	Too Long to Queue Time	-2.48	Lowest
2	Spacious, Clean and Neat Room	-0.43	Highest

To observe the dimensions of service quality, the gap value must be added by grouping the number of each dimension which can be seen in the Table 6.

Table 6. Gap Value for Dimensions Quality

No.	Dimensions Quality	Gap Value
1	Tangible	-4.53
2	Reliability	-8.22
3	Responsiveness	-9.77
4	Assurance	-8.69
5	Empathy	-8.30

Based on Table 6 the lowest value gap dimension quality is Responsiveness that is -9.77 and the highest is the value gap dimension quality from Tangible which is -4.53. The gap value must be reduced by increasing all the attributes that exist in the service provider.

Based on the value of perception shows the quality of service felt by respondents while using the P3T. Defuzzification value after sorting in data processing is known that the attribute that has the greatest value is a spacious, clean and tidy room (P1) with a value of 3.99. Meanwhile, the smallest value, the worst attribute, is the queue time that is too long (P11) with a value of 2.10. From the acknowledgment of some respondents of a spacious, clean and tidy room, P3T is indeed good and students also feel happy because the cleaning staff always do cleaning to keep the P3T room clean, while regarding the queue time that is too long indeed the P3T uses a queue number for each transaction at each counter, but with the queue number not accompanied by the speed to conduct transactions, especially at the financial counter so there is a build up at the financial counter so students feel bored and disappointed while waiting too long.

While the respondent's expectation value indicates the level of desire of the respondent to the quality of service that should be provided by P3T. Defuzzification value after the highest ranking is too long queuing time (H11) with a value of 4.58, while the lowest value in the respondents' expectations is a spacious, clean and tidy (H1) room with a value of 4.42. Queuing time is too long a person's top priority in choosing services and this is also desired by students. Service providers must minimize queuing time for students too long so that the value of the service providers is higher and trusted. Spacious, clean and tidy room turned out to have a fairly low value. This shows that the spacious, clean and tidy room turned out not to be the main priority in the service. But it is better if the service provider provides a spacious, clean and tidy room so that students feel comfortable when entering the service room.

4. Conclusion

The results of the analysis that using Fuzzy and Servqual method reveals that the level of service quality at P3T Universitas Malahayati based on the level of visitor satisfaction is not good, hence, service improvements are needed to increase customer satisfaction.

References

[1] Kotler P and Keller K L 2012 *Marketing Management* Vol **13** (New Jersey : Pearson Prentice Hall, Inc).
[2] Khikmawati E Wibowo H Farla Z 2019 *Jurnal Teknik Industri* **5 1** pp 56-61.
[3] Deveci M, Öner SC, Canıtez F, Öner M. 2019 *Research in Transportation Business & Management* 100387.
[4] Achrol, R.S., Kotler, P., Ahmed, P.K., Rafiq, M. and Albrecht, K. 2017 In *Internal Marketing: Another Approach to Marketing for Growth* (Vol. **55**, No. 4, pp. 1-3).
[5] Liu R, Cui L, Zeng G, Wu H, Wang C, Yan S and Yan B 2015 *Applied Soft Computing* **26**:508-12
[6] Tjiptono F 2006 *Manajemen Jasa* Vol **1**
[7] Kusumadewi S dan Purnomo H 2004 *Aplikasi Logika Fuzzy Untuk Pendukung Keputusan* Vol **2**
[8] Kusumadewi S .2010. Analisa dan Desain Sistem Fuzzy Menggunakan Tool Box Matlab Vol **2**
[9] Perçin S.2018 Journal of Air Transport Management **68**:48-60.
[10] Herdiyanti A, Adityaputri AN, Astuti HM 2017 *Procedia Computer Science* **124**:60, 1-7.

The Effect of Orientation Fibres on Flexural and Tensile Properties of Arenga Pinnata Fibres Reinforced Polyester Composites

R D Widodo[1]*, W Robi[1], Rusiyanto[1], A Nugroho[1] and D H Al-Janan[1]

[1] Department of Mechanical Engineering, Faculty of Engineering, Universitas Negeri Semarang (UNNES), Kampus Sekarang Gunungpati, Semarang, Indonesia

Corresponding e-mail: rahmat_doni@mail.unnes.ac.id

Abstract. This paper presents flexural and tensile properties of sugar palm (Arenga pinnata) fibres-reinforced polyester composites study. The fibres were treated by alkaline solution with 5% NaOH solution for 2 hours of soaking time. The composites were prepared with different orientations of fibres (i.e. 0° + 45°, 0° + 90°, 45° + 90°, and 45° + (-45°)), while the ratio between Arenga pinnata fibres and polyester is 30 percent volume. Hand lay-up method was employed to produce the specimens. ASTM D-790 03 and ASTM D-3039 standard were employed to characterize the specimens in good sequent. As the result, the specimen with 45°+90° orientation generated maximum important values both in flexural strength (24.03 MPa) and modulus (4.01 GPa), tensile strength (23.84 MPa) and modulus elasticity (0.97 GPa). This is due to palm fibres as reinforcement that forms an angle of 90 ° or increasingly upright to the load crossing field (load of bending) and tensile load giving the effect of maximum reinforcement compared to other fibres orientations.

1. Introduction

The use of synthetic fibres as a composite reinforcement has a negative impact on the environment due to non-recyclable waste material issue. Therefore, the use of natural fibres is a solution to overcome this problem. Natural fibre composites have other advantages compared to glass fibres because they are more abundant, environmentally friendly because they can degrade naturally and are cheaper than glass fibres [1-11]. One of the widely used natural fibres is palm fibre or locally known as *ijuk* from Arenga pinnata tree.

The use of palm fibre is widely use in household needs and building materials such as brooms, doormat, ropes and roofs. Palm fibre has the advantage of being resistant to seawater and acid, slowing down the weathering of wood and preventing termite attacks [12-13]. Mechanical properties especially the tensile strength of palm fibres, have similarities with the tensile strength of coir, kenaf, bamboo and hemp fibres, which are between 138.7 MPa to 270 MPa [13]. For the tensile strength of single fibres, it is 190.29 MPa, tensile modulus 3.69 GPa, and strain 19.6% [14]. With these advantages, palm fibre is used as one of the reinforcements in potential composite materials. Palm fibre applications as reinforcement in composite materials such as automotive components [2], portable tables, safety helmets, and composite boats [12], [15-16]. In addition to palm fibre has the advantages mentioned earlier, it also has drawbacks, namely moisture uptake, quality variations, low thermal stability, and their lack of adequate adhesion between the fibre and the matrix [8], [10], [17-20].

In order to overcome these shortcomings or weaknesses, the researchers used physical fabrication (hybrid fabrication, calendaring, thermal treatment and stretching) and chemical treatments (acetylation, benzoylation, treatment by peroxide, treatment by isocyanate, acidic treatments, coupling agents and grafting, polymeric coating and alkalization) on natural fibres before it were then combined with a matrix or other amplifier [3], [21]. This alkalization process is often used or chosen by the treatment of palm fibres [12], [22-27]. The mechanical properties of the single fibres depend on the chemical compositions, size, shape, orientation and thickness of their cell walls [28].

This study aims to determine the effect of variations of palm fibre fibres orientation on composites structures against bending strength, bending elastic modulus, tensile strength, and tensile modulus of elasticity values. Palm fibre is expected to be used as a material for making hull replacing fibre reinforced plastic (FRP) because it is known as that palm fibre is saltwater resistant and has good mechanical value.

2. Research method

The main material in this study is palm fibre which has been soaked in a 5% NaOH solution for 2 hours. The matrix is the Yukalac BQTN-EX 157 polyester with MEXPO hardener. Percentage of volume between fibres: polyester is 30%:70%. Variations in the orientation of fibre angles used were 0° + 45°, 0° + 90°, 45° + 90° and 45° + (-45°). The bending and tensile test uses a 600kN capacity Servo Hydraulic Universal Testing Machine (UTM) tool. The standard bending test (three-point bending method) uses ASTM D-790 03 [29] with dimensions of specimens 150 mm x 15 mm x 6 mm and standard tensile tests using ASTM D-3039 [30] with dimensions of specimens 250 mm x 125 mm x 6 mm. The equation for calculating bending strength is:

$$\sigma_b = \frac{3\,P\,L}{2\,b\,d^2} \quad (1)$$

Description:
σ_b = Bending strength (MPa)
P = Pressure force (N)
L = Distance between 2 span (mm)
b = Specimen wide (mm)
d = Specimen thickness (mm)

The equation to calculate the modulus of elasticity of bending is:

$$E_b = \frac{L^3 m}{4bd^3} \quad (2)$$

Description:
E_b = Bending modulus of elasticity (GPa)x
m = Initial tangent slope on load deflection curve (N/mm)

The equation for calculating the maximum tensile strength is:

$$\sigma_t = \frac{F^{max}}{A} \quad (3)$$

Description:
σ_t = Maximum tensile strength (MPa)
F_{max} = Maximum tensile force (N)

A = Cross section area (mm^2)

Modulus of elasticity of attraction can be calculated by equation:

$$E_t = \frac{\Delta\sigma}{\Delta\varepsilon} \tag{4}$$

Description:
E_t = Tensile modulus of elasticity (GPa)
$\Delta\sigma$ = Stress difference (MPa)
$\Delta\epsilon$ = Strain difference

3. Result and Discussion

3.1. Flexural strength and bending modulus elasticity

The results of testing of composite samples reinforced palm fibre with variations in fibre orientation to bending strength (flexural strength) are shown in Figure 1.

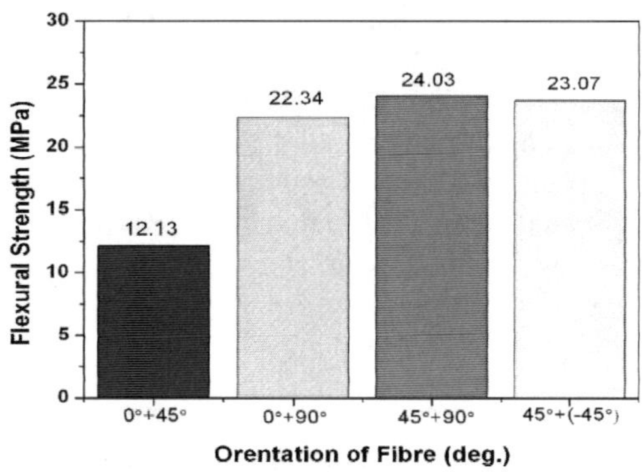

Figure 1. Graph of composites flexural strength.

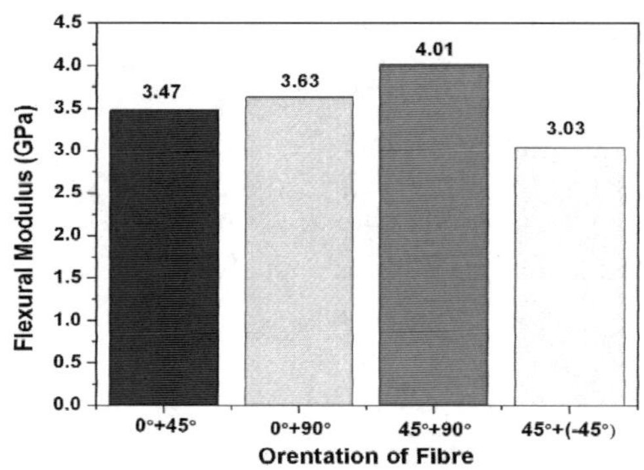

Figure 2. Graph of composites flexural modulus.

Figure 1 shows the relationship between the effect of palm fibre orientation on bending strength, where the orientation of palm fibre fibres 45 ° + 90 ° gets the highest average bending strength value of 24.03 MPa. Figure 2 is a graph that is made based on the results of the calculation of flexural modulus on variations in the fibres.

Based on Figure 2, the maximum flexural modulus value is obtained from the 45° + 90° angle orientation, where the value is 4.01 GPa. The results of the analysis of flexural properties based on Figures 1 and 2 are if the fibres of fibres as reinforcements form an angle of 90° or increasingly perpendicular to the bending load, then the value of strength and bending modulus elasticity will be higher. This is because the load of bending received by the matrix is then forwarded to palm fibre and each fibre of fibres function as a maximum reinforcement, where the load is distributed throughout the fibre [5], [14], [21].

3.2. Tensile strength and tensile modulus elasticity

Tensile testing on composite samples reinforced palm fibre with variations in fibre orientation to tensile strength is shown in Figure 3.

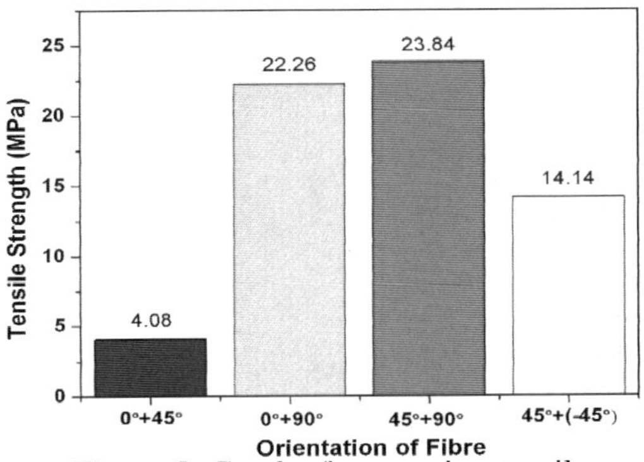

Figure 3. Graph of composites tensile strength.

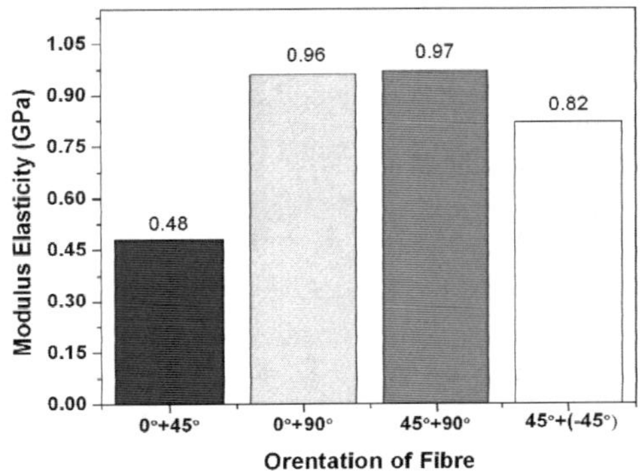

Figure 4. Graph of composites modulus elasticity.

Based on Figure 3 shows that the value of maximum tensile strength occurs at the orientation of fibres 45° + 90° which is 23.84 MPa, while from the results of calculations, the tensile modulus of elasticity is shown in Figure 4.

The calculation of maximum tensile modulus is in the composite with an orientation of fibres of 45° + 90° fibres which is equal to 0.97 GPa. Based on Figures 3 and 4 show that in composites with an orientation of 45° + 90° palm fibre not only has maximum tensile strength but also generates a maximum tensile modulus as well. In composites with fibre orientation upright to the direction of the tensile load, the maximum tensile strength and modulus will be produced. This is because the two composite components, both the matrix and the amplifier, will work together to hold the tensile load given [12].

Conclusions

Based on the conducted research, it can be concluded that:

1. Composite with the orientation of fibres of 45° + 90° fibres with polyester shows the maximum flexural and tensile properties. This is due to palm fibre as reinforcement that forms an angle of 90 ° or increasingly upright to the load crossing field (load of bending) and tensile load giving the effect of maximum reinforcement compared to other fibre orientations.
2. The reinforced composite fibres with polyester-fibre fibres show that the increase in bending and tensile strength is identical to the increase in bending and tensile modulus elasticity.

References

[1] Leman Z, Sapuan S M, Azwan M, Ahmad M M H M and Maleque M A 2008 *J. Polymer-Plastics Technology and Engineering* **47** 606-612

[2] Faris M A O and Sapuan S M 2014 *J. Cleaner Production* **66** 347-354

[3] Ahmad Y A M and Naser A H 2019 *J. Composite Science* **3 27.**

[4] Shahinur S and Hasan M 2019 Reference Module in Materials Science and Materials Engineering 1-9

[5] Begum K and Islam M A 2013 *Res. J. Engineering Sciences* **2** 3 46-53

[6] Sahari J, Sapuan S M, Zainudin E S and Maleque M A 2013 *J. Materials and Design* **49** 285-89

[7] Huzaifah M R M, Sapuan S M, Leman Z and Ishak M R 2019 *J. Fibre and Polymers* **20** 51 1077-84

[8] Rashid B, Leman Z, Jawaid M, Ghazali M J and Ishak M R 2016 *Intl. J. Precision Engineering and Manufacturing* **17** 8 1001-08

[9] Leman Z, Sastra H Y, Sapuan S M, Hamdan M M and Maleque M A 2005 *Journal of Applied Technology* **3**.1 14-19

[10] Sastra H Y, Siregar J P, Sapuan S M and Hamdan M M 2006 *J. Polymer-Plastics Technology and Engineering* **45** 149-155

[11] Ishak M R, Leman Z, Sapuan, S M, Salleh M Y and Misri S 2009 *International Journal of Mechanical and Materilas Engineering* **4** 3 316-20

[12] Ishak M R, Sapuan S M, Leman Z, Rahman M Z A, Anwar U M K and Siregar J P 2013 *J. Carbohydrate Polymers* **91** 699-710

[13] Norizan M N, Abdan K, Sapuan S M and Mohamed R 2017 *J. Physical Science* **28** 3 115-36

[14] Bachtiar D, Sapuan S M, Zainudin E S, Khaline A and Dahlan K Z M 2010 *IOP Conf. S: Materials Science and Engineering* **11** 012012

[15] Ishak M R, Sapuan S M, Leman Z, Rahman M Z A and Anwar U M K 2012 *J. Thermal Analysis and Calorimetry* **109** 981-989

[16] Misri S, Leman Z, Sapuan S M and Ishak M R 2009 *IOP Conf. Ser.: Materials Science and Enineering* **11** 012015

[17] Leman Z, Sapuan S M, Saifol A M, Maleque, M A and Ahmad M M H M 2008 *J. Material and Design* **29** 1666-70

[18] Pulungan M A, Sutikno and Sani M S M "Analysis of Bulletproof Vest Made from Fiber Carbon Composite and Hollow Glass Microsphere (HGM) in Absorbing Energy due to Projectile Impact" IOP Conf. Ser. Mater. Sci. Eng. 206 012001, 2019

[19] Sani M S M, Abdullah N A Z, Zahari S N, Siregar J P and Rahman M M "Finite element model updating of natural fibre reinforced composite structure in structural dynamics" MATEC Web of Conferences 83 03007, 2016

[20] Yaacob R M, Hashim M A H and Sani M S M "Finite element modeling and updating of the composite plate structure" IOP Conf. Ser. Mater. Sci. Eng. 100 01201, 2019

[21] Sonar T, Patil S, Deshmukh V and Achary R 2015 *IOSR J. Mechanical and Civil Engineering* 142-7

[22] Bledzki A K and Gassan J 1999 *Progressin Polymer Science* **24** 221–274

[23] Bachtiar D 2008 *Mechanical Properties of Alkali-Treated Sugar Palm (Arenga Pinnata) fibre-Reinforced Epoxy Composites*.Thesis. Universiti Putra Malaysia.

[24] Bachtiar D, Sapuan S M and Hamdan M M H M 2008 *J. Materials and Design* **29** 1285–90

[25] Mohammed A A, Bachtiar D, Siregar J P and Rejab M R M 2016 *J. Mechanical Engineering and Sciences* **10** 1 1765-77

[26] Bachtiar D, Sapuan S M and Hamdan M M 2009 *Polymer-Plastics Technology and Engineering* **48** 379–83

[27] Bachtiar D, Sapuan S M, Zainudin E S, Khalina A and Dahlan K Z H M 2011 *J. BioResources* **6** 4815–23

[28] Nurazzi M N, Khalina A, Sapuan S M, Laila D A H A M, Rahman M and Hanafee Z 2017 *Pertanika Journal Science and Technology* **25(4)** 1085-102.

[29] ASTM D 790. Standard Test Methods for Flexural Properties of Unreinforced and Reinforced Plastics and Electrical Insulating Materials West Conshohocken, United States, ASTM International

[30] ASTM D 3039. Tensile Properties of Polymer Matrix Composite Materials West Conshohocken, United States, ASTM International Standard

IOP Conf. Series: Materials Science and Engineering **807** (2020) 012032 doi:10.1088/1757-899X/807/1/012032

Analysis of productivity and the need for heavy equipment on grade B aggregate foundation layers work (case study: Pekan Gedang road section, Batang Asai – Muaro Talang District, Sarolangun)

Annisaa Dwiretnani[1], **Henriadi**[2]

[1]Lecturer at the Study Program of Civil Engineering, University Batanghari, Jambi, Indonesia
[2]Student at the Study Program of Civil Engineering, University Batanghari, Jambi, Indonesia

Corresponding e-mail: annisaa.dwiretnani@gmail.com

Abstract. Grade B aggregate foundation layers are part of the road foundation that is located between subgrade and grade A foundation layers. The function of grade B aggregate foundation layers is to minimize the spread of loads to the subgrade. Road infrastructure is one of the factors that support an area development. Road improvements aimed to improve the social economy and open access to isolated areas of considerable potential for the development of natural resources. This research analyses productivity and the need for 5 types of heavy equipment on grade B aggregate foundation layers work (case study: Pekan Gedang road section, Batang Asai – Muaro Talang District, Sarolangun). Noted that heavy equipment specifications were used in project site same as the planned. The results showed that based on time schedule and project volume in this research can be completed properly and on time if using 5 units of dump trucks and 1 unit each for wheel loader, vibratory roller, motor grader and water tank truck, with productivity for wheel loader is 125.42 m^3, dump truck is 2.69 m^3, vibratory roller is 87.15 m^3, motor grader is 57.24 m^3 and water tank truck is 71.14 m^3.

1. Introduction

Roads are defined as all parts of the road, including auxiliary buildings and their equipment intended for public traffic, which are on the surface of the land, above the surface of the land, below the surface of the land and/or water and above the water level, except railroads and roads cable [1]. While local roads are public roads that serve the function of local transportation with the characteristics of short distance travel, low average speed and the number of entry roads is not limited [2]. Pekan Gedang road section in Batang Asai District Sarolangun is an old road which in 2018 has been upgraded to a local road with a road length of 1 km, width of the road body is 6 meters, type of road is 1/2 UD and can be burdened by vehicle loads heaviest up to 8 tons.

The increase in the Pekan Gedang road section is related to the existing conditions in the form of macadam pavement layers which are often badly damaged due to annual landslides due to very high rainfall. Repairing parts of the damaged road continually cost so much that the Ministry of Public Works of the Jambi Province in the field of Bina Marga initiated the road improvement activities for Region VIII in the Sarolangun Regency, namely opening new roads by cutting hills.

Cost, quality, time and work safety factors are benchmarks of project performance in achieving success [3]. Grade B aggregate foundation layers work as part of this project has a total work volume of 1,200 m^3. The definition of aggregate is a major component of highway pavement material which contains 90 – 95% of the rupture area, it comes from nature or through processing [4]. The foundation layers' work is assisted by the use of 5 types of heavy equipments, namely wheel loader, dump truck, vibratory roller, motor grader and water tank truck. Heavy equipment is an important factor in construction work that has a large scale of work. The purpose of using heavy equipment is to facilitate humans in doing work, so that maximum results can be achieved in a short time. Errors in the allocation of heavy equipment can cause work to become problematic, have an impact on the quality of work and scheduled implementation of work [5].

The work sequence of the combination of heavy equipments above in carrying out work in a mechanical way starts with:

i. Wheel loader; function to loads grade B aggregates into a dump truck at base camp or road project in the site.
ii. Dump truck; function to transports grade B aggregates to the site.
iii. Vibratory roller; function to compacts on grade B aggregates that have been soaked in water by a water tank truck.
iv. Motor grader; function to extends grade B aggregates on the road that has been prepared.
v. Water tank truck; function to flushes grade B aggregates evenly that have spread across the road site.

The infrastructure development is an important aspect for the overall development of country. The major problem frequently faced by contractor in the selection of most suitable equipment [6]. "Construction Equipment (CE)" or "Heavy Equipment" refers to heavy-duty self-propelled vehicles, specially designed for executing construction tasks. Its use has a signifcant importance in the successful realisation of civil projects. Therefore, represents a major capital investment for the construction industry [7]. In highway projects most of the activities nowadays are equipment oriented and require only very meagre amount of participation from the unskilled and semi-skilled labourers. Only high skilled manpower and operators of the equipments are required to execute the job and the productivity of the operators can be determined from the productivity of the respective equipments [8]. It's important to have a proper management for maintenance, work schedule, fuel consumption and idle time [9]. As for each type of machine having a completely different cycle time phase. Therefore, studying cycle time statistics and cycle time frequency distribution curves is important [10].

The location of the work, the field of work, the type of work and the time of completion of the work are considered in the process of selecting the type of equipment needed [11]. In addition, the type of material, the volume of work and the ability to work of heavy equipment also influence [12]. The use of heavy equipments that is not right with the conditions and employment situation will affect the low productivity of the equipment and not achieving the predetermined schedule or target [13]. Many found a significant increase in the use of heavy equipment at construction sites as a form of business to meet the demand in the past few decades [14].

By paying attention to the problems that occur, it is concluded that the heavy equipment used in a project is influenced by the productivity of the equipment to the volume of work to be carried out, while the amount of equipment needed depends on the time available, site conditions, weather conditions, equipment efficiency, equipment capacity, number equipment and equipment operator capabilities [3]. Therefore, the authors make a research of the formulation of how much productivity (production capacity) per hour of heavy equipment and how much the need for heavy equipment used in grade B aggregate foundation layers work with a total volume of 1,200 m^3.

2. Method of Research

This type of research is quantitative data study, which is a type of data that can be measured (measurable) or calculated directly as a variable number or number that describe a case or object of research, about productivity and the need for heavy equipment on grade B aggregate foundation layers work (case study: Pekan Gedang road section, Batang Asai – Muaro Talang District, Sarolangun) in 2018.

Data sources in this research consisted of primary data and secondary data. The primary data were obtained using the approach and direct observation in the project site by field observations and interviews, while the secondary data obtained of important documents from owner, contractor, supervising consultant and other sources related to this research. Secondary data collection aimed to obtain information, such as complete project implementation data, heavy equipments data and grade B aggregate foundation layers work data.

The research site is shown in Figure 1:

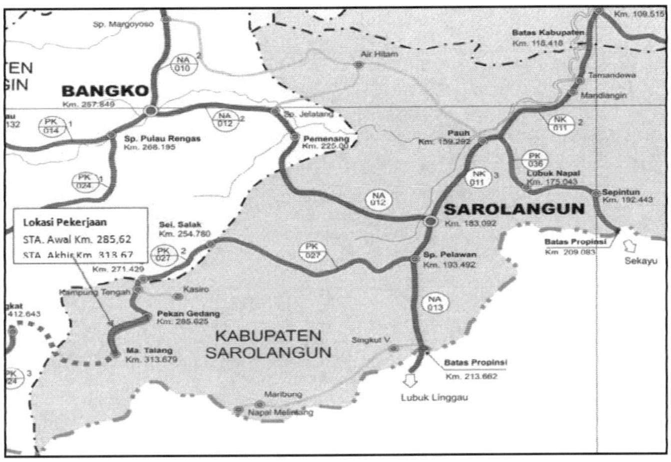

Figure 1. Research Site

2.1 *Preparation Phase*

There are several steps taken at preparation phase, namely:

i. Formulating research, research objectives, define the methods to be used and explore the literature, do literature study by reading the lecture subjects, reference books related to a research report.

ii. Collected project data that will be made as object of research, such as site data from contractor, relevant government institutions as project owners, consultant and other sources related to the execution of the project.

iii. To support the research conducted interviews with parties related to the project.

In processing data to determine productivity and the need for heavy equipment on grade B aggregate foundation layers work, there are formulations and specifications that are used, namely:

2.1.1 *Wheel Loader*

The calculation for production capacity per hour can be calculated according to equation (1) and the specifications of wheel loaders used are as tabulated in Table 1.

$$Q = \frac{V \ x \ F_b \ x \ F_a \ x \ 60}{T_s \ x \ Bip/Bil} \tag{1}$$

The following notation is used to formulate the production capacity of wheel loader:

Q = The production capacity per hour (m^3)
V = Capacity of bucket (m^3)
F_b = Bucket factor

F_a = Equipment efficiency factor
T_s = Cycle time (minute)

Table 1. Specifications of wheel loaders used.

Specification	Description
Brand of heavy equipment	Caterpillar
Type of heavy equipment	Cat 914 G
Heavy equipment power (HP)	95 HP
Bucket capacity (Cp)	1.4 m^3
Heavy equipment age (A)	5 years
Working hours for 1 year (W)	2000 hours

2.1.2 Dump Truck

The calculation for the production capacity per hour can be calculated based on equation (2) and the specification of dump truck used is as displayed in Table 2.

$$Q = \frac{V\ x\ F_a\ x\ 60}{D\ x\ T_s} \tag{2}$$

The following notation is used to formulate the production capacity of dump truck:

V = Capacity of tub (ton)
F_k = Materials development factor
F_a = Equipment efficiency factor
60 = Conversion from hours into minute
D = Weight of material (loose materials) (ton/m^3)
T_s = Cycle time (minute)
= $T_1 + T_2 + T_3 + T_4$

Table 2. Specifications of dump truck used.

Brand of heavy equipment	Hino
Type of heavy equipment	Dutro 130 HD
Heavy equipment power (HP)	130 HP
Dump capacity (Cp)	6 m^3
Heavy equipment age (A)	5 years
Working hours for 1 year (W)	2000 hours

2.1.3 Vibratory Roller

The production capacity per hour can be calculated based on equation (3) and specifications of vibratory rollers used is as shown in Table 3.

$$Q = \frac{(b_e\ x\ v\ x\ 1000)\ x\ t\ x\ F_a}{n} \tag{3}$$

The following notation is used to formulate the production capacity of vibratory roller:

b_e = Effective width of compaction wheel
= $b - b_o$ (overlap) (m)
b = Effective width of compaction (m)
b_o = Overlap width (m)

t = Compaction thickness (m)
v = Equipment average speed (km/hour)
n = Number of tracks (track)
F_a = Equipment efficiency factor (use 0.83 for good condition)
1000 = Conversion from km into m

Table 3. Specifications of vibratory rollers used.

Brand of heavy equipment	Sakai
Type of heavy equipment	Sakai SV525D
Heavy equipment power (HP)	82 HP
Equipment capacity (Cp)	7.1 tons
Heavy equipment age (A)	5 years
Working hours for 1 year (W)	2000 hours

2.1.4 Motor Grader

The production capacity per hour can be calculated as in equation (4) and specifications of motor graders used is tabulated in Table 4.

$$Q = \frac{L_h \, x \, (n \, (b - b_o)) \, x \, F_a \, x \, 60}{N \, x \, n \, x \, T_s} \tag{4}$$

The following notation is used to formulate the production capacity of motor grader:

L_h = Length of overlay (m)
b_o = Overlap width (m)
F_a = Equipment efficiency factor
n = Number of tracks (track)
N = Number of lane (lane)
b = Effective width of blade
60 = Conversion from hours into minute
T_1 = One time track = *(LH x 60) / (v x 1000)* (minute)
T_2 = Other time (minute)
T_s = Cycle time (minute)

Table 4. Specifications of motor graders used.

Brand of heavy equipment	Caterpillar
Type of heavy equipment	Cat 120 K
Heavy equipment power (HP)	125 HP
Heavy equipment age (A)	5 years
Working hours for 1 year (W)	2000 hours
Brand of heavy equipment	Caterpillar

2.1.5 Water Tank Truck

The production capacity per hour can be calculated as in equation (5) and the specifications of water tank truck used is as displayed in Table 5.

$$Q = \frac{P_a \, x \, F_a \, x \, 60}{W_c \, x \, 1000} \tag{5}$$

The following notation is used to formulate the production capacity of water tank truck:

W_c	= The need of water per m^3 solid materials (m^3)
P_a	= Capacity of water pump (use 100 liters/minute)
F_a	= Equipment efficiency factor
60	= Conversion from hours into minute
1000	= Conversion from km into m

Table 5. Specifications of water tank truck used.

Brand of heavy equipment	Mitsubishi
Type of heavy equipment	Canter HD 125 PS
Heavy equipment power (HP)	125 HP
Tank capacity (V)	4000 liters
Heavy equipment age (A)	5 years
Working hours for 1 year (W)	2000 hours

2.1.6 Total Heavy Equipment Need

To determine the need allocation of necessary heavy equipment in road construction project in this research, there are elements that need to be considered which are allocation assumption of heavy equipments (units), volume of work (m^3), production capacity of heavy equipments (m^3), and working hours for 1 year (days)

In this research, working hours specified for grade B aggregate foundation layers work on time schedule is a guideline for allocating working day of each machine. First, determined working days based on largest to smallest production capacity of heavy equipment with assumption there is only one heavy equipment can be calculated according to equation (6).

$$t = \frac{Q}{V} \tag{6}$$

where t is the working days (hour), Q is the production capacity (m^3) and V is the volume of work (m^3). Once obtained each working days then calculate the summary of heavy equipment working days. The number of working days must be less than or equal to the working days on time schedule. In this case, heavy equipment with smallest production capacity must be considered because that smallest one needs more working days to fit same working days in time schedule.

2.2 Process Phase

There are several steps taken at process phase, namely:

i. Calculating the productivity (production capacity) of heavy equipment on grade B aggregate foundation layers work.
ii. Calculating the demand of heavy equipment on grade B aggregate foundation layers work.

IOP Conf. Series: Materials Science and Engineering **807** (2020) 012032 doi:10.1088/1757-899X/807/1/012032

2.3 Final Phase
At this last stage, the research procedure is make conclusion from analysis results of productivity and the need for heavy equipment on grade B aggregate foundation layers work (case study: Pekan Gedang road section, Batang Asai – Muaro Talang District, Sarolangun).

3. Result and Discussion

3.1 Project Volume Data
The implementation of work for the improvement of Pekan Gedang Road (Batang Asai) – Muaro Talang District, Sarolangun, in 2018 by Ministry of Public Works and Public Housing in Jambi Province consists of several types of work as set out in Table 6 below:

Table 6. Project Volume Data

No	Type of Work	Unit	Volume	%
1	Mobilization	Ls	1.00	0.85
2	Excavation for drainage channel and water channel	m^3	59.00	0.06
3	Ordinary excavation	m^3	321.80	0.30
4	Soft stone excavation	m^3	1,120.00	2.76
5	Selection fill from excavation	m^3	263.00	0.31
6	Road set up	m^3	6,000.00	0.38
7	Grade B aggregate foundation layers	m^3	1,200.00	16.94
8	Concrete quality: fc' 25 Mpa (K.300)	m^3	31.10	1.35
9	Concrete quality: fc' 20 Mpa (K.240)	m^3	972.00	48.47
10	Concrete quality: fc' 15 Mpa (K.180)	m^3	50.40	1.77
11	Concrete quality: fc' 10 Mpa (K.120)	m^3	7.00	0.20
12	Reinforcing steel: BJ 24 plain	kg	56,599.50	26.61
	Total			100.00

In this research, discussion is limited for grade B aggregate foundation layers work with a total volume of 1,200 m^3.

3.2 Wheel Loader
The production capacity calculation of wheel loader can be seen in the Table 7.

Table 7. The Production Capacity Calculation Recapitulation of Wheel Loader

No	Data	Code	Coefficient	Quantity	Unit
	Data of Project				
1	Solid weight of grade B aggregate	Bip	–	1.80	ton/m^3
2	Loose weight of grade B aggregate	Bil	–	1.60	ton/m^3
3	Effective working hours per day	T_k	–	7.00	hour
	Data of Wheel Loader				
1	Capacity of bucket	V	–	1.40	m^3
2	Bucket factor	F_b	0.85	–	–
3	Equipment efficiency factor	F_a	0.83	–	–
4	Cycle time	T_s	–	0.45	minute
5	Production capacity per hour	Q	–	125.42	m^3

	$Q = \frac{V \, x \, F_b \, x \, F_a \, x \, 60}{T_s \, x \, Bip/Bil}$				
6	Production capacity per day $Q_h = T_k \, x \, Q$	Q_h	–	877.96	m^3

Based on the calculation results in Table 2 and specifications of the heavy equipment used, wheel loader production capacity is 125.42 m³/hour and 877.96 m³/day. Within 1 day the wheel loader works with effective working hours of 7 hours.

3.3 Dump Truck

The production capacity calculation of dump truck can be seen in the Table 8.

Table 8. The Production Capacity Calculation Recapitulation of Dump Truck

No	Data	Code	Coefficient	Quantity	Unit
	Data of Project				
1	Capacity of dump truck	Q_1	–	118.37	m^3
	Data of Dump Truck				
1	Dump capacity (on site)	V	–	4.80	ton
2	Equipment efficiency factor	F_a	0.80	–	–
3	Average speed of loads vehicle	v_1	–	10.00	km/hour
4	Average speed of empty vehicle	v_2	–	15.00	km/hour
5	Loading time: $(V \, x \, 60) / (Q_1 \, x \, Bil)$	T_1	–	1.52	minute
6	Loads vehicle time: $(L / v_1) \, x \, 60$	T_2	–	30.00	minute
7	Empty vehicle time: $(L / v_2) \, x \, 60$	T_3	–	20.00	minute
8	Etc, like unloading material time	T_4	–	2.00	minute
9	Cycle time: $T_1 + T_2 + T_3 + T_4$	T_s	–	53.52	minute
10	Production capacity per hour	Q	–	2.69	m^3
11	Production capacity per day $Q_h = T_k \, x \, Q$	Q_h	–	18.83	m^3

Based on the calculation results in Table 3 and specifications of dump truck used, the production capacity of dump truck is 2.69 m³/hour and 18.83 m³/day, in 1 day the dump truck works with effective working hours is 7 hour.

3.4 Vibratory Roller

The production capacity calculation of vibratory roller can be seen in the Table 9.

Table 9. The Production Capacity Calculation Recapitulation of Vibratory Roller

No	Data	Code	Coefficient	Quantity	Unit
1	Average speed of vehicle	v	–	1.50	km/hour
2	Effective width of compaction wheel	b_e	–	0.70	m
3	Overlap width	b_o	–	0.30	m
4	Effective width of compaction	b	–	1.00	m
5	Equipment efficiency factor	F_a	0.83	–	–
6	Number of tracks	n	–	2.00	track
7	Production capacity per hour	Q	–	87.15	m^3
8	Production capacity per day $Q_h = T_k \, x \, Q$	Q_h	–	610.05	m^3

Based on the results of calculations in Table 4 and specifications of vibratory roller used, vibratory roller production capacity is 87.15 m³/hour and 610.05 m³/day, in 1 day vibratory roller working with effective working hours is 7 hour.

3.5 *Motor Grader*

The production capacity calculation of vibratory roller can be seen in the Table 10.

Table 10. The Production Capacity Calculation Recapitulation of Motor Grader

No	Data	Code	Coefficient	Quantity	Unit
1	Length of overlay	L_h	–	1,000.00	m
2	Effective width of blade	b	–	1.00	m
3	Overlap width	b_o	–	0.30	m
4	Equipment efficiency factor	F_a	0.83	–	–
5	Average speed of vehicle	v	–	2.50	km/hour
6	Number of tracks	n	–	6.00	track
7	Number of lane	N	–	1.00	lane
8	Cycle time	T_s	–	29.00	minute
9	Leveling 1 track: $(L_h \times 60) / (v \times 1000)$	T_1	–	24.00	minute
10	Etc	T_2	–	5.00	minute
11	Production capacity per hour	Q	–	57.24	m^3
12	Production capacity per day $Q_h = T_k \times Q$	Q_h	–	400.69	m^3

Based on the calculation in Table 5 and the specifications of the motor grader used, the production capacity of motor grader is 57.24 m³/hour and 400.69 m³/day, in 1 day the motor grader work with effective working hours is 7 hours.

3.6 *Water Tank Truck*

The production capacity calculation of water tank truck can be seen in the Table 11.

Table 11. The Production Capacity Calculation Recapitulation of Water Tank Truck

No	Data	Code	Coefficient	Quantity	Unit
1	Tank capacity	V	–	4.00	m^3
2	The need of water per m^3 solid materials	W_c	–	0.07	m^3
3	Capacity of water pump	P_a	–	100.00	liter/ minute
4	Equipment efficiency factor	F_a	0.83	–	–
5	Production capacity per hour	Q	–	71.14	m^3
6	Production capacity per day $Q_h = T_k \times Q$	Q_h	–	498.00	m^3

Based on the calculation in Table 6 and the specifications of the water tank truck used, the production capacity of the water tank truck is 71.14 m³/hour and 498.00 m³/day, in 1 day the water tank truck works with effective working hours is 7 hours.

3.7 *Total Heavy Equipment Needed*

The calculation of total heavy equipment needed can be seen in the Table 12.

Table 12. The Calculation Recapitulation of Total Heavy Equipment Needed

No	Data	Wheel Loader	Dump Truck	Vibratory Roller	Motor Grader	Water Tank Truck
1	Heavy equipment needed (unit)	1	5	1	1	1
2	Volume of work (m^3)	1,200	1,200	1,200	1,200	1,200
3	Production capacity (m^3/day)	877.96	86.05	610.05	400.69	498.00
4	Working day (day)	1.37	13.94	1.97	2.99	2.41
5	Working day (day) (rounded)	2	14	2	3	3

The results of the calculations in Table 7 can be seen that total heavy equipment on grade B aggregate foundation layers work (case study: Pekan Gedang road section, Batang Asai – Muaro Talang District, Sarolangun) in 2018 with a total volume of 1,200 m^3 based on time schedule consist of wheel loader needs 1 unit, dump truck needs 5 units, vibratory roller needs 1 unit, motor grader needs 1 unit and water tank truck needs 1 unit.

4. Conclusion

The use of heavy equipment is required for large-scale construction work. The goal is to make it easier for humans to do work in order to obtain maximum good results within a certain period of time. In addition, the accuracy in choosing the type and combination of heavy equipment greatly affects the quality of the work. Heavy equipment used in a project is also influenced by two important things, namely the productivity of the heavy equipment to the volume of work to be done and the amount of heavy equipment needed in the field. The number of heavy equipment required depends on the time available, field conditions, weather conditions, heavy equipment efficiency, heavy equipment capacity, number of heavy equipment and the ability of the heavy equipment operator.

From the results of the analysis, the results obtained that the productivity of heavy equipment on grade B aggregate foundation layers work (case study: Pekan Gedang road section, Batang Asai – Muaro Talang District, Sarolangun) consists of production capacity per hour for wheel loader is 125.42 m^3, production capacity per hour for dump truck is 2.69 m^3, production capacity per hour for vibratory roller is 87.15 m^3, production capacity per hour for motor grader is 57.24 m^3 and production capacity per hour for water tank truck is 71.14 m^3.

From the results of the analysis, the results obtained that the productivity of heavy equipment on grade B aggregate foundation layers work (case study: Pekan Gedang road section, Batang Asai – Muaro Talang District, Sarolangun) consists of production capacity per hour for wheel loader is 125.42 m^3, production capacity per hour for dump truck is 2.69 m^3, production capacity per hour for vibratory roller is 87.15 m^3, production capacity per hour for motor grader is 57.24 m^3 and production capacity per hour for water tank truck is 71.14 m^3. The results showed that production capacity per hour of each heavy equipment on grade B aggregate foundation layers work in accordance with project time schedule.

In calculating the needs for heavy equipment on grade B aggregate foundation layers work (case study: Pekan Gedang road section, Batang Asai – Muaro Talang District, Sarolangun), all data needed are the aggregate volume of class B, the production capacity of the heavy equipment and the time to carry out the work specified in the work schedule (time schedule). Time schedule for grade B aggregate foundation layers work in this research is 4 weeks or 24 work days or 168 hours, where 1 effective working day is 7 hour. From the calculation results, it can be seen that total heavy equipment needed on grade B aggregate foundation layers work for 1,200 m^3 project volume which consists of Wheel loader needs 1 unit, Dump truck needs 5 units, Vibratory roller needs 1 unit, Motor grader needs 1 unit and Water tank truck needs 1 unit.

IOP Conf. Series: Materials Science and Engineering **807** (2020) 012032 doi:10.1088/1757-899X/807/1/012032

References

[1] Pemerintah Indonesia 2009 Undang-Undang Republik Indonesia Nomor 22 Tahun 2009 Tentang Lalu Lintas dan Angkutan Jalan *Lembaran Negara RI Tahun 2009* **3** (Jakarta: Sekretariat Negara)

[2] Pemerintah Indonesia 2004 Undang-Undang Republik Indonesia Nomor 38 Tahun 2004 Tentang Jalan *Lembaran Negara RI Tahun 2004* **3** (Jakarta: Sekretariat Negara)

[3] Husen A 2009 Manajemen Proyek Perencanaan, Penjadwalan dan Pengendalian Proyek (Yogyakarta: Andi)

[4] Sukirman S 2003 Perkerasan Lentur Jalan Raya (Bandung)

[5] Rostiyanti S F 2008 Alat Berat Untuk Proyek Konstruksi (Jakarta: Rineka Cipta)

[6] Nilesh D Chinchore, Khare P R 2014 Planning and Selection of Heavy Construction Equipment in Civil Engineering *Int. Journal of Engineering Research and Applications* **4**(12)-1 29-31

[7] Naskoudakis I, Petrousatou K 2016 A Thematic Review of the Main Research on Construction Equipment over Recent Years *Periodica Polytechnica Architecture* **47**(2) 110-118

[8] Singh N S 2018 Construction Equipment Productivity and Cost Economics in Highway Projects: Case Study Part II (Seasonal Variation Study) *International Journal of Advanced Research in Engineering and Technology* **9**(3) 163-171

[9] Mithilesh V V S, Nagavinothini R 2016 A Critical Study on Latest Heavy Equipment and Latest Technologies Used In Road Works and Recommending the Types of Equipment and Its Effectiveness for Indian Roads Equipment Recommendation for Indian Roads – A Critical Study *International Journal of Innovative Research in Science, Engineering and Technology* **5**(2)

[10] Ozdogan M, Ozdogan H 2017 Cycle Time Segments and Cycle Time Distribution Curves of Mining Size Wheel Loaders – A Case Study *Madencilik Scientific Mining Journal* **56**(1) 13-21

[11] Wilopo D 2009 Metode Konstruksi dan Alat-alat Berat (Jakarta: Universitas Indonesia)

[12] Tenriajeng A T 2003 Pemindahan Tanah Mekanis (Jakarta: Gunadarma)

[13] Yadam R W, Diputra G A, Sudipta I G K 2015 Optimalisasi Penggunaan Alat Berat Pada Pekerjaan Galian Tanah *Jurnal Ilmiah Elektronik Infrastruktur Teknik Sipil*

[14] Son H J, Cho S Y, Kim Y S, Kim C W 2007 Optimization of Heavy Equipment Operations on A Construction Site Using Simulation *Journal of The Korean Society of Civil Engineers* **27** 579-588

IOP Conf. Series: Materials Science and Engineering **807** (2020) 012033 doi:10.1088/1757-899X/807/1/012033

Anti-adjacency and Laplacian spectra of inverse graph of group of integers modulo *n*

Murni[1], Agung Efriyo Hadi[2*], Ifkra Febry[3], and Abdussakir[4]

[1] Department of Mathematics Education, Faculty of Teacher Training and Education, Universitas Abulyatama, Jl. Blangbintang Lama KM. 8.5, Kabupaten Aceh Besar 24415, Indonesia
[2] Department of Mechanical Engineering, Faculty of Engineering, Universitas Malahayati, Jl. Pramuka No. 27 Kemiling, Kota Bandar Lampung 35153, Lampung, Indonesia
[3] Department of Mathematics, Faculty of Science and Technology, Universitas Islam Negeri Maulana Malik Ibrahim Malang, Jl. Gajayana 50, Kota Malang 65144, Indonesia
[4] Department of Mathematics Education, Graduate Program, Universitas Islam Negeri Maulana Malik Ibrahim Malang, Jl. Raya Ir. Soekarno 34 Dadaprejo, Kota Batu 65233, Indonesia

Corresponding e-mail: *efriyo@malahayati.ac.id

Abstract. Research on the spectra of a graph still attracts the attention of many researchers over the last decades. In addition, research related to graphs obtained from an algebraic structure such as groups and rings is also growing. This paper determines the spectrum of the anti-adjacency and Laplacian matrices of inverse graph of a finite commutative group, namely the addition group of integers modulo *n*. It can be concluded that all eigenvalues of anti-adjacency and Laplacian matrices of the inverse graph of addition group of integers modulo *n* are integer

1. Introduction

Since Norman Bigg [1] first introduced the spectrum adjacency or spectrum concept of a graph in 1974, researchers continue to develop various other spectrum concepts. At present, several spectrum concepts have been developed and studied, such as Laplacian [2-9], signless Laplacian [10-16], detour [17], distance [18-23], distance Laplacian [24], distance signless Laplacian [25], detour distance Laplacian [26], color Laplacian [27] and color signless Laplacian [28] spectra of various types of graphs. This shows that the spectrum topic of a graph is still in great demand by researchers. Laplacian spectrum and its variation received more attention than the adjacency spectrum and detour spectrum from the researchers.

On the other hand, the study of graphs obtained from an algebraic structure also continues to develop and produces various types of graphs. Some examples of graphs obtained from a group are Cayley graph [29], subgroup graph [30], commuting graph [31,32], non-commuting graph [33], identity graph [29], graph conjugate graph [34] and inverse graph [35]. Inverse graph of a group is first introduced by Alfuraidan and Zakariya [35] in 2017. Suppose that Γ is a finite group and S is a set of non-self-invertible elements in Γ, namely $S = \{u \in \Gamma : u \neq u^{-1}\}$. The inverse graph of Γ is denoted by $G_S(\Gamma)$ and is defined as a graph whose set of vertex is the set Γ and two different elements v and w of Γ will be joined by an edge in $G_S(\Gamma)$ if and only if either vw or wv are elements of S.

Several studies on the spectra of graph obtained from a finite group have been conducted and published. For dihedral group, the spectra of non-commuting and commuting graphs [36], conjugate graph and its complement [37,38] and subgroup graph and its complement [39-41] have been reported. Nevertheless, study on the spectra of inverse graph of the symmetry group and the addition group of integers modulo n has not been found.

Edwina and Sugeng [42] introduced the notion of anti-adjacency matrix. Suppose that graph G is simple (without loops or multiple edges) with order p and $A(G) = [a_{ij}]$ $(1 \le i,j \le p)$ is its adjacency matrix. The anti-adjacency matrix of graph G is matrix $B(G) = [b_{ij}]$ $(1 \le i,j \le p)$ where $b_{ij} = 1$ if $a_{ij} = 0$ and $b_{ij} = 0$ if $a_{ij} = 1$. So, anti-adjacency matrix $B(G)$ can be expressed as $B(G) = J - A(G)$ where J is $p \times p$ matrix whose all entries are one [43]. In other words, the anti-adjacency matrix $B(G)$ is the opposite of the matrix $A(G)$ [44]. Until now, no one has examined the anti-adjacency spectrum of graphs, especially of graphs associated with a group.

This study examines the spectrum of anti-adjacency and Laplacian matrices of the inverse graph of a group. This study focuses on the addition group of integers modulo n where n is a positive integer.

2. Literature Review

Suppose G is a simple and finite graph with order $p = |\boldsymbol{V(G)}|$ and size $q = |\boldsymbol{E(G)}|$. Suppose $V(G) = \{v_i : 1 \le i \le p\}$. The degree $\deg(v_i)$ of a vertex v_i in G is defined as the number of vertex v_j $(j \ne i)$ in G such that v_iv_j is an element of $E(G)$ [45]. The matrix $A(G) = [a_{ij}]$ $(1 \le i,j \le p)$ where $a_{ij} = 1$ if v_iv_j is an element of $E(G)$ and $a_{ij} = 0$ if v_iv_j is not element of $E(G)$ is called the adjacency matrix of a graph G [46]. The matrix $D(G) = [d_{ij}]$ $(1 \le i,j \le p)$ where $d_{ij} = \deg(v_i)$ if $i = j$ and $d_{ij} = 0$ if $i \ne j$ is called the degree matrix of graph G [47]. The matrix $L(G) = D(G) - A(G)$ is called the Laplacian matrix of graph G [2,4]. The characteristic polynomial of $A(G)$ is a polynomial $\sigma(\lambda) = \det(A(G) - \lambda I)$ where I is $p \times p$ identity matrix [48]. The roots of $\sigma(\lambda) = 0$ are eigenvalues of $A(G)$ [49]. Suppose $\lambda_1 > \lambda_2 > \ldots > \lambda_k$ $(k \le p)$ are the distinct eigenvalues of $A(G)$ and $m(\lambda_i)$ is the algebraic multiplicity associated with λ_i $(1 \le i \le k)$. The adjacency spectrum $\boldsymbol{spec_A(G)}$ of a graph G is a $2 \times k$ matrix that contains the distinct eigenvalues of $A(G)$ in the first row and their corresponding multiplicities in the second row [50]. The adjacency spectrum of G can be written as

$$spec_A(G) = \begin{bmatrix} \lambda_1 & \lambda_2 & \cdots & \lambda_k \\ m(\lambda_1) & m(\lambda_2) & \cdots & m(\lambda_k) \end{bmatrix} \quad (1)$$

In a similar way, the Laplacian spectrum $spec_L(G)$ of a graph G is obtained from matrix $L(G)$ [51] while the anti-adjacency spectrum $spec_B(G)$ of a graph G is obtained from matrix $B(G)$. If all of the eigenvalues of $L(G)$ are integer, then graph G is called integral [40].

3. Results

Suppose $(Z_n, +)$ is the addition group of integers modulo n where n is positive integer. It is well known that $Z_n = \{\bar{0}, \bar{1}, \bar{2}, \ldots, \overline{n-1}\}$ and $\bar{0}$ is identity element of Z_n. If n is odd, then $\bar{0}$ is the only self-invertible element of Z_n. Hence, the set of non-self-invertible elements S of Z_n is $S = \{\bar{1}, \bar{2}, \ldots, \overline{n-1}\} = Z_n \backslash \{\bar{0}\}$. If n is even, then $\bar{0}$ and $\overline{\frac{n}{2}}$ are self-invertible elements of Z_n. Therefore, the set of non-self-invertible elements S of Z_n is $S = \{\bar{1}, \bar{2}, \ldots, \overline{\frac{n}{2}-1}, \overline{\frac{n}{2}+1}, \ldots, \overline{n-2}, \overline{n-1}\} = Z_n \backslash \{\bar{0}, \overline{\frac{n}{2}}\}$. According to the definition of inverse graph of Z_n, it is straightforward that $\deg(\bar{0}) = |S| - 1 = n - 1$ if n is odd and $\deg(\bar{0}) = |S| - 1 = \deg(\overline{\frac{n}{2}})$ if n is even. Furthermore, $\deg(v) = |S| - 2 = n - 2$ for $v \in Z_n \backslash \{\bar{0}\}$ if n is odd.

The results of the present study on the inverse graph of the addition group Z_n are presented as the following. First, the results will be presented with proof.

Theorem 3.1

The characteristic polynomial of anti-adjacency matrix $B(G_s(Z_n))$ is

$$\sigma(\lambda) = (\lambda - 2)^{\frac{n-1}{2}}(\lambda - 1)(\lambda)^{\frac{n-1}{2}} \tag{2}$$

if n is odd.

Proof

Because n is odd, then the set of all non-self-invertible elements of group Z_n is $S = \{\bar{1}, \bar{2}, \ldots, \bar{k}, \ldots, \overline{n-k}, \ldots \overline{n-2}, \overline{n-1}\}$. In the inverse graph $G_s(Z_n)$, the vertex $\bar{k}$ is not adjacent to vertex $\overline{n-k}$ $(1 \leq k < n)$ and is adjacent to all other vertices. Therefore, the adjacency matrix of $G_s(Z_n)$ is $A(G_S(Z_n)) = [a_{ij}]$ where

$$a_{ij} = \begin{cases} 1, \text{ if } i = j \text{ or } i = t \text{ and } j = n - (t-2)\ (1 < t \leq n) \\ 0, \text{ otherwise} \end{cases}$$

and the anti-adjacency matrix of $G_s(Z_n)$ is $B(G_S(Z_n)) = [b_{ij}]$

$$b_{ij} = \begin{cases} 0, \text{ if } i = j \text{ or } i = t \text{ and } j = n - (t-2)\ (1 < t \leq n) \\ 1, \text{ otherwise} \end{cases}$$

The matrix $B(G_S(Z_n))$ can be presented as

$$B(G_s(Z_n)) = \begin{array}{c} \\ \bar{0} \\ \bar{1} \\ \bar{2} \\ \vdots \\ \bar{k} \\ \vdots \\ \overline{n-k} \\ \vdots \\ \overline{n-2} \\ \overline{n-1} \end{array} \begin{array}{c} \begin{array}{ccccccccccc} \bar{0} & \bar{1} & \bar{2} & \ldots & \bar{k} & \ldots & \overline{n\text{-}k} & \ldots & \overline{n\text{-}2} & \overline{n\text{-}1} \end{array} \\ \begin{bmatrix} 1 & 0 & 0 & \ldots & 0 & \ldots & 0 & \ldots & 0 & 0 \\ 0 & 1 & 0 & \ldots & 0 & \ldots & 0 & \ldots & 0 & 1 \\ 0 & 0 & 1 & \ldots & 0 & \ldots & 0 & \ldots & 1 & 0 \\ \vdots & \vdots & \vdots & \ddots & \vdots & \ddots & \vdots & \ddots & \vdots & \vdots \\ 0 & 0 & 0 & \ldots & 1 & \ldots & 1 & \ldots & 0 & 0 \\ \vdots & \vdots & \vdots & \ddots & \vdots & \ddots & \vdots & \ddots & \vdots & \vdots \\ 0 & 0 & 0 & \ldots & 1 & \ldots & 1 & \ldots & 0 & 0 \\ \vdots & \vdots & \vdots & \ddots & \vdots & \ddots & \vdots & \ddots & \vdots & \vdots \\ 0 & 0 & 1 & \ldots & 0 & \ldots & 0 & \ldots & 1 & 0 \\ 0 & 1 & 0 & \ldots & 0 & \ldots & 0 & \ldots & 0 & 1 \end{bmatrix} \end{array} \tag{3}$$

By using Gaussian elimination on $B(G_s(Z_n)) - \lambda I$, an upper triangular matrix U will be obtained. Therefore, $\sigma(\lambda) = \det(B(G_s(Z_n)) - \lambda I)$ can be obtained by multiplying entries along the main diagonal of U. Finally, by simplifying the results of multiplication, it will be found that $\sigma(\lambda) = (\lambda - 2)^{\frac{n-1}{2}}(\lambda - 1)(\lambda)^{\frac{n-1}{2}}$.

Corollary 3.1

The anti-adjacency spectrum of $G_S(Z_n)$ is

$$spec_B(G_s(Z_n)) = \begin{bmatrix} 2 & 1 & 0 \\ \frac{n-1}{2} & 1 & \frac{n-1}{2} \end{bmatrix} \tag{4}$$

if n is odd.

Proof

Based on Theorem 3.1, the different eigenvalues of $B(G_S(Z_n))$ are $\lambda_1 = 2,\ \lambda_2 = 1$ and $\lambda_3 = 0$ and their multiplicities are $m(\lambda_1) = \frac{n-1}{2},\ m(\lambda_2) = 1$ and $m(\lambda_3) = \frac{n-1}{2}$, respectively. By the definition of the spectrum of a graph, it is obvious that

$$spec_B(\mathrm{G_s(Z_n)}) = \begin{bmatrix} 2 & 1 & 0 \\ \frac{n-1}{2} & 1 & \frac{n-1}{2} \end{bmatrix} \tag{5}$$

Theorem 3.2

The characteristic polynomial of $L(G_s(Z_n))$ is

$$\sigma(\lambda) = (\lambda - n)^{\frac{n-1}{2}}(\lambda - n + 2)^{\frac{n-1}{2}}(\lambda) \tag{6}$$

if n is odd.

Proof

Previously explained that $\deg(\bar{0}) = |S| - 1 = n - 1$ and $\deg(v) = |S| - 2 = n - 2$ for $v \in Z_n \backslash \{\bar{0}\}$ if n is odd. Hence, the degree matrix of $G_s(Z_n)$ is $D(G_S(Z_n)) = [d_{ij}]$ where

$$d_{ij} = \begin{cases} n-1, \text{ if } i=j=1 \\ n-2, \text{ if } i=j\neq 1 \\ 0, i\neq j \end{cases}$$

Thus, the matrix $L(\Gamma_s(Z_n)) = D(\Gamma_s(Z_n))$ - $A(\Gamma_s(Z_n))$ is

$$\mathrm{L}(\Gamma_s(\mathrm{Z_n})) = \begin{array}{c} \\ \bar{0} \\ \bar{1} \\ \bar{2} \\ \vdots \\ \bar{k} \\ \vdots \\ \overline{n-k} \\ \vdots \\ \overline{n-2} \\ \overline{n-1} \end{array} \begin{array}{c} \begin{array}{cccccccccc} \bar{0} & \bar{1} & \bar{2} & \dots & \bar{k} & \dots & \overline{n-k} & \dots & \overline{n-2} & \overline{n-1} \end{array} \\ \begin{bmatrix} n-1 & -1 & -1 & \dots & -1 & \dots & -1 & \dots & -1 & -1 \\ -1 & n-2 & -1 & \dots & -1 & \dots & -1 & \dots & -1 & 0 \\ -1 & -1 & n-2 & \dots & -1 & \dots & -1 & \dots & 0 & -1 \\ \vdots & \vdots & \vdots & \ddots & \vdots & \ddots & \vdots & \ddots & \vdots & \vdots \\ -1 & -1 & -1 & \dots & n-2 & \dots & 0 & \dots & -1 & -1 \\ \vdots & \vdots & \vdots & \ddots & \vdots & \ddots & \vdots & \ddots & \vdots & \vdots \\ -1 & -1 & -1 & \dots & 0 & \dots & n-2 & \dots & -1 & -1 \\ \vdots & \vdots & \vdots & \ddots & \vdots & \ddots & \vdots & \ddots & \vdots & \vdots \\ -1 & -1 & 0 & \dots & -1 & \dots & -1 & \dots & n-2 & -1 \\ -1 & 0 & -1 & \dots & -1 & \dots & -1 & \dots & -1 & n-2 \end{bmatrix} \end{array} \tag{7}$$

Through Gaussian elimination on $L(G_s(Z_n)) - \lambda I$ and some computation will be obtained $\sigma(\lambda) = \det(L(G_s(\mathrm{Z_n})) - \lambda I) = (\lambda - n)^{\frac{n-1}{2}}(\lambda - n + 2)^{\frac{n-1}{2}}(\lambda)$.

Corollary 3.2

The Laplacian spectrum of $G_S(Z_n)$ is

$$spec_L(G_s(Z_n)) = \begin{bmatrix} n & n-2 & 0 \\ \frac{n-1}{2} & \frac{n-1}{2} & 1 \end{bmatrix} \tag{8}$$

if n is odd.

Proof

It is obvious from Theorem 3.2,

If n is even, then $n \equiv 0 \pmod 4$ or $n \equiv 2 \pmod 4$. Henceforth, the main results of this study are presented without proof.

Theorem 3.3

The characteristic polynomial of $B(G_s(Z_n))$ is

$$\sigma(\lambda) = (\lambda - 3)^{\frac{n-4}{4}}(\lambda - 2)^2(\lambda - 1)^{\frac{n-4}{2}}(\lambda)^2(\lambda + 1)^{\frac{n-4}{4}} \tag{9}$$

if $n \geq 8$ and $n \equiv 0 \pmod 4$.

Corollary 3.3

The anti-adjacency spectrum of $G_S(Z_n)$ is

$$Spec_B\big(\mathrm{G_s}(\mathrm{Z_n})\big) = \begin{bmatrix} 3 & 2 & 1 & 0 & -1 \\ \frac{n-4}{4} & 2 & \frac{n-4}{2} & 2 & \frac{n-4}{4} \end{bmatrix} \tag{10}$$

if $n \geq 8$ and $n \equiv 0 \pmod 4$.

Theorem 3.4

The characteristic polynomial of $L(G_s(Z_n))$ is

$$\sigma(\lambda) = (\lambda - n)^{\frac{n}{4}}(\lambda - n + 2)^{\frac{n}{2}}(\lambda - n + 4)^{\frac{n-4}{4}}(\lambda) \tag{11}$$

if $n \geq 8$ and $n \equiv 0 \pmod 4$.

Corollary 3.4

The Laplacian spectrum of $G_S(Z_n)$ is

$$spec_L\big(\mathrm{G_s}(\mathrm{Z_n})\big) = \begin{bmatrix} n & n-2 & n-4 & 0 \\ \frac{n}{4} & \frac{n}{2} & \frac{n-4}{4} & 1 \end{bmatrix} \tag{12}$$

if $n \geq 8$ and $n \equiv 0 \pmod 4$.

Theorem 3.5

The characteristic polynomial of $B(G_s(Z_n))$ is

$$\sigma(\lambda) = (\lambda - 3)^{\frac{n-2}{4}}(\lambda - 2)(\lambda - 1)^{\frac{n-2}{2}}(\lambda)(\lambda + 1)^{\frac{n-2}{4}} \tag{13}$$

if $n \equiv 2 \pmod 4$.

IOP Conf. Series: Materials Science and Engineering **807** (2020) 012033 doi:10.1088/1757-899X/807/1/012033

Corollary 3.5

The anti-adjacency spectrum of $G_S(Z_n)$ is

$$spec_B\big(G_s(Z_n)\big) = \begin{bmatrix} 3 & 2 & 1 & 0 & -1 \\ \frac{n-2}{4} & 1 & \frac{n-2}{2} & 1 & \frac{n-2}{4} \end{bmatrix} \quad (14)$$

if $n \equiv 2 \pmod 4$.

Theorem 3.6

The characteristic polynomial of $L(G_s(Z_n))$ is

$$\sigma(\lambda) = (\lambda - n)^{\frac{n-2}{4}}(\lambda - n + 2)^{\frac{n}{2}}(\lambda - n + 4)^{\frac{n-2}{4}}(\lambda) \quad (15)$$

if $n \equiv 2 \pmod 4$.

Corollary 3.6

The Laplacian spectrum of $G_S(Z_n)$ is

$$spec_L\big(G_s(Z_n)\big) = \begin{bmatrix} n & n-2 & n-4 & 0 \\ \frac{n-2}{4} & \frac{n}{2} & \frac{n-2}{4} & 1 \end{bmatrix} \quad (16)$$

if $n \equiv 2 \pmod 4$.

4. Conclusion

According to the results of this study, it can be seen that all eigenvalues of anti-adjacency and Laplacian matrices of the inverse graph of addition group of integers modulo n are integer. So, the inverse graph of this group is integral. The next research can be done to examined the other spectrum of inverse graph of this group or other groups.

References

[1] Biggs N 1974 Algebraic graph theory (Cambridge: Cambridge University Press).

[2] Mohan B 1991 The Laplacian spectrum of graphs. *Graph Theory, Comb Appl* **2** 871–898.

[3] Grone R, Merris R 1994 The Laplacian spectrum of a graph II. *SIAM J Discret Math* **7** 221–229.

[4] Newman MW 2000 *The Laplacian Spectrum of Graphs* (Minatoba: The University of Manitoba)

[5] Jamakovic A & Van Mieghem P 2006 The Laplacian spectrum of complex networks. *Proceedings of the European Conference on Complex Systems, Oxford, September 25-29, 2006.*

[6] Bapat RB, Lal AK & Pati S 2008 Laplacian spectrum of weakly quasi-threshold graphs. *Graphs Comb* **24** 273–290.

[7] Lu M, Liu H & Tian F 2005 Bounds of Laplacian spectrum of graphs based on the domination number. *Linear Algebra Appl* **402** 390–396.

[8] Lange SC De, Reus MA De & Heuvel MP van den 2014 The Laplacian spectrum of neural networks. *Front Comput Neurosci* **7** 1–12.

[9] Agliari E & Tavani F 2017 The exact Laplacian spectrum for the Dyson hierarchical network. *Nat Publ Gr* 1–21.

[10] Zhang XD 2009 The signless Laplacian spectral radius of graphs with given degree sequences. *Discret Appl Math* **157** 2928–2937.

[11] Chang TJ & Tam BS 2010 Graphs with maximal signless Laplacian spectral radius. *Linear*

Algebra Appl **432** 1708–1733.

[12] Yu G, Wu Y & Shu J 2011 Signless Laplacian spectral radii of graphs with given chromatic number. *Linear Algebra Appl* **435** 1813–1822.

[13] Guo G & Wang G 2013 On the (signless) Laplacian spectral characterization of the line graphs of lollipop graphs. *Linear Algebra Appl* **438** 4595–4605.

[14] Cui S-Y & Tian G-X 2017 The spectra and the signless Laplacian spectra of graphs with pockets. *Appl Math Comput* **315** 363–371.

[15] Abdussakir & Khasanah R 2018 Spektrum signless-Laplace dan spektrum detour graf konjugasi dari grup dihedral. *J Kubik* **3** 45–51.

[16] Cui S-Y & Tian G-X 2017 The spectra and the signless Laplacian spectra of graphs with pockets. *Appl Math Comput* **315** 363–371.

[17] Ayyaswamy SK & Balachandran S 2010 On detour spectra of some graphs. *Int J Math Comput Phys Electr Comput Eng* **4** 1038–1040.

[18] Indulal G & Gutman I 2008 On the distance spectra of some graphs. *MATCH Commun Math Comput Chem* **13** 123–131.

[19] Ramane HS, Gutman I & Revankar DS 2008 Distance equienergetic graphs. *MATCH Commun Math Comput Chem* **60** 473–484.

[20] Stevanović D & Indulal G 2009 The distance spectrum and energy of the compositions of regular graphs. *Appl Math Lett* **22** 1136–1140.

[21] Indulal G & Stevanovi D 2015 The distance spectrum of corona and cluster of two graphs ☆. *AKCE Int J Graphs Comb* **12** 186–192.

[22] Gopalapillai I 2009 Sharp bounds on distance spectral radius and distance energy of graphs. *Linear Algebr Its Appl Multilinear Algebr* **430** 106–113.

[23] Renteln P 2011 The distance spectra of Cayley graphs of Coxeter groups. *Discrete Math* **311** 738–755.

[24] Aouchiche M & Hansen P 2013 Two Laplacians for the distance matrix of a graph. *Linear Algebra Appl* **439** 21–33.

[25] Xing R & Zhou B 2013 On the distance and distance signless Laplacian spectral radii of bicyclic graphs. *Linear Algebra Appl* **439** 3955–3963.

[26] Kaladevi V & Abinayaa A 2017 On detour distance Laplacian energy. *J Informatics Math Sci* **9** 721–732.

[27] Shigehalli VS & Betageri KS 2015 Color Laplacian Energy of Graphs. *J Comput Math Sci* **6** 485–494.

[28] Bhat PG & D'Souza S 2017 Color signless Laplacian energy of graphs. *AKCE Int J Graphs Comb* **14** 142–148.

[29] Kandasamy WBV & Smarandache F 2009 *Groups as graphs*. (Judetul Olt, Romania: Editura CuArt).

[30] Anderson DF, Fasteen J & Lagrange JD 2012 The subgroup graph of a group. *Arab J Math* **1** 17–27.

[31] Vahidi J & Talebi AA 2010 The commuting graphs on groups D2n and Qn. *J Math Comput Sci* **1** 123–127.

[32] Ali F, Salman M & Huang S 2016 On the commuting graph of dihedral group. *Commun Algebr* **44** 2389–2401.

[33] Darafsheh MR 2009 Groups with the same non-commuting graph. *Discret Appl Math* **157** 833–837.

[34] Erfanian A & Tolue B 2012 Conjugate graphs of finite groups. *Discret Math Algorithms Appl* **04** 1–8.

[35] Alfuraidan MR & Zakariya YF 2017 Inverse graphs associated with finite groups. *Electron J Graph Theory Appl* **5** 142–154.

[36] Abdussakir, Elvierayani RR & Nafisah M 2017 On the spectra of commuting and non commuting graph on dihedral group. *Cauchy-Jurnal Mat Murni dan Apl* **4** 176–182.

[37] Abdussakir 2017 Spektrum graf konjugasi dan komplemen graf konjugasi dari grup dihedral. *Pros Semin Nas Teknol Informasi, Komun dan Ind 9* 670–674.
[38] Abdussakir, Muzakir & Marzuki CC 2018 Detour spectrum and detour energy of conjugate graph complement of dihedral group. *J Phys Conf Ser* **1028** 1-6
[39] Abdussakir, Akhadiyah DA, Layali A, et al. The adjacency spectrum of subgroup graphs of dihedral group. *IOP Conf Ser Earth Environ Sci*; 243. Epub ahead of print 2019. DOI: 10.1088/1755-1315/243/1/012042.
[40] Abdussakir, Akhadiyah DA, Layali A, Putra AT 2018 Q-spectral and L-spectral radius of subgroup graphs of dihedral group. *J Phys Conf Ser* **1114** 1–6.
[41] Abdussakir, Susanti E, Turmudi, Jauhari MN & Ulya NM 2018 On the distance spectrum and distance energy of complement of subgroup graphs of dihedral group. *J Phys Conf Ser* **1114** 1-6
[42] Edwina M & Sugeng KA 2017 Determinant of antiadjacency matrix of union and join operation from two disjoint of several classes of graphs. *AIP Conf Proc* **1862** 1–5.
[43] Widiastuti L, Utama S & Aminah S 2018 Characteristic Polynomial of Antiadjacency Matrix of Directed Cyclic Wheel Graph (W →n). *J Phys Conf Ser* **1108** 1–7.
[44] Wang J, Lu M, Belardo F, Randic M 2018 The anti-adjacency matrix of a graph: Eccentricity matrix. *Discret Appl Math* **251** 299–309.
[45] Chartrand G, Lesniak L & Zhang P 2016 *Graphs and digraphs*. 6th ed. (Florida: CRC Press)
[46] Bondy JA & Murty USR 2008 Graph theory (New York: Springer)
[47] Harary F 1969 *Graph theory*. (California: Addison-Wesley Publishing Company)
[48] Elvierayani RR & Abdussakir 2013 Spectrum of the Laplacian matrix of non-commuting graph of dihedral group D2n. *Proceeding of International Conference The 4th Green Technology*. Malang: Faculty of Science and Technology 321–323.
[49] Biggs N 1993 *Algebraic graph theory*. 2nd ed. (New York: Cambridge University)
[50] Yin S 2008 Investigation on spectrum of the adjacency matrix and Laplacian matrix of graph Gl. *WSEAS Trans Syst* **7** 362–372.
[51] Abdussakir, Ikawati DSE & Sari FKN 2017 On the Laplacian and Signless Laplacian Spectra of Complete Multipartite Graphs. *Proceedings of the International Conference on Green Technology*. Malang, Indonesia, 79–82.

IOP Conf. Series: Materials Science and Engineering **807** (2020) 012034 doi:10.1088/1757-899X/807/1/012034

The Effectiveness of Corncob Activated Carbon in Reducing Chromium (Cr), Cadmium (Cd), Copper (Cu), and Zinc (Zn) Levels in Electroplating Wastewater

Ergantara, R.I[1*], N. Panisean[1], Dewi, S. N[1], Sidiq, I[1]

[1]Department of Environmental Engineering, Universitas Malahayati

Corresponding e-mail: ergantararani@yahoo.com

Abstract. The heavy metal from industrial electroplating wastewater can directly affect the environment even at very low concentration. The application of agricultural waste as low-cost adsorbents of removing heavy metal has been developed. This study aimed to determine the effectiveness of corncob activated carbon in reducing levels of Cu, Zn, Cr, and Cd in electroplating wastewater. The adsorbent from corncob dried in the sun, carbonized for 24 hours until carbon was formed. Then it was activated by HCl 9% for 24 hours, dried, crushed, and sieved with 80 mesh size sifter. The variation of the discharge are 15ml/min, 20 ml/min, and 25ml/min, the sample was taken at 0, 5, 10, 15, and 20 minutes. The results show that the highest reduction of Cr, Cd, Cu, and Zn occurred at 15 ml/minute discharge in the 5th minute with the total removal efficiency of Cr, Cd, Cu, and Zn were 25.75%, 43.42%, 50.39%, and 53.97%. Furthermore, it indicates that corncob activated carbon was found to be most effective in reducing of Zinc (Zn).

1. Introduction

The development of the Indonesian economy is characterized by various industries like electroplating industry. Their activities not only providing benefits, but also has a negative impact due to the discharge of heavy metal contaminated wastewater, such as Cd, Cr, Zn Cu, Ni, As, Ag, and Pb. These include electroplating, coating, cleaning, electroless deposition, etc. All of these metals are discharged into the environment directly or indirectly, reaching the water, air, and land. Then these metals can be absorbed by living organism and accumulated in the vital organ on the human being, plants, and another living organism. Besides, these can lead to various health disorder [1- 2]. It is necessary to treat metal-contaminated wastewater as well as reduction the environment effect.

Recently, numerous researches have been studied for the cheaper and effective technologies for the dissolved and suspended heavy metals from wastewater. These include ion exchange, adsorption, electrochemical treatment technologies, chemical precipitation, membrane filtration, electro dialysis, photocatalysis [3- 10]. Adsorption is one of the alternative treatments that has intensified increase. Many adsorbents may be used from agricultural raw materials and residues, such as corncobs, nutshells, rice husks, fruit pits, soybean, rice straw, sugarcane bagasse [11],[12]. The choice of those adsorbents based on the low-cost adsorbents, easy operating conditions, having a wide pH range, and high metal binding capacities [13].

Lampung province is the 3rd largest corn-producing after Jawa Timur and Jawa Tengah [14]. Its followed by a large amount of corncob waste with the estimate of waste 18 kg per 100 kg of corn grain production[15]. The utilization of corncob as the activated carbon being economical and eco-friendly because of their high adsorptive capacities based on pore volume, porosity, and surface area. Due to the activation process, all of that's developed and increase the absorption. Furthermore, temperature, pH, and contact time also affect carbon absorption. The process of activated carbon divided two types, the chemical and physical activation. The chemical activation method involves the co-carbonization of feedstock with a chemical, such as zinc chloride (ZnCl2), phosphoric acid (H3PO4), and potassium hydroxide/carbonate. The physical activation method involves carbonization of the raw material and subsequent activation in high temperature in carbon dioxide or steam.

Several studies on the various method have been done. The preparation of activated carbons from corncob with chemical activation with $ZnCL_2$, H_2SO_4, and HCl [16- 18]. But it can be pointed out that ZnCl2 and H3PO4 are the most widely used as chemical agents. Hence, the objective of the present study was to analyze the effectiveness of corncob activated carbon by chemical activation (HCl) with discharge and contact time variation for Chromium (Cr), Cadmium (Cd), Copper (Cu), and Zinc (Zn) contaminants wastewater.

2. Methods

2.1. Materials
Industrial electroplating wastewater was obtained from CV Bintang Chroom and the corncob was obtained from the farmer in East Lampung.

2.2. Sample Preparation
Ten kilograms corncob washed, cleaned, and dried in the sun for 3 days until completely dry. Subsequently, air-dried corncob was carbonized in artificial furnace from stacked bricks for 24 hours. In this process, the rice husk sprinkled as the bottom layer the furnace and on the corncob mound to make the combustion process easier.

2.3. Sample Activation
After the carbon is formed, then the activation is carried out by soaking the carbon into HCl with a concentration of 9% in a 1000 ml beaker glass for 24 hours. The separating carbon then dried in an oven at 100°C for 2 hours. The carbon was inserted to desiccator so that the carbon temperature was stable and washed with distilled water to release the acid. The carbon then dried with aerating it with the wind. Finally, the activated carbon is finely ground and it sieved with 80 mesh size sifter.

2.4. Filtration Process
A schematic diagram of the experimental for the sample filtration as shown in Figure 1.

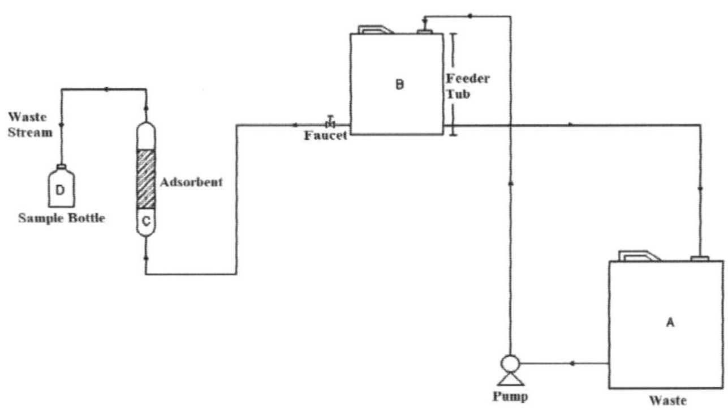

Figure 1. A Schematic Diagram Sample Filtration

A is an industrial electroplating wastewater container, wastewater from A then pumped to B which is a feeder container that serves to keep the discharge stable. C is a reactor tube with corncob activated carbon. Treated wastewater then collected in the bottle sample (D) for further laboratory testing.

3. Results and Discussion

3.1. The Concentration of Adsorption Cr, Cd, Cu, and Zn with discharge 15 ml/minute, 20 ml/minute, and 25 ml/minute

The concentration of corncob activated carbon adsorption can be seen in the Table 1, 2 and 3.

Table 1. The Concentration of Adsorption Cr, Cd, Cu, and Zn with discharge 15 ml/min

No	Minutes	Cr (mg/l) (Before)	Cr (mg/l) (After)	Cd (mg/l) (Before)	Cd (mg/l) (After)	Cu (mg/l) (Before)	Cu (mg/l) After	Zn (mg/l) (Before)	Zn (mg/l) (After)
1	0		171.56		25.24		20.54		7.17
2	5		161.43		17.66		12.40		4.02
3	10	171.56	145.10	25.24	16.51	20.54	11.42	7.17	3.89
4	15		127.39		14.28		10.19		3.30
5	20		136.96		16.06		11.01		3.70
6	25		157.53		17.88		11.30		3.79

Table 2. The Concentration of Adsorption Cr, Cd, Cu, and Zn with discharge 20 ml/min

No	Minutes	Cr (mg/l) (Before)	Cr (mg/l) (After)	Cd (mg/l) (Before)	Cd (mg/l) (After)	Cu (mg/l) (Before)	Cu (mg/l) After	Zn (mg/l) (Before)	Zn (mg/l) (After)
1	0		171.56		25.24		20.54		7.17
2	5		163.30		18.20		13.52		4.35
3	10	171.56	147.07	25.24	17.16	20.54	12.59	7.17	4.12
4	15		130.63		15.53		11.80		3.78
5	20		138.92		17.77		12.53		4.00
6	25		159.89		17.88		12.56		4.08

Table 3. The Concentration of Adsorption Cr, Cd, Cu, and Zn with discharge 25 ml/min

No	Minutes	Cr (mg/l) (Before)	Cr (mg/l) (After)	Cd (mg/l) (Before)	Cd (mg/l) (After)	Cu (mg/l) (Before)	Cu (mg/l) After	Zn (mg/l) (Before)	Zn (mg/l) (After)
1	0		171.56		25.24		20.54		7.17
2	5		165.90		19.18		15.05		5.04
3	10	171.56	149.79	25.24	18.77	20.54	14.55	7.17	4.70
4	15		133.23		16.06		13.78		4.12
5	20		141.52		18.50		13.99		4.32
6	25		162.77		18.98		14.20		4.70

Table 1, 2, and 3 showed the concentration of Cr, Cd, Cu, and Zn have changed. Generally, all of the heavy metals concentration decrease at the same time 0, 5, 10 minutes, and have a turning point at 15, 20, 25 minutes. At 0, 5, 10 minutes 5th minute the adsorption process was occurred. It happened because the absorption of Cr, Cd, Cu, and Zn occurred in the surface layer or between phases where ions/molecules of Cr, Cd, Cu, and Zn are fill the cavity and collected in corn activated carbon. Then, at the 15th to 25th minutes, the concentration of Cr, Cd, Cu, and Zn gradually increased because the process of releasing back the ions/molecules Cr, Cd, Cu, and Zn that have bonded with the corn activated carbon.

The ability of corncob activated carbon in 4 heavy metals process is most effective at 5th minute because the cavity of corncob can absorb well and there is a balance between adsorption to desorption in the 15th minute.

3.2. The Effectiveness of Adsorption

The effectiveness measurement of adsorption in this study were all carried out using discharge 15 ml/minute. That's based on the experiment results from the variation of discharge, the highest velocity of adsorption occurred at 15 ml/minute. The effectiveness of corncob activated carbon adsorption can be seen in Figure 2.

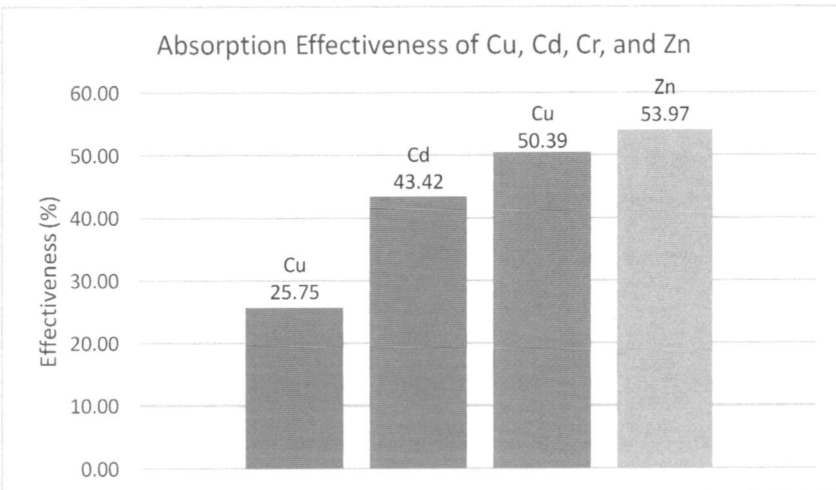

Figure 2. Absorption Effectiveness

Based on the data, it can be concluded the total removal efficiency of Cr, Cd, Cu, and Zn were 25.75%, 43.42%, 50.39%, and 53.97%. Furthermore, it indicates that corncob activated carbon was found to be most effective in reducing of Zinc (Zn).

4. Conclusion

The most significant finding showed that the corncob activated carbon was most effective in reducing Zn (53.97%) than Cr (25.75%), Cd (43.42%), Cu (50.38%) at 15 ml/min of discharge.

Acknowledgements

The authors would like to express the sincere gratitude for Universitas Malahayati for aiding this research financially.

References

[1] Hu, H. 2002 *MIT Press* (Cambridge, MA, USA) p. 65
[2] Jaishankar M, Tseten T, Anbalagan N, Mathew B B and Beeregowda N K 2014 *Interdisciplinary toxicology* 7(2) p. 60-72
[3] Barakat M 2011 *Arabian journal of chemistry* **4**(4) p. 361-377
[4] Said N I 2018 *Jurnal Air Indonesia* **6**(2).
[5] Abu-El-Halawa R and S A Zabin 2017 *Journal of Taibah University for Science* **11**(1) p. 57-65.

[6] Leung W C, Wong M F, Chua H, Lo W, Yu, P H F and Leung C K 2000 *Water science and technology* **41**(12) p. 233-240
[7] Kurniawan T A, Chan G Y S, Lo W H and Babel S 2006 *Chemical Engineering Journal* **118**(1) p. 83-98
[8] Mohammadi T, Moheb A, Sadrzadeh M and Razmi A 2005. *Separation and Purification Technology* **41**(1) p. 73-82
[9] Qdais H A and H Moussa *Desalination* **164**(2) p. 105-110
[10] Chen G 2004 *Separation and Purification Technology* **38**(1) p. 11-41
[11] Marshall W E and E T Champagne 1995 *Journal of Environmental Science & Health Part A* **30**(2) p. 241-261
[12] Abdul A and F Aberuagba 2005 *AU J Technol* **9** p. 59
[13] Kurniawan T A, Chan G Y, Lo W H and Babel S *Science of the total environment* **366**(2-3) p. 409-426
[14] Lampung, B P S P 2016 Lampung: *BPS-Statistics of Lampung Province*
[15] Bagby M and N Widstrom 1987 *Biomass uses and conversions*
[16] Tsai W T, C Y Chang and S L Lee 1998 *Bioresource Technology* **64**(3) p. 211-217
[17] Rizky, I P 2015 (UNIVERSITAS NEGERI SEMARANG)
[18] El-Sayed, G O, M M Yehia, and A A Asaad 2014 *Water Resources and Industry* **7-8** p. 66-75

IOP Conf. Series: Materials Science and Engineering **807** (2020) 012035 doi:10.1088/1757-899X/807/1/012035

Investigation on the dynamic properties of propeller structure with different number of blades

N.A.Z. Abdullah[1], P. Brapakaran[1], M.S.M. Sani [1,2*], M.W. Marcel[3] and E.H. Agung[3]
[1]Faculty of Mechanical and Automotive Engineering Technology, Universiti Malaysia Pahang, 26600 Pekan, Pahang, Malaysia
[2]Automotive Engineering Centre, Universiti Malaysia Pahang, 26600 Pekan, Pahang, Malaysia
[3]Faculty of Engineering, Universitas Malahayati, Lampung, Indonesia

Corresponding e-mail: *mshahrir@ump.edu.my

Abstract A propeller is mainly used for propelling the movement for an aircraft or ship. As it gives major contribution towards the performance of an aircraft or ship, study on propeller structure has become popular. In this study, the dynamic or modal characteristics of propeller structure with different number of blade of the same design are investigated. Modal properties of two blades, three blades and four blades propeller that are obtained through computational analysis are compared after validation of three blades propeller is conducted. It has been observed that the four blades propeller has the lowest nominal value of natural frequencies and shows most mode shapes deflection. The results confirm that as the propeller has more number of blades, it will produce more energy and hence has lower natural frequency.

1. Introduction

A propeller is a structure equipped with a rotating fan, which is used to propel a ship or a plane by generating thrust force from a power transmitted by the main engine. As the structure plays an important role in deciding the performance of modern ships and planes, topic of propeller design and performance has become a perennial subject of research. Most of the researches are focusing on the how the design of propeller blade can affect the performance of the propeller in creating thrust for more powerful movement [1–7]. For example, the study on several different configuration of propeller blades were conducted and the pressure field and thrust created from the propeller were analysed.

On the other hand, it has also been discussed that the dynamic behaviour of the blades may cause dangerous phenomenon such as resonance to the whole propelling system [5]. Generally, study of vibration on structure has always been an important topic of research as knowledge on the dynamic characteristic of a structure contributes to better structural design process. Specifically, for propeller structure, resonance may occur if the operating frequency of the propeller match its natural frequency. It also has been mentioned by reference [8] that propeller induces a pressure field due to the rotation movement and the thrust generated by the blades. Propeller also may be exposed to unstable vibration at a certain rotation speed as energy is absorbed from its surrounding [9]. Therefore, it is crucial to understand the effect of blades on the propeller towards the modal characteristics of the propeller.

Moreover, in practical application for designing propeller for airplane, modal properties of the propeller are crucial knowledge because of safety issues [10-11].

Consequently, there are studies that presented the analysis of propeller's modal properties under several approach. For example, the modal behaviour of propeller structure under different medium was investigated [12]. Hong et.al compared the free vibration characteristics of a marine propeller in air and water, and the influence of fluid inertial effect on the propeller was analysed. A study on modal characteristics of propeller blades of different material (metal and composites) was also investigated [13]. Other study on propeller structure that also concern the materials and damping element toward the modal characteristics of the propeller was also conducted by Morad et.al [14].

However, when concerning the effect of number of blades on propeller towards the performance of propeller, very little study was conducted. The study by Wang et.al shows how same number of blades with different skew angle contributes to the wake propelled by the propeller [7]. Similarly, study by Felli et.al. compares how different number of blades contributes towards the wake propulsion [15]. The subject of choosing number of blades on propeller mostly changed based on the design requirement of the propeller itself as there are many parameters that need to be considered as well.

This paper will provide the analysis and comparison of the modal properties of a propeller structure with different number of blades. The design of blades in terms of size and skew for each propeller remains constant.

2. Validation of modal properties of three blades propeller structure

2.1. Finite element analysis on three blades propeller structure

Initially, the propeller blade structure was modelled in SolidWorks software (see Figure 1) before being imported into a finite element analysis software (see Figure 2). Analytical frequency analysis was performed using MSC. Nastran/Patran software by utilizing the SOL103 which is the normal mode analysis. Finite element model of the propeller blade structure was created using 8-node tetrahedral elements which were selected during the meshing process. Isotropic material properties of steel were assigned to the model. The nominal values of the assigned material properties on the structure are as follows; Young's Modulus is 69 GPa, density is 2800 kg/m^3, and Poisson's ration is 0.3. No constraints were applied to the model which means the analysis was conducted under free-free boundary condition. The frequency range of interest was set between 10 Hz to 8000 Hz in order to avoid the calculation of rigid body modes during the analysis. In this study also, only first six modes were observed and analysed.

Figure 1. CAD model of three blades propeller structure.

Figure 2. FE model of three blades propeller structure.

2.2. Experimental modal analysis on three blades propeller structure

IOP Conf. Series: Materials Science and Engineering **807** (2020) 012035 doi:10.1088/1757-899X/807/1/012035

Modal testing on the three blades propeller structure was conducted using impact hammer testing. One tri-axial accelerometer was used during the testing. The impact hammer test was performed by roving accelerometer method where the accelerometer was roved around to all measurement point while providing excitation on one single point only using an impact hammer [16]. The three blades propeller structure was labelled with 36 measurement points. The experimental model was sketch in a wire-frame model for the purpose of displaying the response collected by the structure at each of the measurement point during the testing (see Figure 3).The structure was placed in hanging position as shown in Figure 4 in order to portray the free-free boundary condition that was set during computational analysis [17].

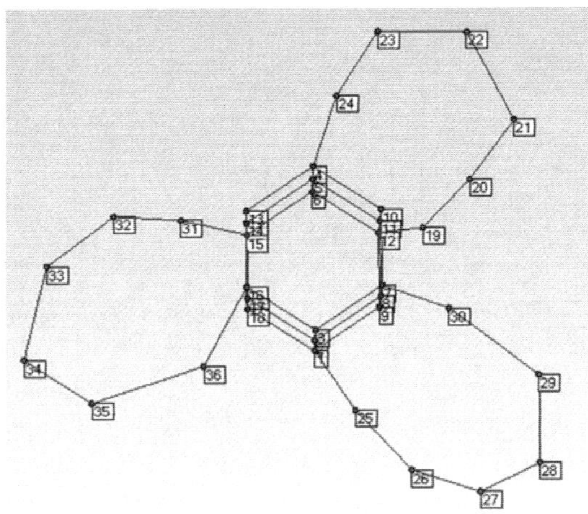

Figure 3. Experimental model of the three blades propeller.

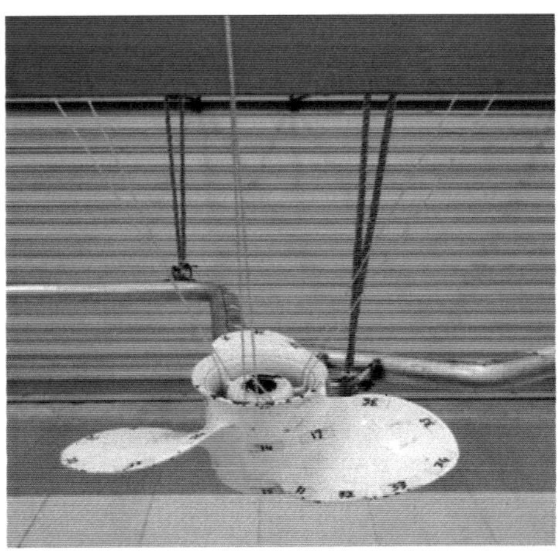

Figure 4. Three blades propeller hanging from test rig.

2.3. Correlation of natural frequencies of three blades propeller structure

Correlation of the output gathered in both analysis (experiments and analytical) intends to validate the data obtained through computational analysis. This is owing to the reason that data obtained through computational analysis is based from assumption and need to be validated experimentally before trusting its reliability for providing accurate results [18]. Figure 5 provides the curve of natural frequencies of the three blades propeller structure for the first six modes, that are obtained through impact hammer test (experimental) and normal mode analysis (analytical/computational). Based on the figure, the values of natural frequency obtained computationally are well correlated with the values obtained during experimental works. It is clearly shown that in most modes such as 2nd, 3rd, and 4th mode, the values of natural frequency are very well correlated and shows almost no error. Therefore, the model of the three blades propeller structure is considered as validated and reliable enough for further prediction analysis.

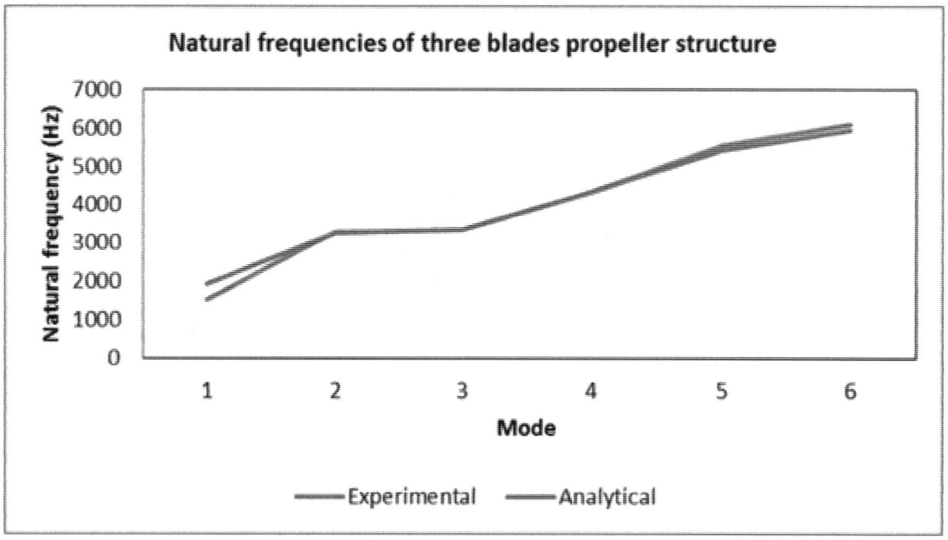

Figure 5. Correlation of natural frequencies obtained through experimental and computational analysis of the three blades propeller structure.

3. Computational modal analysis on propeller structure with different number of blades

Normal mode analysis was performed on the model of three propellers with different number of blades (2 blades, 3 blades and 4 blades) of the same design as shown in Figure 6. The set up for analysis for all three models of propeller were the same with the one conducted for the three blades propeller as described in previous section. The data for material properties remains unchanged as well as the boundary condition and the mode of interest.

The nominal value of the first six natural frequencies for propeller of different blade number is as shown in Table 1. The computational data presented in Table 1 indicates that when propeller has lower number of blade, the natural frequencies of the propeller is significantly higher. Figure 7 shows the natural frequencies curve for propeller models with different blade number. It shows that the two blades propeller has higher value of natural frequencies in each mode. This is owing to the reason that more number of blades contribute to more mass to the propeller structure, and therefore decrease its frequency.

On the other hand, the mode shapes of the first six natural frequencies for two blades, three blades and four blades propeller is tabulated in Table 2, Table 3 and Table4 respectively. Understanding the mode shapes is important in order to understand the behaviour of the propeller under resonance. Therefore, enhancement such as stiffeners may be added in order to fortify the propeller design for better operation [19]. Based on the computed mode shapes, the higher the number of blades on the propeller, the more it experiences deflection on each blades compared to the propeller with smaller number of blades. The two blades propeller shows deflection in all blades only in the first mode. The three blades propeller shows deflection in all blades in third and sixth mode. The four blades propeller shows deflection in all blades in the first, second, fourth and sixth mode. This agrees with some findings that discussed that more blades on propeller produced more thrust and therefore contains more energy [20]. Thus, resonance is more likely to occur in propeller with more blades.

(a) (b) (c)

Figure 6. Finite element model of propeller structure with (a) two blades, (b) three blades and (c) four blades.

Table 1. Comparison of first six natural frequencies for propeller with different number of blades

Mode		1	2	3	4	5	6
Natural	2 blades	2072.1	4451.2	5451.0	5510.9	7606.3	8086.8
frequencies of	3 blades	1926.2	3233.1	3360.1	4353.1	5425.6	5953.2
propeller (Hz)	4 blades	572.95	2636.2	2677.4	3096.8	4346.3	4793.1

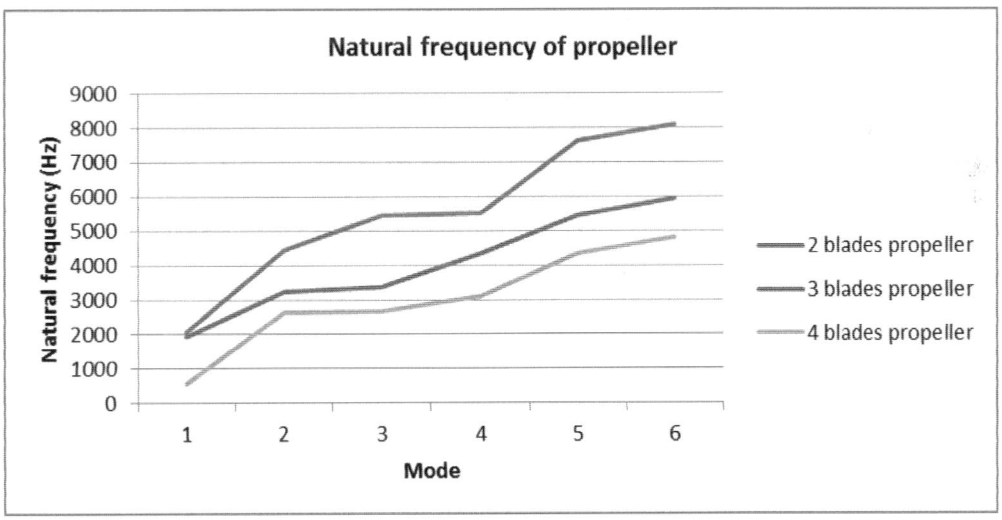

Figure 7. The natural frequencies curves of propeller with different number of blade.

Table 2. Comparison of mode shapes for two blades propeller

Mode	Mode shapes		
1	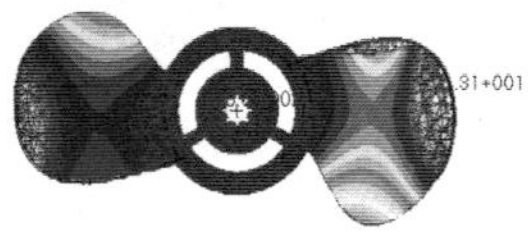	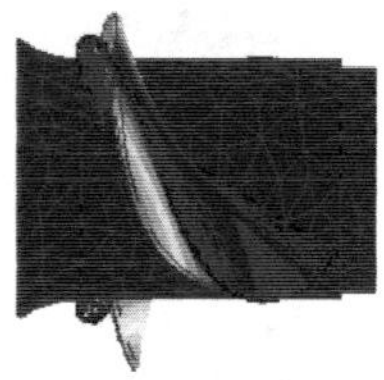	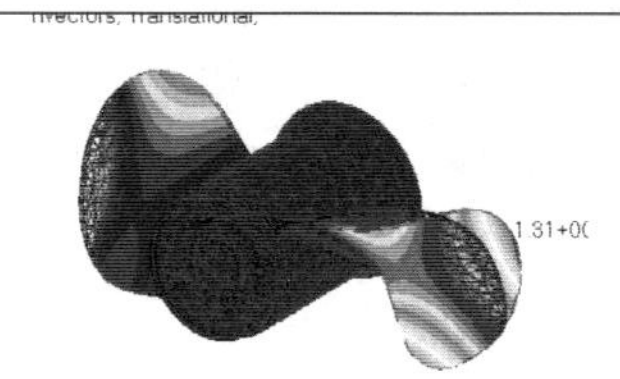
2		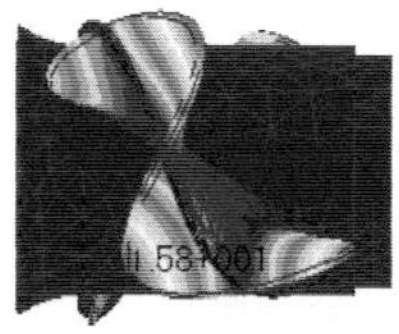	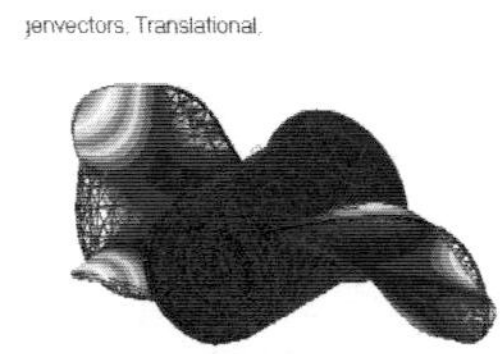
3		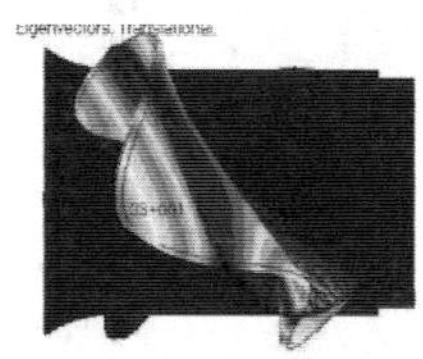	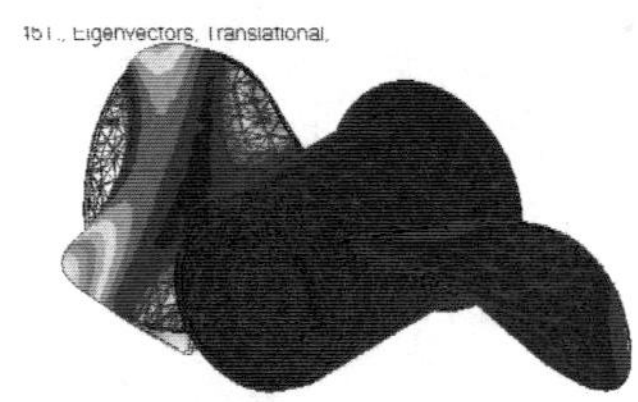
4		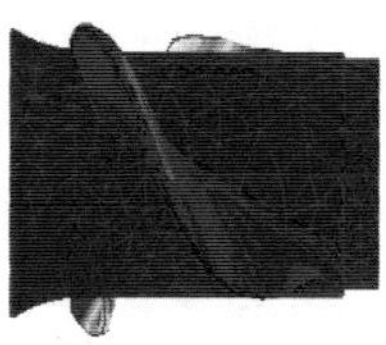	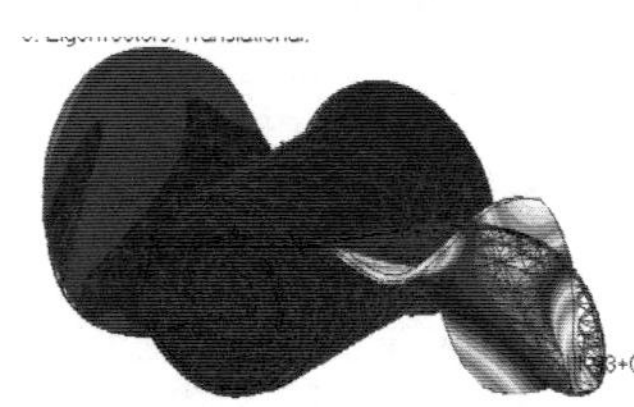
5			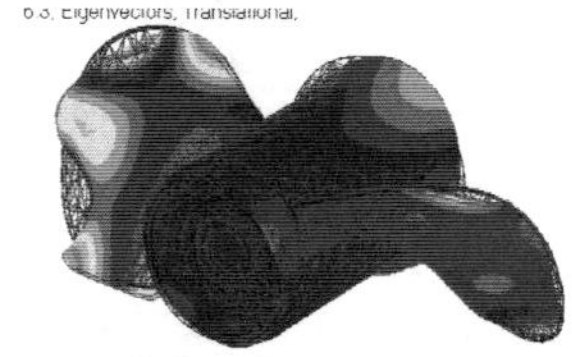
6		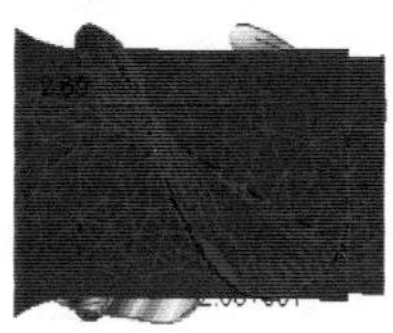	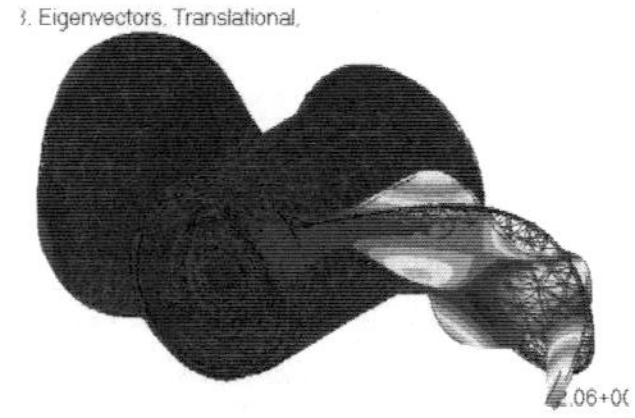

Table 3. Comparison of mode shapes for three blades propeller

Mode	Mode shapes		
1	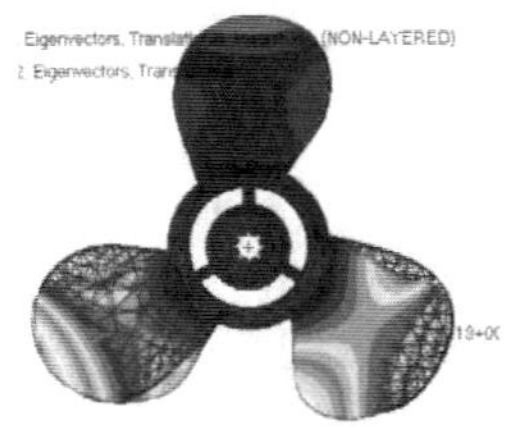	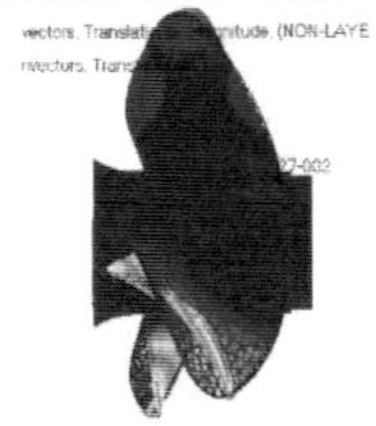	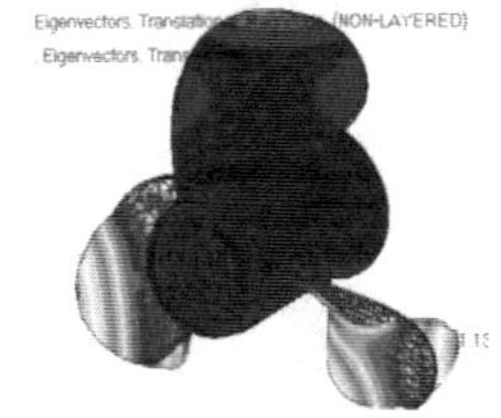
2	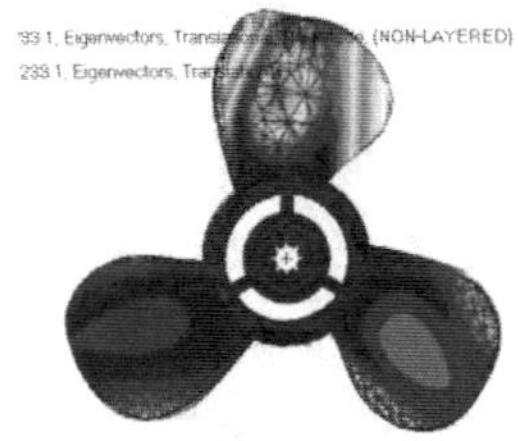	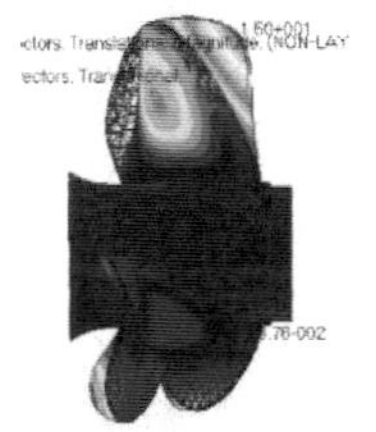	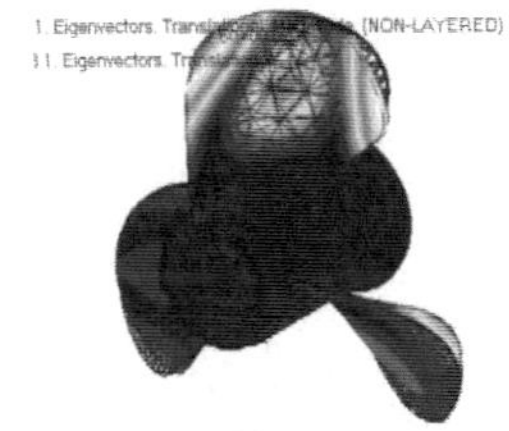
3	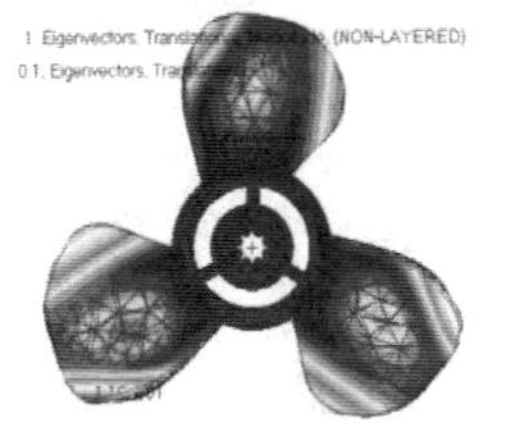	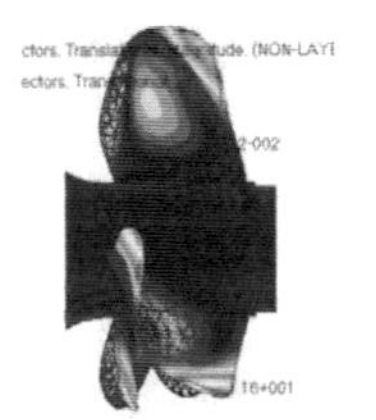	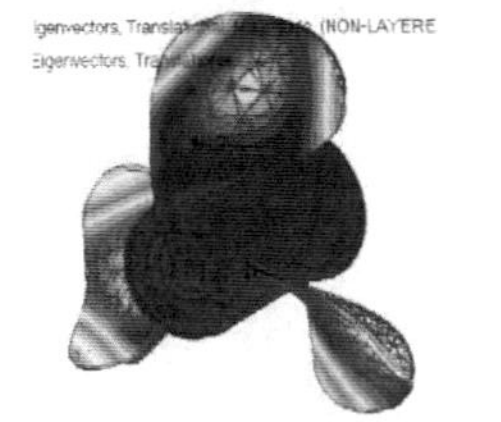
4	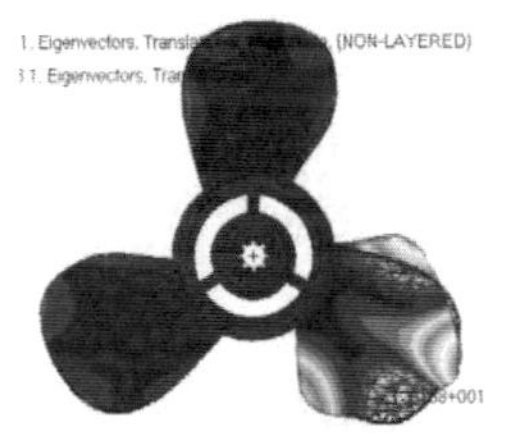	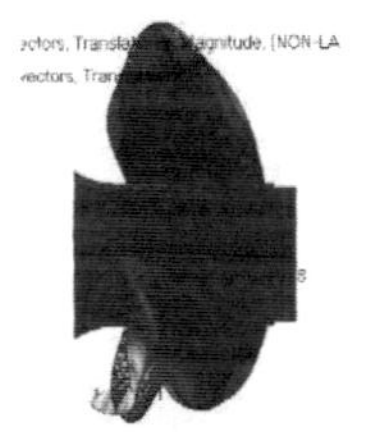	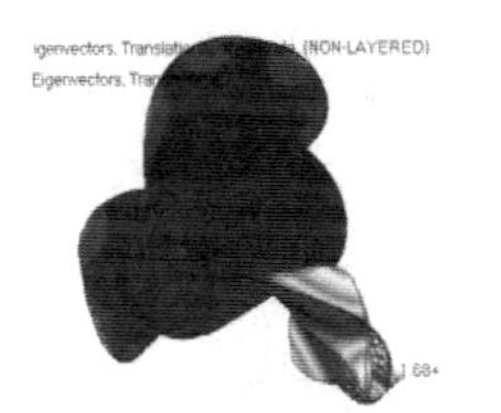
5	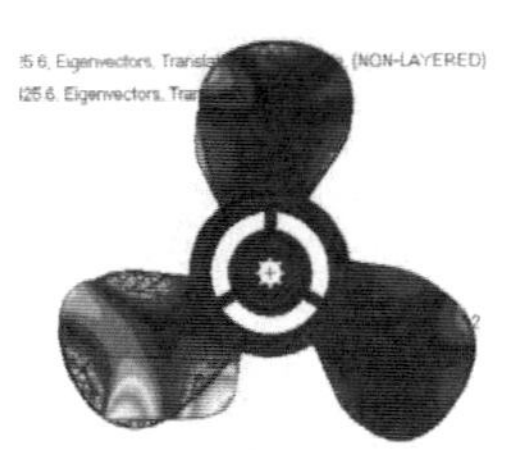	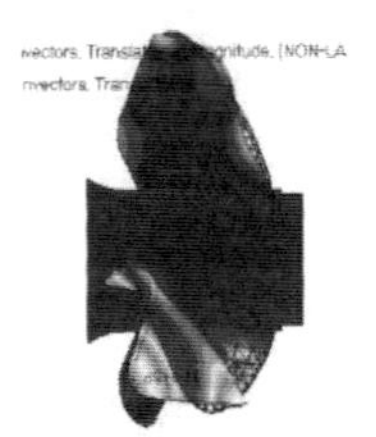	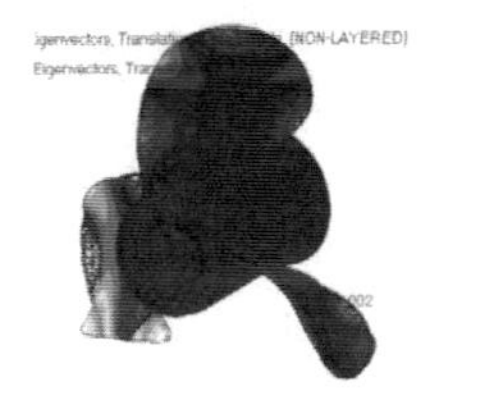
6	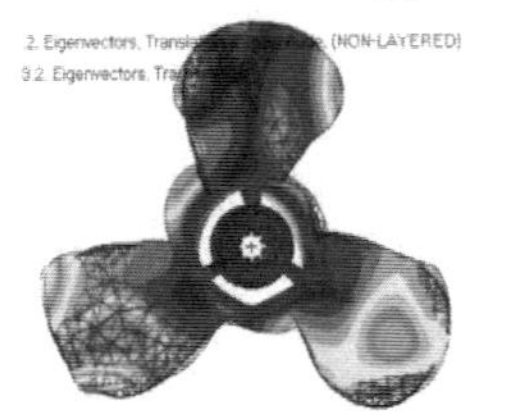	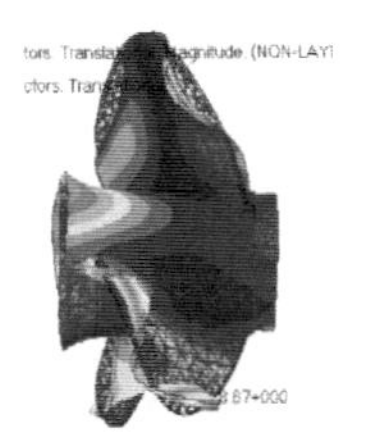	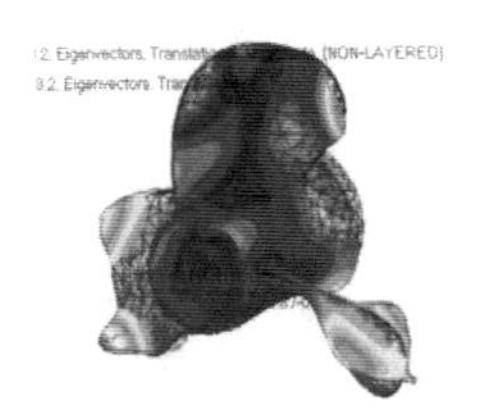

Table 4. Comparison of mode shapes for four blades propeller

Mode	Mode shapes		
1	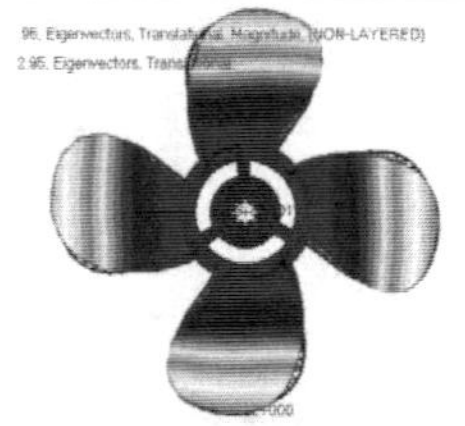	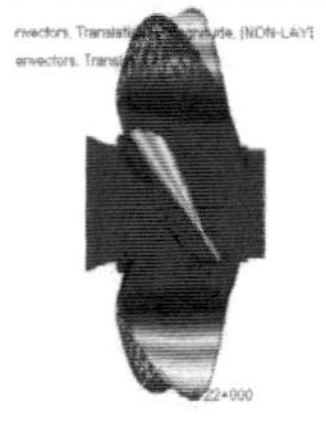	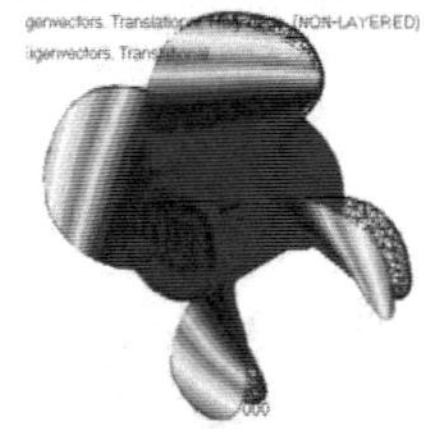
2	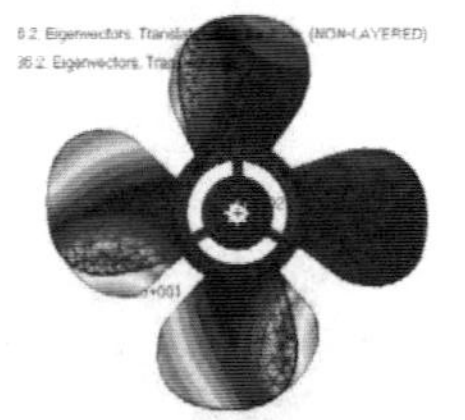	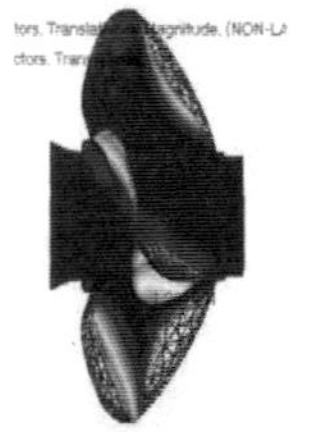	
3	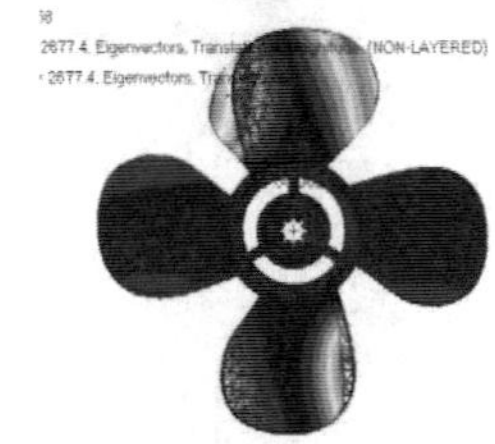	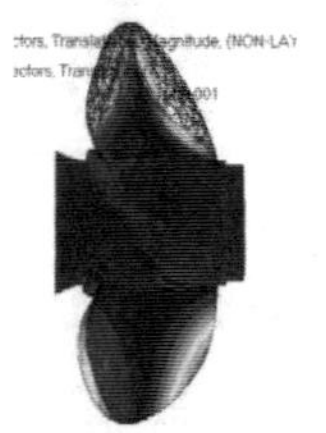	
4		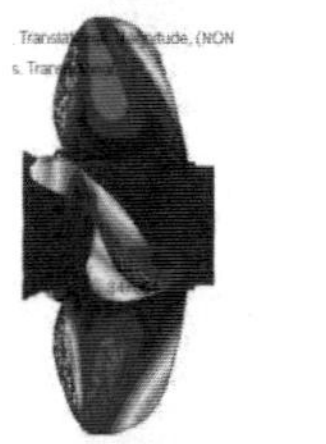	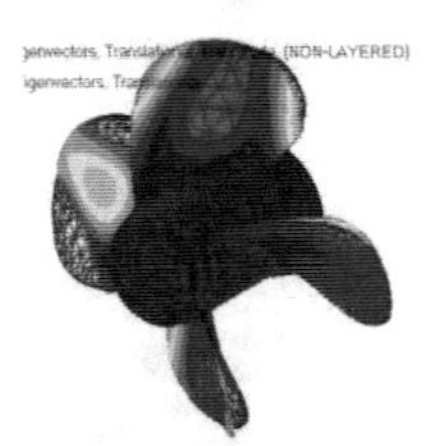
5	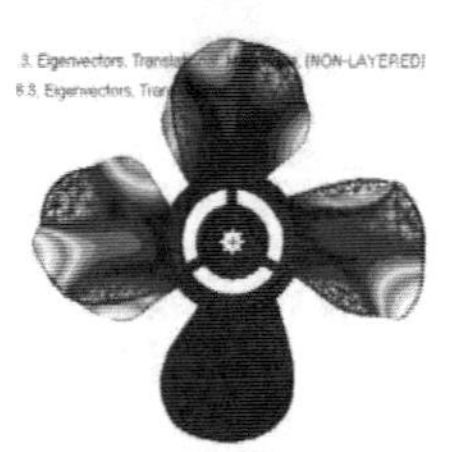		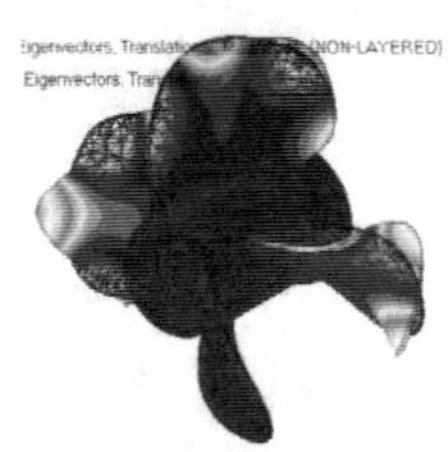
6	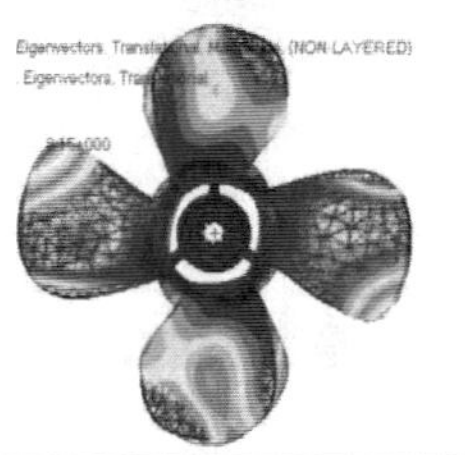	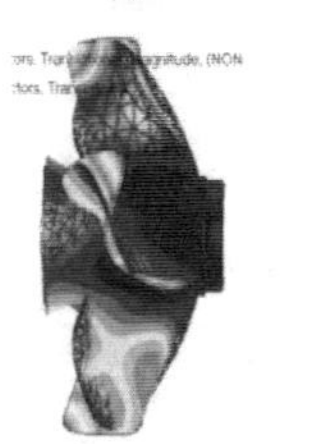	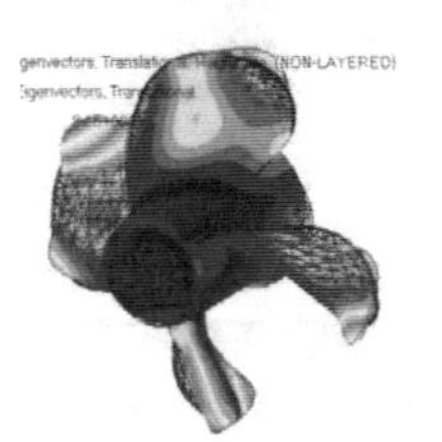

However, in determining the right number of propeller blades that should be used, it depends on several other factors such as the engine power, operating frequency for the propeller, performance requirements and many more. As more blades produce higher thrust with drag which may contributes to the loss of fuel efficiency. Therefore, in determining the optimum number of blades a propeller should have depends on combination of many factors which will vary depending on where and with how much power the propeller will work [8].

4. Conclusion

The modal characteristics of propeller with different number of blades, which are two blades, three blades, and four blades were investigated. The value of natural frequency for each propeller together with their mode shapes were obtained computationally and analysed later on. The following findings were pointed out:

- Computational and experimental analysis on the three blades propeller structure shows good correlation in most of the mode. Even though the results is well correlated, there are still room for improvement. A method call model updating is very recommended for reducing the available discrepancies between analytical and experimental results.
- When the propeller has less number of blades shows higher value of natural frequency. The two blades propeller has the highest value of natural frequency in every first six mode than the three blades and four blades propeller. This is owing to the fact that less blade means less mass. As the value of natural frequency is inversely proportional to the mass of a structure, the two blades propeller which has the lowest mass has the highest value of natural frequency.
- The mode shapes for all propellers were computed and analysed. For the four blades propeller was observed to show the most deflection in its mode shape. This situation suggested that , resonance is more likely to occur in propeller with more blades.

References

[1] Kurniawan R, Himawanto D A and Widodo P J 2019 The performance of numbers of blade towards picohydro propeller turbine *IOP Conference Series: Materials Science and Engineering*

[2] Liu Z, Liu P, Hu T and Qu Q 2018 Experimental investigations on high altitude airship propellers with blade planform variations *Proc. Inst. Mech. Eng. Part G J. Aerosp. Eng.*

[3] Magionesi F, Dubbioso G, Muscari R and Di Mascio A 2018 Modal analysis of the wake past a marine propeller *J. Fluid Mech.*

[4] Sinha S K and Zylka R P 2017 Vibration analysis of composite airfoil blade using orthotropic thin shell bending theory *Int. J. Mech. Sci.*

[5] Feng C, Yong C and Hongxing H 2016 Local modes analysis of a rotating marine ship propeller with higher order harmonic elements *Journal of Physics: Conference Series*

[6] Felli M, Camussi R and Di Felice F 2011 Mechanisms of evolution of the propeller wake in the transition and far fields *J. Fluid Mech.*

[7] Wang L Z, Guo C Y, Su Y M and Wu T C 2018 A numerical study on the correlation between the evolution of propeller trailing vortex wake and skew of propellers *Int. J. Nav. Archit. Ocean Eng.*

[8] Bertram V 2012 Chapter 5 - Vibrations *Pract. Sh. Hydrodyn. (Second Ed.*

[9] Radkowski S and Szulim P 2015 Analysis of vibration of rotors in unmanned aircraft *Adv. Intell. Syst. Comput.*

[10] R.G. Manley Grad. R.Ae.S. B S 1944 Vibrations in Aircraft *Aircr. Eng. Aerosp. Technol.*

[11] Abdullah N A Z, Sani M S M, Hussain N A, Rahman M M and Zaman I 2017 Dynamics properties of a Go-kart chassis structure and its prediction improvement using model updating approach *International Journal of Automotive and Mechanical Engineering (IJAME)* 14 3887-

97
[12] Ren H, Li F C and Zhao T Y 2014 Modal analysis of marine propeller submerged in fluid *Advanced Materials Research*
[13] Prasad L M and Kumar S 2008 Structural and Vibration Analysis of propeller blade in small aircraft *Int. Res. J. Eng. Technol.* **595**
[14] Morad A M, Elzahaby A, Abdallah S, Kamel M and Khalil M K 2015 Application of Piezoelectric Materials for Aircraft Propeller Blades Vibration Damping *Int. J. Sc ientific Eng. Researc h*
[15] Felli M, Guj G and Camussi R 2008 Effect of the number of blades on propeller wake evolution *Exp. Fluids*
[16] Jelani K M, Razip M H M, Nazri N A, Sani M S M and Yasar M 2019 Dynamics investigation on motorcycle chassis based on Finite Element (FE) modelling and updating *IOP Conf. Ser. Mater. Sci. Eng.* **469** 012103
[17] Fouzi M S M, Jelani K M, Nazri N A and Sani M S M 2018 Finite Element Modelling and Updating of Welded Thin-Walled Beam *Int. J. Automot. Mech. Eng.* **15** 5874–89
[18] Fouzi M S M, Sani M S M and Muchlis Y 2019 Finite Element Modelling and updating of welded joint for dynamic study of exhaust structure *IOP Conf. Ser. Mater. Sci. Eng.* **469** 012099
[19] Qu Y, Su J, Hua H and Meng G 2017 Structural vibration and acoustic radiation of coupled propeller-shafting and submarine hull system due to propeller forces *J. Sound Vib.*
[20] Zhang C and Huang Q 2017 Experimental research on the characteristics of ship propulsion system under dynamic excitations *2017 4th International Conference on Transportation Information and Safety, ICTIS 2017 - Proceedings*

IOP Conf. Series: Materials Science and Engineering **807** (2020) 012036 doi:10.1088/1757-899X/807/1/012036

The Application of Non-deterministic Model Updating method of a Complex Jointed Structure using Central Composite Design based meta-model

M.A.S. Aziz Shah[1], M.A. Yunus[1*], M.N. Abdul Rani[1], M.S.M. Sani[2]
[1] Structural Dynamics Analysis and Validation, Faculty of Mechanical Engineering, Universiti Teknologi MARA (UiTM), 40000 Shah Alam, Selangor, Malaysia
[2] Faculty of Mechanical Engineering, Universiti Malaysia Pahang (UMP), 26600 Pekan, Pahang Malaysia.

*E-mail: mayunus@salam.uitm.edu.my

Abstract. Finite element (FE) model of a structure is highly depended on idealising and simplification of model, however, the model may not truly represent the physical structure. The application of FE model updating is advocated to modify FE model of structure in order to acquired better correlation between the predicted and measured structure response. Nevertheless, for complex jointed structure, a FE model is containing a very large degree of freedoms that may contribute to high computational time. Therefore, this paper presents the application of response surface method in non-deterministic model updating using central composite design (CCD) sampling for improving the efficiency of a finite element model of a laser stitch welded structure. In this study, FE model of the structure is developed using CQUAD4 shell elements and ACM2 element connectors been used to representing the laser stitch weld joints. For the measured data, experimental modal analysis was performed using LMS SCADAS and conducted under free-free boundary conditions. In the model updating, the minimisation of uncertainties parameters in the FE model is based on the objective function that is formed from the residuals between the FE and experimental natural frequencies. The results show that the response surface method using CCD sampling is efficient to be used in FE model updating because it is capable of improving the accuracy of the initial FE model.

1. Introduction

In the simulation analysis for structural dynamics problem, finite element (FE) method is an important tool to simulate the dynamic behaviour of complex engineering structures. However, engineers or scientist usually design the FE models by making engineering assumptions. These engineering assumptions generally may not truly embody all the characteristics of a physical structures [1-4]. Consequently, the predicted responses from FE models may extremely inaccurate and differ went comparing with actual structure. The divergences that are produced from FE models basically are originated from the uncertainties in simplifying assumptions of the structural geometry, material properties, mechanical joints and boundary conditions. To improve the accuracy of FE models, the optimisation procedure such as FE model updating is required to adjust uncertainty in parameters that can increase the accuracy of predictions results [5-6].

Inversed method such as FE model updating where used responses such as natural frequency and mode shape to reducing uncertainties in the initial FE model [7]. However, Ren et. al has highlight that setting up of an objective function, selecting updating parameters and applying robust optimisation algorithm are the most vital procedure in improving the predicted results via model updating [8]. Meanwhile, study showed that, it is very important to ensure the reliability of updated parameter in order to accurately mimic the physical structure [9].

There are a lot of research that used the sensitivity based model updating [10-14]. For example, Russo et. al had mentioned the complexity of the sensitivity based model updating for complex structure in constructions the sensitivity matrices. This is because the FE models must be tuned and recomputed iteratively during optimisation process [15]. Besides, FE model of a complex assembled structure is containing with high degree of freedoms and frequently causing convergence difficulty, ill-conditions and requiring very high computational process [16]. The mentioned problems show that the ineffectiveness of deterministic model updating. Thus, alternative methods such as the used of response surface method in development of meta-model are more efficient in improving the predicted model.

Response surface (RS) methodology is a method to replace FE model using surrogate model. Initially, surrogate model is a simplified form of a FE model. The FE model been replaced statically using design of experiment (DOE) approach. Therefore, RS method is more preferable due to the advantages such as low computation memory needed and low processing time. In RS method, significant parameters are included in predicting the dynamic behaviour and make this method more efficient [17-18].

In this paper, the procedure of FE model updating using RS methodology to improve the correlation between initial predicted in the light of measured data is presented. The structure used to demonstrate the procedure stated is assembled laser stitch welds structure. The structure is selected because of the complexity in the joint development which contribute to the structural uncertainty and thus, non-deterministic procedure can be applied.

2. Structural Testing via Experimental Modal Analysis (EMA)

For the EMA, the structure assembled using laser stitch welds as shown in Figure 1 were tested. The structure was selected as to replicate a sub-structure of car body-in-white. There are 20 laser stitch welds with 10 mm length for each. The structure was fabricated using mild steel.

In this work, frequency of interest was 1 – 1000 Hz. As shown in Figure 2, the structure was setup under free-free boundary conditions using four sets of strings and springs. The free-free boundary conditions setup is essential in this work as to reduce the uncertainty due to the boundary conditions. Meanwhile, impact hammer and roving accelerometers technique was applied to the structure to measure the dynamic behaviour such as natural frequencies and mode shapes (Figure 3). Finally, data acquisition system (DAQ) such as LMS SCADAS were used to processes the output data obtained.

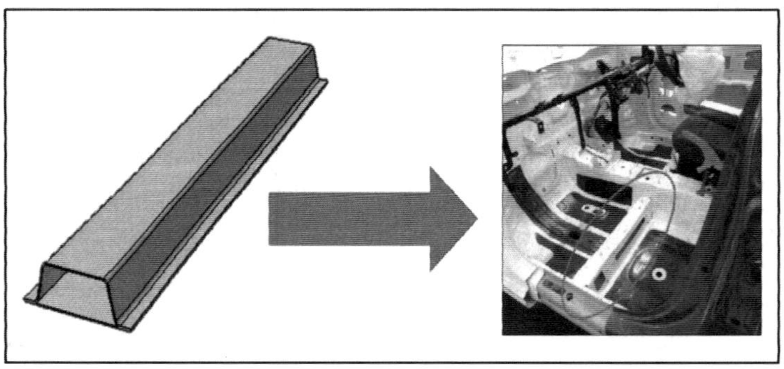

Figure 1. Simplified model of a car body-in-white

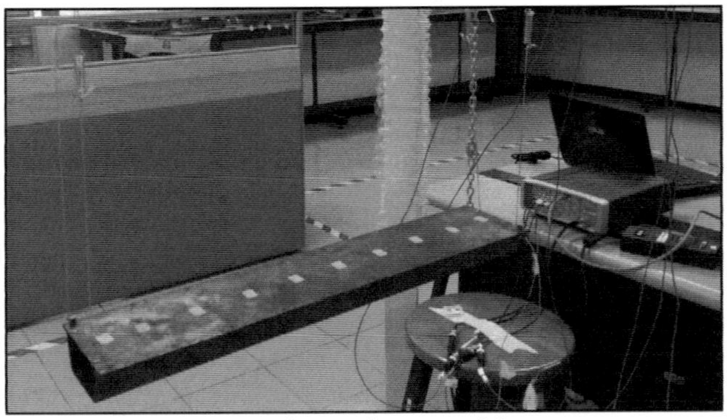

Figure 2. Experimental setup of the structure

3. Finite element (FE) modelling and Analysis

The process of FE analysis for the structure are distinguished into 3 stages which are pre-processing (input model), solver (normal mode analysis) and post-processing (output data). In the pre-processing, the initial FE model was constructed using MSC software. The FE model was constructed using shell elements with 5 mm meshing size. To represent laser stitch weld joints in the FE model, ACM2 element were used [19-21]. Standard properties of mild steel were used as input properties of the FE model.

Table 1. Properties of the welded structure [21]

	Parameter	Value	Unit
Mild steel	Young's Modulus	210	GPa
	Poisson's Ratio	0.30	Unitless
	Mass Density	7700	kg/m^3
Laser stitch welds (ACM2)	Young's Modulus	210	GPa
	Poisson's Ratio	0.30	Unitless
	Mass Density	7700	kg/m^3

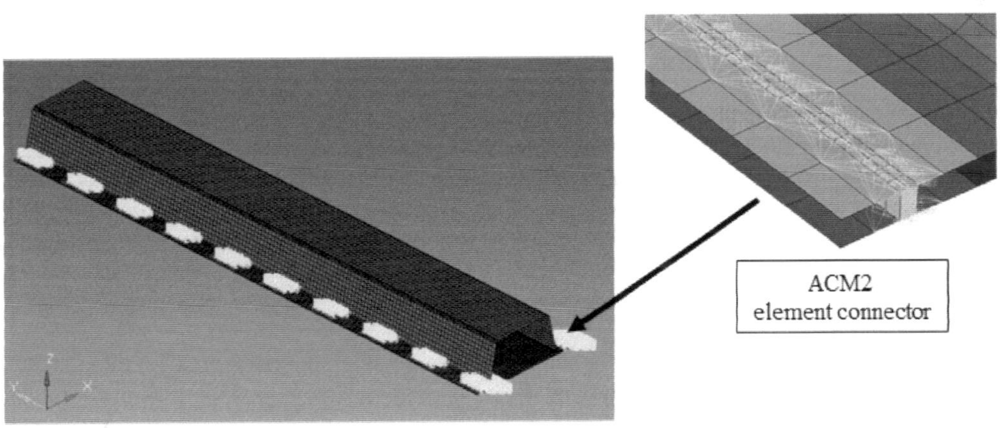

Figure 3. The FE model of the welded structure

For the solver stage, SOL 103 solver was used to solve the equation of motion. SOL 103 is a solver for normal modes analysis which is to calculate the dynamic behaviour of FE model. The equation of motion is given as

IOP Conf. Series: Materials Science and Engineering **807** (2020) 012036 doi:10.1088/1757-899X/807/1/012036

$$(\mathbf{K} - \omega^2\mathbf{M})\phi = 0 \quad (1)$$

where **K** and **M** are symmetric matrices of stiffness and mass. Meanwhile ω and ϕ are the natural frequency and mode shape of the system.

4. RS based FE Model Updating

RS methodology based model updating is a method that creating functional evaluation in the design space to globally approximate the response of the structure with comprises objectives and contains. Generally, this method often comprises with a statistical methods to enhancing structures responses [22]. The flowchart of the FE model updating based on RS methodology is shown in Figure 4.

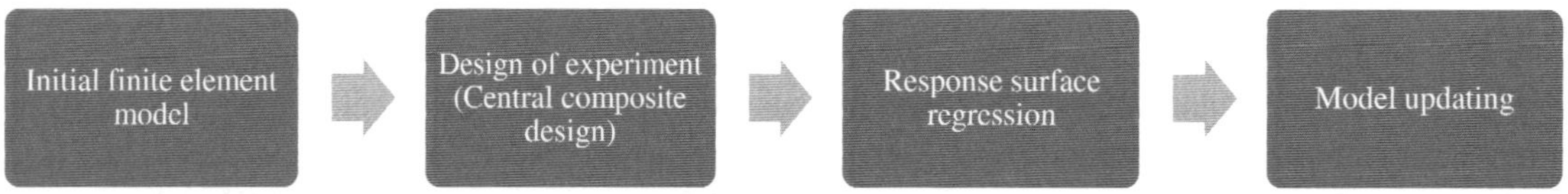

Figure 4. The process of model updating using response surface methodology

4.1. Central Composite Design (CCD) method

One of the basic process to develop a meta-model using response surface (RS) is to calculate predicted response features at numerous points in the design space by solving Eq (1) at selected points using design of experiment (DOE). The values that obtain for solving the Eq (1) at various points are fit with a RS and serve as sampling. However, it is important to understand that, the efficiency and the accuracy of a RS are depending on the selection of sample points. This is because less sampling points may reduce the RS accuracy, meanwhile, high sampling points may improve the response surface accuracy but with high computational time. In this study, CCD method is used in constructing the RS. This method has been found as the most accurate and simple DOE for creation of polynomial surfaces [23-24].

4.2. Parameters selection and identification

In the model updating, the selection of updating parameters to improve the correlation of the predicted data is a crucial part. The potential updated parameters should be selected properly and able to maintain the physical significant of the structure. In the RS model updating, the selected parameters are been used to construct RS and further, the parameters been adjusted to the satisfactory level of accuracy. The parameters such as material properties of geometry or joints can be considered as updated parameters, however it is important to make sure the responses such as natural frequencies are sensitive to the selected parameters.

4.3. RS regression

It is important to develop a meta-model that can represent the dynamic behaviour efficiently. For the structural dynamic problems, second-order polynomials are the most appropriate forms to representing a RS because the calculations are simple and the resulting function is closed-form algebraic expression with less complicity [25]. Polynomials are also capable to approximate the map function between physical parameters and response and thus optimising the responses.

4.4. FE model updating

The intelligent method to increase a confident of predicted responses is FE model updating method. The aim of the method is to improve the correlation of initial FE model in the light of experimental data. The improvement of the initial predicted responses is achieved by altering the assumptions of the predicted model to an acceptable level. Natural frequencies were used as objective response in this research. The objective function is formulated as in Eq (6).

$$J = \sum_{i=1}^{n} W_i \left(\frac{\lambda_i^{\text{fe}}}{\lambda_i^{\text{exp}}} - 1 \right)^2 \tag{6}$$

where, λ_i^{exp} is the i-th experimental eigenvalue and λ_i^{fe} is the i-th predicted eigenvalue from the FE model and n is the number of eigenvalues involved in the updating procedure.

5. Results and Discussion

Table 2 illustrations the results of the initial FE model and experimental modal analysis (EMA) of laser stitch welded structure in term of natural frequencies. From the results, a vast error was recognised in the initial FE model which is 50.41 percent for the first 6th modes. Meanwhile, the comparison also reveals that the high contribution to the total error is in the mode 2rd, mode 4th, mode 5th, and mode 6th.

The discrepancies in the natural frequencies between FE analysis and EMA were arisen of the inability of the initial FE model to replicate the physical structure accurately. In this work, the initial FE model was developed by idealisation of geometry, nominal material properties, and simplified weld joints [26]. Hence, alteration on the initial FE model must be done to accurately represent the structure.

Parameters such as Young's modulus of mild steel and Young's modulus of weld joints were investigated using sensitivity analysis, to identify significant parameters that need to be updated. However, this study only included the global material properties of the structure since the objective of this work is improving initial FE model by using low computational time based model updating method in which by reducing the complexity of the FE model using meta-model (as shown in Figure5).

RSM based model updating was effectively conducted to the initial FE model. From the result in Table 2, the total error has been managed to reduce from 50.1 percent to 26.84 percent. The result in Table 2 also shows that, the minimisation in the individual error of every mode particularly for the 4th mode, from 15.11 percent to 9.71 percent. Meanwhile, Table 3 shows the comparison of initial and updated value of influential parameters (sensitive) of the structure where the Young's modulus of mild steel and Young's modulus of weld joints were updated from 200 GPa to 206 GPa and 231 GPa respectively. In this work, it was found that, the proposed method has managed to improve the predicted result reasonably.

Table 2. Comparison between EMA, FE and RSM.

Mode	I. EMA (Hz)	II. FE (Hz)	III. Error between I & II (%)	IV. RSM (Hz)	V. Error between I & II (%)
1	521.25	514.11	1.37	526.11	0.93
2	590.04	533.05	9.66	563.20	4.55
3	596.88	561.07	6.00	574.20	3.80
4	672.43	570.82	15.11	607.14	9.71
5	681.01	613.39	9.93	649.40	4.64
6	693.44	635.56	8.35	671.20	3.21
	Total Error		50.41		26.84

Table 3. Comparison values of influential parameters of the structure

Parameter	Initial value (GPa)	Updated value (GPa)	Difference (%)
Young's modulus of mild steel	200	206	3.1
Young's modulus of laser welds	200	231	15.5

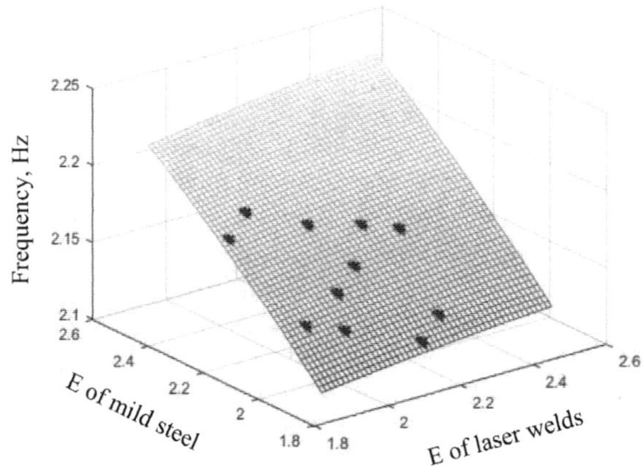

Figure 5. Meta-model of welded structure

6. Conclusions

The procedure of FE model updating using response surface (RS) method for structures that assembled by laser stitch welds has successfully presented in this paper. To improve the correlation of predicted responses with measured data, the meta-model based on RS using second-order polynomial was constructed with central composite design (CCD) as FE statistical sampling. This research show that, the initial FE model can be efficiently and successfully optimise using mentioned procedures.

Acknowledgments

The authors would like to acknowledge the Malaysia Ministry of Higher Education (MOHE) and Universiti Teknologi MARA (UiTM) for funding this research through the fundamental research grant scheme (FRGS) 600-IRMI/FRGS 5/3 (172/2019). The authors would also like to extend their sincere gratitude for the support and help given by Research Management Institute (RMI), Prof. Hadariah Bahron and all SDAV members.

References

[1] S. E. Abdel Raheem, A. K. Abdel Zaher, and A. M. Taha, "Finite element modeling assumptions impact on seismic response demands of MRF-buildings," *Earthq. Eng. Eng. Vib.*, vol. 17, no. 4, pp. 821–834, 2018.

[2] M. A. Yunus, M. A. S. Aziz Shah, M. N. Abdul Rani, W. I. I. Wan Iskandar Mirza, and S. Kushairi, "Experimental and Finite Element Analysis of Laser Stitch Welded Structure," *J. Mech. Eng.*, vol. 4, no. 3, pp. 155–162, 2017.

[3] M. Zhan, Q. Guo, L. Yue, and B. Zhang, "Modeling and Stochastic Model Updating of Bolt-Jointed Structure," *Shock Vib.*, vol. 2018, pp. 1–12, 2018.

[4] M. A. S. Aziz Shah, M. A. Yunus, M. N. Abdul Rani, N. A. Tormidi, A. A. Mat Isa, and W. I. I. Wan Iskandar Mirza, "Closed Mode Separation of The Symmetrical Structure," *J. Mech. Eng.*, 2018.

[5] S. Daouk, F. Louf, O. Dorival, L. Champaney, and S. Audebert, "Uncertainties in structural dynamics: overview and comparative analysis of methods," *Mech. Ind.*, vol. 16, no. 4, p. 404, 2015.
[6] M. A. Yunus, M. N. A. Rani, M. S. M. Sani, and M. A. S. A. Shah, "Finite element model updating of riveted joints of simplified model aircraft structure," *AIP Conf. Proc.*, vol. 1952, no. 1, p. 20013, 2018.
[7] M. Friswell and J. E. Mottershead, *Finite Element Model Updating in Structural Dynamics*. Springer Netherlands, 1995.
[8] W. X. Ren and H. B. Chen, "Finite element model updating in structural dynamics by using the response surface method," *Eng. Struct.*, vol. 32, no. 8, pp. 2455–2465, 2010.
[9] J. E. Mottershead and M. I. Friswell, "Model Updating In Structural Dynamics: A Survey," *Journal of Sound and Vibration*, vol. 167, no. 2. pp. 347–375, 1993.
[10] M. S. Mohd Zin, M. N. Abdul Rani, M. A. Yunus, W. I. I. Wan Iskandar Mirza, A. A. Mat Isa, and Z. Mohamed, "Modal and FRF Based Updating Methods for the Investigation of the Dynamic Behaviour of a Plate," *J. Mech. Eng.*, vol. 4, no. 3, pp. 175–189, 2017.
[11] Y. Yuan and H. Liu, "An iterative method for solving finite element model updating problems," *Appl. Math. Model.*, vol. 35, no. 2, pp. 848–858, 2011.
[12] J. E. Mottershead, M. Link, and M. I. Friswell, "The sensitivity method in finite element model updating: A tutorial," *Mech. Syst. Signal Process.*, vol. 25, no. 7, pp. 2275–2296, 2011.
[13] M. A. Yunus, H. Ouyang, M. N. A. Rani, and A. A. M. Isa, "Modal test and model updating for a welded structure made from thin steel sheets," *20th Int. Congr. Sound Vib. 2013, ICSV 2013*, no. 7, pp. 7–11, 2013.
[14] M. N. Abdul Rani, S. Kasolang, M. H. Othman, M. A. Yunus, W. I. I. Wan Iskandar Mirza, and H. Ouyang, "Finite Element Modelling and Modal Based Updating for the Dynamic Behaviour of a Laser Spot Welded Structure," *ICSV 2016 - 23rd Int. Congr. Sound Vib. From Anc. to Mod. Acoust.*, pp. 1–8, 2016.
[15] G. Boscato, S. Russo, R. Ceravolo, and L. Zanotti Fragonara, "Global Sensitivity-Based Model Updating for Heritage Structures," *Comput. Civ. Infrastruct. Eng.*, vol. 30, 2015.
[16] L. Deng and C. S. Cai, "Bridge Model Updating Using Response Surface Method and Genetic Algorithm," *J. Bridg. Eng.*, vol. 15, no. 5, pp. 553–564, 2009.
[17] T. Marwala, *Finite-element-model updating using computional intelligence techniques: Applications to structural dynamics*. 2010.
[18] S. Sehgal and H. Kumar, "Structural Dynamic Model Updating Techniques: A State of the Art Review," *Arch. Comput. Methods Eng.*, vol. 23, no. 3, pp. 515–533, 2016.
[19] M. Palmonella, M. I. Friswell, J. E. Mottershead, and A. W. Lees, "Guidelines for the implementation of the CWELD and ACM2 spot weld models in structural dynamics," *Finite Elem. Anal. Des.*, vol. 41, no. 2, pp. 193–210, 2004.
[20] R. M. N. . Abd Halim. N.I, Yunus.M.A, Kasolang.S, "Finite Element Model of the Spot Welded Joints of Door in White (DIW)," *Eng. Appl. Sci.*, vol. 11, no. 11, pp. 2474–2478, 2016.
[21] M. S. M. Zin, M. N. A. Rani, M. A. Yunus, M. S. M. Sani, W. I. I. Wan Iskandar Mirza, and A. A. Mat Isa, "Frequency response function (FRF) based updating of a laser spot welded structure," *AIP Conf. Proc.*, vol. 020055, p. 020055, 2018.
[22] M. A. Yunus, S. Nazri, M. N. Abdul Rani, A. Tormodi, and S. Kasolang, "Response surface reconciliation method of bolted joints structure," *MATEC Web Conf. AiGEV 2016*, vol. 90, p. 01013, 2017.
[23] G. G. Wang, "Adaptive Response Surface Method Using Inherited Latin Hypercube Design Points," *J. Mech. Des.*, vol. 125, no. 2, p. 210, 2003.
[24] S. W. Doebling, F. M. Hemez, J. F. Schultze, and A. L. Cundy, "A Metamodel-Based Approach to Model Validation for Nonlinear Finite Element Simulations Model Validation – Beyond Test / Analysis Correlation," *Simulation*, 2002.
[25] C. Mares, J. E. Mottershead, and M. I. Friswell, "Stochastic model updating of a spot welded

structure," *Int. Conf. Noise Vib. Eng.*, vol. 1–8, pp. 1885–1898, 2005.
[26] M. S. M. Zin, M. N. A. Rani, M. A. Yunus, M. A. Ayub, M. S. M. Sani, and M. A. S. A. Shah, "Modal based updating for the dynamic behaviour of a car trunk lid," in *AIP Conference Proceedings*, 2019, p. 020001.

IOP Conf. Series: Materials Science and Engineering **807** (2020) 012037 doi:10.1088/1757-899X/807/1/012037

Ergonomic Design for Tempe Production Tool Based on User Voice

E W Asih[1], A Mawadati[1] and N Lestari[1]

[1]Industrial Engineering Department, Faculty of Industrial Technology, AKPRIND Yogyakarta Institute of Science & Technology, Jl. Kalisahak 28 Yogyakarta

Corresponding e-mail: endang.akprind@gmail.com

Abstract. The process of tempe production in small industry are considering not optimal yet. The process does not consider the ergonomic aspects and is not environment friendly. It causing complaint at some body parts and pollution. For the solution, this research aims to design a multifunctional tool that considering ergonomic and eco- friendly aspect. The design will be based on Ergonomic and the Quality Function Deployment (QFD) methods. Based on evaluation using the Nordic Body Map, it shows that body parts with the most complaints are the waist, back, neck, right and left shoulder, wrist and leg. Complaints on these body parts are caused by wrong postural when doing work which is not ergonomic. And by using Quality Function Deployment (QFD) the tool designed by considering user expectation and desire. it was found that the user expect for the tool to be made based on worker anthropometry, moved with a semi-manual machine and the tool model is suitable for work. Overall the design shows that the Ergonomic Tempe Maker Tool is a multifunctional tool that easy, flexible, safe and comfortable to be operated which focused on anthropometric size, uses a semi-manual machine and can be arranged up or down.

1. Introduction

In one of small industry that produce tempe, the production process performance and work environment are considering not optimal yet. It has low productivity and efficiency. The reason is that production process is not ergonomic and unhealthy. The non-ergonomic production process can be seen at the process of breaking soybeans and overall they have to go back and forth from one process to another, also their waste management process is not environmentally. In the process of breaking soybeans, workers work by bending and squatting (can be seen in Figure 1). Tools used in the process of breaking soybeans are also not in accordance with the dimensions of the body of the worker, so workers must bend to do their work. It can cause pain in the back and neck, as well as the risk of muscle and spinal back disorders.

At the washing and cracking process include a repetition and back and forth process. This process starts by cooking the soybean, then after being boiled in the kitchen the soybean taken to the well for washing, after being washed, then brought back to close to the kitchen for the soybean cracking process. In the process, workers need to go back and forth from the kitchen to the well then returning back to the kitchen, this is done 5 times a day, because for boiling 20 kg they divided it into 5 washing parts, so that the production process becomes long, and cause the results are not optimal (low worker productivity).

In this small industry it also can be seen that the work environment is not environmentally friendly, where the disposal of waste from the washing process is disposed to sewers and river (see Figure 2), which polluting the environment. As a result of environmental pollution, causing air odor, sewers or streams become smelly, and the soil becomes muddy. Pollutants are the amount of pollutant weight in a certain time unit, which is the result of the multiplication of pollutant levels with liquid waste discharge [1]. The parameters used to measure the content of pollutants include BOD (Biological Oxygen Demand) and COD (Chemical Oxygen Demand),.

To increase the productivity of small-scale industries and reduce the impact of waste, these influential factors must get serious attention from small industry owners, especially factors such as; ergonomic risk factors, awkward posture while working and work layout [2]–[3] and waste. Previous studies have shown that workers in small industries are suspectible to disease due to work posture errors [4–8]. In addition to potentially causing disease, this non-ergonomics work posture and work environment also has the potential to cause productivity to decline. Surely this can affect the output of the business. The problem of increasing productivity is done as an effort to get alternative ways of working that are good, effective, efficient, healthy, comfortable and safe.

A good, efficient and comfortable alternative for small industries Rasimun is making a mulfunctional tool that considering the ergonomic and eco friendly aspect [9]. This tool is designed so can be used for washing and cracking the soybeans and the residual liquid can be stored in the container so that can be processed into something useful. For this reason, it is necessary to design a multifunctional tool that can be used for washing, splitting and filtering (to separate soybeans from liquid waste) while being ergonomic and environmentally friendly.

Figure 1. Process of cracking soybeans.

Figure 2. Disposal of waste in a ditch.

2. Methodology

2.1 Ergonomics

Ergonomics comes from the word ergo which means work and nomos which means rules, principles or rules. So the word ergonomics means a study of the relationship between humans and their work. In its development, the notion of ergonomics is a term that is widely used in terms of human engineering or human factor, which is a study of interface devices and forms of interaction between humans and objects used with the work environment. By using a more comprehensive approach, ergonomic is carried out through three main things, namely; focus, purpose and ergonomics [10]. The focus of ergonomics is humans and their interactions with products, equipment, facilities, procedures and work environments and daily life. The goal of ergonomics is to increase the effectiveness and efficiency of work, improve safety, reduce fatigue and stress, increase comfort, receive greater user acceptance, improve job satisfaction and improve quality of life. The approach taken in ergonomics is a systematic application of relevant information about abilities, limitations, characteristics, behavior and motivation humans to product design and procedures used for the environment in which they are used.

Based on the above approach, Chaffin summarizes the definition of ergonomics as a science that explores and applies information about behavior, abilities, limitations, and other human characteristics to design equipment, machinery, systems, jobs and environments to improve productivity, safety, comfort and the effectiveness of human work [11].

Ergonomics also can be define as a systematic branch of knowledge to utilize information about the nature, abilities and limitations of humans to design a work system so that people can live and work in the system by both achieving the desired goals through work effectively, safely and comfortably [12]. In ergonomics, one of the principles that must always be used is the principle of fitting the task / job to man. This implies that work must be adapted to the capabilities and limitations of humans, so that the results achieved can be better. The absence of ergonomist can result in undesireable product design [13]. For the example the tools that be used by the worker in Rasimun. The machine is not made based on the anthropometryand biomechanics of the body of the worker. Thus, there are many complaint regarding pain in the back and neck. In this paper, we develop a new design of tempe production tool that minimize physical discomfort and risk of of muscle and spinal back disorders.

2.2 Quality Function Deployment (QFD)

Quality Function Deployment (QFD) is a methodology used by companies to anticipate and prioritize consumer needs and desires. and combining the needs and desires of consumers in the product or service provided to consumers [14]. The use of QFD in the product design process will help the management to gain competitive advantage through the process of creating the characteristics and attributes of quality products or services that can increase customer satisfaction. In addition, the implementation of the QFD methodology is also able to guarantee that information on consumer needs obtained in the initial stages of the planning process is applied to all stages of the product cycle, from the design concept, component planning, process and production planning, to the product for consumers.

The key to competitive ability is the ability to respond to challenges in producing a product and / or service quickly [15]. But there are some obstacles for companies to produce products and / or services quickly, such as: low understanding of customer needs, strategic mistakes in determining priorities, the desire to take risks that can be returned, there is a tendency towards the use of designs that are not developing or unable to provide services, determination of specifications that are too high, and the wrong testing scenario in determining the main error. QFD is used for three reasons: to focus on customer need and satisfaction, to save development and design time, also to improve communication in all parts of the organization [16].

Many research has done using QFD as their tool to develop a new product [17–20] or for evaluation of system [21–23]. The use of QFD has also been carried out in research on the design of ergonomic products [13,24,25]. In this study QFD is used to identify the desires and needs of worker regarding the tool they will use at tempe production.

3. Results and Discussion

This study begin by collecting initial data (qualitative data), was carried out with a questionnaire on workers and home industry owners, with the aim of obtaining questionnaire attributes. Attributes related to the use of tools in the production process, that are ergonomic and environmentally friendly. The questionnaire is divided into 3 parts, namely; section 1 contains the satisfaction of workers on the existing tools used for the production process, part 2 contains the importance of the tool and part 3 contains the complaints of the biomechanical workers on the existing tools. The questionnaires distribut to 25 workers and tempe industry owners.

The results of the questionnaire were calculated using the Importance to Customer value and satisfaction level. This data processing reflects the level of interest of workers on existing indicators based on the questionnaire. Measuring the level of customer satisfaction of product users is obtained based on the results of a questionnaire regarding how far the workers feel satisfied with the quality of the tempe production tool used at this time. Determination of the level of importance and level of satisfaction of worker indicators based on the average value of the weight or answers of respondents in processing the questionnaire.

Ergonomics is a verbal discussion of consumers which is then classified into the primary level. From the primary language, the ergonomic meaning is translated into the level of secondary language, which is easy, flexible, safe, comfortable and of appropriate size. Tertiary language is a sorting and secondary language translation which is a detailed result of the wishes of workers.

The final result of the QFD method is a product development plan. The development plan can be seen in the target value located at the bottom of the House of Quality (HOQ). The QFD method includes complete processes ranging from identifying problems to achieving targeted development projects through the birth of design specifications. For details of the overall results of House of Quality seen in Figure 3.

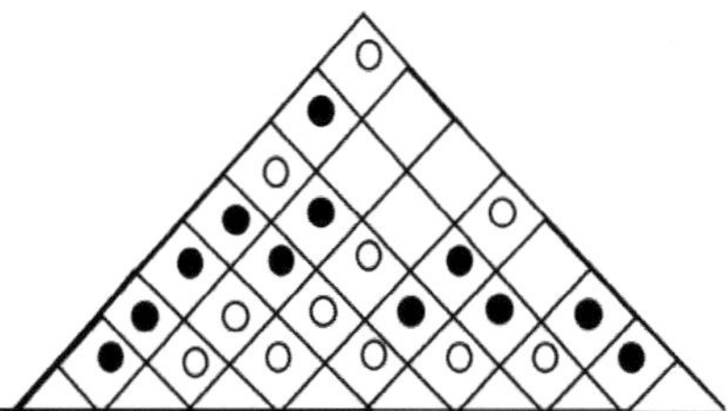

	Customer need \ Technical respond		Machine	Material Weight	Material Strength	material quality	according to anthropometry	can be move up and down	suitable to work	security
ERGONOMICS	Easy	Easy to be operated	20	4	16	20	20	12	20	
	Flexible	The ability to be used variously	25		15	15	25	20	25	
		The rotation speed can be adjusted	20	12	16	16	12		16	
		To work standing or sitting	20		12	12	25	25	15	
		Use inside or outside the room	9	15	9	9	15		19	
	Right Size	Easy to be moved (lightweight)	12	15	9	9	15		3	
		match size with body dimensions	25		15	15	25	20		
	Secure	The condition of the tool when used is clean	20		15	15	25		15	25
		security of equipment when used	12		9	9	9	9	20	20
	Comfort	Comfort to be used	20			12	25	25	25	15
	Total		**183**	**26**	**132**	**132**	**187**	**111**	**148**	**60**

Figure 3. House Of Quality .

From the HOQ matrix (house quality) can be known the level of the relationship between the attributes of workers' needs and the characteristics of the technique. Technical characteristics are used to measure or quantify consumer needs that are still qualitative. From the level of importance, it can be seen that the characteristics that have a value of 5, show that the most important thing is that the tool has the ability to use various processes, the condition of the equipment when used cleanly, and its measurement tools according to the dimensions of the human body. The level of importance with a value of 4, shows that it is very important for workers that the Tempe maker is easy to operate or control and security of use. It is important for consumers that the tool can be used inside or outside the room (can be moved),

In the relationship matrix between the needs of workers with technical characteristics shows workers prefer the tool to be easy to operate, the tool can be used in a variety of tempe making processes namely to wash, break and filter soybeans, convenient tools to use, tools can be used to work standing and or sitting, inside the meaning of the tool can be adjusted worker's work when standing by using a semi-manual machine and the tool can be used comfortably. This can be seen in the largest number found in the relationship matrix. From the target value of Figure 4. HOQ shows that the characteristics of the measurement technique with anthropometry, that is, the size of the tool made by Tempe is based on the anthropometry of workers, has the greatest value. The second biggest characteristic is that the tool is moved by a semi-manual machine and the tool model is in accordance with the work. This shows that the tool that makes ergonomic Tempe is an easy, flexible, safe and comfortable tool when operated focused on anthropometry size, the tool uses a semi-manual machine and the tool can be raised and lowered.

After House of Quality, the next step is design planning. The design phase of the ergonomic Tempe maker design tool starts from the largest target value generated from HOQ (House of Quality). Based on the target value, the order of the technical characteristics that are the top priority is the size of the tool in accordance with anthropometry, the tool is operated by the machine, the equipment can be raised and lowered, the tool must be made of strong material, good quality material and the seat is in accordance with tools and work. At this design, the ergonomic Tempe maker has many functions, such as washing, breaking and filtering soybeans based on priority characteristics as in Figure 3. And the drawing of the design based on QFD can be seen in Figure 4.

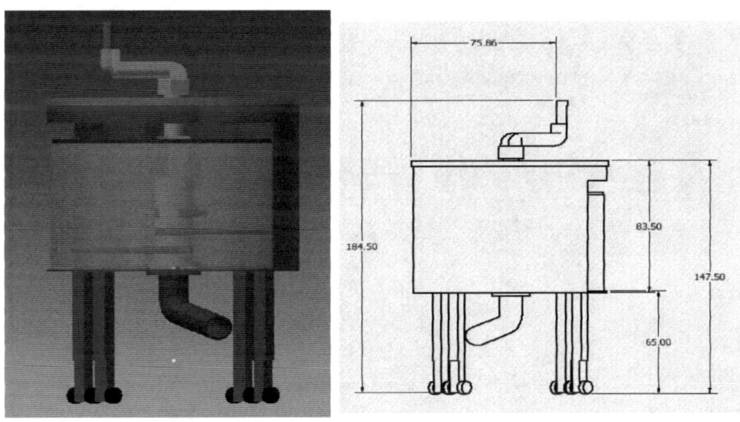

Figure 4. Cicahring Tool Design based on HOQ results

Figure 3 is a complete picture of the design product which we called is Cicahring. The design of Cicahring had considered several aspects of ergonomics so that the product is comfortable to be operated. The anthropometric measurement used in this design adjusts the body posture of Indonesian men, Javanese, and age ranges from 20-50 years.

This product consists of 1 core component, 3 subcomponents, and 1 supporting component. Each component is divided again into several components. The core component is a tub component that functions as a container of soybeans and water to be ground. Material used for this component is mild steel so it is easy to move. The diameter for the tub is 98 cm, which is the body dimension of the reach of the hand going forward. This tub has a hole that has function as a drainage center for soybeans after being ground. After the epidermis is separated and the washing water has been removed, the tub can be removed and the results of the mill are taken. This core component can be seen in Figure 5.

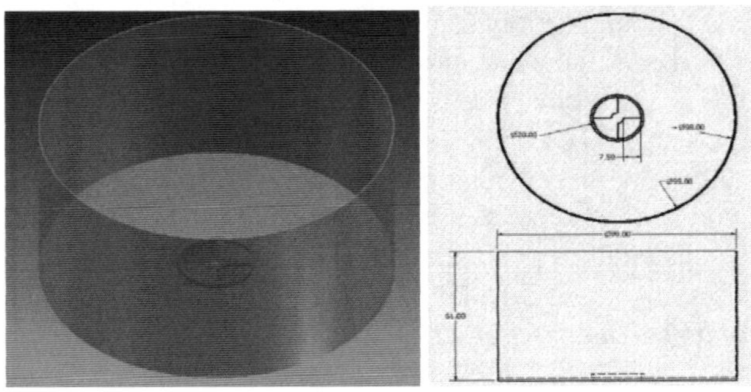

Figure 5. Core Component

The subcomponents consist of subcomponent 1, subcomponent 2 and subcomponent 3. Sub component 1 is a container and its grinding blade. In this component there are legs to support the engine load, also wheels so the product could be move easily. The wheel is equipped with a lock to prevent the product from move around being used. This subcomponents 1 can be seen at Figure 6.

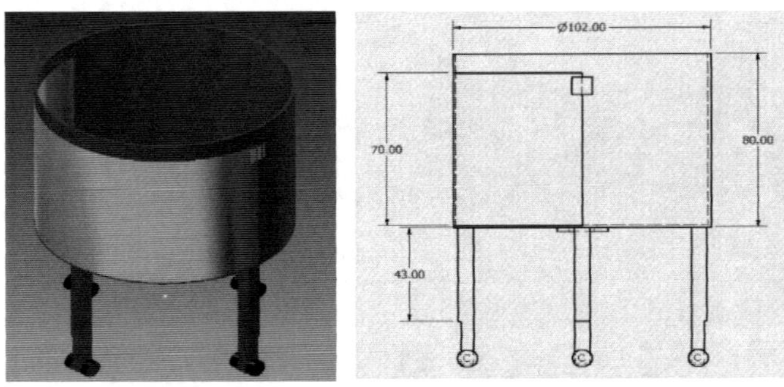

Figure 6. Sub Component 1

Sub component 2 is a combination of several components, such as the base of the blade, blade, bolt, filter, filter holder, and cover. At the base of the blade equipped with a rubber that serves as a lock and water barrier to prevent leakage. The blade is adjusted to the height of the filter, which later can be opened after the grinding process is complete. The filter can be opened and closed. During the grinding process the filter is opened, after finished the filter is closed to remove the soybean husk. Then the head of the blade base can be removed thus the blade and filter can be removed. This component can be seen at Figure 7.

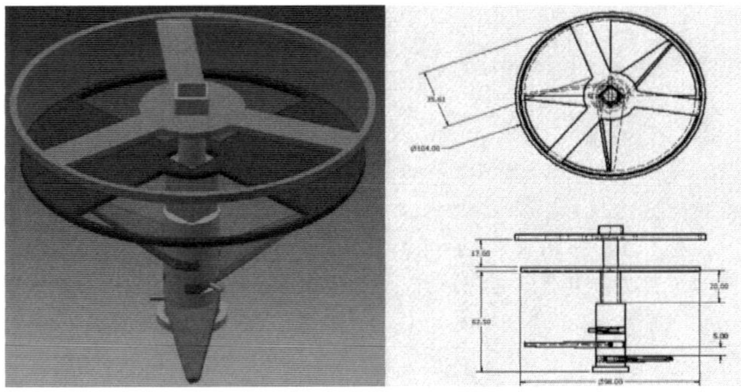

Figure 7. Sub Component 2

Sub component 3 is a component that functions as a blade grinder lever. For the length of the grinder lever uses body dimensions shoulder-hand length forward = 61cm, which is the total addition of the radius of the tub plus the length of the lever body. The lever consists of two components, namely the main lever and the hand handle. This subcomponent can be seen at Figure 8.

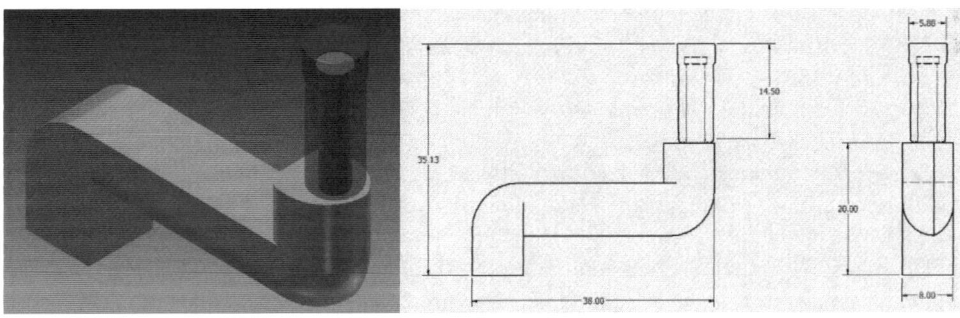

Figure 8. Sub Component 3

Supporting component is additional components used to dispose the disposal water so it does not splatter. This component has a pipe that functions to close and open the channel through the tap. The design of this supporting component can be seen at Figure 9.

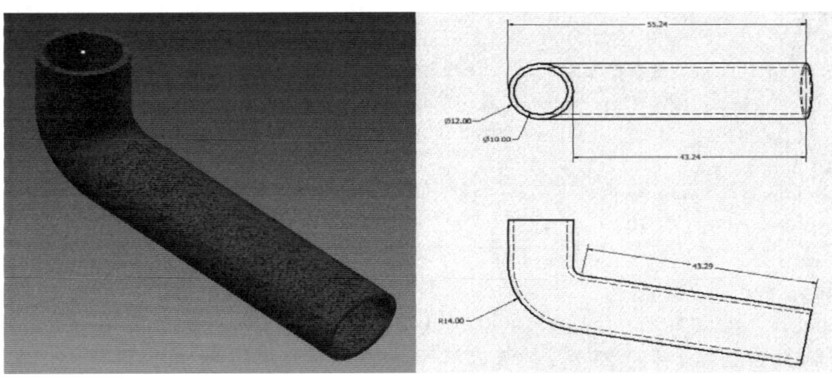

Figure 9. Additional Component

4. Conclusion

Based on previously discussed outcomes, it can be concluded that:

i. The results of the questionnaire with Nordic Body Map are that most respondents (85%) stated that work tools can affect work performance. For work posture 76% stated that work posture can affect work performance
ii. The body parts that have the most complaints are the waist, back, neck, right and left shoulder, wrist and leg. Complaints on the waist and back are caused by postural errors when doing work.
iii. From the results of the QFD through HOQ the tool designed is a multi-functional tempeh process tool for various processes, namely the process of washing, breaking and filtering ergonomic Cicahring with an easy, flexible, safe and comfortable tool when focused on anthropometric measurements. machine tools, can be used for a variety of processes and tools can be raised and lowered

References

[1] Anon 1999 Surat Keputusan Gubernur no 61
[2] Asih E and Oesman T 2015 Usulan Perancangan Fasilitas Kerja Yang Ergonomis Guna Meningkatkan Kinerja Pekerja Industri Kecil Mozaik *Proceeding 11th Natl. Conf. Indones. Ergon. Soc.*
[3] Asih E W, Parwati C I, Widyastuti N, Industri J T, Sains I, Teknologi & and Yogyakarta A 2015 Analisis Produktivitas Pada Proses Penyepuhan dengan Metode Green Productivity *Prosiding IENACO* (Surakarta: Universitas Muhamadiyah Surakarta)
[4] Mulyati Invironmental Politeknik Kesehatan Kementerian Kesehatan Bengkulu Bengkulu S 2019 *The Relationship Between Work Posture And Musculoskeletal Disorders (Msds) In Laundry Workers In The Area Of Puskesmas Sukamerindu Bengkulu*
[5] Gde I K, Suarbawa J, Adiputra N, Pangkahila A, Dewa I, Sutjana P, Ketut I and Suarbawa G J 2016 Work Posture Improvement Using Ergonomic Approach Decreases Subjective Disorders of Perapen Workers on the Process of Nguwad Gamelan in Bali *Int. Res. J. Eng.*
[6] Splittstoesser R E, Yang G, Knapik G G, Trippany D R, Hoyle J A, Lahoti P, Vatan Korkmaz S, Sommerich C M, Lavender S A and Marras W S Spinal loading during manual materials handling in a kneeling posture
[7] Gangopadhyay S, Ghosh T, Das T, Ghoshal G and Das B 2010 Effect of Working Posture on Occurence of Musculoskeletal Disorders among the Sand Core Making Workers of West Bengal Postural *Cent Eur J Public Heal.* **18** 38–42
[8] Vandergrift J L, Gold J E, Hanlon A and Punnett L 2012 Physical and psychosocial ergonomic risk factors for low back pain in automobile manufacturing workers *Occup. Environ. Med.* **69** 29–34
[9] Asih E W 2009 Perancangan Alat Pemecah Kedelai Yang Ergonomis Dengan Pendekatan Integrasi Model Kano & Quality Function Deployment
[10] Sanders M and McCormick E 1987 Human factors in engineering and design
[11] Don B. Chaffin, Gunnar B. J. Andersson B J M 2006 *Occupational Biomechanics* (England: John Wliley and Son Ltd)
[12] Sutalaksana I, Anggawisastra R and JH Tjakraatmadja 1970 *Teknik tata cara kerja* (Bandung: Jurusan Teknik Industri ITB)
[13] Marsot J 2005 QFD: A methodological tool for integration of ergonomics at the design stage *Appl. Ergon.* **36** 185–92
[14] Mazur G 2003 Voice of the customer (define): QFD to define value *Proc. 57th Am. Qual.*
[15] Cohen L *Quality Function Deployment: How to Make QFD Works for You* (Michigan: Addison Wesley-Publishing Company)
[16] Akbar R, Noor S and Shah W 2010 Qfd As a Tool for Improvement of Car Dashboard *J. Qual. Technol. Manag.* **VI** 1–22
[17] Wu X, Hong Z, Li Y, Zhou F, Niu Y and Xue C 2020 A function combined baby stroller design

method developed by fusing Kano, QFD and FAST methodologies *Int. J. Ind. Ergon.* **75** 102867

[18] Kuijt-Evers L F M, Morel K P N, Eikelenberg N L W and Vink P 2009 Application of the QFD as a design approach to ensure comfort in using hand tools: Can the design team complete the House of Quality appropriately? *Appl. Ergon.* **40** 519–26

[19] Moghimi V, Jusan M B M, Izadpanahi P and Mahdinejad J 2017 Incorporating User Values into Housing Design through Indirect User Participation using MEC-QFD Model *J. Build. Eng.* **9** 76–83

[20] Kurtulmuşoğlu F B and Pakdil F 2017 Combined analysis of service expectations and perceptions in lodging industry through quality function deployment *Total Qual. Manag. Bus. Excell.* **28** 1393–413

[21] Pramanik D, Haldar A, Mondal S C, Naskar S K and Ray A 2017 Resilient supplier selection using AHP-TOPSIS-QFD under a fuzzy environment *Int. J. Manag. Sci. Eng. Manag.* **12** 45–54

[22] Cebi S, Ozkok M and Demirci E 2014 Evaluation of Design Parameters for Vessel Engine Room by Using A Modified QFD Technique *J. Mult. Log. Soft Comput.* **23** 559–87

[23] Bulut E, Duru O and Huang S T 2018 A multidimensional QFD design for the service quality assessment of Kansai International Airport, Japan *Total Qual. Manag. Bus. Excell.* **29** 202–24

[24] Hashim A M and Dawal S Z M 2012 Kano Model and QFD Integration Approach for Ergonomic Design Improvement *Procedia - Soc. Behav. Sci.* **57** 22–32

[25] Zadry H R, Rahmayanti D, Susanti L and Fatrias D 2015 Identification of Design Requirements for Ergonomic Long Spinal Board Using Quality Function Deployment (QFD) *Procedia Manuf.* **3** 4673–80

Identification of the delaying factors on construction work completion in Jambi city according to the contractor perception

Elvira Handayani, Kiki Rizky Amalia, Fakrul Rozi Yamali

Civil Engineering Department, Universitas Batanghari

Corresponding e-mail : Elvira.handayani2@yahoo.co.id

Abstract. One of the successes in the Implementation of Construction Works is marked by the implementation of timely work. However, in the process of implementing the construction work often encountered various problems, one of which is work delays. the number of factors is the cause of delays that are often faced by parties involved, especially contractors who act as executors. This study aims to determine the main factors of the delay in the Construction Work. The questionnaire was distributed to 63 Middle-class contractors with the value of the company's assets qualified into the M1-M2 based on the Business Entity Registration List (DRBU) of the Jambi City Construction Services Development Agency (LPJK). The factors of delay in construction work are grouped into 7 factors, each of which consists of several sub-factors. The results showed that the most dominant factor in the construction work in Jambi City was environmental factors with sub-factors Heavy rain / waterlogged work location with a percentage of 15.73%, while the Time Factors namely Sub-factors Too late shop drawing agreement is the 7th most influential factor in work delay has the smallest percentage value of 13.41%.

1. Introduction

Every Implementation of a Construction Work, Wants to be successful in implementing project completion on time. In the development process it is very common to encounter various problems causing work delays, the number of factors that cause delays is a problem that is often faced by parties involved, especially contractors who act as executors.

With the existence of problems or constraints in the process of work will result in moral and material losses for parties directly related to construction activities, as the implementing party, the contractor can be said to be the party most disadvantaged for the delay in the work. Delays are things that greatly affect the success of a job, delays will be detrimental to all parties involved in a construction industry, this condition requires a good handling of the project so that the project delays can be minimized or avoided.

Delay in the completion of the project is a problem that is very undesirable by the contractor and the owner because it will always cause adverse consequences for both parties [1]. The impact of delays is conflict and debate about what and who is the cause plus the demands of time and additional costs. Not only for the contractor and owner, the loss is also experienced by the society around the project location due to activity disruption caused by the late completion of the project [2,3].

This research is only aimed at Contractors who play a role in Construction Work in Jambi city and are categorized as "Medium" class contractors with the value of the company's assets qualified into the M1-M2 based on the Business Entity Registration List (DRBU) of Jambi City Construction Services Development Institute (LPJK) period January to December 2018, focusing on the Contractor's perception based on their experience of the most influential factors or those that dominate the delay in the work.

With the research, it is hoped that it can become a reference material for the owner or contractor in preparing a more careful project planning and scheduling, in an effort to avoid and or control the delay in project implementation.

1.1 Purposes of the Research

The Purposes of this research are:

i. Knowing what factors influence work delays in construction projects in Jambi city in 2018
ii. Looking for the most dominant factor in the delay in construction work in Jambi city in 2018

2. Literature Review

2.1. Construction Activities

Construction activities are an organized effort to achieve important goals, objectives and expectations by using the budget of funds and available resources, which must be completed within a certain period of time [4].

Construction activities are a series of activities which are only one time carried out and are generally short-term [5]. Construction activities can be interpreted as a temporary activity that takes place in a limited period of time, with the allocation of certain resources and intended to carry out tasks whose goals have been clearly outlined [6]. These tasks can be in the form of building factories, making new products or conducting research and development.

Construction activities are activities that must go through a long process and in which many problems must be solved. In addition, there are sequential and related sequences in construction activities [7]. Usually the series starts from the birth of an idea that arises from a need (need), the thought of possible implementation (feasibility study), a decision to build and make a more detailed explanation of the formulation (briefing), pouring in the form of a preliminary design (preliminary design), making a more detailed and definite design (design development and detail design), administrative preparation for the implementation of development by selecting procurement candidates, then implementing construction on the location provided (construction), as well as maintenance and preparation for use the building (maintenance, start-up, and implementation) [8]. The building activity ends when the building is put into use.

2.2. Delay of Construction Work

Important parameters in the implementation of projects are often used as project targets, namely budget, schedule, and quality. Success in carrying out projects on time, cost, and quality that has been planned is one of the most important goals for owners and contractors [6]. Project implementation that is not in accordance with the plan, can result in project delays. In the implementation of construction projects, project delays often occur, which can cause various kinds of losses for service providers and service users. For contractors, delays in addition to causing project cost overruns due to increased time of project implementation can also lead to a decrease in the credibility of contractors for the future. Whereas for the owner, the delay in the use or operation of the results of the construction project and often has the potential to cause disputes and claims between the owner and the contractor [9].

The delay in construction projects means increasing the time of completion of project completion that has been planned and listed in the contract document [10]. In general, in development activities there are often delays due to lack of attention to the resources involved in it. The active role of management is one of the main keys to the success of a job.

IOP Conf. Series: Materials Science and Engineering **807** (2020) 012038 doi:10.1088/1757-899X/807/1/012038

2.3. The Causes of Work's Delay

Delays can be caused by different parties [11], namely:

i. Owner or representative (delay caused by owner or his agent). If the owner or deputy causes a delay, for example due to late work drawings or delays in giving approval to the image, the contractor will generally be permitted to obtain an extension of time and may also submit legitimate claims to obtain extra compensation.

ii. Allowable delay by third parties (excusable triedparty delay). There are often delays caused by different strengths beyond the control range of the owner or contractor. Examples that are generally not questioned include fire, floods, earthquakes and other things. Other things that are often a matter of disputes include strikes, embargoes for transportation, accidents and delays in submitting understandable ones. Also included are those that cannot be included in conditions that existed at the time the offer was made and the weather conditions were bad. In case this can be agreed upon, the type of delay of these types generally results in an extension of time but is not accompanied by additional compensation.

iii. Contractor-caused delay. Such delays will generally result in no extension of time and no additional compensation. If in extreme situations these things will cause a contract break.

2.4. Criteria of Project Delays

Criteria for project delays in this study can be seen in Table 1 below.

Table 1. Criteria of Project Delay

No	Factor	Sub Factor
1	Materials	Lack of Construction Materials
		Slow Material Delivery
		Material Provision on Location
		Delay in Ordering Material
		The amount of Material sent by the Supplier is not appropriate
		Material Damage due to Storage
		Material Management Mistakes by the Contractor
2	Environment	Heavy Rain / Waterlogged Location
		Social Aspects (War, Riots) on Location
		Project locations that are difficult to reach
		Community Response That Does Not Support the Project
		Land conditions that are different than expected
3	Finance	Insufficient Contractor's Capital
		Financial Administration to Contractors
		Late Payment by Client Owner
4	Changed	Change in Method of Work by the Contractor
		Errors in Interpreting Images or Specifications
		Changes in Plumbing, Structure and Electrical Workshops
5	Labor/Worker	The Quality of Bad Workers
		Work Strikes Conducted by the Contractor
		Workforce who lack expertise and experience
		Inaccuracy of Workforce Planning
		Lack of Labor by the Contractor
6	Time and Control	Poorly planned scheduling by the contractor
		Lack of Control of Work Safety

No	Factor	Sub Factor
		Late Approval of Shop Drawing by a Consultant
		Less Optimal Productivity from the Contractor
		Repairing Completed Work
		Implementation of Poor Stages of Work by the Contractor
		Repairing Work Damage Due to Strikes
7	Tools	Equipment Damage
		Equipment Not Available On Time
		Productive Equipment

3. Research Methodology

In this chapter we will explain the research methods used to look for factors that cause delays in construction project work.

3.1 Data Collection

T.he method used in this study has several stages, starting from the study of literature to the collection of primary and secondary data. During literature study, the researcher seeks and collects supporting theories and formulas according to experts related to this research, as a reference for making questionnaires. In data collection phase, researchers use primary data and secondary data. For primary data, namely data obtained by conducting a direct survey of space which will later be carried out by researchers by distributing questionnaires to respondents as a measuring instrument. For secondary data, namely field data sourced from related institutions or institutions, here the researcher obtained secondary data through the Jambi City Construction Services Provider (LPJK) Institute regarding the list of names and addresses of contractors in Jambi City who would later be made as respondents. Next, researched data were managed and analyzed before making a summary discussion and make conclusions from the results of the study.

3.2. Sample

Simple random sampling is a method of selecting samples from populations chosen one by one randomly or randomly. All members of the population get the same opportunity to be chosen and if it has been chosen it cannot be chosen again [12].

In determining the number of samples researchers used the Slovin formula as shown in equation (1).

$$kn = \frac{N}{1 + Ne^2} \tag{1}$$

where n is the number of samples, N is the population, and e is the error tolerance, used 10%. Data in this step data is operated using simple random sampling technique as a sampling technique, because the population is homogeneous. Based on the size of the population as many as 206 companies that are registered and registered in the Jambi Province LPJK data, using the Slovin formula to get the number of samples that will be used as respondents is 63 samples of Contractor respondents in the city of Jambi.

3.3. Questionary Design

The questionnaire was designed in two groups as below:

i. The profile of the respondent, namely the respondent's data regarding the name of the company where the respondent works, the respondent's education, the duration of experience of the respondent working and the position or position of the respondent in the company.
ii. Delay factors, namely bullet points about factors that often cause delays.

3.4. Method Of Data Analysis

In this study the measurement scale uses an ordinal scale that shows the differences in subject levels quantitatively, such as data expressed in the form of rank or ranking. The respondent's perception can be sorted into: No effect; rather influential; take effect; and very influential. Then the quantitative data is converted into qualitative data as follows:

i. No effect is indicated by the number 0.
ii. Quite influential is indicated by number 1.
iii. The effect is indicated by number 2.
iv. Very influential is indicated by number 3.

The analysis of this study uses qualitative research methods that are transformed into quantitative, to find out how many factors are estimated to influence the delay in construction work and the factors that influence and most determine based on the ranking ranking in each assessment of each respondent. After all the data has been obtained from the respondents collected, then the descriptive statistical analysis is carried out using the excel program.

i. Mapping respondents and respondent's answers.
ii. Conduct an analysis of central tendencies.

4. Result and Discussion

4.1. Result

Based on data from questionnaires as many as 63 respondents of construction service companies, the following is an analysis of the sub-factors causing delays in construction work, including: Material and material factors, environmental factors, financial factors, change factors, labor factors, time factors and controls, and equipment factor. The outputs were tabulated in Table 2 until Table 10 below.

Table 2. Interpretation of Mean Values, Modes

Value	Interpretation
0	No effect
0,1 – 1	Quite influential
1,1 – 2	Take effect
2,1 – 3	Very influential

Table 3. Results of Material and Material Analysis

No.	Factor	Total	Mean	Modus	Percentage %	Delay Rating
1	Lack of Construction Materials	120	2,26	3	16,48	1
2	Slow Material Delivery	104	1,96	2	14,29	4
3	Material Provision on Location	107	2,02	3	14,70	3
4	Delay in Ordering Material	92	1,74	2	12,64	7
5	The amount of Material sent by the Supplier is not appropriate	113	2,13	2	15,52	2
6	Material Damage due to Storage	94	1,77	2	12,91	6
7	Material Management Mistakes by the Contractor	98	1,85	2	13,46	5
	Total	728			100	

Table 4. Results of Analysis of Environmental Factors

No.	Factor	Total	Mean	Modus	Percentage %	Delay Rating
1	Heavy Rain / Waterlogged Location	129	2,43	3	25,00	1
2	Social Aspects (War, Riots) on Location	98	1,85	2	18,99	3
3	Project locations that are difficult to reach	94	1,77	1	18,22	4
4	Community Response That Does Not Support the Project	91	1,72	2	17,64	5
5	Land conditions that are different than expected	104	1,96	2	20,16	2
	Total	526			100	

Table 5. Results of Financial Factor Analysis

No.	Factor	Total	Mean	Modus	Percentage %	Delay Rating
1	Insufficient Contractor's Capital	101	1,91	3	33,78	2
2	Financial Administration to Contractors	85	1,60	2	28,43	3
3	Late Payment by Client Owner	113	2,13	3	37,79	1
	Total	299			100	

Table 6. Results of Change Factor Analysis

No.	Factor	Total	Mean	Modus	Perscentage %	Delay Rating
1	Change in Method of Work by the Contractor	84	1,58	2	27,72	2
2	Errors in Interpreting Images or Specifications	111	2,09	2	36,6	1
3	Changes in Plumbing, Structure and Electrical	108	2,04	2	35,6	3
	Total	303			100	

Table 7. Results of Labor Factor Analysis

No.	Factor	Total	Mean	Modus	Percentage %	Delay Rating
1	The Quality of Bad Workers	125	2,36	3	22,28	1
2	Work Strikes Conducted by the Contractor	110	2,08	3	19,61	2
3	Workforce who lack expertise and experience	109	2,06	2	19,43	4
4	Inaccuracy of Workforce Planning	107	2,02	2	19,07	5
5	Lack of Labor by the Contractor	110	2,08	2	19,61	3

No.	Factor	Total	Mean	Modus	Percentage %	Delay Rating
	Total	561			100	

Table 8. Results of Time and Control Factor Analysis

No.	Factor	Total	Mean	Modus	Percentage %	Delay Rating
1	Poorly planned scheduling by the contractor	98	1,85	2	14,24	4
2	Lack of Control of Work Safety	107	2,02	2	15,55	2
3	Late Approval of Shop Drawing by a Consultant	110	2,08	3	15,99	1
4	Less Optimal Productivity from the Contractor	95	1,79	2	13,8	6
5	Repairing Completed Work	74	1,40	2	10,76	7
6	Implementation of Poor Stages of Work by the Contractor	106	2,00	2	15,41	3
7	Repairing Work Damage Due to Strikes	98	1,85	2	14,24	5
	Total	688			100	

Table 9. Results of Equipment Factor Analysis

No.	Factor	Total	Mean	Modus	Percentage %	Delay Rating
1	Equipment Damage	112	2,11	3	34,04	1
2	Equipment Not Available On Time	109	2,06	2	33,13	2
3	Productive Equipment	108	2,04	2	32,83	3
	Total	329			100	

Table 10. Data Comparison of the Causes of Delay Factors

No.	Factor	Total	Mean	Persentage %	Delay Rating
1	Heavy rain / waterlogged work location	129	2,43	15,73	1
2	Poor quality of labor	125	2,36	15,24	2
3	Lack of Construction Materials	120	2,26	14,63	3
4	Late Payment by Client Owner	113	2,13	13,78	4
5	Equipment Damage	112	2,11	13,66	5
6	Errors in Interpreting Images or Specifications	111	2,09	13,54	6
7	Late Approval of Shop Drawing by a consultant	110	2,08	13,41	7
	Total	820		100	

5. Conclusion

From the results of the research, the causes of the delay in construction work in the city of Jambi according to the perceptions of the contractor, some conclusions can be drawn including:

i. Environmental Factors namely Heavy Rain Sub-factors / locations of waterlogged work are the most influential factors in work delay with a percentage of 15.73%
ii. Labor Factors namely Sub-factors Poor quality of labor is the second most influential factor in work delays with a percentage of 15.24%
iii. Material / material factors, namely Sub-factors The lack of construction materials is the third most influential factor in work delay with a percentage of 14.63%
iv. Financial factor, namely Sub-factor Delay in payment by the client / owner is the fourth most influential factor in work delay with a percentage of 13.78%
v. Equipment Factors namely Sub-factor damage Equipment is the most influential factor to 5 in the work delay with a percentage of 13.66%
vi. Change Factors, namely Sub-factors Errors in interpreting images / specifications are the most influential factors to 6 in work delays with a percentage of 13.54%
vii. Time Factors namely Sub-factors Too late shop drawing agreement is the 7th most influential factor in work delay with a percentage of 13.41%

References

[1] Purvitasari, Evi, Azis, Subandiyah, Tiong I 2018 Evaluation of Factors Affecting the Delay in a Construction Project of Dr. Iskak's Hospital Pavilion of Tulungagung District *International Journal of Scientific Engineering and Science* 2(9)
[2] Messah Y A, Widodo, Adoe M L 2013 Kajian penyebab keterlambatan pelaksanaan proyek konstruksi gedung di kota Kupang *Jurnal Teknik Sipil Nusa Cendana* **II**(2)
[3] Atkinson R 1999 Project management: Cost, time and quality, two best guesses and a phenomenon, Its time to accept other success criteria *International Journal of Project Management* **17**(6)
[4] Dipohusodo, Istimawan 1996 Manajemen proyek & konstruksi **2** (Yogyakarta: Kanisius)
[5] Wulfram I, Ervianto 2005 *Manjemen Proyek Konstruksi Edisi Revisi* (Yogyakarta :Andi)
[6] Soeharto, Imam 1995 Manajemen proyek (Jakarta: Erlangga)
[7] Wulfram I, Ervianto 2004 Teori – Aplikasi Manajemen Proyek Konstruksi (Yogyakarta :Andi)
[8] Husen A 2010 Manjemen Proyek (Yogyakarta :Andi)
[9] Dipohusodo, Istimawan 1996 Manajemen proyek & konstruksi **1** (Yogyakarta: Kanisius)
[10] Amperawan R 1999 Studi Keterlambatan Kontraktor dalam Melaksanakan Proyek Konstruksi di Daerah Istimewa Yogyakarta
[11] Barrie, Donald, Poulson, Boyd C 1984 Manajemen Konstruksi Profesional - Alih Bahasa Sudinarto **2** (Jakarta: Erlangga)
[12] Kountur, Ronny 2007 Metode Penelitian untuk penulisan Skripsi dan Tesis (Jakarta : PPM)

IOP Conf. Series: Materials Science and Engineering **807** (2020) 012039 doi:10.1088/1757-899X/807/1/012039

New Topology Multilevel Inverter Type Diode Clamped Five Level Single Phase

Hendi Matalata[1]*and Venny Yusiana[1]

[1] Department of Electrical and Engineering, Batanghari Jambi University, SlametRiyadi-Broni Road, Lake Sipin, Jambi County, Jambi, Indonesia.

*Corresponding e-mail: hendi.matalata@unbari.ac.id

Abstract. Multilevel inverter is a converter that converts DC power sources into AC power sources with the voltage output higher than two levels. Conventional multilevel inverter topology for five levels that is currently developing generally uses eight components of power switches. In this paper, a five-level multilevel inverter topology is proposed by reducing the power switch component to four and aided with two pin diodes and capacitors as voltage couplings. The simulation with MATLAB that has been done shows various advantages of this topology compared to conventional multilevel inverter topologies. Results and discussion This topology compare the switching techniques used between Pulse Width Modulation (PWM) and Sinus Pulse Width Modulation (SPWM) for five-level multilevel inverter. Cohesiveness in switching techniques topology is proposed and validated by using passive filter. The harmonic analysis performed by the proposed topology could reduce the output voltage with the low harmonics.

1. Introduction

Developments of power electronics allows it to produce the generation, the processing and the use of electrical energy and economic efficiency. Power electronics is defined as an electronic application that focuses on regulating electrical equipment by making changes to electrical parameters (current, voltage, frequency and electrical power). Today multilevel inverter has attracted great interest in the high usage power industry with numerous advantages. The power electronics equipment is widely applied such as utilization of DC resources, uninterruptible power supplies, direct voltage power transmission (HVDC), variable frequency drive systems, air conditioners and others [1].

Multilevel inverter (MLI) is a converter that converts DC power sources into AC power sources where the output voltage forms are more than two levels where there are three levels, five levels, seven levels and so on. Those numerous output levels can be adjusted by the number of semiconductor switches power circuit. Multilevel inverters that are currently developing can be recognized based on their architecture (topology) such as diode clamped, flying capacitor and cascade H-bridge. Those are the three multilevel inverter structures known in industrial applications with separate DC sources. The advantage of diode clamped, and flying capacitor is to use a voltage balancer using a capacitor so that for many voltage levels on DC sources could be minimized

Conventional five-level multilevel inverter requires eight switches to enumerate the two DC sources separately while the next topology is proposed using six switches and two DC sources apart from being given with two diode clamped and two capacitors as a voltage balancers that have the form of cascade waves with smaller harmonic distortions compared to the conventional multilevel inverters [2]. Then the topology is then reduced by six switches and two separate DC sources without using clamped and capacitor diodes as a voltage balancer where the wave cascade forms with harmonic distortion reduced compared to the previous topology [3].

By analysing the advantages and disadvantages of existing topologies, a new topology is proposed in reducing multilevel inverter switches using four switches to count two DC sources for positive polarity and negative polarity assisted by two pin diodes and capacitors as voltage couplings. Results and discussion This topology compare the switching techniques used in Pulse Width Modulation (PWM) and Sinus Pulse Width Modulation (SPWM) to analyse the resulting harmonics [4]. The discussion would be on overcoming the weaknesses of the existing topology.

2. Multilevel inverter five levels type diode clamped

Conventional topology in the clamped multilevel inverter was first proposed in 1981 then also known as the neutral point. Many studies were then published that discussed on the advantages and disadvantages. Modified sinusoidal waveforms in a five-level multilevel inverter are shown in Figure 1. Unlike the cascaded H-bridge inverters, this topology requires diodes to be used as clamping devices. This DC voltage is further divided through the capacitor coupling. The components which needed by the multilevel inverter topology diode type clamped for *n* levels are listed as below [4].

- number of switches = 2 (*n*-1),
- number of diodes = (*n*-1) (n-2),
- number of capacitors = (*n*-1)

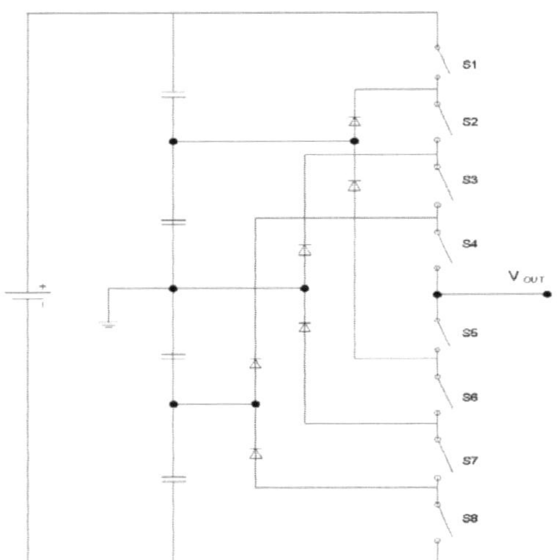

Figure 1. MLI five levels of type diodes clamped

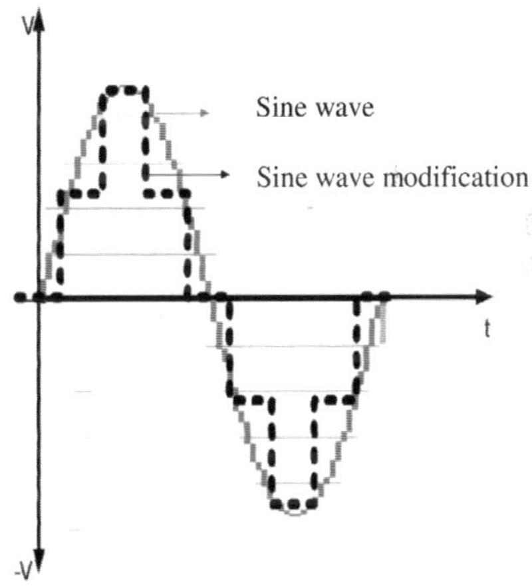

Figure 2. Sinusoidal wave modifications the five-level multilevel inverter

2.1. *Proposed Topology*

Topology proposed in a five-level multilevel inverter is built using two DC sources, four switches, four diodes and two capacitors, in this topology is to show in the reduction of the number of switches used in conventional topologies, validation of the output of the proposed topology can be formed five-level output voltage, as for the proposed topology shown in Figure 3.

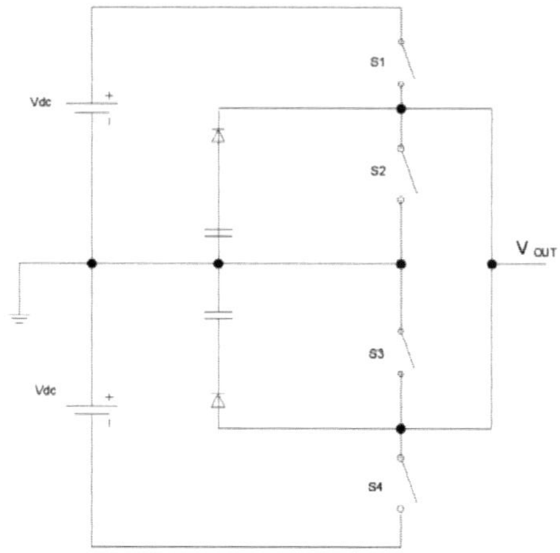

Figure 3. Multilevel inverter five topology the proposed level

From Figure 3 above can be explained the work of the proposed topology where there are two DC sources namely at the upper limit and at the lower limit, the upper DC source is used for positive cycle polarity while the lower limit DC source is used for negative cycle polarity, two capacitors used as a coupling at the voltage level of a large DC source while the two diodes are used as clamps from each DC source. The switching topology of the four switches is shown in Table 1.

Table 1. MLI Switching Topology Five Levels Using Four Switches

No	Condition Switch				Output (Vdc)
	S1	S2	S3	S4	
1	1	0	1	0	0.5
2	1	0	0	0	0.0
3	1	1	0	0	0.5
4	0	1	0	0	0.0
5	0	1	0	1	-0.5
6	0	0	0	1	-0.5
7	0	0	1	1	-0.5
8	0	0	1	0	0.0

2.2. Pulse Width Modulation (PWM)

The basic principle of a PWM switching technique is to compare a triangular wave ($V_{carrier}$) with a sine wave ($V_{reference}$) which is called a modulation technique, this process produces a pulse wave at the basic frequency of a sine wave. This PWM wave is in the form of a box with a certain frequency in the ratio of the width of the high and low pulses that are different

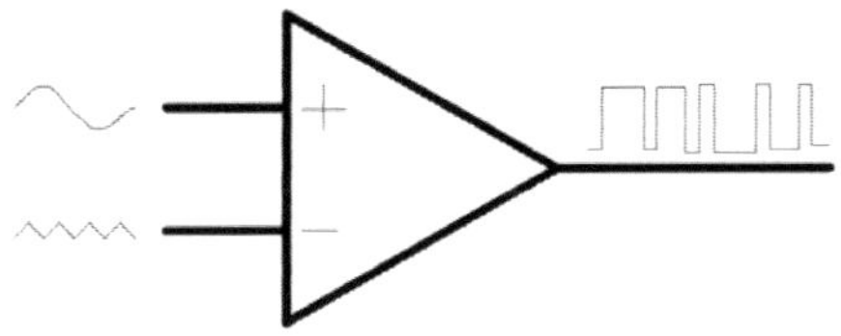

Figure 4. Processed PWM

From the above description the PWM duty cycle can be explained in the following equation (1) and (2) [4].

$$\frac{V_{reference}}{V_{carrier}} = \frac{t_{on}}{(t_{on} + t_{off})} \quad (1)$$

$$D = \frac{t_{on}}{(t_{on} + t_{off})} \quad (2)$$

Where;
D = duty cycle

2.3. Total Harmonic Distortion (THD)
Research on Multilevel Inverter technology to get the output voltage is more than two levels has been developed by various researchers. The important technology in developing multilevel inverters is how to modify the shape of the output level nearing sinusoidal waves. With Fourier series, the square wave can be expressed as the sum of harmonic components and expressed in the following equation

$$V_o(\omega t) = V_o + \sum_{n=1}^{\infty} (a_n \sin n\, \omega t) + \sum_{n=1}^{\infty} (b_n \cos n\, \omega t) V_o(\omega t) = V_o + \sum_{n=1}^{\infty} (a_n \sin n\, \omega t) + \sum_{n=1}^{\infty} (b_n \cos n\, \omega t) \quad (3)$$

THD states the amount of distortion or deviation of a wave containing harmonics expressed in percent (%) formulated as follows:

$$THD = \frac{\sqrt{V^2{}_2(rms) + V^2{}_3(rms) + V^2{}_4(rms) + \cdots + V^2{}_n(rms)}}{V_1(rms)} \quad (4)$$

Where:
$V_{1\,(rms)}$ = effective basic harmonic voltage
$V_{n\,(rms)}$ = harmonic effective voltage to n

3. Results and analysis

The simulation conducted is intended to show that for the harmonic level of output voltage generated from topology multilevel inverter five conventional levels using eight switches (MOSFETs) compared to the proposed topology.

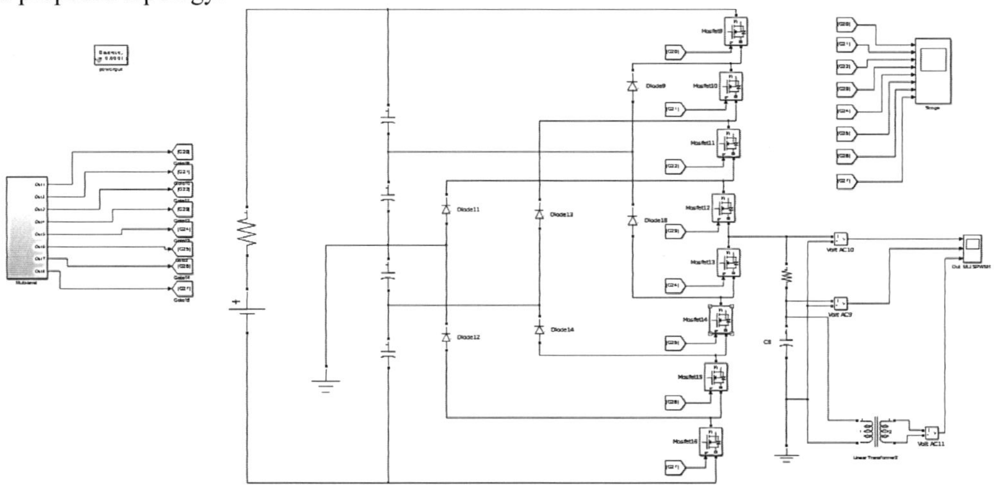

Figure 5. Simulation of conventional multilevel five-level inverter model

Simulation carried out from Figure 5 above use a 24 Volt DC source, five-level multilevel inverter voltage output shown in Figure 5, to reduce harmonic magnitude from multilevel voltage output passive filter R = 15 Ohm and C = 25mF, for the purpose of increasing the voltage, a step up transformer with a ratio of 1:34 is given, that the output voltage is shown in Figure 6. The harmonic voltage of each output uses analysis of Fast Fourier Transform (FFT) at the fundamental frequency of 50 Hz as in Figure 8 and 9.

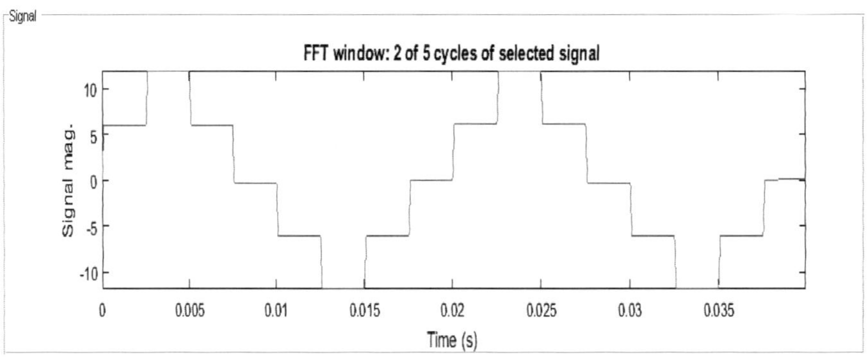

Figure 6. MLI five-level conventional MLI five level conventional

IOP Conf. Series: Materials Science and Engineering **807** (2020) 012039 doi:10.1088/1757-899X/807/1/012039

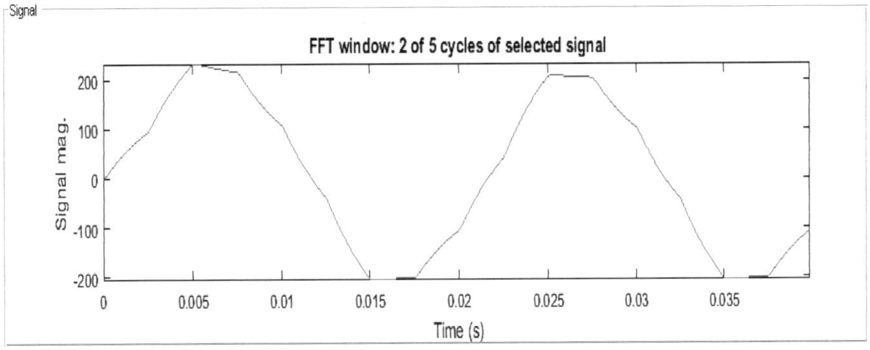

Figure 7. Step up MLI five level conventional mode

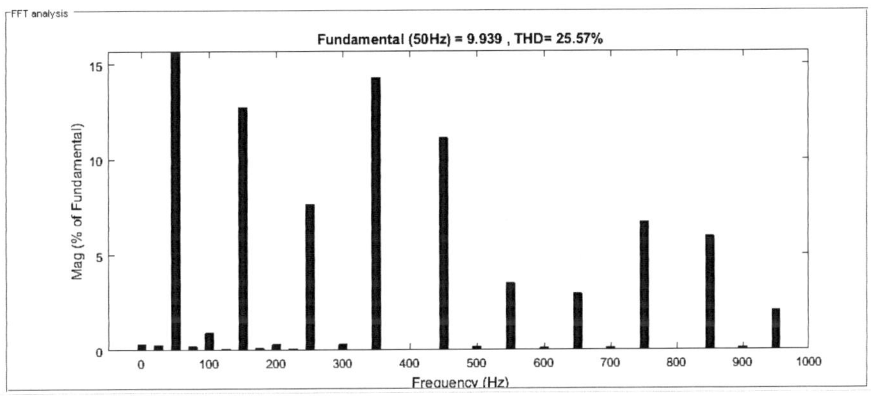

Figure 8. Conventional five-level MLI FFT

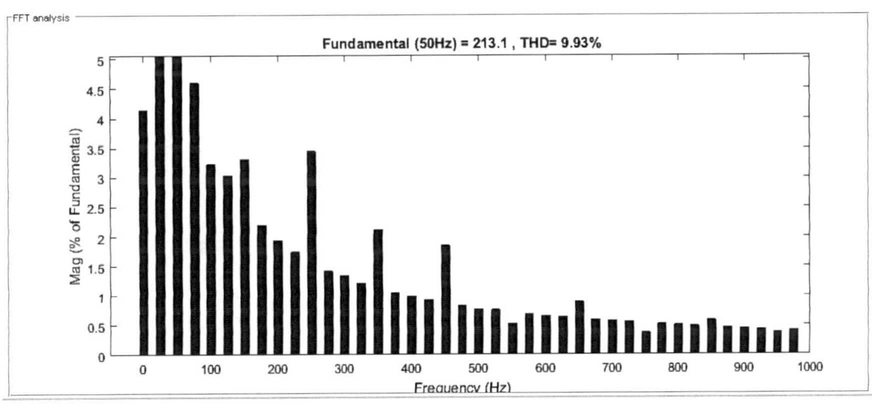

Figure 9. Step up MLI five level conventional mode

Figure 7 shows the output voltage harmonic of large MLI that uses FFT the level obtained is 25.57% while Figure 8 uses passive filter and the large harmonic step up mode MLI output voltage obtained is 9.93%. From the conventional simulation, the topology is proposed by reducing the number of switches by validation using the PWM and SPWM techniques, to obtain the optimum efficiency.

IOP Conf. Series: Materials Science and Engineering **807** (2020) 012039 doi:10.1088/1757-899X/807/1/012039

3.1. The Five Level MLI Simulation Model proposed by the PWM Technique

Proposed as a comparison discussed by using the PWM technique. Figure 9 below shows the proposed five-level multilevel inverter simulation model which uses four switches (MOSFETs). And consists of two DC sources.

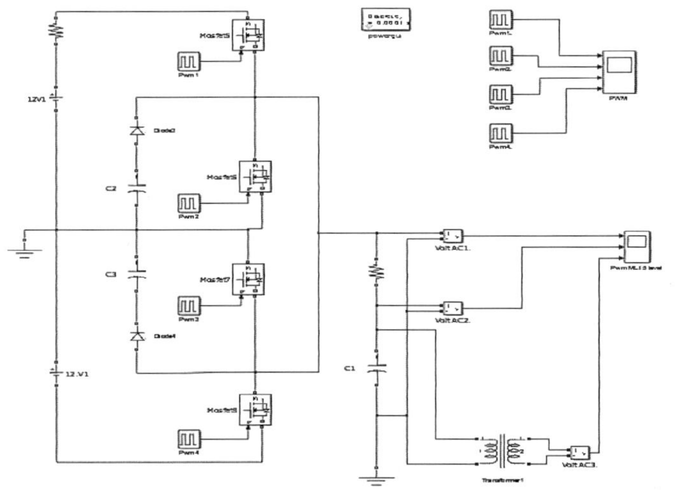

Figure 10. Five-level multilevel inverter simulation model with PWM technique

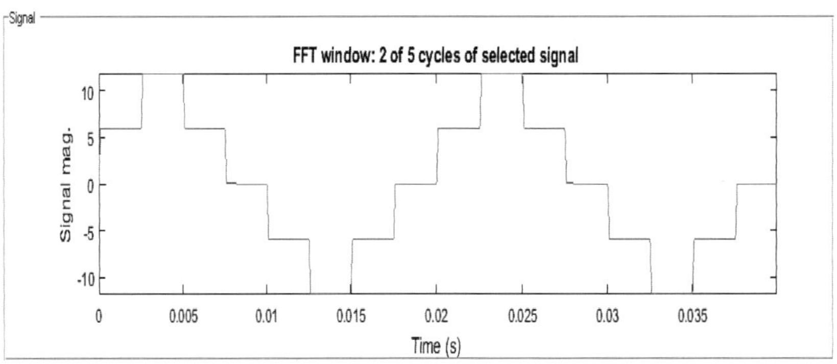

Figure 11. MLI proposed five levels with PWM

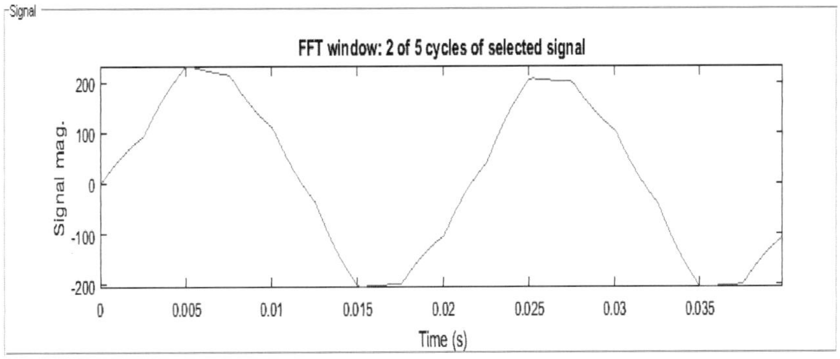

Figure 12. Proposed step up mode for five-level MLI with PWM

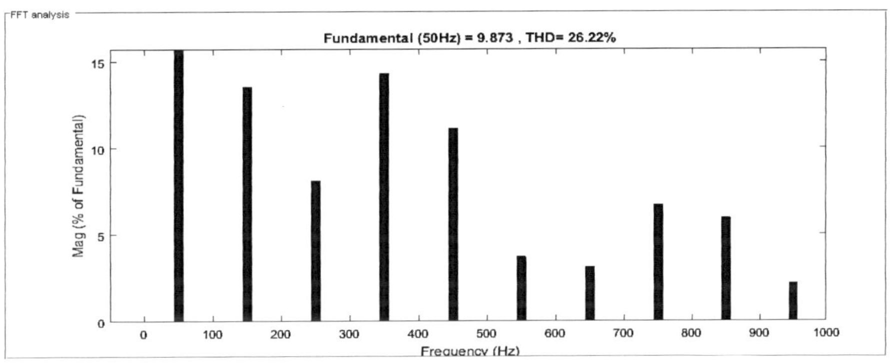

Figure 13. Proposed five-level MLI FFT with PWM

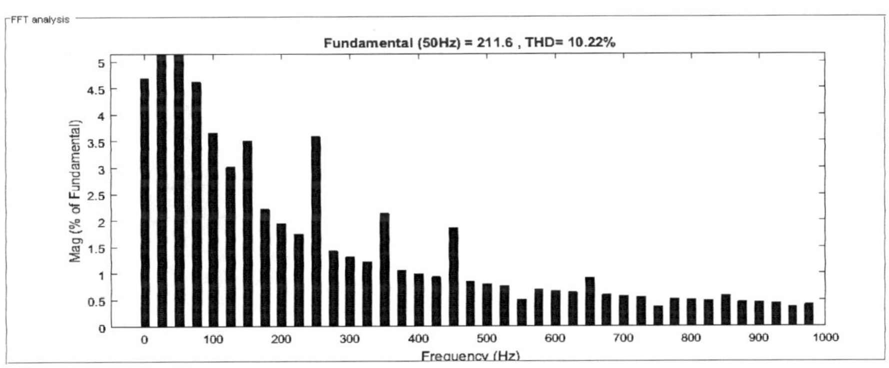

Figure 14. Proposed step up mode five level MLI FFT with PWM

As shown in Figure 13 that use the PWM technique, the harmonic output of the MLI output using FFT is 26.22%, while from Figure 14, it uses a passive filter and a large step-up harmonic mode. The output voltage is 10.22%.

The proposed topology simulation results show that to multilevel the five-level inverter by reducing the number of switches can be done, then to reduce the proposed Topology harmonic voltage, the SPWM technique is one of the first steps in multilevel inverter devices to produce low harmonics.

IOP Conf. Series: Materials Science and Engineering **807** (2020) 012039 doi:10.1088/1757-899X/807/1/012039

3.2. The Five Level MLI Simulation Model proposed by the SPWM Technique

Modulation of the PWM signal pulse width generated is a reference voltage reference (sine wave) and carrier voltage (triangle wave), in this paper the voltage frequency used is 50 Hz and 1.125 kHz respectively. The simulation model shown is shown in Figure 15 below.

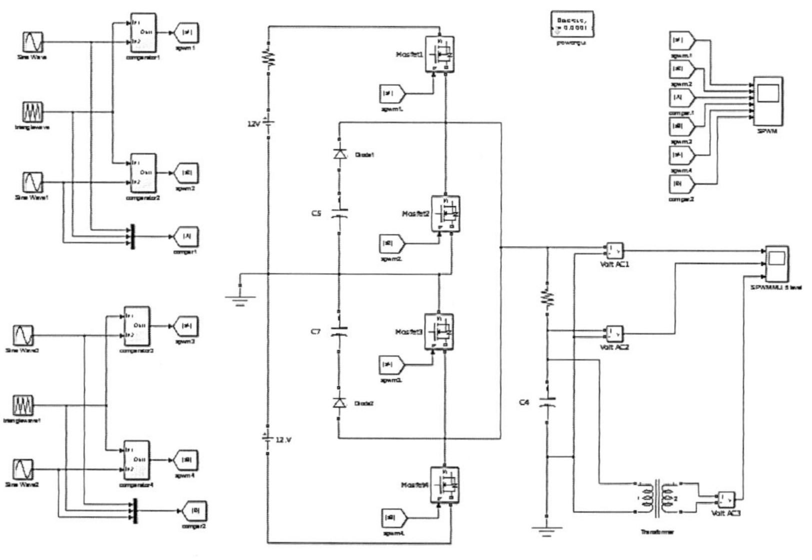

Figure 15. Simulink five-level model of proposed SPWM technic multilevel inverter

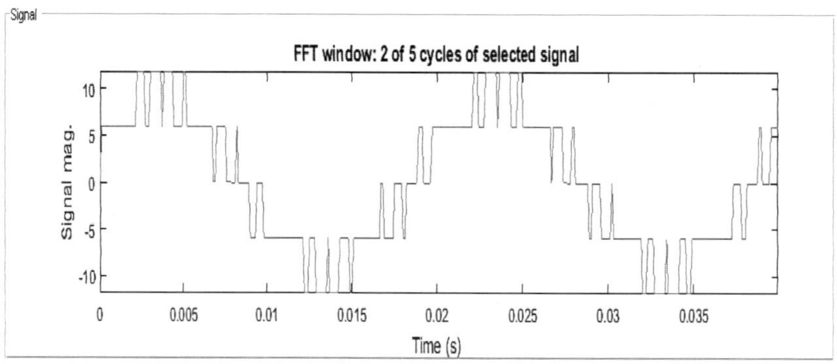

Figure 16. MLI proposed five levels with SPWM

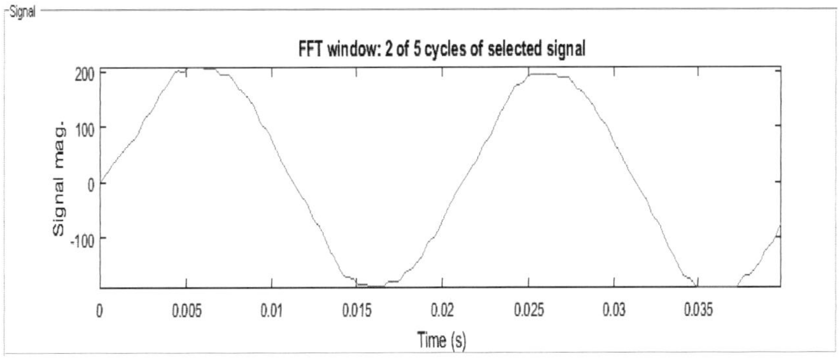

Figure 17. Proposed step up mode MLI five levels with SPWM

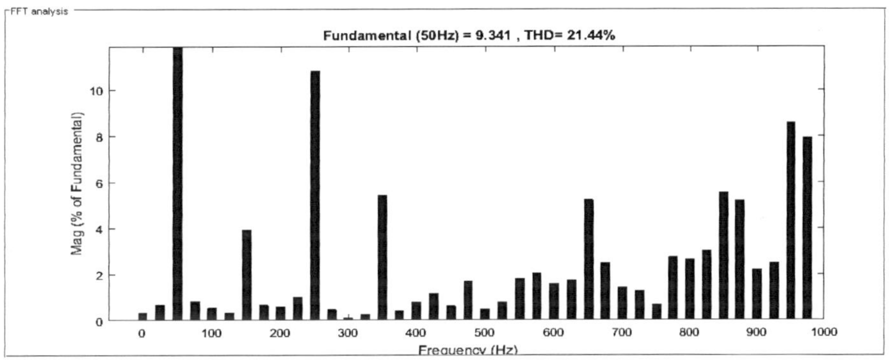

Figure 18. Proposed MLI FFT five levels with SPWM

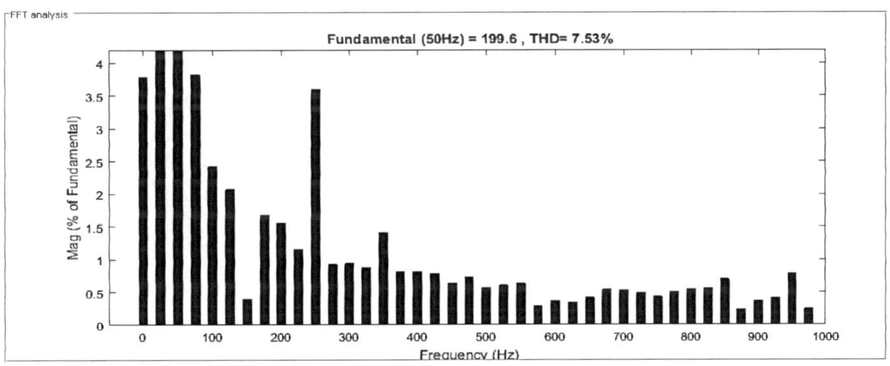

Figure 19. Proposed step up mode MLI FFT five levels with SPWM

As shown in Figure 17 the harmonic output voltage of MLI using FFT is 21.55%, while from Figure 19 that uses a passive filter and a large step-up harmonic mode the output voltage is 7.53%.

3.3. THD Comparison
Table 2 shows the comparison values of the THD

Table 2. Comparison values of THD

Harmonic mode	Conventional PWM		Proposed PWM		Proposed SPWM	
Voltage of five levels (V)	MLI	Filter	MLI	Filter	MLI	Filter
	26.22	9.93	25.75	10.22	21.55	7.53

4.0 Conclusion

The proposed topology has been able to redirect component switches on a five-level multilevel inverter. MATLAB simulations performed on the conventional five-level multilevel inverters in reducing harmonic magnitude can be obtained using the topology that uses only four switch components (MOSFETs) in PWM and SPWM switching modes, the magnitude of the output voltage harmonics is 27.75% in PWM mode and 21.55% in SPWM mode A better THD using a passive filter provides a harmonic reduction in output voltage of 9.93% in PWM mode and 7.53% in SPWM mode. Thus, the proposed topology can overcome the disadvantages of conventional topology in the cost of using switch components. The implementation of hardware can also be made so that it is applied in various aspects of industrial needs.

References

[1] Palanisamy R, Kuppusamy S, Devadoss S, Dhandapani K 2018 Simulation of single phase 3-level z-source NPC inverters with PV system *Telecomm. Comp. Elec. Ctrl.* **16** 2293-301

[2] Devarajan N and Reena A 2015 Reduction of switches and DC sources in cascade multilevel inverters *Bull. Ele. Info.* **4** 186-95

[3] Patil V and Patil V V 2016 Five level inverter to reduce Harmonics *Intl. J. Engin. Res. Technol.*.**3** 2000-204

[4] Mathapati B S and Landge S S 2017 A single phase five level inveter based on switched DC source with reduce number of total component count by novel topology *Int. J. Rec. I. Tr. Comp. Comm.* 5 218-23

[5] Khanfara M, Mohammed B, Bachtiri R E and Hammoumi K 2018 A multicarrier PWM technique for five levels of connected to grid inverters *Intl. J. Pwr. Elect. Dr. Sys.* **9** 1774-83

[6] Karthik K, Narsimharaju B L and Srinivasa R S 2015 Five-level using POD PWM technique *Intl Conf. Electrics Electronics Signals Communication and Optimization.*

[7] Yusof N A, Sapari N M, Mokhlis H, Selvaraj J 2012 A comparative study of 5-level and 7-level multilevel inverter connected to the grid *Intl. Conf. on Power and Energy.* **2** 542-7

[8] Rasyid M 2014 Power electronics (India: Prentice Hall of India)

[9] Kale P J, Kulkarni V D and Bhad S 2015 Modelling and simulation of five levels of diode clamped inverter for reduction of common voltage modes in induction motor drive *Intl. J. Engineering Resources Technology.* **4** 622-5

IOP Conf. Series: Materials Science and Engineering **807** (2020) 012040 doi:10.1088/1757-899X/807/1/012040

The design programming simulation as a reference for the adjustment of the relay GFR on the protection of the airway medium voltage 20 KV

Umar Djufri, Fadli Eka Yandra

Department of Electrical Engineering, Faculty of Engineering, Batanghari University

Corresponding e-mail: umar.djufri@unbari.ac.id

Abstract. Short circuit fault phase to ground and phase to phase is one of the problems that may arise in the operation of the power transformer in a Substation. The interference caused by the presence of a short circuit caused a lot of losses, losses on the transmission system of electricity and losses on the part of consumers of electric energy. One way to overcome this disorder is with safety protection relay OCR and GFR that works with the Power Breaker (Circuit Breaker). This research will create a Programming-Based Object (PBO) using the GUI of MATLAB, where programs are designed is one answer to cope with the calculation of the adjustment GFR (Ground Fault Relay) attached on the network, the validation of the programming that is done using armed forces data from the company, which for reasons of privacy mentioned as XYZ. Great protection Setting of the interference 1 phase ground occurs along the feeder can be concluded that the working time of GFR on the side of the outgoing and incoming have a difference of 0.4 seconds, thus the design of programming windows in this study can be used as a reference as setting tms on GFR which is mounted on an equipment of protection of electric power because it has been in accordance with the applicable standards.

1. Introduction

The system of electric power distribution from power plant to consumer is very important to learn because the process of distribution through several stages starting from the power plants that produce electrical energy, transmitted to the network transmission of Air Channel High Voltage (SUTT) directly to the substation, from the substation the electric power is distributed to the primary distribution network Voltage Air Channel Medium (SUTM), usually channel this electrical energy is utilized industry-large capacity. Through a network of SUTM this electric power is distributed through substations secondary distribution Air Ducts Low Voltage (SUTR) and through the SUTR electric power is delivered to consumers [1].

The Distribution network is part of the electric power systems that deliver electricity from the substation, wherein the distribution using the conductor of the cable channel through the air duct. Disturbances experienced by the network installed on the airway medium voltage 20 KV, is the disturbances caused by lightning, animals and trees that exposed a network of electric power that can cause interference 1 phase ground, interference 2 phase and interference 3 phase, as a result the distribution of electric power is interrupted or is experiencing outages [2]. Analysis of short circuit is a part of the power system, in order to maintain the security of the distribution of electrical energy. need count large short circuit current interruption for the purposes of tuning protection on the side of the feeder.

In this research carried out the study setup protection on the side of the feeder caused by the fault short circuit single phase to ground. To calculate the large short circuit current is the basic formula used is OHM's law as shown in equation (1), namely:

$$I = \frac{V}{Z} \tag{1}$$

where I is the short circuit current (Ampère), V is the voltage source (Volts), Z is the impedance equivalent from the source to the point disorders (Ohm). Short circuit fault single phase to ground using the following equation [1]:

2. Setting OCR and GFR

Relay over current and relay ground fault is a protection relay that works with the power Breaker (Circuit Breaker), wherein the power Breaker will work if you get a response from the relay are attached. The characteristics of the OCR Inverse work if the flows are perceived to exceed the current setting, the relay current is this will open the child's contacts with time delays have been determined [3]. To setup the Current relay is installed in the feeder is calculated based on the magnitude of current from each type of disorder.

Relay setting relay:
Based on the current
- Relay Definite = 1.2 x I Load
- Relay Inverse = 1.05 x I Load

Based on time
- Relay Definite = Directly on the TAP
- Relay Inverse = Calculated based on fault current

The Calculation of The Coordination Relay Inverse
I Set Primary = 1.05 x I Load
I Set Secondary = I Primary x 1/Ratio of CT

For the calculation of the settings on the relay, theoretically described in the equation (2) below

$$tms = \frac{t_{set} \text{ x} \left(\left(\frac{I_f}{I_{Set}} \right)^{0,02} - 1 \right)}{0,14} \tag{2}$$

where: tms is time multiplier setting, t_{set} is the time setting on relay incoming and outgoing usually set of the section downstream of the first (outgoing) with the provisions of 0.3 seconds, for setting the incoming 0.4 seconds longer, If is the fault current (Ampere), and I_{set} is the current setting (Amperes). Setting the relay as the protection starts from the outgoing Substation, to get a longer trip time of the PMT used the equation (3) below.

$$t_{set} = \frac{0,14 \text{ x tms}}{\left(\left(\frac{I_f}{I_{Set}} \right)^{0,02} - 1 \right)} \tag{3}$$

3. Methodology

Ground Fault Relay (GFR) which serves to detect the existence of a short circuit to the ground while overcurrent relays (OCR) to detect short circuit interruption between phases.

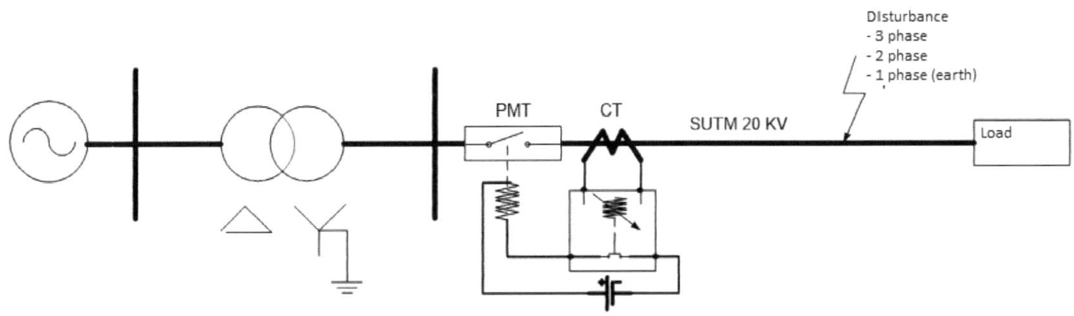

Figure 1. Single Line Interference on The Side of the Feeder

To determine the setting of the protection GFR on the side of the feeder then first calculate the fault current of short circuit. The purpose of the calculation of the short circuit fault is to calculate the magnitude short-circuit current maximum and minimum, which occurs along the feeders [2].

3.1. Model Flow of Research

The research methods used in this study are generally structured as the flow chart shown in Figure 2.

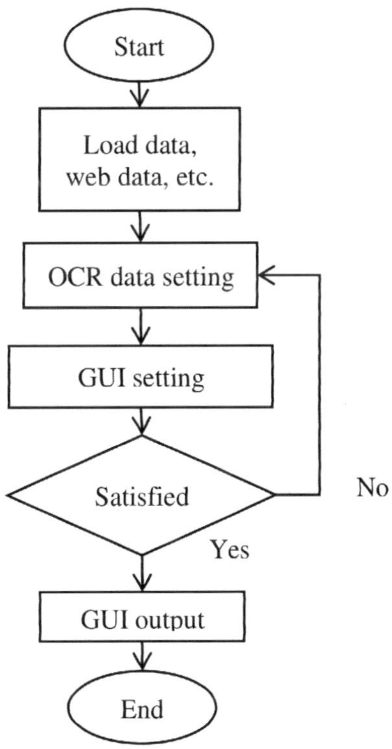

Figure 2. The Flow of Research

3.2. Parameters of the Research

The data assumptions of this research study is as shown in Table 1.

IOP Conf. Series: Materials Science and Engineering **807** (2020) 012040 doi:10.1088/1757-899X/807/1/012040

Table 1. Data Assumptions of the Study Research

Power of 80 MVA	Power of 80 MVA
The Secondary voltage of 150 KV	The Secondary voltage of 150 KV
Primary voltage 20 KV	Primary voltage 20 KV
The ratio of CT Power Transformers 2000/5	The ratio of CT Power Transformers 2000/5
The Relationship of Delta Transformer There	The Relationship of Delta Transformer There
The capacity of the Entanglement Delta 20 MVA	The capacity of the Entanglement Delta 20 MVA
Reactance (%) 10	Reactance (%) 10
Ground Resistor 40	Ground Resistor 40
The ratio of CT (current transformer) 300/5	The ratio of CT (current transformer) 300/5
Data Feeders (SUTM 20 KV)	
The Length of The Feeder 20 Km	The Length of The Feeder 20 Km
AAC 240 mm2	AAC 240 mm2
AAAC 240 mm2	AAAC 240 mm2
Generator	
MVA Short Circuit 7547 MVA	MVA Short Circuit 7547 MVA
Load Current 320 Ampere	Load Current 320 Ampere
Data attached	
GFR Incoming GFR Outgoing	GFR Incoming GFR Outgoing
Tms = 0.9	Tms = 0.9

4. Results and Discussion

The main Menu in programming windows GUI is a display that leads to the simulation of network impairments, in this view in the program menu there is a choice of the type of interference that is 1 phase to ground. The appearance of the GUI is as shown in Figure 3 below.

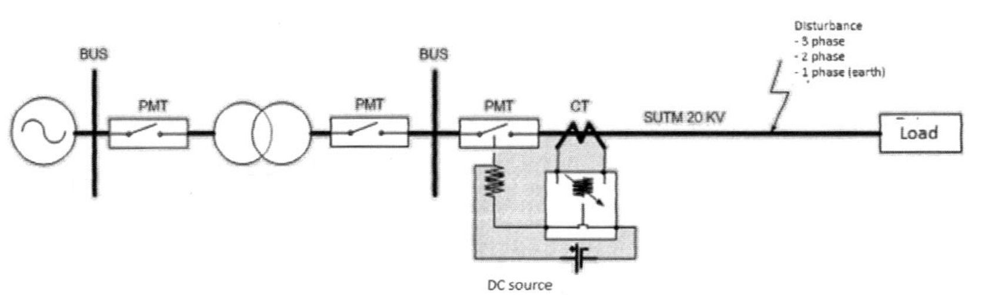

Figure 3. The appearance of the GUI

Every choice of the type of disturbance on the menu the program will provide information about the data input disorders, where this information should be input based on the channel data attached.

4.1 Display Data Input Interruption 1 Phase to Ground

To simulate disorders 1 phase to ground occurs on the channel, then the parameters study discussed in the previous chapter to be a reference research for analyzing the short circuit fault and setting GFR. The simulation results of the parameter study shown in Figure 4

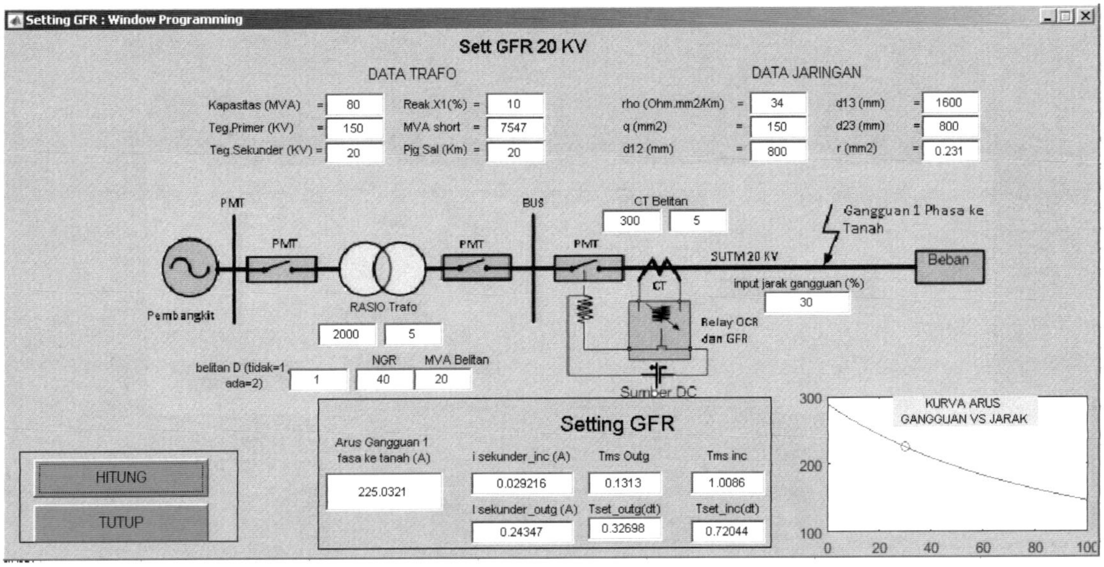

Figure 4. Display data input interruption 1 phase to ground

From Figure 4 above can be explained that the disorder occurs in 30% of the length of the channel where the channel length is 20 km, the fault current obtained 225.03 Amperes where each setting tms outgoing and tms incoming obtained by 0.1313 seconds and 1.0086 seconds
the chart of working time on the side of the outgoing when the interference occurs along the feeder shown in figure 5 below.

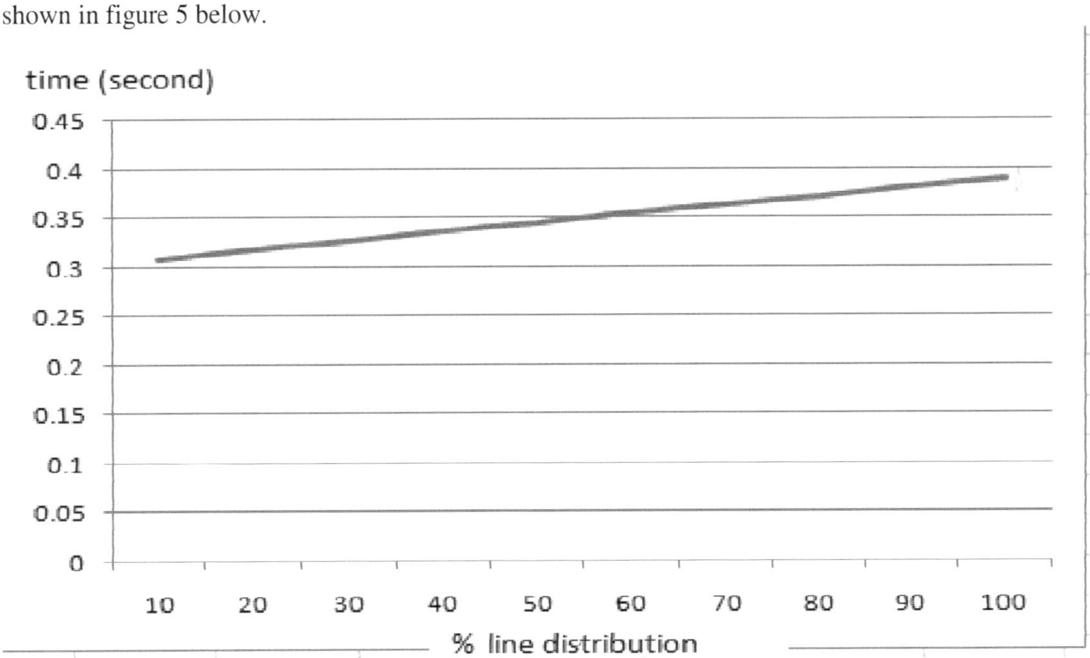

Figure 5. The chart Setting GFR outgoing

while the chart of working time on the side of the incoming when an interruption occurs along the feeders is shown in Figure 6 below.

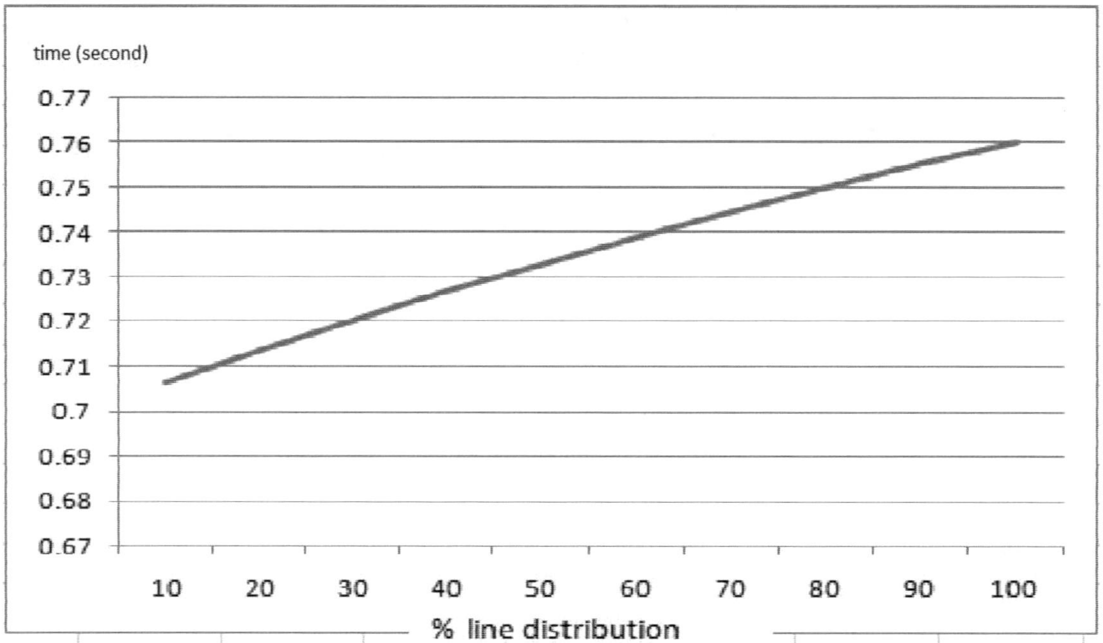

Figure 6. The chart Setting GFR incoming

From Figure 5 and Figure 6 above, the difference in working time of GFR from the side of the outgoing and incoming when an interruption occurs along the channel shown in Figure 7 below

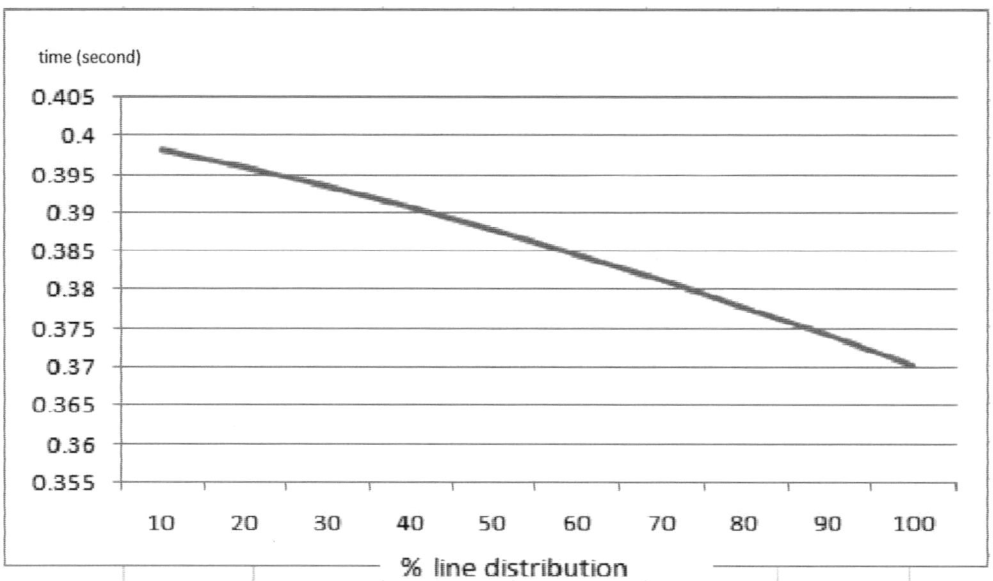

Figure 7. The difference in working time of GFR on the side of the outgoing and incoming

5. Conclusion

From the analysis of the calculation of the simulation GUI is made large, the protection Setting of the interference 1 phase ground occurs along the feeder can be concluded that the working time of GFR on the side of the outgoing and the incoming have the difference between the average of 0.38 seconds, thus the design of programming windows in this study can be used as a reference as setting tms on GFR which are attached to a protection equipment of electric power.

References

[1] Pribadi K, Wahyudi.S 2005 Pelatihan Proteksi Distribusi, Udiklat Padang (Persero)
[2] Hendi M 2018 Analisa KedipTegangan Pada Saluran Udara 20 KV Akibat Gangguan Satu Phasake Tanah Berdasarkan Jenis Konduktor Penyulang *Jurnal Momentum* **20**(2)
[3] Made I Y D, Wayan I R, Made I S 2016 Analisa KedipTegangan Akibat Gangguan HubungSingkat Pada Penyulang Abang Di KarangAsem *E-Journal SPEKTRUM* **3**(2)
[4] Impedansi Kawat Penghantar Menurut SPLN 64:1995
[5] MATLAB User's Guide, The Mathwork, inc, 1989.
[6] Nasution M W, Teguh Y, Ontoseno P 2012 Pengaruh Kedip Tegangan dan Koodinasi Rele Arus Lebih pada Pabrik Semen *Jurnal Teknik ITS* **1**(1)

IOP Conf. Series: Materials Science and Engineering **807** (2020) 012041 doi:10.1088/1757-899X/807/1/012041

The Nature of Coconut Fibre Fly Ash-Based Mechanical Geopolymer

Ria Zulfiati[1], Saloma[2], and Yakni Idris[2]

[1]Civil Engineering Department, Faculty of Engineering, Batanghari University, Jl. Slamet Riyadi, Broni, Jambi, Indonesia.
[2]Civil Engineering Department, Faculty of Engineering, Sriwijaya University, Jl. Palembang-Prabumulih KM. 32, OganIlir District, South Sumatera, Indonesia.

Corresponding author: riazulfiatiyati@yahoo.co.id

Abstract. In order to reduce carbon emissions in the atmosphere, one of the steps is to use environmentally friendly concrete. Environmentally friendly concrete uses industrial waste as a substitute for cement, such as rice husk ash, palm burning ash, coal burning ash or fly ash. Today, the development of the use of fly ash as a substitute for cement has been widely used, especially in geopolymer concrete. This study uses natural fibres such as coconut fibre. Coconut fibre is also an alternative material in the use of natural fibres as geopolymer mortar composites. This research carried out compressive strength and flexural strength test by doing several variations such as the length of coconut fibres 10 mm, 20 mm, and 30 mm and the percentage of fibre used was 0, 0.25% and 0.50% of the weight of the mortar. Geopolymer mortar made from 100% fly ash. The concentration of NaOH solution used was 14 M and 16 M. The results of this study indicate that the fibre length of 30 mm with a percentage of 0.50% fibre at a concentration of 16 M NaOH solution gives a compressive strength of 40.016 MPa and the greatest flexural strength of 8.799 MPa at 28 days mortar age.

1. Introduction

Concrete is a construction material that is widely used in infrastructure development such as bridges, dams, buildings and other infrastructure. The use of concrete structures continues to increase with the increasing need for facilities and infrastructure needed by humans. This is because concrete has advantages such as being able to withstand press, fresh concrete is easy to print as desired, in the process of repairing concrete, fresh concrete can be added to the cracked surface. Increasing the use of concrete structure materials, there are two important factors that must be considered, namely durability and environmental friendliness.

The Efforts to obtain environmentally friendly concrete is through the development of concrete using inorganic binding materials such as alumina-silicate polymers or known as geopolymers which are synthesized from natural geological materials or industrial by-products such as fly ash which is rich in silica content and alumina [1].

Flying ash is one of the industrial by-products that can be used in the manufacture of binders in geopolymer concrete. Fly Ash is a very fine material like dust, originating from the unused coal combustion. Fly Ash has high levels of cement material and has pozzolanic properties, which can react with free lime released by cement during the hydration process and form compounds that are binding at normal temperatures in the presence of water. Fly ash is activated by an alkaline solution in the form of sodium hydroxide (NaOH) and sodium silicate (Na2SiO3) as the catalyst.

This combination is used as a material for the construction of infiltration construction in the form of a cube. In the process of geopolomerization of alkaline solutions which are often used is a combination of sodium hydroxide (NaOH) or potassium hydroxide (KOH) and calcium silicate (K2SiO3). The type of alkaline solution plays a very important role in the polymerization process. The use of sodium (Na) material which is sodium hydroxide or sodium silicate will be cheaper so that costs can be reduced and the technology used is more flexible [2].

Fine aggregate used ranges from 70 - 75% of the total volume of mortar. Therefore, the quality of the aggregate has a major influence on the quality of the mortar to be produced. The choice to use a good aggregate will make the mortar workable, strong, durable and economical. Sodium silicate and sodium hydroxide are used as alkaline activators [2]. Sodium silicate serves to accelerate the polymerization reaction, while sodium hydroxide functions to react the Al and Si elements contained in the fly ash so that it can produce strong polymeric bonds. Alkaline activator used in this research is a combination of sodium silicate solution and sodium hydroxide.

Geopolymer concrete is an environmentally friendly concrete that is divided into several types, one of which is fibrous geopolymer concrete. Fibre concrete is a composite material consisting of two or more materials that have stronger mechanical properties than its forming material, which is normal concrete and fibre-added material as a composite filler.

Waste from coconut plants, namely coconut fibre, is fibre taken from the outer cover of coconuts and is a plant native to the tropics. Coir is also considered a seed fibre, although it looks similar to fibre from tree bark with cellulose (about 44%), lignin (45%), pectin and related compounds (3%), and water (5%). Coconut coir fibre has an average diameter of 236μm, coconut coir fibre has the most resilient properties (Sunariyo, 2008). Coconut fibre has the most ductile properties. Meanwhile, the strength of valuable coconut fibre is lowest when compared to other natural fibres. One of the causes of the low strength of coconut fibre is its single fibre structure and its large diameter is 236 μm. However, coconut fibre has a high elongation to break> 31%. Composites that use fibre as reinforcement can be classified into two parts, namely short fibre composite (short fibre composite) and long fibre composite (long fibre composite). Long fibre is stronger than short fibre. Long fibre (continuous fibre) is more efficient in its use than short fibre but short fibre is easier to place than long fibre. Fibre length can affect the processing capability of fibre composites. Judging from his theory, long fibres can continue the load and the voltage from the point of tension towards the other fibres [3-6].

High and low strength of the composite is very dependent on the fibre used because the voltage applied to the first composite is received by the matrix and passed to the fibre, so the fibre will withstand the load to the maximum load. Therefore, fibre must also have a higher tensile stress and modulus of elasticity than the composite matrix [7-12].

2. Materials and methods

This geopolymer mortar study uses 100% fly ash as its base material. The combination of sodium silicate and sodium hydroxide is used as an alkaline activator.

The material used in this study is sand, fly ash, sodium silicate, sodium hydroxide, water or aquades, and coconut fibre. The sand that is used comes from the Tanjung Raja area, South Sumatra. Before being used, sand is examined for specific gravity, sludge content, and organic content.

Fly ash used in this study is class F fly ash from PT. Bukit Asam, Muara Enim, South Sumatra. The fly ash used has passed the filter no. 200 which was carried out in the concrete material laboratory of Sriwijaya University, Palembang and SEM tests have been carried out in the geological survey center laboratory, Bandung.

Table 1. Composition of *fly ash*

Oxida	SiO_2	TiO_2	Al_2O_3	Fe_2O_3	MnO	CaO	MgO	Na_2O	K_2O	P_2O_5	SO_3
Amount (%)	53,490	0,782	29,350	5,980	0,063	3,330	1,350	0,680	0,509	0,424	0,506

Use of sodium silicate (Na2SiO3) or water glass as polymerization. Sodium hydroxide (NaOH) is a strong alkaline solution when it is dissolved in water. The water used must be clean, non-greasy and free of harmful organic substances, in this study aquades are used.

This study uses natural fibres such as coconut fibre. Coconut fibre are obtained from Blitar farmers in ready-to-use conditions. Coconut fibre fibres before use are treated with immersion in a 5% NaOH solution for 6 hours. After soaking the fibres are washed clean and carried out drying at room temperature. After the fibres are dry, the fibres are cut into 10 mm, 20 mm and 30 mm lengths [13-14].

Mixed composition with NaOH concentrations of 14 M and 16 M. Variants of Na2SiO3 and NaOH ratios of 2. Activator variants and fly ash precursors are 0.417. Fine aggregate variants and fly ash precursors are 2. Using fibre lengths of 10 mm, 20 mm and 30 mm. For the percentage of fibre use of 0%, 0.25%, and 0.50% fibre by weight of the mixture.

Table 2. Composition of the mixture/m³

Composition	NaOH Concentration	*Fly ash*	Fine aggregate	NaOH	Na_2SiO_3	Fibre percentage	Coconut fibre
		(kg)	(kg)	(kg)	(kg)	%	(kg)
K0-14 M	14 M	577,216	1154,448	120,256	240,512	-	-
K0-16 M	16 M	577,216	1154,448	120,256	240,512	-	-
K10-14 M	14 M	577,216	1154,448	120,256	240,512	0,25%	5,224
K20-14 M	14 M	577,216	1154,448	120,256	240,512	0,25%	5,224
K30-14 M	14 M	577,216	1154,448	120,256	240,512	0,25%	5,224
K10-16 M	16 M	577,216	1154,448	120,256	240,512	0,25%	5,224
K20-16 M	16 M	577,216	1154,448	120,256	240,512	0,25%	5,224
K30-16 M	16 M	577,216	1154,448	120,256	240,512	0,25%	5,224
K10-14 M	14 M	577,216	1154,448	120,256	240,512	0,50%	10,464
K20-14 M	14 M	577,216	1154,448	120,256	240,512	0,50%	10,464
K30-14 M	14 M	577,216	1154,448	120,256	240,512	0,50%	10,464
K10-16 M	16 M	577,216	1154,448	120,256	240,512	0,50%	10,464
K20-16 M	16 M	577,216	1154,448	120,256	240,512	0,50%	10,464
K30-16 M	16 M	577,216	1154,448	120,256	240,512	0,50%	10,464

3. Mortar compressive strength

The sample for compressive strength testing consists of ten samples with 0% fibre variation of 14 M and 16 M, thirty samples with a variation of 0.25% fibre 14 M and 16 M, thirty samples with 0.50% variation of 14 M fibre and 16 M. Compressive strength testing was carried out in the material and concrete laboratory at Inderalaya University, Sriwijaya University.

Figure 1 shows the relationship between the percentage of 14 M fibre with compressive strength and the results show that the fibre length of 30 mm with a percentage of fibre 0.50% compressive strength of 25.472 MPa. Figure 2 shows the relationship between the percentage of 16 M fibre and compressive strength with the result that the fibre length is 30 mm and the percentage of fibre 0.50% results in compressive strength of 40.016 MPa.

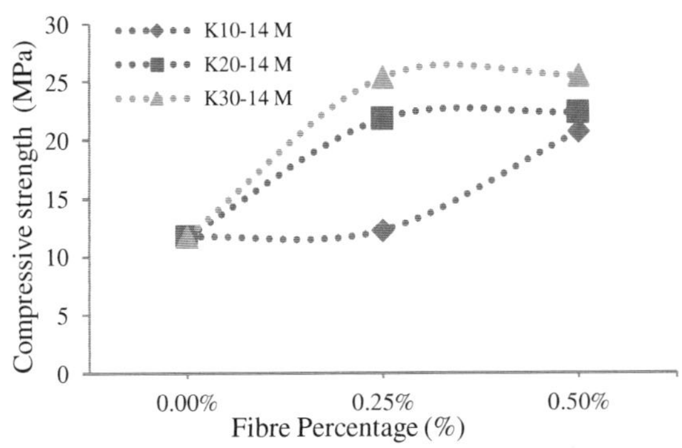

Fig 1. Fibre (%) and Compressive Strength 14M

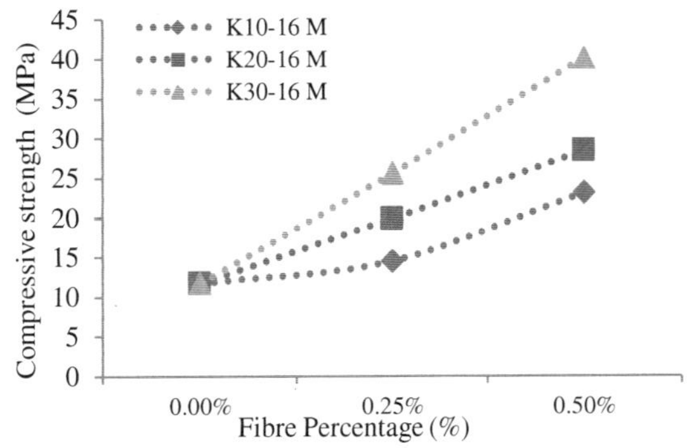

Fig 2. Fibre (%) and Compressive Strength 16M

From figure 1 and figure 2 show that the mixture with a percentage of 0.5% fibre at 14 M NaOH concentration and 16 M NaOH concentration results in higher compressive strength compared with 0.25% fibre percentage. This proves that a greater percentage of fibre use will increase the compressive strength value of geopolymer mortars but it must also be known that not all fibre usage with a greater percentage can increase the compressive strength value because it can happen with the use of a greater percentage of fibres which will result in bonding between Other fibres and mortar constituents do not interact properly so that the bonding matrix is weak and the compressive strength value is low. Therefore, it is important to pay attention to the percentage of fibre usage, if too much fibre usage is not balanced with the matrix volume, the strength will be weakened. So the percentage of fibre use must also be limited so that a high compressive strength mortar is obtained, therefore in this study the percentage of fibre is limited to 0.50%.

Figure 3 shows the relationship between the length of the fibre and the compressive strength value with the results of the compressive strength test that the fibre length of 30 mm using a percentage of 0.50% fibre and 14 M NaOH concentration showed a compressive strength result of 25.472 MPa. Figure 4 shows the relationship between fibre length and compressive strength with 16 M NaOH concentration. The results show that the fibre is 30 mm and the percentage of fibre is 0.50% the compressive strength value is 40.016 MPa.

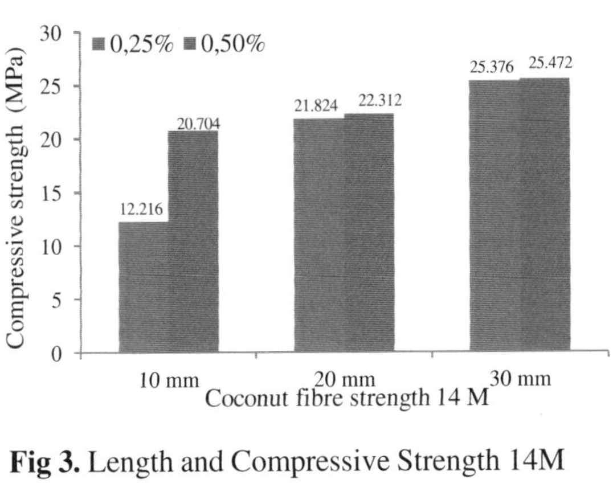

Fig 3. Length and Compressive Strength 14M

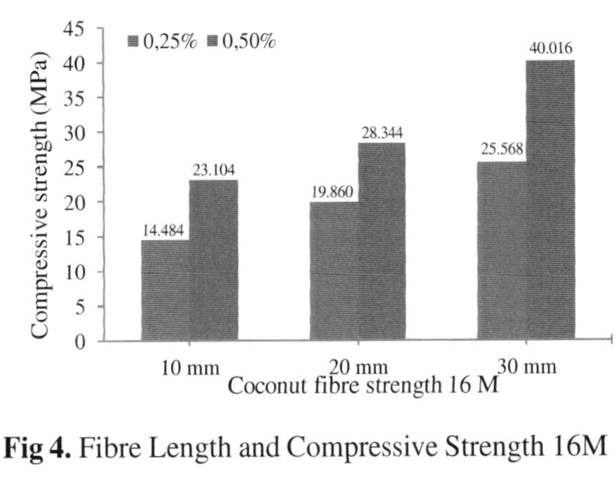

Fig 4. Fibre Length and Compressive Strength 16M

So Figure 3 and Figure 4 show that the fibre length of 30 mm at 16 M NaOH concentration with a percentage of fibre 0.50% compressive strength can be higher than the fibre length of 10 mm and 20 mm. The difference in compressive strength is very significant, this shows that the length of the fibre in the manufacture of fibre composites on the matrix is very influential on its strength. Judging from his theory, long fibres can flow loads and stresses from voltage points to other fibres. In an ideal continuous fibre structure, the fibre will be stress free or have the same stress. During fabrication, some fibres will receive high stress and others may not be exposed to stress so that the above conditions cannot be achieved [3].

Figure 5 shows the relationship between molarity and compressive strength. Fibre length of 30 mm percentage of 0.50% fibre with 16 M NaOH concentration shows a compressive strength value of 40.016 MPa.

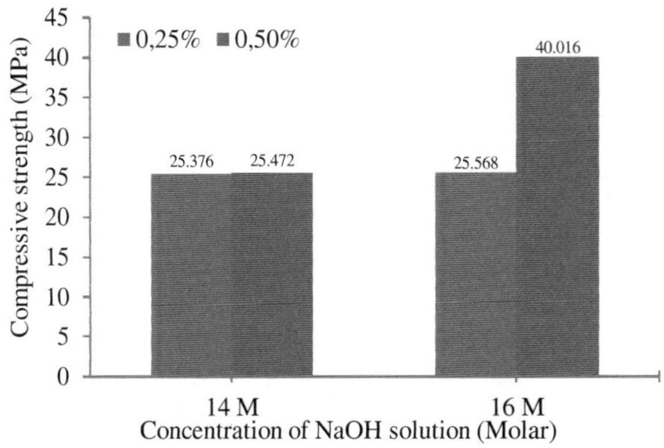

Figure 5. Molarity and Compressive Strength

Figure 5 shows that the concentration of NaOH solution can also affect the compressive strength value, an increase in compressive strength value with increasing concentration of NaOH solution. Concentration of 16 M NaOH solution can make the matrix binding process faster and with increasing time the polymerization process will be more maximal, so that the bond formed can be more solid.

4. Mortar flexural strength

The flexural strength test sample consists of ten specimens with 0% variation of 14 M and 16 M fibres, thirty samples with variation of 0.25% 14 M and 16 M fibres, thirty samples with 0.50% variation of 14 M fibre and 16 M. This flexural strength test was carried out in the material and concrete laboratory of the Inderalaya campus of Sriwijaya University.

Figure 6 shows the relationship between the percentage of 14 M fibre and flexural strength. Fibre with a fibre length of 30 mm and a percentage of 0.50% fibres showed flexural strength results of 7,576 MPa. Figure 7 shows the relationship between the percentage of 16 M fibre and the value of flexural strength. Fibre with a fibre length of 30 mm with a percentage of 0.50% fibre shows a flexural strength of 8.799 MPa.

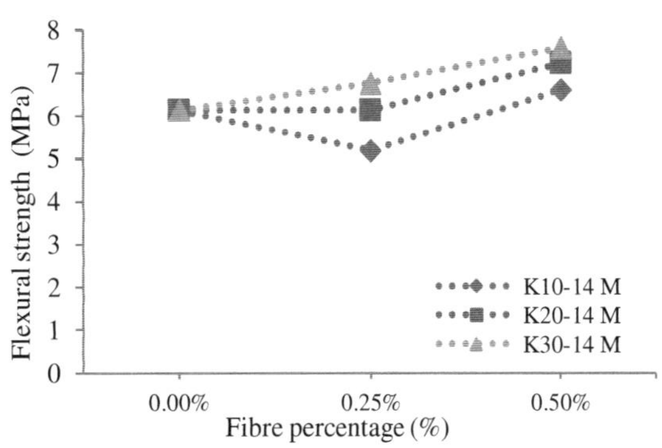

Fig 6. Fibre (%) and Flexural Strength 14M

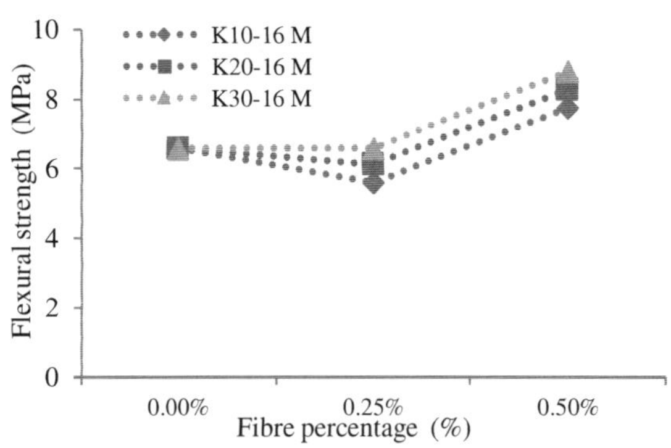

Fig 7. Fibre (%)and Flexural Strength 16M

From the picture of the flexural test results above shows that if the fibre is more fibre then the bending stress will increase. The increasing strength of the flexural strength is due to the greater composite dimensions. The more fibre usage, the bigger the composite dimension will be. However, it should be noted that if the comparison of the percentage of fibre usage is not balanced with the volume of the matrix it can cause the bending value to weaken because the composite only relies on the fibre alone, if the increase in fibre, will automatically reduce the binding material which causes the binding material to weaken as well.

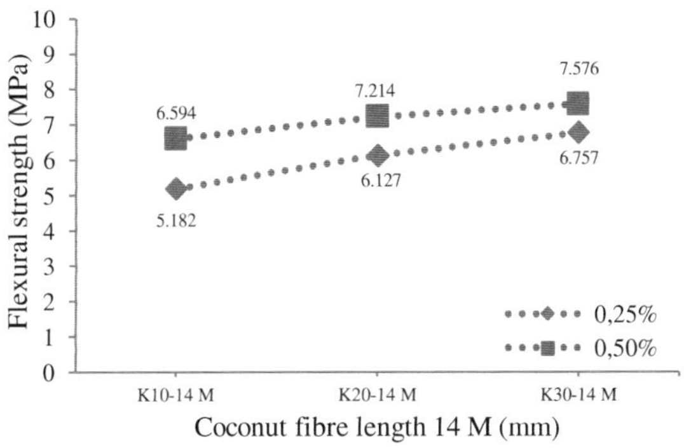

Fig 8. Fibre and Flexural Strength 14M

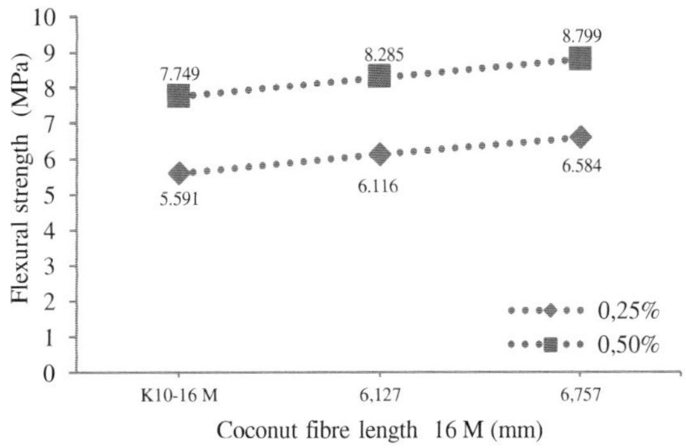

Fig 9. Fibre and Flexural Strength 16M

Composites with 30 mm fibre length have higher flexural strength values than composites with fibre lengths of 10 mm and 20 mm. This is because the 30 mm long fibre can be distributed well and evenly during the process of making composites, so that the bond between the reinforcement ie coconut fibre fibres with the matrix can take place perfectly, so that it will directly increase the value of the flexural strength of the coconut fibre reinforced composites. In the 30 mm long fibre composite with a percentage of 0.50% fibre reaches maximum strength because the amount of fibre is more than the percentage of fibres 0.25% so that it will be better in dividing the load then the matrix bond with the fibre will get better too which causes the value of flexural strength will get higher.

Figure 10 shows the relationship between molarity and flexural strength. The results show the flexural strength of 30 mm fibre length and percentage of fibre 0.50% with a concentration of 16 M molarity of 8,799 MPa.

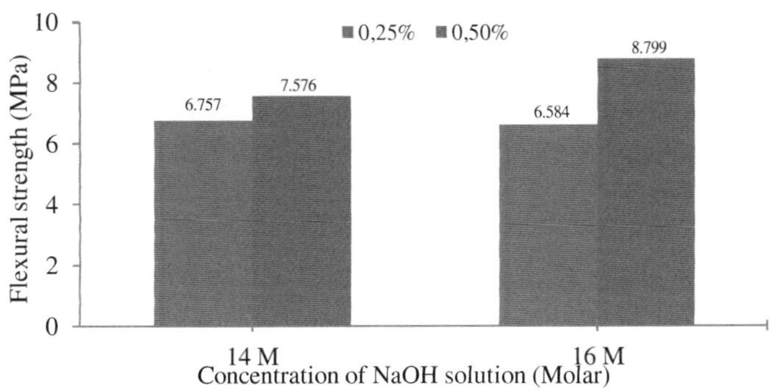

Figure 10. Molarity and Flexural Strength

The results of flexural strength tests show that the use of fibre with a percentage of 0.50% and the concentration of 16 molar NaOH solution will give good flexural test results for natural fibrous geopolymer mortar compared to other variations. This is directly proportional to the compressive strength test results. Increasing the concentration of NaOH solution also affects the increase in the value of flexural strength.

IOP Conf. Series: Materials Science and Engineering **807** (2020) 012041 doi:10.1088/1757-899X/807/1/012041

5. Microstructure test

Geopolymer particle morphology analysis was carried out using SEM (Scanning Electron Microscopy) test. The purpose of SEM analysis is to determine the formation of a geopolymer matrix, unreacted material, pores and the entry of coconut fibre fibres into the geopolymer matrix. Micrographs show geopolymers with part or all of the unreacted mortar forming matrix and cracks drawn as lines. The test was carried out at the Bandung geological survey center laboratory. SEM test results of fibre length of 30 mm 16 M with a percentage of 0.50% shows the distribution of coconut fibre fibres more evenly compared to the results of SEM test fibre lengths of 10 mm and 20 mm. The morphology of geopolymer particles which is dominated by solid and homogeneous geopolymer matrices on fibre length 30 mm 16 M 0.50% fibre indicates that the geopolymerization reaction is more perfect and has a higher compressive strength. This is directly proportional to the results of the compressive strength and flexural test results.

Figures 11 through 16 show that both a 0.25% fibre sample and a 0.50% fibre fibre geopolymer matrix are formed. Matrices in geopolymers are the result of reactions of aluminina and silica monomers that form polymers [15]. Figures 11-16 show that examples with 0.25% fibre and 0.50% fibre were observed as unreacted material. The existence of this unreacted material greatly affects the compressive strength and flexural strength. In addition to the unreacting material, the presence of pores also affects the compressive strength and flexural strength values. Pores contained in geopolymers occur due to air bubbles trapped in geopolymer paste during the printing process.

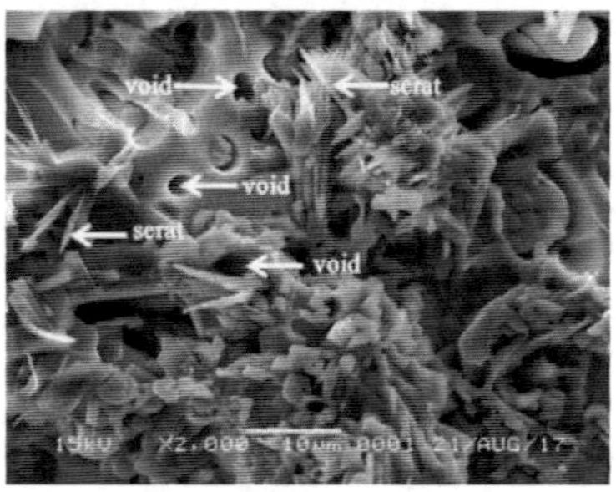

Figure 11. K10-14 M

Figure 12. K20-14 M

Figure 13. K30-14 M

Figure 14. K10-16 M

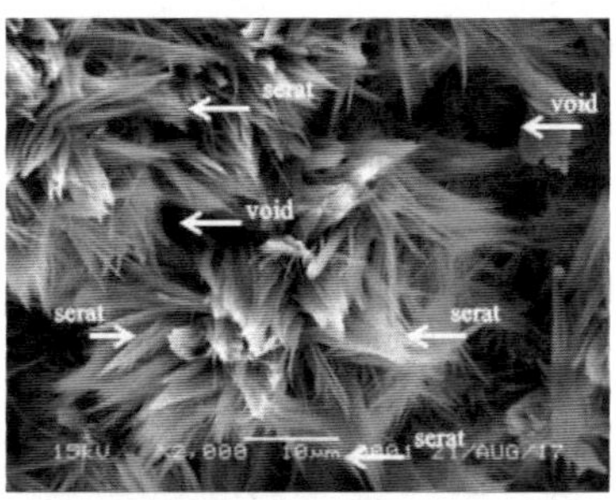

Figure 15. K20-16 M

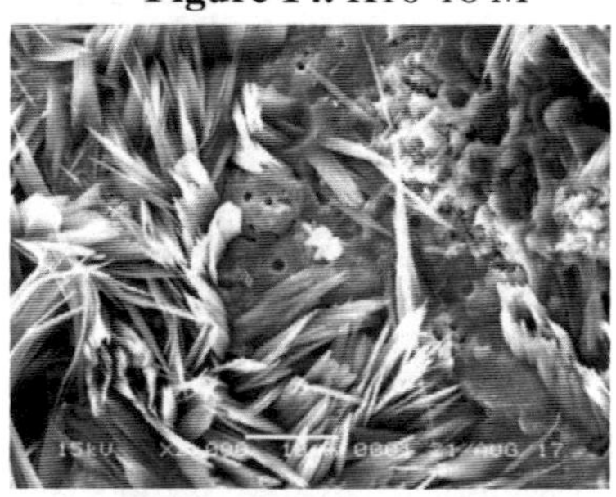

Figure 16. K30-16

6. Conclusion

There are several conclusions obtained from this study, namely:

i. From the variation of fibre length shows that the fibre length is 30 mm with a percentage of fibre 0.50% and the concentration of 16 M NaOH solution has a higher compressive strength value of 40.016 MPa and a flexural strength value of 8.799 MPa at 28 days.

ii. An increase in compressive strength value from the use of 0% 16 M fibre to the use of 0.50% 16 M fibre is 12,168 MPa to 40,016 MPa.

iii. An increase in the value of flexural strength from the use of 0% 16 M fibre to the use of 0.50% 16 M fibre is 6.584 MPa to 8.799 MPa.

iv. An increase in compressive strength and flexural strength value from the use of 0% fibre to the use of 0.50% fibre, this shows that the fibre has a very good influence for improving the quality of fly ash-based geopolymer mortar.

References

[1] Davidovits, J., 1999. *Chemistry of Geopolymer Systems*. Terminology Geopolymer '99 International Conference, Saint-Quentin, France, pp. 9-22.

[2] Hardjito, D., Wallah, S.E and Rangan, B.V., 2004.*Factor Influencing the Compressive Strength of Fly Ash-Based GeopolymerConcrete*.Civil Engineering Dimension. 6. Issue:2, pp. 88.

[3] Schwart, M.M., 1984, *Composite Materials Handbook, McGraw-Hill Book Co*., New York.

[4] Doraiswarmy et al. (1993). Pineapple Leaf Fibres, Textile Progress Vol. 24 Number 1,Textile Institute.

[5] Lerry, M., N., Gerung., 2012. *The Effect of Pineapple Leaf Fibre with 0.075% Fibre Concentration and Fibre Length Variation 0.5 cm; 1.0 cm; 1.5 cm; towards Compressive Strength of Normal Concrete*. Scientific Journal of Media Engineering vol. 2.

[6] Rangan, B.V, S. Wallah, D. Sumajouw and D. Hardjito, 2006.*Heat-cured, Low Calcium Fly Ash – Based GeopolymerConcrete*. The Indian Concrete Journal, pp. 47 – 52.

[7] R Zulfiati, Saloma, Y Idris. 2019 "Mechanical Properties of Fly Ash-Based Geopolymer with Natural Fibre". Journal of Physics: Conferense series, 2019.

[8] Saloma, Hanafiah, Ilma Pratiwi, K. 2016 "Effect NaOH Concentration on Bagasse Ash Based Geopolymerization". MATEC Web of Conferences 78,01025.

[9] Saloma, Iqbal, M.M., Aqil, I., 2017 "Sulfate resistance of fly ash-based geopolymer mortar", AIP Conference Proceedings, 1885, 020038.

[10] Pulungan M A, Sutikno and Sani M S M "Analysis of Bulletproof Vest Made from Fiber Carbon Composite and Hollow Glass Microsphere (HGM) in Absorbing Energy due to Projectile Impact" IOP Conf. Ser. Mater. Sci. Eng. 206 012001, 2019

[11] Sani M S M, Abdullah N A Z, Zahari S N, Siregar J P and Rahman M M "Finite element model updating of natural fibre reinforced composite structure in structural dynamics" MATEC Web of Conferences 83 03007, 2016

[12] Yaacob R M, Hashim M A H and Sani M S M "Finite element modeling and updating of the composite plate structure" IOP Conf. Ser. Mater. Sci. Eng. 100 01201, 2019

[13] ASTM C 109/C109-07 (2008). *Standard Test Methods for Compressive Strength of Hydraulic Cement Mortar ASTM Annual Books*.USA: ASTM.

[14] ASTM C 78 – 02 (2003). *Standard Test Methods for Flexural Strength of Concrete (Using Simple Beam with Third-Point Loading)*.ASTM International, West Conshohocken, PA.

[15] Rattanasak, U., Kanokwan P., & Prinya, C.2011. *Effect of chemical admixtures on properties of high-calcium fly ash geopolymer*.International Journal of Minerals:Metallurgy and Materials, 18: 364-369

IOP Conf. Series: Materials Science and Engineering **807** (2020) 012042 doi:10.1088/1757-899X/807/1/012042

Road Improvement Overlay Pavement Flexible at Cibomo – Terumbu Road Serang City

Telly Rosdiyani, Nila Prasetyo Artiwi and Ogi Suratman

Department of Civil Engineering, Universitas Banten

Corresponding e-mail: prasetyonila2@gmail.com

Abstract. Cibomo - Terumbu road to makes an alternative as connecting road (Axis) residential area in Kasemen District, Serang City. There are some damage conditions on the road structure and there is absence of adequate drainage channel with heavy traffic through the road, which lead congestion and inconvenience in driving. It is necessary to solve these problems. This research was conducted to determine the thickness of the pavement layer, and to design the dimensions of the drainage channel. The method used in this planning was the calculation of Bina Marga 1987. Geometric control of the road based on the Road Geometric Planning Basics 1999 , while drainage planning was used in the Procedure for Road Surface Drainage Planning (SNI-03-34241994) . Planning period for 15 (fifteen) years. The final results of the plan require an additional layer of 6 (six) cm Laston (Concrete Asphalt Surface Coating) MS 590. The size of the drainage channel in accordance with the calculation, channel height (h) is 0.40 m, the channel width (b) is 0.30, and the watch height (w) is 0.45.

1. Introduction

Road as one of the transportation infrastructure is an important element in the development of national and state life. Many roads are the main means of connecting that experience crisis conditions often occur before reaching the age of the plan such as Jalan Cibomo – Terumbu [1]. This could have happened due to several causes because the data of road pavement calculation in the planning period until the implementation was not in accordance with the parameter specifications that had been given and determined by the regulations and road planning guidelines issued by the regulations and road planning guidelines by Bina Marga [2, 3].

Physical conditions Jalan Cibomo – Terumbu has many holes, in which case this road is an alternative road as well as a connecting road (Poros) residential area in Kasemen District, Serang City, Banten Province [4-6]. The road was repaired in 2015. Visually, the road has damaged the structure of the road in several places, and does not have adequate drainage channels [7-9]. Population growth and vehicle ownership will spur increased population activity. The heavy traffic flow through the road section causes traffic jams and inconvenience in driving, so it is deemed necessary to address improvements in road conditions and services with flexible pavement overlay (flexible pavement). Thus the authors are interested in conducting to determine the thickness of the pavement layer, and to design the dimensions of the drainage channel [10, 11].

2. Research Method

2.1. Data Collection Technique
In road improvement overlay the flexible pavement on Cibomo – Terumbu Serang city, as supporting requiring data.To is done the collection of primary data and secondary. Primary data obtained from vehicle survey / interviewing related parties, while secondary data in the or map, technical specifications, implementation guidelines, NSPM (a norm, standard, guide and manual).

2.2. Analysis Method
As previously mentioned, this paper is about improving road pavement with the method of calculation spurring to Bina Marga to know the thickness pavement of overlay and dimensions of the drainage channel as for steps are planned as the flowchart shown in the Figure 1 below.

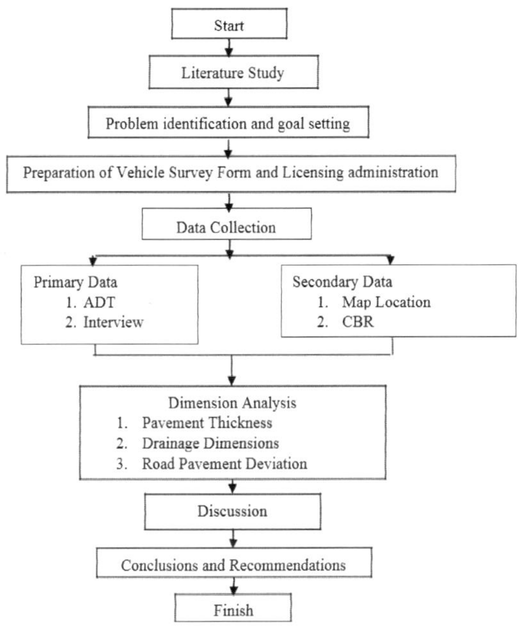

Figure 1. Flow Chart of Pavement Improvement Planning for Cibomo - Terumbu Road.

From Figure 1 above can be explained that the data obtained, primary data and secondary data, the next step is analyzing the data to plan the thickness of the asphalt layer for increasing Overlay and planning the dimensions of drainage channels, this calculation is also reviewed for the security stability set in the method calculation of Bina Marga. And the designed dimensions do not meet the security stability, the calculation will return or review the stage of determining the design of the layer thickness and dimensions of the road drainage channel.

3. Results and Discussion

3.1. Field Data
The data that the author can get from the Public Works Office of the City of Serang are as follows:

i. Pavement thickness for 1 lane and 2 directions.
ii. Population growth rate = 5%.
iii. The planned road is class III (Local road).

iv. Age of plan is 15 years.
v. CBR value is 4.
vi. ITP 4.78

The data obtained through the results of research in the field are the number of vehicles passing on Cibomo – Terumbu Road are as shown in Table 1.

Table 1. Number of vehicles passing in 1 day.

No	Types of Vehicles in Weight	Number of Vehicles
1	Light ride	230 Vehicles
2	Bus 8 ton	35 Vehicles
3	2 axles 10 ton truck	57 Vehicles
4	2 axles 13 tons trucks	10 Vehicles
	Total	332 Vehicles

3.2. Average Daily Trafic (ADT)
This data is for calculating the average daily traffic with the age of 15 while the results are shown in Table 2 below.

Table 2. Calculated members' stiffness

ADT	Light ride	Bus 8 Ton	2 axles 10 ton truck	2 axles 13 tons trucks
	478.15	72.76	118.50	20.79

3.3. Calculation Of The Equivalent Number Of Axle Loads
The value of the equivalent number of axle loads are as shown in Table 3.

Table 3. Value the equivalent number of axle loads

E	Light ride	Bus 8 Ton	2 axles 10 ton truck	2 axles 13 tons trucks
	0.0004	0.1593	0.3500	1.0648

3.4. Calculation Of The Equivalent Number Of Axle Loads
This calculation is to know the cross equivalent to the method:

Light Ride (1+1): 230 x 1.0 x 0.0004 = 0.092
Bus 8 ton (3 + 5): 35 x 1.0 x 0.1593 = 5.576
2 axles 10 ton Truck (4+6): 57 x 1.0 x 0.3500 = 19.95
2 axles 13 ton Truck (5+8): 10 x 1.0 x 1.0648 = 10.64
Total Perminal Equivalent = 36.266

3.5. Final Equivalence Calculation
It is the sum of the average daily equivalents of a single axis as heavy as 8.16 (18000 lb). On the path of the plan:

On the path of the plan:
Light Ride (1+1): 478.15 x 1.0 x 0.0004 = 0.191
Bus 8 ton (3 + 5): 72.76 x 1.0 x 0.1593 = 11.591
2 axles 10 ton Truck (4+6): 118.49 x 1.0 x 0.3500 = 41.475

2 axles 13 ton Truck (5+8): 20.789 x 1.0 x 1.0648 = 22.136
Total Final Equivalent = 75.393

3.6. Final Equivalence Calculation
Cross-calculation is equivalent 15 year obtained:

Central Cross Equivalent 15: = ½ x (Perminal Equivalent + Final Equivalent
Bus 8 ton (3 + 5): = 0.5 (36.266 + 75.393) = 55.829

3.7. Calculation of Equivalent Cross Plan
Calculation of equivalent cross plan is as follows:

Equivalent Cross Plant = Final Equivalent x Central Cross Equivalent where is the value CCP
= (age of plan /10)
= 55.829 x (15/10) = 83.74

3.8. Determine The Value of Regional Factors (FR)
Determine the value of FR with consideration of less than 6% and the regional climate is less than 900 year so regional factors can be used on the Cibomo-Terumbu road which is 1.0 - 1.5

3.9. Determine The Value Of The Surface At The End Of Age Plan (IP)
Determine the value of index see of the value of traffic equivalent plan is 83.74 and Roads are local roads and surface value index of 1.5

3.10. Determine The Value Of CBR
Segments can be determined using analytical or graphical methods, known data in the CBR field = 8, 6, 5, 4, 5, 5, 4, 5, and 3 which is shown in Table 4 below. Based on Table 4, the CBR value is taken by CBR 4 which shows 90%.

Table 4. Value of CBR

CBR	Same or More Amount	Percent (%) Equal Or Greater	
3	9	9/9 x100 = 100	100
4	8	8/9 x 100 = 100	88,9
5	6	6/9 x 100 = 100	66.7
6	2	2/9 x 100 = 100	22.2
8	1	1/9 x 100 = 100	11.1

3.11. Determine the Carrying Capacity of the Land
After setting the value CBR so capacity land was obtained through interesting way a straight line on diagram Soil support so that the CBR 4 found soil support value of 4.3 tons.

3.12. Determine the value of ITP
The pavement thickness index was obtained using a nomogram, determining Cross Equivalent at 83.74, using the Laston surface type and value 3.9 - 3.5, the surface index value at the end of the 1.5 plan, the writer determines the Nomogram number 5 as shown in Figure 2 below.

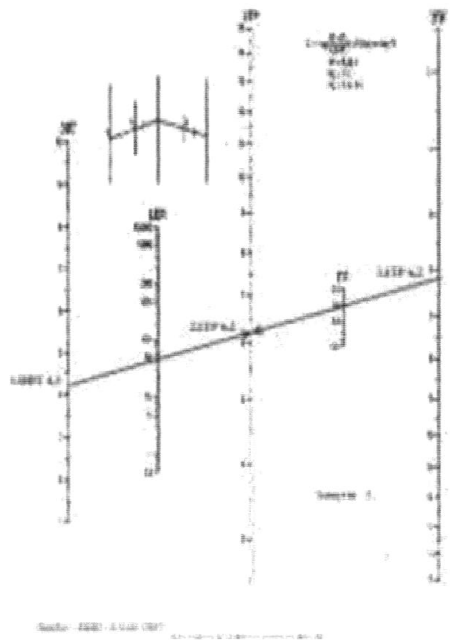

Figure 2. Flow Chart of Pavement Improvement Planning for Cibomo - Terumbu Road.

3.13. Determine the add Layer Thickness

Surface layer thickness in road conditions is assumed to be 25% reduced, conditions in the field with a thickness of 4 cm using laston MS 590 which have a relative strength coefficient of 0.35, the thickness of the upper foundation layer does not experience conduction in the field 10 cm thick and uses stone broken (CBR 100) which has a relative strength coefficient of 0.14. Lower foundation thickness in road conditions 20 cm thick. So that the Additional Layer Thickness:

Land Strength
Laston (MS 590) 4 cm = 75% x 4 x 0.35 = 0.78
Broken stone class A = 100% x 0.14 =1.40
Broken stone class B = 100% x 20 x 0.13 = 2.60
ITP = 4.78

IPT = ITP_{15} – ITP
= 2.11

1.41 = 0.35 x D_1 , so D_1 = 6.04 or 6 cm

3.14. Road Drainage Planning

3.14.1. Calculation of Debit Flow Plans (Q)

The calculation of the debit flow plans are as follows:

Length of channel plan : 400 m
Width of drainage area
I_1 Road Pavement = 2.25 ; I_2 Roadside = 0.5 m ; I_3 Outside road = 10 m
Area of Jetting;.
A_1 Road Pavement = 2.25 x 400 = 900 m^2
A_2 Roadside = 0.5 x 400 = 200 m^2
A3Outside road = 10 x 400 = 4000 m^2 km^2
Total Large =5.100 m^2 = 0.00510 km^2

3.14.2. Calculation the Average Flow Coefficient

Use the average flow coefficient

Table 5. Flowing coefficient

Soil Surface Conditions	Coefficient Streaming (c)	Runoff Factor (fk)
Ingredients		
1. Concrete road and asphalt road	0.70 -0.95	
2. Gravel road & dirt road	0.40 -0.70	
3. Roadside		
- Fine-grained soil	0.40 -0.65	
- Coarse grained soil	0.10 -0.20	
- Massive rock hard	0.70 -0.85	
- Soft massive rock	0.60 -0.75	
Land use		
1. Urban area	0.70 -0.95	2.0
2. Suburban area	0.60 -0.70	1.5
3. Industrial Area	0.60 -0.90	1.2
4. Dense settlement	0.60 -0.80	2.0

3.14.3 Counting The Time Of Concentration

The time of concentration or Tc is the longest time needed by the service area in channeling water flow.

$$Tc = T1 + T2 \quad (3)$$

Where :

Tc = Time of concentration

T_1 = The time needed to reach the start of the channel from the point Farthest

T2 = Time of flow in channels along L from the upstream end

I_0 = Distance from the farthest point to the drainage facility (m)

L = Channel length (m)

nd = Coefficient of resistance

Is = The channel slope is elongated

V = Average speed in the drainage channel (m/second)

Then the time of the road concentration is equal to:

T_1 = T_1 Road Pavement + T_1 Roadside + T_1 Outside road

Where:

I_0 = 2.25 make

= $(\ x\ 3.28\ x\ I_0\ x\ )^{0,167}$ = 0.784 minutes

T_1 Roadside

Where:

I_0 = 0.5 make T_1 Roadside as to 0.93 minutes

T_1 Outside road

Where

I_0 = 10 make T_1 Outside road as to 1.783 minutes

Total T_1 is 3.531 minutes and can be known value T_2 as 3.333 minutes then the amount of time for the

IOP Conf. Series: Materials Science and Engineering **807** (2020) 012042 doi:10.1088/1757-899X/807/1/012042

Cibomo -Terumbu road concentration is:

Tc $= T_1 + T_2$

Tc $= T_1 + T_2$

3.14.4 Rainfall Data Analysis

The maximum mean data is taken from the BMKG class 1 Serang Station in 2016 as follows:

Table 6. Maximum Average Rainfall Data

No	Month	Rainfall (mm)	Day Rain
1	Jan	125.9	19
2	Feb	271.9	26
3	Mar	228.8	23
4	Apr	88	19
5	May	143.3	18
6	Jun	93	13
7	Jul	134.8	16
8	Aug	86.6	16
9	Sep	158.9	24
10	Oct	164.8	24
11	Nov	138	22
12	Dec	178.8	22
Total		1812.8	242
Average		151,07	
Maximal		271.9	

The results of the calculation of the rainy day are as follows:

Table 7. Result of Calculation of Rain reoccurs

No.	T (Year)	Rainfall (mm)	Rounding (mm)
1	2	151.07	151
2	5	197.97	198
3	10	222.54	222
4	15	234.06	234

3.14.5 Rainfall Intensity Analysis

The calculation method is used as follows:

$$I = \{(R/24)\} \times \{(24/Tc)\}^{2/3} \quad (4)$$

Where:

I is the intensity of rain in mm / hour

Tc is the Time of Concentration in Hours

R is design rainfall in mm

is the Time of Concentration in Hours

3.14.6 Calculate Flow Discharge (Q)

Calculate the flow rate based on the Pd T-02-2006 guideline on road drainage system planning

$I = \{(R/24)\} \times \{(24/Tc)\}^{2/3}$

Where:

Q = flow (m^3/sec.)

C = Average Flowing Coefficient

I = Rainfall intensity (mm/hr)

A = area service (km^2) consists of A_1, A_2, A_3

then value Q obtained 0.165 m^3/sec

3.14.7 Calculating Channel Dimensions

Planned in the form of a square with concrete materials with the following conditions:

Allowable flow rate	= 1.5 m3 / sec
Lengthened channel slope	= 1.5%
Coefficient of material roughness according to the stone pair channel	= 0.020 from the SNI data
Flow discharge	= 0.165 m3 / sec
Channel base	= 0.75
	= 0.75 h^2
Wet circumference of the channel Ps	= b + 2 h
	= 2.75 h
Hidraulic Radius	= Fs/Ps
	= 0.273 h
Formula Manning V	= $1/n \times Rs^{2/3} \times I^{1/2}$
	= 2.577 $h^{2/3}$
Make Q	= Fs. V
0.165 m^3/sec	= 0.75 h^2 x 2.577 $h^{2/3}$
$h^{8/3}$	= 0.085
h	= $0.085^{3/8}$ = 0.40 ; b = 0.3 m
Hence the dimensions of the drainage channel h	= 0. 40 m and b = 0.30 m

3.14.8 Security Height (W)

Using formulas:

W = (11)

Then the value of w is obtained 0.45 m

3.14.9 Control Channel Dimensions

A = Flow speed

= 0.12 m^2 ;

wet roving = 1.1 m

Hydraulic radius = 0.109 m

Flow speed = 1.86 m/ sec.

Q = A.V= 0.223 m^3/ sec. $\geq$ 0.165 m^3/ sec

4. Conclusions

From the description above, this study can be concluded as follows:

i. Total overlay of asphalt concrete for Cibomo – Terumbu road with a thickness of 6 (six) cm, uses laston MS 590.

ii. The size of the drainage channel in accordance with the calculation, the channel height (h) is 0.40 m, the channel width (b) is 0.30, and the watch height (w) is 0.45.
iii. Pavement Deviation Bending Cibomo – Terumbu road in the results of the review have a difference of 2 cm in the age of the plan of 15 years.

To support conclusions, the suggestions that can be conveyed by the author are as follows:
i. In road planning, it is expected that the planner must also plan road drainage to avoid flooding or standing water on the pavement so that the road is more durable.
ii. In connection with the field conditions that affect the construction of the pavement layer supporting the spread of the load, the type of material used must be in accordance with the local soil conditions

References

[1] Hardiyatmo H C 2017 Road Pavement Design and Soil Survey (UGM Press, Yogyakarta)
[2] Government Regulation Number 34 concerning Roads 200 (Jakarta)
[3] Law of the Republic of Indonesia Number 38 concerning Roads 2004 (Jakarta)
[4] Ministry of Public Works Directorate General of Highways 2006 Drainage System Planning, Road Surface Drainage Design Guidelines (Jakarta)
[5] National Standardization Board, Procedure For Drainage Design
[6] Perdani N 2016 Analisis of flexible Pavement thickness Design PD T-01-2002 Manual Method Pavemen Design (MDP) and Metod Nottingham of Road I Gusti Ngurahrai Palu *Jurnal Fropil* **4**(20)
[7] Public Works Departement 1987 Directions for Planning the Thickness of The Flexible Road Pavement Using The Component Analysis Method (Jakarta)
[8] Silvia S 1995 Flexible Pavement Road (Bandung: Nova Publisher)
[9] Silvia S 1999 Basics of Road Geometric Planning (Bandung: Nova Publisher)
[10] Silvia S 2003 Hot Mix Asphalt Concrete (Jakarta: Granite Publisher)
[11] Suwardo, Iman H 2016 Geometric Road Design (Yogyakarta: UGM Press)

IOP Conf. Series: Materials Science and Engineering **807** (2020) 012043 doi:10.1088/1757-899X/807/1/012043

Influence of Temperature on The Yield and Characteristic of Bio Oil from Pyrolysis of Pine Sap

Wim Widyo Baskoro[1], Muhammad Zainal Arifin[1] and Samsudin Anis[1*]

[1]Department of Mechanical Engineering, Universitas Negeri Semarang, Kampus Sekaran, Gunungpati, 50229 Semarang, Indonesia, Tel. +62 24 8508101, Fax. +62 24 8508101

Corresponding e-mail: samsudin_anis@mail.unnes.ac.id

Abstract. The purpose of this work is to investigate the influence of temperature on the yield and characteristic of bio oil produced from pyrolysis of pine sap. The batch pyrolysis processes were conducted in an electric furnace at various temperature ranging from 350°C to 450°C. In each experiment, about 1 kg sample was pyrolyzed to obtain bio oil under inert environment by flowing nitrogen gas at 0.4 lpm. The produced bio oils were then characterized to obtain their physical and chemical properties. The result exhibited that the bio oil yield increased as the pyrolysis temperature increases. The highest yield of bio oil of 23.03 wt.% was achieved at 450°C. At this condition, the bio oil had specific gravity of 0.998, degree of acidity (pH) of 3.83, heating value of 5673.05 cal/g, and viscosity of 1.97 mm^2/s. The bio oil composed of hydrocarbon component with carbon chain of C_{10}–C_{21}. It was found that more than 70 wt.% bio oil occupied by hydrocarbon fuel of >C_{14} that could be categorized as biodiesel fuel. In the meantime, bio oil produced at 350°C composed of hydrocarbon with carbon chain of C_{10} only that could be utilized as gasoline fuel.

1. Introduction

Energy needs in Indonesia, especially in the fields of transportation, industry and households continue to increase in line with economic and population growth. Based on the data from the state electricity company, most of Indonesia's energy needs are still supplied by fossil energy. Petroleum still ranks highest, which is 51.66%. The second level is natural gas that occupy about 28.57%. The rest is supplied from oil energy at 15.34% and renewable energy 4.43% [1]. Ironically, conventional energy sources in the form of fossil energy, which is the main energy source in Indonesia, have increasingly limited reserves. This encourages efforts to find alternative energy as a substitute for fossil energy as an energy supply.

One of the alternative energies is biomass, both in the form of products and waste. In Indonesia, biomass is a very important natural resource with various primary products as fiber, wood, oil, foodstuffs and so on, besides being used for primary purposes biomass is also used as an energy source (fuel). Biomass is the only source of carbon that can be renewed and also be processed into good gas, liquid and solid fuels [2].

Pyrolysis is an irregular decomposition process of organic materials caused by the heating process without the presence of air [3]. This process produces organic vapors, pyrolysis gases, and charcoal where the organic vapors produced contain carbon monoxide (CO), methane (CH_4), carbon dioxide

(CO_2), tar and water (H_2O). Organic vapor is then condensed into liquid that is known as bio-oil. Several parameters that affect the product of pyrolysis are temperature, moisture or water content, type of material, carrier gas flow rate, residence time, particle size of the sample, and type of pyrolysis [4],[5]. The temperature affects the product to be produced because the decomposition reaction of the chemical bond in the material occurs due to high temperature heating. Pyrolysis temperature plays an important role in the process of bio oil production, both the yield and composition of bio oil [6]. Unlike other alcohol-based fuels, pine oil is an oxygen-containing fuel obtained from pine sap. Pine oil is a renewable energy that has a higher heating value, viscosity, and boiling point than diesel [7]. A previous study also found that by mixing pine oil with diesel fuel, HC and CO content can be reduced by 30% and 65%, respectively. Nevertheless, in order to obtain good pine oils for fuel application, information about the properties of pine oil produced at different pyrolysis temperatures is necessary to be evaluated. Some important physical and chemical properties of bio oil that commonly considered for fuel application are viscosity, specific gravity, heating value, and chemical composition.

Based on the description above, this study focused on the investigation of the effect of temperatures on the yield and characteristic of bio oil produced from pyrolysis of pine sap.

2. Materials and Methods

2.1. Materials

Pine sap (mercury) originating from several pine trees in Central Java Indonesia was used as the feedstock. It was obtained from Perum. Perhutani of Central Java Regional Division. Nitrogen gas with a purity of 99.99% was also used as a carrier gas during the experiment to ensure there is no oxygen in the system and to carry the vapor product into the condenser.

2.2. Experimental Setup

Figure 1 shows the schematic diagram of the experimental setup. The electrically-heated pyrolysis reactor constructed from stainless steel 340 with a thickness of 2 mm, a diameter of 100 mm, and a height of 500 mm was used in this work. The inside and wall temperatures of the reactor were monitored by using K-type thermocouples. The cooling system consisted of a 300 mm length of Liebig condenser and a 250 ml glass pot for collecting the condensed bio oil.

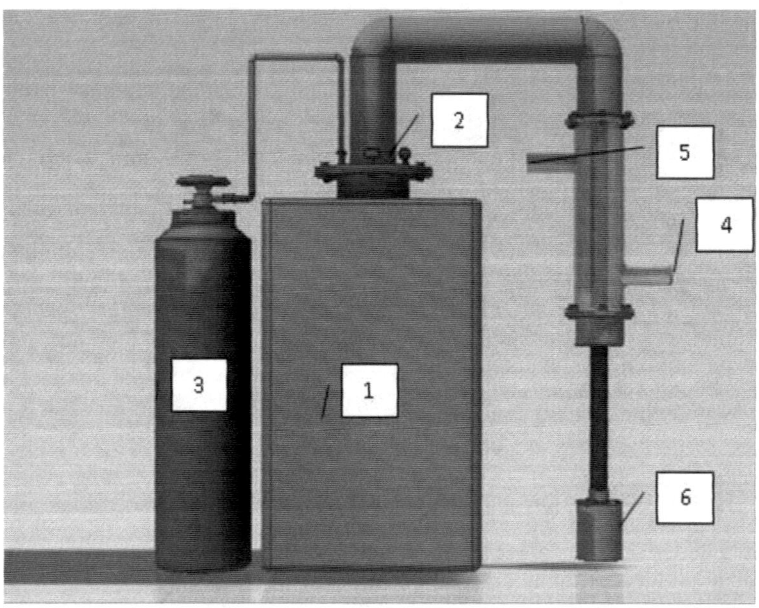

1. Reactor
2. Thermocouple and safety valve
3. Nitrogen gas cylinder
4. Water cooler input

Figure 1. Schematic representation of the pyrolysis experimental setup

2.3. Pyrolysis Experimental Procedure

In each experimental run, about 0.4 lpm high purity nitrogen gas was supplied into the reactor and others equipment for ensuring inert environment and for sweeping the vapor product from the pyrolysis reactor into the condenser.

The experiments were initially performed by put about 1 kg of pine sap into the reactor and then heated to a predetermined pyrolysis temperature. The pyrolysis reactor was designed and operated in batch processes. In this study, the sample was treated at several pyrolysis temperatures ranging from 350°C to 450°C. During the process, the bio oil products were collected in the collector. There are two types of bio oil samples for each pyrolysis temperature, namely bio oil collected before reaching the specified pyrolysis temperature, called sample A, and bio oil collected after reaching the specified pyrolysis temperature until the completion of pyrolysis process, called sample B. Both types of bio oil were then subjected to determine the specific gravity, heating value, viscosity, acidity (pH), and chemical composition.

3. Results and Discussion

3.1. Yield of Products

The yield of solid, liquid and gas products (% by weight) as a result of the pyrolysis process of pine sap at various temperatures are presented in Figure 2. As shown in the figure, the rate of the decomposition reaction increased with the increase of temperature. This condition caused the increase of the rate of chain breaking of the compounds in the material. A greater amount of water and volatile compounds evaporated rapidly as the temperature increases resulting in the increase of liquid products. The obtained results are in line with a previous study reported by Anis et. al [8] and Kim et. al [9] whose found that the yields of liquid and gas products increase as the temperature rises.

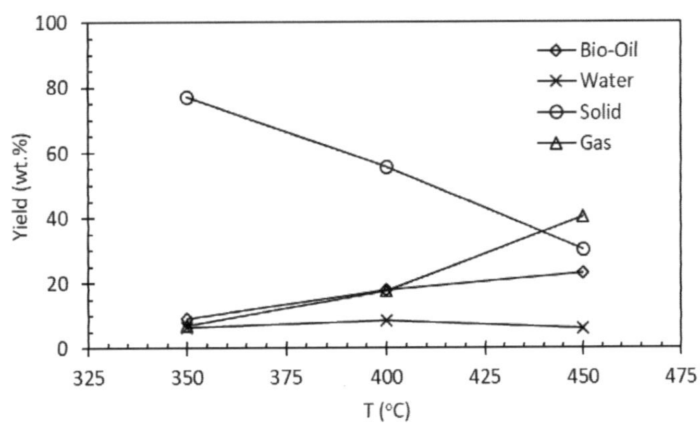

Figure 2. The yield of products from pyrolysis of pine sap as a function of pyrolysis temperature

3.2. Specific Gravity of Bio Oils

Figure 3 shows the specific gravity of bio oils. It could be observed that the specific gravity of bio oil from pyrolysis process of pine sap varied greatly ranging from 0.678 to 0.998. At a temperature of 400°C the specific gravity of bio oils was 0.843 for A sample and 0.886 for B sample, which is almost the same as the specific gravity of biodiesel fuel i.e. 0.815 - 0.860 and avtur fuel of 0.835. Nevertheless, these specific gravity of bio oils from pyrolysis process of pine sap were smaller than that of gasoline. This showed that bio oil generated by pyrolysis of pine sap can be utilized as a mixture of biodiesel or avtur fuel but cannot be used as a mixture of gasoline when viewed from the value of its specific gravity.

Based on the figure, the higher the pyrolysis temperature, the specific gravity of bio oils from pine pyrolysis was also higher. A higher pyrolysis temperature caused the decomposition of pine sap to be more intense. As a result, more smoke was formed, then condensed to become volatile. Volatile material produced at higher pyrolysis temperatures has the highest specific gravity because the ratio of mass to volume was greater [6],[10].

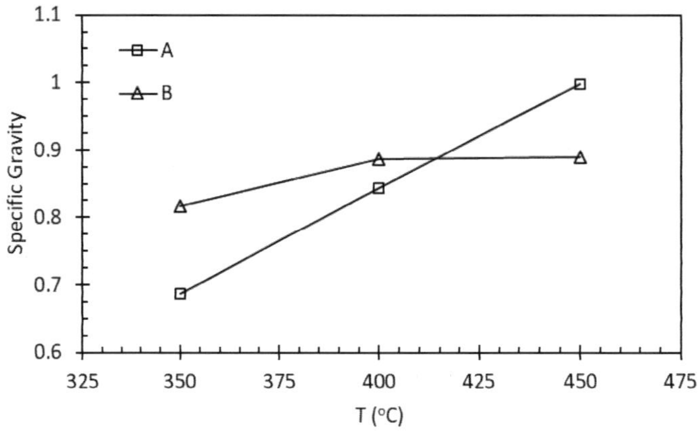

Figure 3. Specific gravity of bio oils from pyrolysis of pine sap as a function of pyrolysis temperature

3.3. Heating Value of Bio Oils

Figure 4 shows the heating value of bio oils at various pyrolysis temperatures. It could be observed that heating value of bio oils increased with the increase of temperature for both A and B bio oil samples. The biggest increase in the heating value of bio oil was obtained in B sample which was equivalent to 50.19%. This is due to bio oil produced at high temperatures tended to contain more hydrocarbons compounds which contribute to the calculation of the heating value of bio oil. It was found that the highest heating values of bio oils were obtained at 450°C of about 7,638.23 cal/g for A sample and of about 7,461.91 cal/g for B sample. These heating values of bio oil are comparable to diesel.

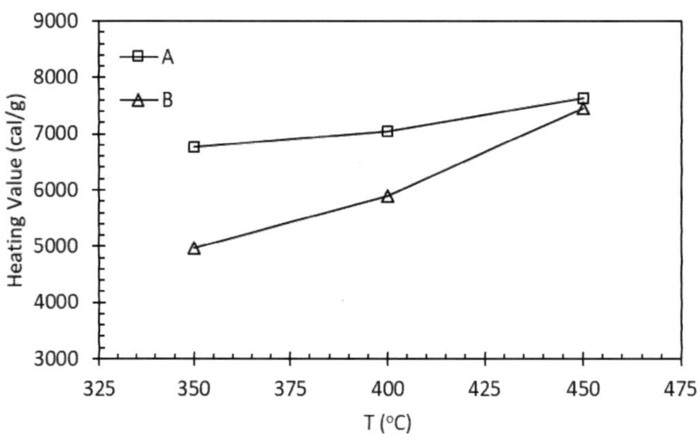

Figure 4. Heating value of bio oil from pyrolysis of pine sap as a function of pyrolysis temperature

The heating value of a fuel is equal to the absolute combustion enthalpy value where the high heating value of the fuel is influenced by the high volatile matter content which usually burns longer. High fuel heating value is not only influenced by high volatile matter and fixed carbon content, but also by low ash and moisture content in the fuel [11- 13].

3.4. Viscosity of Bio Oils

Figure 5 shows the viscosity of bio oils at various pyrolysis temperatures. The viscosity of bio oils from pyrolysis process of pine sap increased with increasing temperature. It was found that the highest viscosity of bio oil was obtained at 450°C of about 1.966 mm^2/s and the lowest viscosity of bio oil at

350°C of about 1.128 mm²/s. In general, the viscosity of various pyrolysis oils is influenced by various factors including the type of raw material, pyrolysis temperature, water content in pyrolysis oil, and the pyrolysis process used [14, 15].

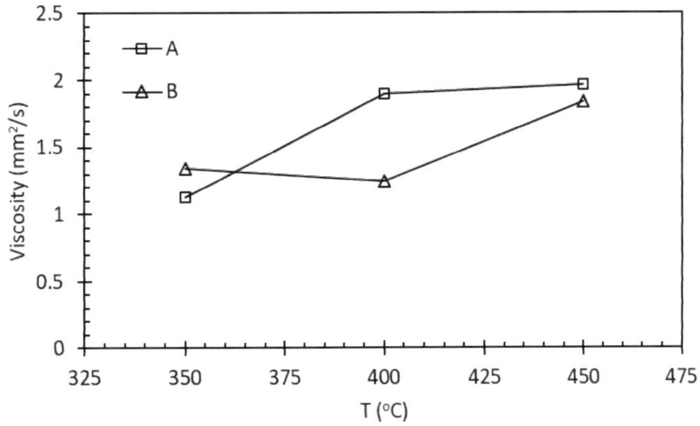

Figure 5. Viscosity of bio oil from pyrolysis of pine sap as a function of pyrolysis temperature

Figure 5 also indicates that the value of the viscosity of bio oils from pyrolysis of pine sap varied greatly between 1.1 mm²/s to 1.9 mm²/s. It was found that viscosity of bio oils from pyrolysis of pine sap was comparable than the viscosity of diesel and avtur fuels. This shows that for bio oil produced at higher pyrolysis temperature of 450°C, it can be directly applied in diesel engine without further purification or as a mixture of diesel and aviation fuels. In the meantime, for bio oil of pine sap produced at lower pyrolysis temperature of 350°C, it can be utilized as a mixture of gasoline fuel.

3.5. Degree of Acidity (pH) of Bio Oils

Figure 6 shows the degree of acidity (pH) of bio oil particularly the B sample of bio oil at various pyrolysis temperatures. It could be seen that the bio oil pH increased as the increase of pyrolysis temperature. This identifies that the higher the temperature process, the pH of bio oil tended to close to neutral condition. In this work, the highest bio oil pH produced from pyrolysis of pine sap was obtained at 450°C of about 3.83. It should be noted that the value of acidity (pH) is influenced by the temperature and type of raw material [15].

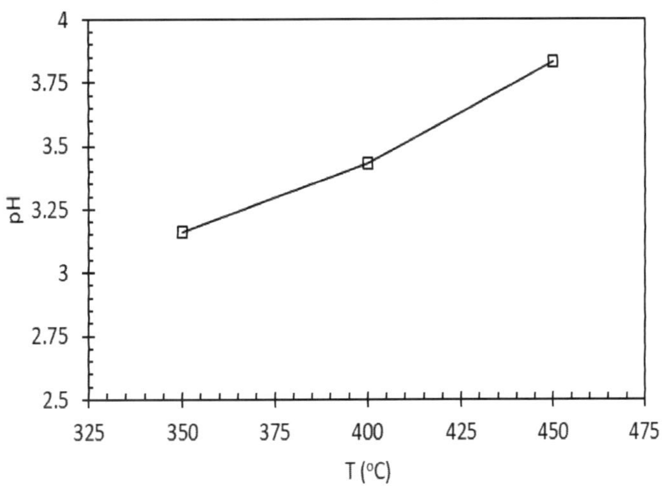

Figure 6. Acidity (pH) of bio oil as a function of temperature

3.6. Chemical Composition of Bio Oils

Chemical compositions of bio oil particularly the B sample of bio oil at various temperatures of pyrolysis are tabulated in Tables 1 to 3. Compositions of bio oil was analyzed by using a Gas Chromatography–Mass Spectrometry (GC-MS) analyzer. At a temperature of 350°C, the bio oil was dominated by the C_{10} bond of hydrocarbon compounds while at higher temperatures of 400°C and 450°C, the bio oil was dominated by the C_{10}–C_{21} bonds of hydrocarbon compounds. This result indicated that the higher the temperature of pyrolysis, the more bio oil compounds produced. It was found that at a temperature of 450°C, there were 21 peaks of compounds in which Methyl pimar was found to be the largest compound in bio oil with 19.66%. Meanwhile at a lower temperature of 350°C, there are only 10 peaks of compounds with α-Pinene of 31.82% as the major compound in bio oil. These results proved that the bio oil is mainly composed of lighter hydrocarbon components.

Table 1. Chemical composition of bio oil produced at 350°C

No	Compound	Molecular Formula	% Area
1	1R-α-Pinene	$C_{10}H_{16}$	31.82
2	Camphene	$C_{10}H_{16}$	6.41
3	Bicyclo[3.1.1]heptane, 6,6-dimethyl-2-methylene-, (1S)-	$C_{10}H_{16}$	1.49
4	Bicyclo[3.1.0]hexane, 6-isopropylidene-1-methyl-	$C_{10}H_{16}$	1.57
5	3-Carene	$C_{10}H_{16}$	14.79
6	1,3-Cyclohexadiene, 1-methyl-4-(1-methylethyl)-	$C_{10}H_{16}$	3.49
7	Cyclohexene, 1-methyl-4-(1-methylethenyl)-, (S)-	$C_{10}H_{16}$	28.57
8	1,4-Cyclohexadiene, 1-methyl-4-(1-methylethyl)-	$C_{10}H_{16}$	3.08
9	Terpinolen	$C_{10}H_{16}$	5.66
10	2,4,6-Octatriene, 2,6-dimethyl-, (E,Z)-	$C_{10}H_{16}$	3.14

Table 2. Chemical composition of bio oil produced at 400°C

No	Compound	Molecular Formula	% Area
1	1,3-Cyclohexadiene, 1,3,5,5-tetramethyl-	$C_{10}H_{16}$	10.54
2	α-Pinene	$C_{10}H_{16}$	1.38
3	Camphene	$C_{10}H_{16}$	1.44
4	1,3,6-Octatriene, 3,7-dimethyl-	$C_{10}H_{16}$	9.20
5	Cyclohexene, 1-methyl-4-(1-methylethylidene)-	$C_{10}H_{16}$	8.27
6	Cyclohexene, 1-methyl-4	$C_{10}H_{16}$	39.25
7	1,4-Cyclohexadiene, 1-methyl-4-(1-methylethyl)-	$C_{10}H_{16}$	2.43
8	2,4,6-Octatriene, 2,6-dimethyl-, (E,Z)-	$C_{10}H_{16}$	14.48
9	7-Acetyl-6-ethyl-1,1,4,4-tetramethyltetralin	$C_{18}H_{26}O$	3.30
10	Methyl pimar-7-en-18-oate #	$C_{21}H_{34}O_2$	6.90
11	Methyl abietate	$C_{21}H_{34}O_2$	2.81

Table 3. Chemical composition of bio oil produced at 450°C

No	Composition	Molecular Formula	% Area
1	1,3-Cyclohexadiene, 1,3,5,5-tetramethyl-	$C_{10}H_{16}$	1.05
2	α-Pinene	$C_{10}H_{16}$	0.86
3	Camphene	$C_{10}H_{16}$	1.08
4	1,3-Cyclohexadiene, 1,3,5,5-tetramethyl-	$C_{10}H_{16}$	3.12
5	3-Carene	$C_{10}H_{16}$	3.66
6	Cyclohexene, 1-methyl-4-(1-methylethenyl)-, (S)-	$C_{10}H_{16}$	14.73
7	Cyclohexene, 1-methyl-4-(1-methylethylidene)-	$C_{10}H_{16}$	1.96
8	2,4,6-Octatriene, 2,6-dimethyl-, (E,Z)	$C_{10}H_{16}$	2.37
9	Naphthalene, 1,2,3,4-tetrahydro-1,1,6-trimethyl-	$C_{13}H_{18}$	1.07
10	Androst-2,16-diene	$C_{19}H_{28}$	3.23
11	Methyl pimar-7-en-18-oate	$C_{21}H_{34}O_2$	19.66
12	Phenanthrene, 7-ethenyl-1,2,3,4,4a,4b,5,6,7,8,10, 10a - dodecahydro-4a,7-dimethyl-1-methylene-, [4aS-(4aα,4aβ,7β,10aβ)]-	$C_{19}H_{28}$	2.35
13	Androstan-17-one, 3-ethyl-3-hydroxy-, (5α)-	$C_{21}H_{34}O_2$	2.40
14	7-Acetyl-6-ethyl-1,1,4,4-tetramethyltetralin	$C_{18}H_{26}O$	10.95
15	Methyl abietate	$C_{21}H_{32}O_2$	9.60
16	Androst-2,16-diene	$C_{19}H_{28}$	3.28
17	1-(Hydroxymethyl) - 4a-methyl-7-vinyl -1,4,4a,4b,5,6,7,8, 10,10a-decahydro-3(2H)-phenanthrenone #	$C_{18}H_{26}O_2$	1.81
18	10,18-Bisnorabieta-8,11,13-triene	$C_{18}H_{26}$	3.10
19	Pimaric acid	$C_{20}H_{30}O_2$	7.48
20	Dehydroabietic acid	$C_{20}H_{28}O_2$	2.84
21	Abietic acid	$C_{20}H_{30}O_2$	3.39

Table 4 shows bio oil fraction composition based on the carbon atoms number in the bio oil compounds. These bio oil fraction refer to general fuel fraction including gasoline, kerosene, diesel, biodiesel, and heavy fuel fractions. Gasoline fraction is composed of hydrocarbon compounds with carbon atoms ranging from C_5 to C_{10} [16]. Gasoline is one type of fuel oil for internal combustion engines application. Compared to others fuel from hydrocarbon compounds, gasoline has a small number of carbon atoms. Besides gasoline fraction, there is a fraction of diesel from the bio oil achieved from pyrolysis process of pine sap. Diesel is hydrocarbon fuel used for diesel engines. Diesel has a number of carbon atoms ranging from C_{10} to C_{15}. The bio oil contained also heavy fuels fractions. Heavy fuels are heating fuels with carbon atoms of C_{18} to C_{25} [16].

Table 4. Bio oil fraction composition at different pyrolysis temperature

Number of Carbon Atoms	Bio oil Fraction Composition (%)			Fuel Fraction
	350°C	400°C	450°C	
C_5-C_{10}	100	86.99	28.83	Gasoline
C_{11}-C_{12}	-	-	-	Kerosene
C_{10}-C_{15}	-	-	1.07	Diesel
C_{14}-C_{24}	-	13.01	70.09	Biodiesel
C_{18}-C_{25}	-	13.01	70.09	Heavy Fuel

4. Conclusion

Pine sap feedstock was successfully pyrolyzed using an electrically-heated reactor to produce bio oils. The greater the temperature of pyrolysis process, the higher the yield of bio oils was obtained. Chemical compounds of bio oils became vary at high temperature processes making its specific gravity, heating value, viscosity, and degree of acidity (pH) values increased considerably. The results showed that only C_{10} carbon atom structures were produced at a lower pyrolysis temperature of 350°C that is totally gasoline fraction. In the meantime, at a higher pyrolysis temperature of 450°C, the produced bio oil contained more hydrocarbon components and is dominated by the C_{18}–C_{21} carbon atom structures indicating the bio oil has the characteristics of biodiesel fraction. The relatively high gasoline and biodiesel fractions composition illustrates that bio oil generated from pyrolysis process of pine sap has the potential as a source of hydrocarbon fuels.

Acknowledgements

The authors gratefully acknowledge the Applied Research Grant (No. 85.18.3/UN37/PPK.3.1/2019) awarded by the Ministry of Research, Technology and Higher Education of the Republic of Indonesia to carry out this work.

References

[1] Agency for the Assessment and Application of Technology (BPPT) 2017 *Outlook Energi Indonesia: Clean Energy Technology Development Initiatives.* Center for Technology of Energy Resources and Chemical Industry, Jakarta.
[2] Gironès V C, Moret M, Peduzzi E, Nasato M and Marécha F 2017 Optimal use of biomass in large-scale energy systems: Insights for energy policy. *Energy* **137** 789–97
[3] Beneroso D, Bermúdez J M, Arenillas A and Menéndez J A 2014 Influence of the microwave absorbent and moisture content on the microwave pyrolysis of an organic municipal solid waste. *J. Anal. Appl. Pyrol.* **105** 234–40
[4] Guedes R E, Luna A S and Torres A R 2018 Operating parameters for bio-oil production in biomass pyrolysis: A review. *J. Anal. Appl. Pyrol.* **129** 134–49
[5] Tao K, Vladimir S and Tim J E 2016 Lignocellulosic biomass pyrolysis: A review of product properties and effects of pyrolysis parameters. *Renew. Sustain. Energy Rev.* **57** 1126–40
[6] Lyu G, Wu S and Zhang H 2015 Estimation and Comparison of Bio-Oil Components from Different Pyrolysis Conditions. *Front. Energy Res.* **3** 1–11
[7] Vallinayagam R, Vedharaj S, Yang W M, Lee P S, Chua K J E and Chou S K 2013 Combustion performance and emission characteristics study of pine oil in a diesel engine. *Energy* **57** 344–51
[8] Anis S, Lestari S E, Wahyudi and Karnowo 2018 Production of Liquid Oil from Thermolysis of Electrical and Electronic Wastes (E-waste) under Microwave Irradiation. *Journal of Physical Science* **29** 53–62
[9] Kwang H K, Tannon J D, Ryan S, Mark M-W and Robert C B 2019 Recovery of resin acids from fast pyrolysis of pine. *J. Anal. Appl. Pyrol.* **138** 132–6
[10] Anis S and Zainal Z A 2013 Upgrading producer gas quality from rubber wood gasification in a radio frequency tar thermocatalytic treatment reactor. *Bioresour. Technol.* **150** 328–37
[11] Demirbas A 1997 Calculation of higher heating values of biomass fuels. *Fuel* **76** 431–4
[12] Yin C-Y 2011 Prediction of higher heating values of biomass from proximate and ultimate analyses. *Fuel* **76** 431–4
[13] Hosseinpoura S, Aghbashlo M, Tabatabaei and Mehrpooya M 2017 Estimation of biomass higher heating value (HHV) based on the proximate analysis by using iterative neural network-adapted partial least squares (INNPLS). *Energy* **138** 473–9
[14] Nolte M W and Liberatore M W 2010 Viscosity of biomass pyrolysis oils from various feedstocks. *Energy Fuels* **24** 6601–8

[15] Boscagli C, Morgano M T, Raffelt K and Leibold H 2018 Influence of feedstock, catalyst, pyrolysis and hydrotreatment temperature on the composition of upgraded oils from intermediate pyrolysis. Biomass and Bioenergy **116** 236–48
[16] Lam S S, Mahari W A W, Cheng C K, Omar R, Chong C T and Chase H A 2016 Recovery of diesel-like fuel from waste palm oil by pyrolysis using a microwave heated bed of activated carbon. *Energy* **115** 791–9

IOP Conf. Series: Materials Science and Engineering **807** (2020) 012044 doi:10.1088/1757-899X/807/1/012044

Decision Support System Model for The Process of Increase of Position in Hospital Pertamina Bintang Amin (RSPBA), Lampung

E Khikmawati[1]*, M Y Sulfarano[2] and A Purnomo[2]

[1] Department of Industrial Engineering, Universitas Malahayati, Bandar Lampung, 35153, Indonesia
[2,3]School of Management, Faculty of Economics, Universitas Bandar Lampung, Bandar Lampung, 35153, Indonesia

*Corresponding e-mail: emy_khikmawati@yahoo.com

Abstract. Decision making related to property criteria in the office will contribute to both organizational performances. This research intends to construct a model of the support system of the decision making in the process of increasing the performance in the office and to determine the criteria used for an elected employee who be promoted to a new position in the office. Model decision-making are expected to help the decision makers to get the information so that it is easier for the election of worker to be promoted in the organization. Development of decision-making system using an analytic hierarchy process (AHP) were used Hospital Pertamina Bintang Amin, Lampung. This system would yield a weighing criterion that would affects to the decree of the election of workers. An alternative model decision making is also developed in the process of increasing the efficiency within the organization itself. The working principle of the model system is by analysing and deciding the employees who are worth to be promoted. The employee's promotion system model does provide services for effectively analysing and deciding the decent employees. The analytical hierarchy process provides an assessment of criteria that influence the selection of decisions by the alternative decision-making models. The system does compare the interest between the criteria, the intensity and referring to the assessment of employees results in producing the output results by the end of the system process.

1. Introduction

A health care institution in conducting a perfect manner of health care, the decision-making process regarding on promoting a position of a decent worker requires a long time and the decision depends on policy makers. Decision Support System is a system that helps managers to solve a problem [1]. Determining the criteria and choices are some of the problems that delayed many processes especially in the process of promotion of workers. Decision support system developed do runs by the selection of employees that fit the criteria that exist in the specific position. In this case the task of analysing the employees who fit the criteria of employee's assessment are in accordance with the predetermined criteria that facilitate the process of promotion [2]. Decision making related to the right criteria for a position will make a good contribution towards the organizational performance. Plus, Decision Support

System is a system that is able to provide the ability to solve problems and the ability to communicate with problems with semi-structured and unstructured conditions [3]. The research objective is to build a decision support system model for the promotion process and determine the criteria used for the selection of employees to be promoted for promotion. Selection of the best employees for promotion by using the Analytical Hierarchy Process (AHP) method can help companies in selecting the best employees for promotion in accordance with the criteria desired by the company. By using the AHP method, the most appropriate employees in accordance with the criteria which are desired are selected to be promoted [4]

2. Methods

The research conducted was a descriptive analytic qualitative research. The results of data analysis in the form of exposure to the situation under study are presented in the form of narrative descriptions, the tools used, namely the Analytical Hierarchy Process (AHP). The assessment given in the use of the AHP method gives us flexibility in assessing, namely the AHP shows logically personal values and values [5]. Data collection techniques: Observation, Interview, Documentation and Literature.

In solving the problems, for AHP there are several principles [7]. Among them are making a hierarchy, evaluating the criteria and alternatives as well as determining the priorities and logical consistency. AHP procedures are defining the problems, creating a hierarchical structure, forming a pairwise comparison matrix and normalizing the data. The eigenvector value is also calculated and test for its consistency. If it does not meet with $CR \leq 0.100$ then the assessment must be repeated. The calculation of consistency index, $CI = \frac{(\lambda max - n)}{n}$ is to find out either the output value of CI do meet the good level or not, it is compulsory to know the ratio that is considered good, which is $CR \leq 0.1$ and the equation of consistency ratio is $CR = \frac{CI}{IR}$ [7].

3. Result and Discussion

Decision support system that includes the three criteria which are the objectives, criteria and alternatives. The goal in this study is that the selected employees who could fill the positions promoted for the management level would use those criteria as parameters to be used as a benchmark for making decisions. The criteria that become the determining parameters in the selection of employees in order to occupy the important positions of management level are the skills, loyalty, cooperation and honesty. The explanation of these criteria is based on several factors that are considered as follows [8]:

i. Skills (KP): employee skills in their fields, supported by expertise certificates or workshops.
ii. Loyalty (LY): loyalty that should be carried out in a variety of conditions and binds employees and companies, therefore loyalty is reflected in how long a person is in the organization and it can be in the form of thoughts or attention.
iii. Working period (MK): the duration of employee that has worked from the first time they entered until recent working.
iv. Honesty (KJ): honest behaviour in the workplace by obeying and abiding by applicable rules and ethical standards. The trust of a leader shown to employees in taking risks in relying on people.

The HR department would process the data and provides instructions as well as the selection process. Meanwhile, the assessment team or the leaders are the decision maker. The process begins with filling out forms that match the data that meets the requirements. Then proceed with the initial meeting to discuss and analyse the appropriate data, where the position is needed, then the decision-making system is carried out. The analytical hierarchy process (AHP) proposed in this study aims to provide an assessment of the criteria that influence the selection decisions of alternative decision-making models for the selection of positions. The choice of methodology is based on the characteristics of the problem and consideration of the advantages and disadvantages of other methodologies. The researcher assessed the importance of each criterion according to the value of the pair of criteria being compared. The result of AHP is a ranking or priority weighting of each alternative model.

In this study the focus is on the formulation of an AHP-based model to assess the three alternative decision-making models for the selection of the proposed position and has the best feasibility among the three. Generally, the promotion is done to carry out the process of filling out the promotion form, continue to elect employees who meet the criteria until the appointment letter is issued [6].

Table 1. Calculation calculations for pairwise criteria

Criteria	MS	LY	KP	KJ	Weight Normalization
MS	1.00	3.00	5.00	9.00	0.54
LY	0.33	1.00	5.00	5.00	0.28
KP	0.20	0.20	1.00	5.00	0.12
KJ	0.11	0.20	0.20	1.00	0.04
Total	1.64	4.40	11.20	20.00	0.98

λmax = 4.25
CI = 0.08
CR = 0.08 (accepted)

A consistency ratio of 0.08 is less than the tolerance limit of 0.1. Then the pairwise comparison matrix is said to be consistent For the AHP model, the comparison matrix can be accepted if the ratio value is consistent lesser than 0.1. The value of CR is greater than 0.1 thus it is a good level of consistency and could be accounted for.

3.1. Pairwise Comparison Matrix For alternatives or choices.
Make a pairwise comparison matrix that illustrates the relative contribution of the influence of each element to each of the objectives of the same level above [5].

3.1.1. Years of Service
First, the researcher will assess who of the three alternatives has the longest working period.

Table 2. Computational calculation of alternative MS criteria in pairs

Alternative	KY 1	KY 2	KY3	Weight Normalization
KY 1	1.00	2.00	3.00	0.54
KY 2	0.50	1.00	2.00	0.29
KY 3	0.33	0.50	1.00	0.16

Total	1.83	3.50	6.00	0.99

λmax = 2.96
CI = 0.02
CR = 0.03 (accepted)

3.1.2. Loyalty Criteria

Table 3. Comparative calculation of alternative LY criteria in pairs

Alternative	KY 1	KY 2	KY 3	Weight Normalization
KY 1	1.00	2.00	2.00	0.49
KY 2	0.50	1.00	2.00	0.31
KY 3	0.50	0.50	1.00	0.19
Total	2.00	3.50	5.00	0.99

λmax = 3,043

CI = 0.022

CR = 0.04 (accepted)

3.1.3. Skill Criteria

Table 4. Calculation of alternative comparison of paired KP criteria

Alternative	KY 1	KY2	KY 3	Weight Normalization
KY 1	1.00	2.00	2.00	0.47
KY 2	0.50	1.00	3.00	0.38
KY 3	0.50	0.33	1.00	0.18
Total	2.00	3.33	6.00	1.03

λmax = 3.22
CI = 0.11
CR = 0.1 (accepted)

3.1.4. Honesty Criteria

Table 5. Calculation comparison of alternative LY criteria in pairs

Alternative	KY 1	KY 2	KY 3	Weight Normalization
KY 1	1.00	3.00	2.00	0.54
KY 2	0.33	1.00	2.00	0.27

KY 3	0.5	0.50	1.00	0.19
Total	1.83	4.50	5.00	1.00

λmax = 3.145
CI = 007
CR = 0.1 (accepted).

Table 6. Calculation of pairwise comparisons for each alternative criterion

Criteria / Alternative	Weight Normalization		
	KY 1	KY 2	KY 3
MS (0.54)	0.54	0.29	0.16
LY (0.28)	0.49	0.31	0.19
KP (0.12)	0.47	0.38	0.18
KJ (0.04)	0.54	0.27	0.19

Alternative criteria weight:

i. KY 1 = 0.54 (0.54) + 0.28 (0.49) + 0.12 (0.47) + 0.04 (0.54) = 0.51.
ii. KY 2 = 0.54 (0.29) + 0.28 (0.31) + 0.12 (0.38) + 0.04 (0.27) = 0.29.
iii. KY 3 = 0.54 (0.16) + 0.28 (0.19) + 0.12 (0.18) + 0.04 (0.19) = 0.17

From the above table it can be concluded that KY 1 has the highest score of 0.51, followed by KY 2 with a score of 0.29 and most recently KY 3 with a score of 0.17. Finally, researchers will choose KY 1as the best employee who will occupy the position. The highest alternative weight is used as a material for consideration for promotion to employees. The results of the AHP calculation that is applied will produce the highest priority intensity employee output so that the employee who has the highest value (KY1) is eligible to get the promoted position.

This decision support system provides only a decisive alternative that can serve as a leader's reference in the promotion of the office, while the decision itself is held by the decision maker. The results of the assessment in this case by doing the type of face validity, which is related to what appears in measuring something and not to what should be measured [6]. Job promotions must be based on existing provisions, which can meet the needs of institutions and provide satisfaction to employees. This rationale for promotion promotes accuracy and objectivity in evaluating employees. The requirements for promotion are fulfilling the service period and the assessment requirements for the employee concerned, also having other criteria, namely loyalty, skills and honesty. Job promotion as one of the elements that can cause employee job satisfaction, therefore the implementation must be mature.

4.0 Conclusion

The employee promotion system model provides services for effectively analysing and deciding decent employees. The analytical hierarchy process provides an assessment of the criteria that influence the selection decisions of alternative decision-making models for the selection of positions.

Acknowledgment

Herewith, we are highly appreciated to Universitas Malahayati for providing fund of this research project.

References

[1] Suhud D S 2014 Analysis of supporting decisions in determination of media promotion for new student admissions at Serang Raya University using the AHP Method *Jurnal Pengembangan Riset dan Observasi Sistem Komputer* **3**

[2] Turban E, Jay E A 2005 Decision support system and intelligent system (New Jersey: Hall Education International)

[3] Darmawan P R 2017 Selection of the best employees for the promotion of supervisor position with the analytical hierarchy process method at PT. Karya Indah Alam Sejahtera *J. MATRIC* **17** 2 01-06

[4] Saaty T L 1993 Multicriteria decision making: The analytic hierarchy process. (University of Pittsburgh: RWS Publication)

[5] Khikmawati E, Widya M and Hertiati R 2019 Analysis of wood supplier selection using analytichal hierarchy *Proc Ntl. Sem. Innov. App.Tech. Ind.*

[6] Moedjiono 2016 Position promotion decision making system with the analytical hierarchy process method of a chemical industry company In*tl. Conf. on Information and Communication Technology*

[7] Kusrini 2007 Concept and application of decision support systems

[8] Hasibuan and Malayu S P 2003 Human Resources Management (Indonesia: Earth Literacy Publisher)

IOP Conf. Series: Materials Science and Engineering **807** (2020) 012045 doi:10.1088/1757-899X/807/1/012045

Estimating Rotational FRFs of a Complex Structure Using FE Model Updating, Mode Expansion and FRF Synthesis Method

WII Wan Iskandar Mirza[1], MN Abdul Rani[1*], MA Yunus[1] , Rina Febrina[2] and MSM Sani[3]

[1]Structural Dynamics Analysis & Validation (SDAV), Faculty of Mechanical Engineering, UniversitiTeknologi MARA, Malaysia
[2]Department of Civil Engineering, Malahayati University, Lampung, Indonesia
[3]Advanced Structural Integrity and Vibration Research (ASIVR), Faculty of Mechanical Engineering, Universiti Malaysia Pahang, 26600 Pekan, Pahang, Malaysia

Corresponding e-mail: mnarani@uitm.edu.my

Abstract. Estimating rotational frequency response functions (FRFs) from experimental data is very challenging, and the estimation often leads to poor FRFs due to the presence of spurious resonance peaks. In practice, special equipment is essentially required in the measurement process, unless the rotational FRFs are synthesised using measured modal data. This paper presents an alternative approach for estimating the rotational FRFs of a geometrically complex structure by using finite element modal updating, mode expansion and the FRF synthesis method. The applicability and accuracy of the proposed approach are demonstrated on a car front-end module A simplified finite element (SFE) model of the test structure is introduced and the SFE model is then updated based on the measured modal model obtained from experimental modal analysis. The mode shapes of the updated SFE model are expanded to the test model to obtain the entire translational and rotational modal vector. The rotational FRFs of the expended experimental model is derived via the FRF synthesis method. The derived translational FRF is compared with the measured counterparts for validation purposes. It was found that the derived translational and measured FRF are in a perfect match, and the derived rotational FRF has successfully estimated all the resonance peaks within the frequency of interest. The findings from this work may be useful in improving the accuracy of the experimental rotational FRFs, which are crucial for particular structural dynamics analysis.

1. Introduction

Experimental rotational degrees of freedom (DOFs) are crucial in many areas of dynamic investigations, such as experimental dynamic substructuring [1-2], model reduction/updating [3], structural modification [4], and vibration control [5]. For instance, in the automotive domain, a crash imparts tremendous energy to the occupants, typically in the form of significant rotational inertia. Therefore, the acceleration experienced by the rotating part of the structure is often an essential parameter during system design. It is noteworthy that measuring rotational DOFs data, which usually requires special measurement sensors, is very difficult to be executed and is not a straightforward assignment. There has been a variety of techniques reported by [6-7] employing two spatially separated and sensitivity-matched accelerometers to estimate rotational acceleration. For instance, an X-block attachment was used to allow the attachment of several translational accelerometers to estimate the rotational frequency response functions (FRFs)of a component in the attempt to perform a structural modification of a helicopter tail [8]. Another striking point is a T-block mass was attached to the structure and was de-coupled to expand and obtain the rotational FRFs. This approach is also known as transmission simulator [9]. However, the reported methods only work best particularly when the rotational responses are high.

Recently, rotational accelerometers have been introduced to efficiently allow the measurement of the rotational FRFs of a test structure[10]. However, it was reported in [11] the measurement only limited to force excitation responses which means moment excitation of experimental substructures are required to be expanded based on the modal model to obtain a full DOF of the structure. Although there are several techniques [7,10-11] to directly measure the rotational data, the techniques usually require special equipment such laser vibrometeror rotational accelerometers, which may not be readily available in most laboratories [12].

This paper puts forward an approach of estimating the rotational FRF data of a geometrically complex structure using the updated FE model based mode expansion method. This approach is developed based on model updating, mode expansion and the FRF synthesis method. The process flow of the proposed approach is presented in Figure 1. In this work, a simplified finite element (SFE) model is introduced in order to overcome the difficulties of modelling a complex geometrical structure [13-14]. Later, the updated SFE model is then expanded to the experimental model to obtain all the translational and rotational FRFs.

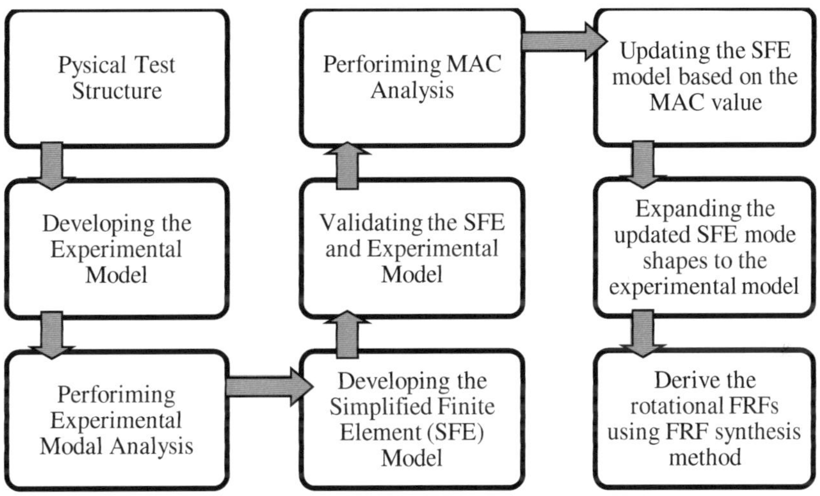

Figure 1. Process flow of the proposed approach

The proposed approach is performed on a car front end module structure as shown in Figure 2. The structure is used as a case study because its configuration has played a crucial contribution to the vibration level of a car. The one end of the test structure as shown in Figure 2 is selected as the reference point because the point covers all modes of interest of the structure.

IOP Conf. Series: Materials Science and Engineering **807** (2020) 012045 doi:10.1088/1757-899X/807/1/012045

Figure 2. The front-end module of a car body in white

2. Development of the SFE model

Previous works [7,10] reported that rotational accelerometers can be used to directly measure the rotational responses of the test structure. However, the measured responses are only for the force-excitation. Normally, the moment-excitation responses are expanded by using modal expansion methods based on the numerical modal model [11] but the modal expansion method relies on the accuracy of the numerical model. For particular cases involving complex structures, developing accurate finite element models are time consuming and difficult. This work employs an approximation modelling approach, where the geometry of the FE model of the structure is simplified. By employing the approach, the FE modelling time is reduced.

Figure 3 depicts the construction process of the SFE model of structure. 2D-Shell elements with an approximate and simplified outline of the experimental model were used in constructing the SFE model. The procedure adopted gives the necessary details of embodying the physical test structure. The use of 2D shell element allows the rotational data to be extracted from the SFE model [13] and its parameters to be correlated with the experimental model [15,16]. To increase the effectiveness of the SFE model, the structure was discretized into seven different parts, allowing more parameters to be used for updating. The updating has led the mode shapes of the updated SFE model to be well correlated with the experimental ones. Subsequently, the updated SFE model was employed in the expanding process of unmeasured rotational data to the experimental model.

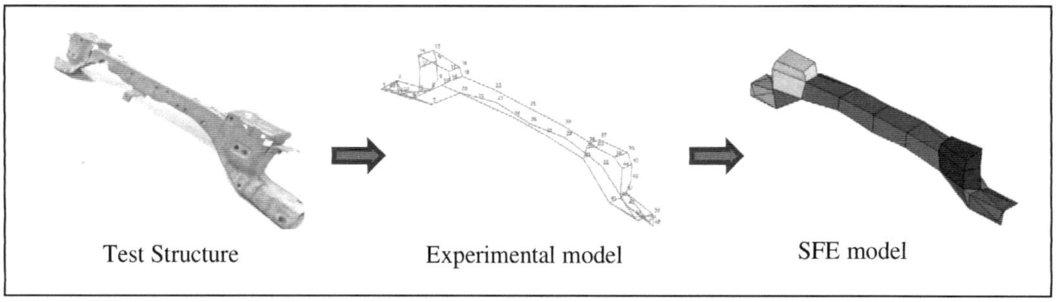

Figure 3. The test structure, experimental model and SFE model

3. Experimental Modal Analysis

The updating of the SFE model relies heavily on the experimentally measured data. In this work, the test structure was tested by using experimental modal analysis to measure the natural frequencies and mode shapes. The test was performed under free boundary conditions as demonstrated in [17]. The experimental configuration was designed as to the finite element model. Figure 4 illustrates the test configuration of the test structure in which three 10mV/g uniaxial accelerometer and a 21.65mV/N impact hammer were used in the measurement. Leuven Measurement System was used to acquire the data. The frequency bandwidth was set between 0 Hz to 512 Hz.

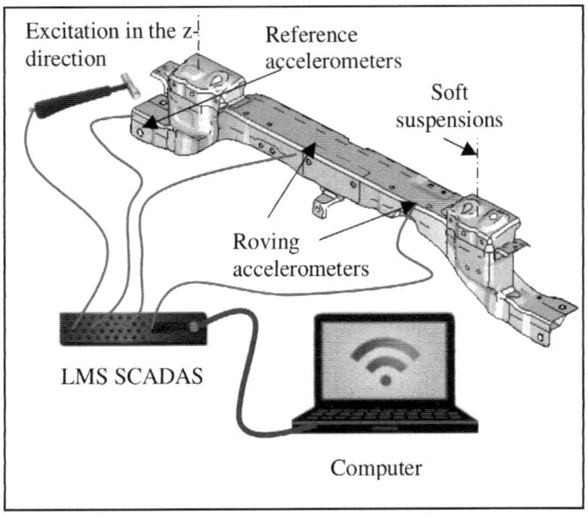

Figure 4. The experimental modal analysis configuration of the test structure

There are eight modes identified within the frequency of interest. The damping values of the modes are also taken into account for FRF synthesis purposes. The natural frequencies and damping values of the structure are tabulated in Table 1. The structure can be considered as a lightly damped structure because the average damping value of each mode is only 0.49 per cent, which the first mode is the highest with 0.84 per cent.

Table 1. The natural frequencies and damping values of the test structure

Mode	Frequency (Hz)	Damping (%)
1	98.78	0.8490
2	162.12	0.5129
3	201.37	0.1553
4	217.89	0.5039
5	302.15	0.5025
6	328.94	0.6566
7	382.65	0.3831
8	483.46	0.4383

4. Model Updating Method

The purpose of updating the SFE model was to improve the accuracy of the model by reconciling certain parameters in the light of the experimental data. The sensitivity analysis can be performed as follows

$$\mathbf{S} = \Phi_i^T \left[\frac{\delta \mathbf{K}}{\delta \theta_j} - \omega_i \frac{\delta \mathbf{M}}{\delta \theta_j} \right] \Phi_i \tag{1}$$

where matrix **S** is the sensitivity matrix, Φ, ω and θ are the eigenvector, eigenvalue and parameter respectively. The j is the j-th parameter and i represents the i-th eigenvalue. The sensitivity analysis is used to quantify the impact of parameters used in modelling on the predicted responses which are the natural frequencies and mode shapes.

Before performing the modal updating method, the accuracy of the predicted mode shapes of the SFE model is quantified using Modal assurance criterion (MAC) analysis. The detailed information on the MAC analysis can be found in [21]. In this work, the parameters of the SFE model were reconciled by referring to the MAC values. This is because only the accurate mode shapes are required for mode expansion purposes. Therefore, the objective function containing only the MAC number is defined in order to reduce the errors between the FE model and EMA mode shapes.

MSC NASTRAN Solution 200 adopted by [22] was used for the updating process and a total of eight mode shapes was employed. The comparison between the EMA and updated SFE mode shapes of the SFE model is presented in Table 2. The MAC values are also included in the table.

Table 2. The comparisons between EMA and updated SFE mode shapes

Mode	EMA Mode Shapes	Updated SFE Mode Shapes	MAC Percentage
1			80.00 %
2	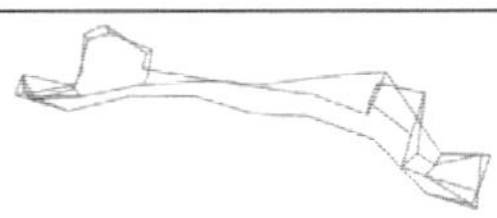	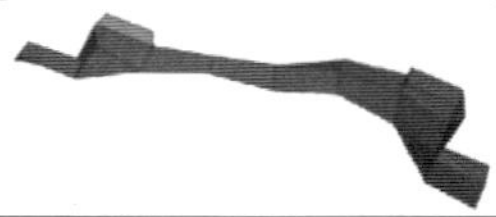	83.90%
3	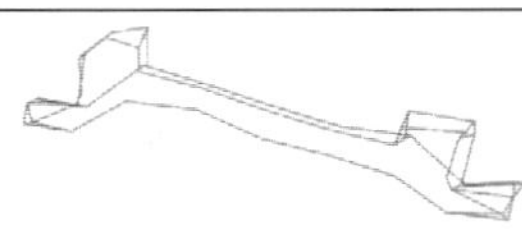	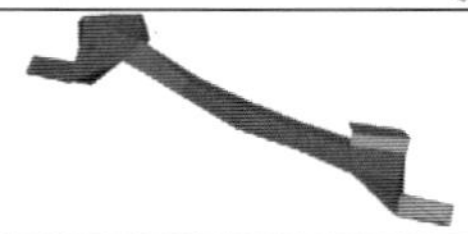	80.90%
4	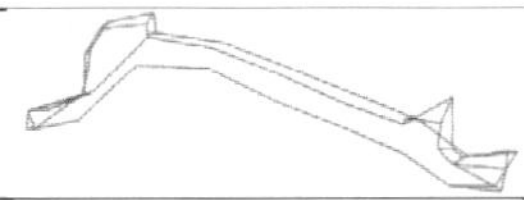		70.60%
5	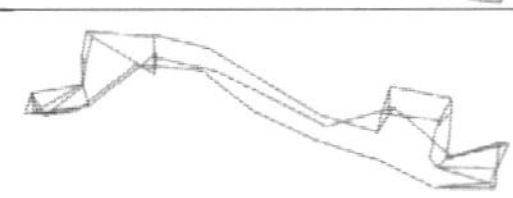		89.60%
6	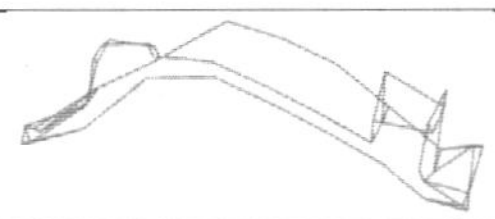		72.70%
7	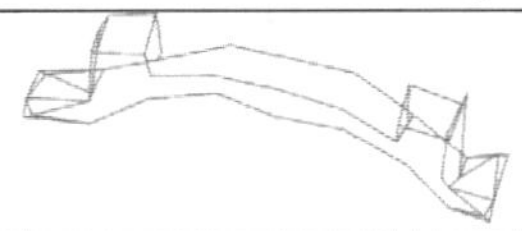	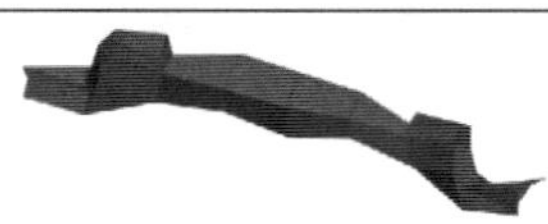	60.10%
8	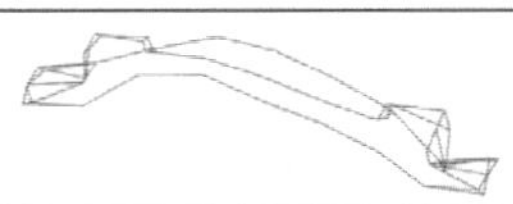		62.40%

Table 2 indicates that the average MAC value between the EMA and updated SFE model mode shapes is 75.02 per cent. The value is within an acceptable range, considering the complexity of the structure [21]. Meanwhile, the 7th mode records the lowest MAC value of 60.10 per cent. The lowest record might be because of the incapability of the developed SFE model of accurately simulating the local deformations of the structure. However, based on engineering judgement and observation, the correlation of the mode shapes between both models is relatively strong.

5. Mode Expansion of the Experimental Model

The purpose of the mode expansion of the EMA mode shapes is to obtain the unmeasured rotational data in the light of the updated SFE mode shapes[23]. In this work, the unmeasured rotational modal vectors in the experimental model were expanded by using the system equivalent reduction expansion process (SEREP) method[24]. The expansion process was carried out based on the modal vectors of the updated SFE model. For this method, the expended experimental modal vector with a full set of DOFs, $\mathbf{E_n}$ is expressed in the form of:

$$\mathbf{E}_n = \begin{bmatrix} \mathbf{E}_a \\ \mathbf{E}_d \end{bmatrix} = \mathbf{T}_u \mathbf{E}_a = [\mathbf{U}_n \mathbf{U}_a]^g \mathbf{E}_a = \begin{bmatrix} \mathbf{U}_a \\ \mathbf{U}_d \end{bmatrix} \mathbf{U}_a{}^g \mathbf{E}_a \tag{2}$$

where **E**,**U** and **T** are experimental modal vectors, analytical modal and transformation matrices. Subscripts a,d and u indicate tested set of experimental DOFs, deleted set of DOFs and modal vectors respectively. Table 3 presents the mode shapes comparisons between the EMA and expanded experimental model. The accuracy of the expanded mode shapes is quantified using MAC analysis.

Table 3. The comparison between EMA and expanded mode shapes

Mode	EMA Mode Shapes	Expanded Mode Shapes	MAC Percentage
1			82.70 %
2	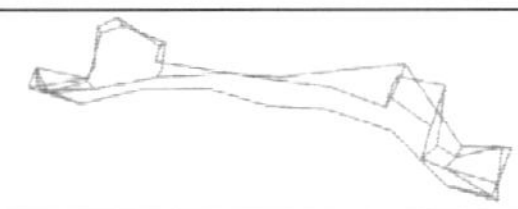	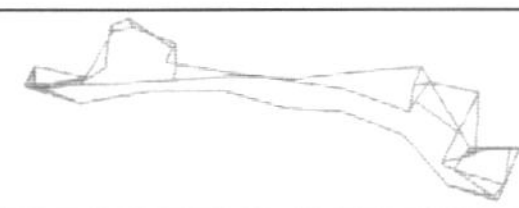	88.70%
3	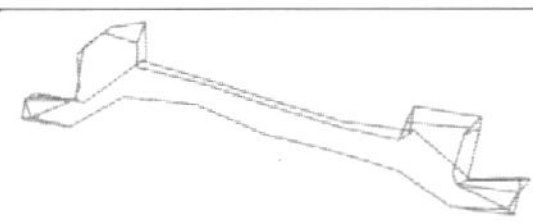	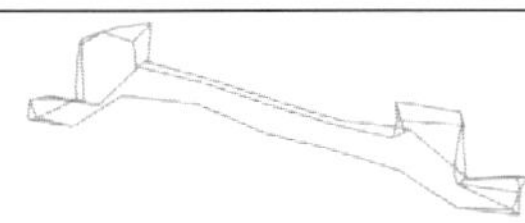	80.90%
4	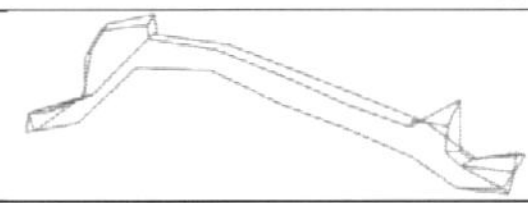	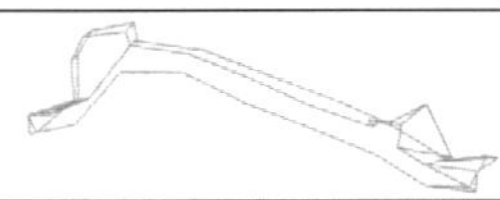	79.60%
5		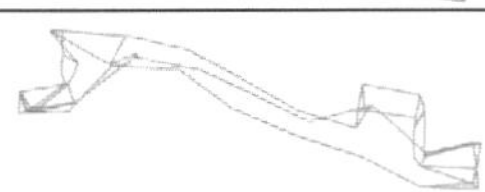	89.60%
6	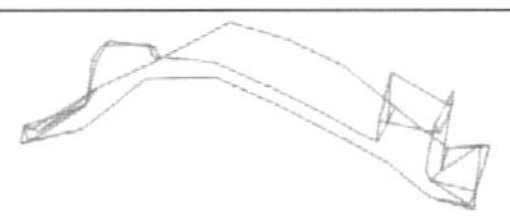	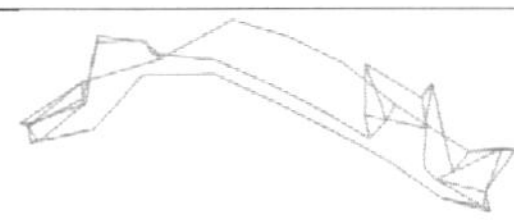	83.90%
7	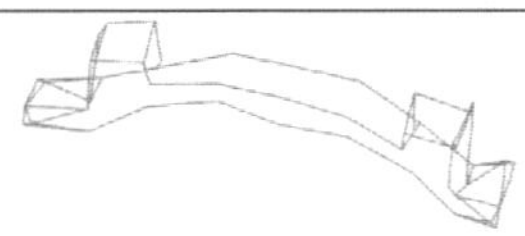	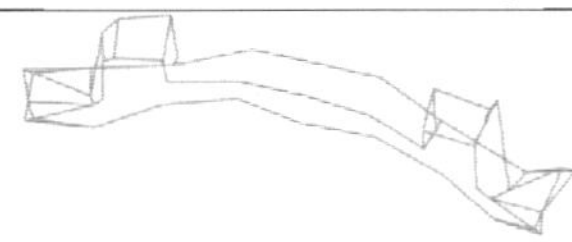	77.50%
8	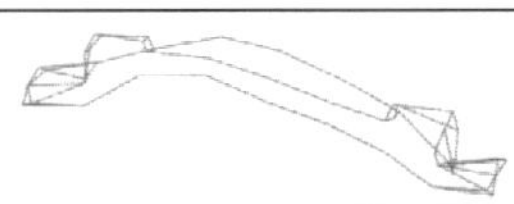	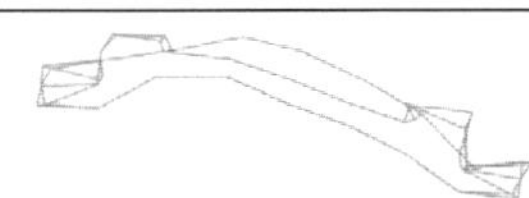	90.80%

From Table 3, it was found that there is a high correlation between the EMA and expanded mode shapes. The average MAC value of the whole mode is 84.18 per cent. The 8th mode of the expanded model records the highest correlation with EMA with MAC value of 90.80 per cent. Therefore, the use of the expanded model in predicting the dynamic behaviour of the physical front-end module is reliable. The 3D view of MAC comparison between EMA and expanded mode shapes presented in Figure 5 also indicates that there are no swapped modes between the two models.

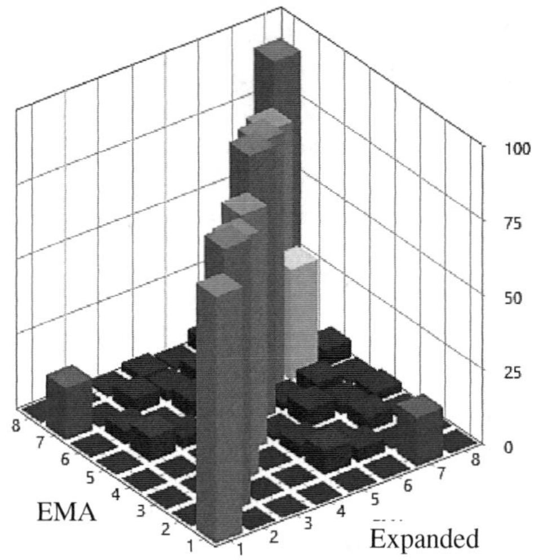

Figure 5. 3D view of MAC comparisons between EMA and expanded mode shapes

6. FRF Synthesis of the Expanded Experimental Model

The objective of the FRF synthesis of the expanded model was to obtain the FRF based on the modal model. The process was carried out for validation purposes. The FRF synthesis method does not require a matrix inversion of a potentially large dynamic stiffness matrix. Only one eigenvalue analysis is needed to determine the natural frequencies and mode shapes[12]. Therefore only one has to be synthesised the FRFs of the DOFs and frequencies of interest. In this work, the x and z-axis translational FRFs at the reference point of the expanded experimental model were derived. The y-axis FRF was not included in the validation process because the sensor cannot be mounted at the reference point of the structure. For this method, the synthesised FRF matrix H_{syn} ($\boldsymbol{\omega_k}$) and mode shapes are expressed in the form of:

$$\mathrm{H_{syn}}(\omega_k) = \sum_{i=1}^{N} \frac{\{\emptyset\}_i \{\emptyset\}_i^T}{\left(\omega_{n_i}^2 - \omega_k^2\right) + j2\xi_i \omega_k \omega_{n_i}} \tag{3}$$

where N represents the number of calculated modes, $\{\emptyset\}_i$ represents the ith mass normalised mode shapes, ω_{n_i} represents ith natural frequency and ξ_i represents the ith modal damping ratio. Figures 6 and 7 show the comparisons between the x and z-axis expanded model FRF (orange) and that of EMA at the reference point.

IOP Conf. Series: Materials Science and Engineering **807** (2020) 012045 doi:10.1088/1757-899X/807/1/012045

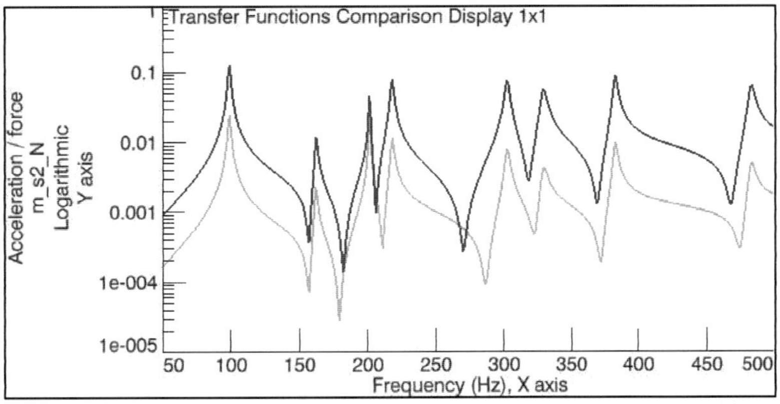

Figure 6. Comparisons between x-axis expanded model (orange) and EMA (blue) FRF at the reference point

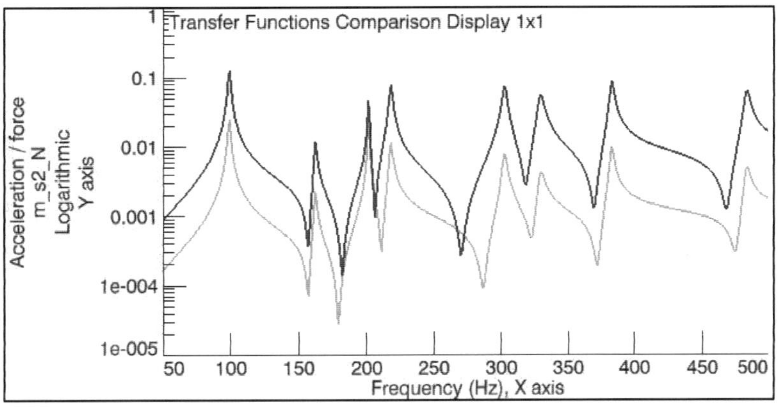

Figure 7. Comparisons between z-axis expanded model (orange) and EMA (blue) FRF at the reference point

Figures 6 and 7 clearly indicate that the resonance frequencies are well matched with the EMA FRF data. In addition, the anti-resonance frequencies have recorded minor discrepancies in the frequency range of 250-300 Hz. The discrepancies observed may be due to the effect of residual modes in the FRF synthesis of the expanded model [25] . On the other hand, the amplitudes of the synthesised FRF are lower than that of the EMA FRF. The low amplitudes may be due to the normalization between the measured and SFE model modal vector during the expansion process. However, based on the pattern of FRFs calculated for the resonances as well as anti-resonances, the SFE model is found to be in good agreement with the experimental model. Since there is a strong correlation between the synthesised and EMA FRFs, the rotational FRFs are then derived and presented in Figures 8 and 9.

Figure 8 presents the FRFs of the force excitation and rotational responses in x,y and z axis , while Figure 9 presents the FRFs of the moment excitation and rotational response in x,y and z axis. In this work, the validation of the rotational FRFs was not performed because the FRFs were not measured during the activities of EMA due to rotational accelerometers were unavailable. From the figures, it can be clearly seen that there are 8 visible resonance peaks appeared in the FRFs. This indicates that each mode experiences rotational acceleration during resonance. It is worth noting that the rotational resonance might introduce damage and unwanted vibration to the structure due to the external forces or moment excitations. Therefore, the rotational responses must be definitely considered in the system design.

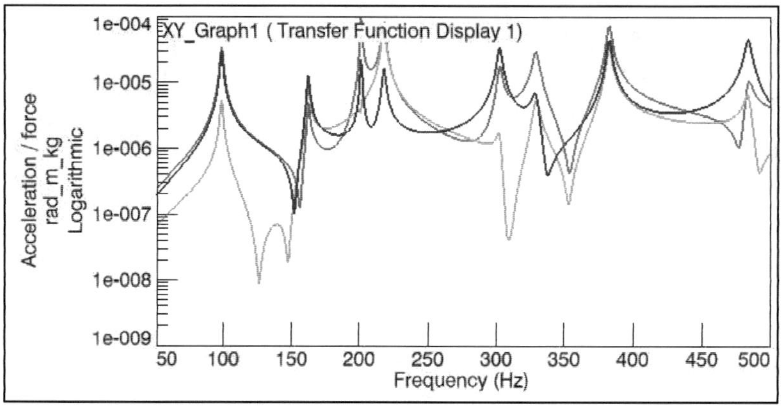

Figure 8. Force x-rotational x (blue), force y-rotational y (orange) and force z-rotational z (green) FRFs.

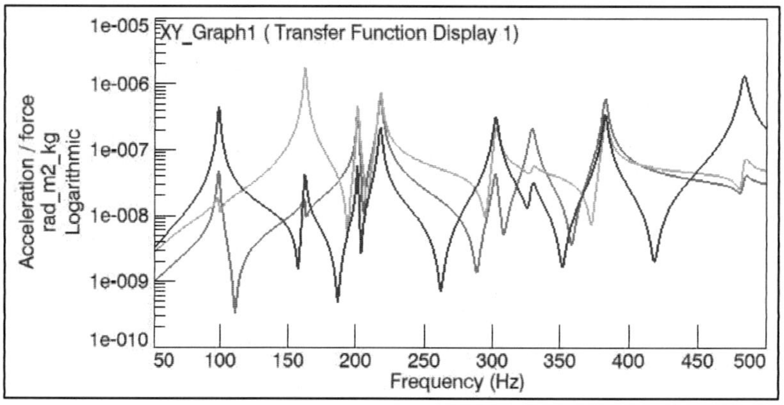

Figure 9. Moment x-rotational x (blue), moment y-rotational y (orange) and moment z-rotational z (green) FRFs.

7. Conclusions

The rotational FRFs estimation approach based on the simplified finite element (SFE) model, model updating, mode expansion and FRF synthesis method has been proposed and presented. The proposed approach has been demonstrated on a car front-end module structure. It was found that the rotational FRFs comprising force and moment excitations have been successfully derived from the expanded experimental model. This finding reveals that the approach used is capable of significantly minimising the heavy reliance on the rotational FRFs measurement, which is very difficult to be performed. The approach used in this research could be employed for other dynamic investigations without a dramatic decline in efficiency and accuracy. Although the rotational responses are less popular in most structural dynamics cases, indeed they are very important especially in the field of experimental dynamic substructuring (modal and frequency based) and structural modification methods.

Acknowledgments

The authors would like to express their sincere gratitude to the Malaysian Ministry of Higher Education (MOHE) for granting financial support for this research through the Geran Inisiatif Penyeliaan(GIP) 600-IRMI 5/3/GIP (012/2019).

References

[1] Law M, Rentzsch H and Ihlenfeldt S 2015 Development of a Dynamic Substructuring Framework to Facilitate in Situ Machining Solutions Using Mobile Machine Tools *Procedia Manufacturing***1** 756–67

[2] Nicgorski D and Avitabile P 2010 Experimental issues related to frequency response function measurements for frequency-based substructuring *Mechanical Systems and Signal Processing***24** 1324–37

[3] Van Der Seijs M V., De Klerk D and Rixen D J 2016 General framework for transfer path analysis: History, theory and classification of techniques *Mechanical Systems and Signal Processing***68–69** 217–44

[4] Shah M A S A, Yunus M A, Rani M N A and Ayub M A 2019 Structural modification and evaluation using finite element reconciliation method Structural Modification and Evaluation Using Finite Element Reconciliation Method **020007**

[5] Liu Z, Li W and Ouyang H 2016 Structural Modifications for Torsional Vibration Control of Shafting Systems Based on Torsional Receptances *Shock and Vibration***2016**

[6] Montalv D 2004 Estimation of the rotational terms of the dynamic response matrix *Shock and Vibration***11** 333–50

[7] Drozg A, Rogelj J, Čepon G and Boltežar M 2018 On the performance of direct piezoelectric rotational accelerometers in experimental structural dynamics *Measurement: Journal of the International Measurement Confederation***127** 292–8

[8] Mottershead J E, Ghandchi Tehrani M, Stancioiu D, James S and Shahverdi H 2006 Structural modification of a helicopter tailcone *Journal of Sound and Vibration***298** 366–84

[9] Konjerla K C 2016 FRF Based Experimental – Analytical Dynamic Substructuring Using Transmission Simulator

[10] Dumont M and Kinsley N 2016 Rotational Accelerometers and Their Usage in Investigating Shaker Head Rotations *Sensor and Instrumentation***5** 85–92

[11] Drozg A, Čepon G and Boltežar M 2018 Full-degrees-of-freedom frequency based substructuring *Mechanical Systems and Signal Processing***98** 570–9

[12] Gibbons T J, Öztürk E and Sims N D 2018 Rotational degree-of-freedom synthesis: An optimised finite difference method for non-exact data *Journal of Sound and Vibration***412** 207–21

[13] Wan Iskandar Mirza W I I, Abdul Rani M N, Yunus M A, Sani M S M and Mohd Zin M S 2019 The Prediction of the Dynamic Behaviour of a Structure Using Model Updating and Frequency Based Substructuring *IOP Conference Series: Materials Science and Engineering***506**

[14] Wan Iskandar Mirza W I ., Abdul Rani M ., Othman M H, Kasolang S and Yunus M A 2016 Reduced order model for model updating of a jointed structure *Journal of Engineering and Applied Sciences***11** 2383–6

[15] Omar R, Rani M N A, Yunus M A, Mirza W I I W I and Zin M S M 2018 Efficient finite element modelling for the investigation of the dynamic behaviour of a structure with bolted joints *AIP Conference Proceedings***020082** 020082

[16] Zin M S M, Rani M N A, Yunus M A, Sani M S M, Wan Iskandar Mirza W I I and Mat Isa A A 2018 Frequency response function (FRF) based updating of a laser spot welded structure *AIP Conference Proceedings***020055** 020055

[17] Wan Iskandar Mirza W I ., Abdul Rani M N, Yunus M A, Ayub M A, Sani M S M and Mohd Zin M S 2019 Frequency Based Substructuringfor Structure with Double Bolted Joints: A Case Study *International Journal of Automotive and Mechanical Engineering***16** 6188–99

[18] Abdul Rani M N, Kasolang S, Othman M H, Yunus M A and Wan Iskandar Mirza W I I 2016 Finite Element Modelling and Modal Based Up- Dating for The Dynamic Behaviour of a Laser Spot Welded Structure *ICSV 2016 - 23rd International Congress on Sound and Vibration: From Ancient to Modern Acoustics* pp 1–8

[19] Abdullah N A Z, Sani M S M, Husain N A, Rahman M M and Zaman I 2017 Dynamics properties of a Go-kart chassis structure and its prediction improvement using model updating approach *International Journal of Automotive and Mechanical Engineering***14** 3887–97

[20] Fouzi M S M, Jelani K M, Nazri N A and Sani M S M 2018 Finite Element Modelling and

Updating of Welded Thin-Walled Beam *International Journal of Automotive and Mechanical Engineering* **15** 5874–89

[21] Pastor M, Binda M and Harčarik T 2012 Modal assurance criterion *Procedia Engineering* **48** 543–8

[22] Mohd Zin M S, Abdul Rani M N, Yunus M A, Wan Iskandar Mirza W I I, Mat Isa A A and Mohamed Z 2017 Modal and FRF based updating methods for the investigation of the dynamic behaviour of a plate *Journal of Mechanical Engineering* **SI 4** 175–89

[23] Avitabile P and O'Callahan J 2003 Frequency response function expansion for unmeasured translation and rotation DOFS for impedance modelling applications *Mechanical Systems and Signal Processing* **17** 723–45

[24] Sastry C V S, Mahapatra D R, Gopalakrishnan S and Ramamurthy T S 2003 An iterative system equivalent reduction expansion process for extraction of high frequency response from reduced order finite element model *Computer Methods in Applied Mechanics and Engineering* **192** 1821–40

[25] Wan Iskandar Mirza W I ., Abdul Rani M N, Yunus M A, Omar R and Mohd Zin M S 1394 Alternative Scheme for Frequency Response Function Measurement of Experimental-Analytical Dynamic Substructuring *Journal of Mechanical Engineering and Science* **13** 4946–57

IOP Conf. Series: Materials Science and Engineering **807** (2020) 012046 doi:10.1088/1757-899X/807/1/012046

Coating of Polyaniline-Titanium Dioxide (PAni-TiO_2) Composite for Corrosion Protection in Low Carbon Steel

E Sulistyaningsih[1], **N Lestari**[2]

[1] Industrial Engineering Department, Faculty of Industrial Technology, Institut Sains & Teknologi AKPRIND Yogyakarta, Indonesia
[2] Mechanical Engineering Department, Faculty of Industrial Technology, Institut Sains & Teknologi AKPRIND Yogyakarta, Indonesia

Corresponding e-mail: sulistyaningsih@akprind.ac.id

Abstract. Carbon Steel is used as an industrial component and an electronic component that has a lifetime or a certain lifetime. Steel is not a rust-resistant material, so it can corrode over time, especially Indonesia is a tropical country that has high rainfall and humidity. Almost all industrial sectors have problems with corrosion. Problems that arise can be in the form of damage, age of use of goods that do not meet expectations until the security factor is inadequate. This study aims to investigate the inhibiting efficiency of coating on Low Carbon Steel with semiconductor materials and morphology changes on Low Carbon Steel. The semiconductor material used is a combination of PAni-TiO_2. Coating of PAni-TiO_2 composite on low carbon steel was carried out by electrodeposition method on acidic medium. The corrosion process is carried out with acid and salt media. Effectiveness is based on inhibition efficiency. The highest inhibition efficiency was produced on plate 3 with a composition of 10^{-2} M PAni-10^{-3} TiO_2 with a corrosion rate 0.003560 mpy. Uncoated specimens have a highest corrosion rate of 0.007359 mpy, so it suffers the most damage. From the results of visual observations conducted samples undergoing corrosion in the form of uniform corrosion.

1. Introduction

Steel is a metal (Fe) alloy with carbon as the main mixing material. In general, steel used as an industrial component and an electronic component that has a lifetime or a certain lifetime. Steel has properties that vary from the softest and easily made to the hardest and sharpest. One of the steels used in making industrial machinery is low carbon steel. However, Carbon steel is not a rust-resistant material so it can corrode over time. Corrosion is a serious problem and detrimental to the industrial sector in Indonesia, because Indonesia is a tropical country that has high rainfall and humidity. Corrosion is a decrease in the quality of metals due to electrochemical reactions with the environment that is directly related to opened air, often referred to as atmospheric corrosion [1]. Therefore, materials made of metal or alloys can be damaged due to corrosion [2]. Because corrosion is a big problem for buildings and equipment that uses metal base materials such as buildings, bridges, machinery, pipes, cars, ships, etc., so a method is needed for inhibiting corrosion.

Various methods have been carried out to reduce corrosion rates, one of which is by using an inhibitor. The use of corrosion inhibitors is intended to change the interface between metals and the corrosion environment by isolating one from the other. The inhibitors used are conductive polymers.

IOP Conf. Series: Materials Science and Engineering **807** (2020) 012046 doi:10.1088/1757-899X/807/1/012046

The use of conductive polymers as corrosion protective materials is one alternative that is being widely studied because conductive polymers have special features compared to other types of polymers.

The most popular conductive polymer is polyaniline. One study on the use of PAni is a study conducted by Yi et al. [3]. Yi et al reported that the inhibition of steel corrosion in 0.5 M HCl of efficiency increased with increasing concentration of PAni compounds. The inhibition efficiency obtained was 92.3% with a concentration of 40 mg / L. TiO_2 is an n-type semiconductor that can inhibit the reduction of oxygen in the mechanism of corrosion. Titania nanostructures in thin film layers act as important nanomaterials from a technological point of view in maintaining properties such as chemical and photoelectrochemical stability. TiO_2 has roles in several industries, including composite coatings for self-cleaning windows, electrolysis of water for energy conversion and anti-material corrosion.

Based on the ability possessed by polyaniline and TiO_2, the manufacture of polyaniline composites (PANi) -TiO_2 is expected to be an alternative corrosion protection material. The use of polyaniline as a composite matrix phase with TiO_2 as a fill phase can slow the rate of corrosion of metals with HCl as corrosion media [4]. Based on these problems, we need a study of the ability of composite polyaniline PAni-TiO_2 as carbon steel coating material, so this research will synthesize PAni-TiO_2 composites by electrodeposition method. PAni-TiO_2 composites are expected to have a high ability to suppress corrosion rates because they combine PAni which suppresses the oxidation rate and TiO_2 which suppresses the rate of reduction. This study varied the composition of nanocomposite PAni-TiO_2 to determine the effect of nanocomposite composition on the effectiveness of inhibitors. Material characterization was carried out before and after corrosion in the form of characterization of visual observations and pore morphology using Scanning Electron Microscope (SEM), while weight loss testing for analyzing Inhibition efficiency.

2. Method

Carbon steel protection is carried out by coating the surface of steel with PAni-TiO_2 composites through an electrodeposition process. This electrodeposition process is based on Abaci & Ness's research [5]. Steel plate samples that have been prepared, coated with PAni-TiO_2 composites are electrochemically carried out by electrode position on the sample for 10 minutes. The sample made amounted to 4 pieces, of which 1 piece was a sample uncoated, while 4 coated samples with composition variations of PAni-TiO_2. The electrolyte solution used is a solution of Sulfuric Acid and LiCl and set up in pH 2. After the sample is electrode positioned, the sample is weighed and recorded as the initial weight. Comparison of the composition of PAni-TiO_2 is presented in Table 1.

Table 1. Composition of PAni-TiO_2 for coating low carbon steel

Samples	PAni (M)	TiO_2 (M)	H_2SO_4
Uncoated Plate	0	0	250 mL
Steel Plate1	10^{-4}	10^{-1}	250 mL
Steel Plate 2	10^{-3}	10^{-2}	250 mL
Steel Plate 3	10^{-2}	10^{-3}	250 mL
Steel Plate 4	10^{-1}	10^{-4}	250 mL

The corrosion process carried out by immersing specimens (coated and uncoated) in sulfuric acid solutions and NaCl for 12 days. The weight of each specimen before and after immersion measured. The Inhibition Efficiency (IE) was calculated using the expression (loss weight method) [6]:

$$IE\% = \left[\frac{W_1 - W_2}{W_2}\right] x100\% \quad (1)$$

Where W_1 and W_2 are the corrosion rates in the absence and presence of the inhibitor, respectively

3. Results and Discussion

Steel plate coating is carried out by electrodeposition when the electrodeposition process, at the anode and the cathode changes due to the potential there is a direct current of electricity. Solution electrolytes used in this study are sulfuric acid and LiCl a solution. Thin Layer of electrodeposition results is affected by some variables including meeting current, temperature, time of deposition, stirring, the acidity of solution (pH), solution concentration, electric potential, and substrate. Synthesis of Polyaniline is carried out by polymerization electrochemically produce products in the form of films covering the steel plate. Use of H_2SO_4 acid as a solution electrolyte also aim to accelerate polymerization. The addition of acid to the polymerization process will form aniline, acid which is more soluble in solution polymerization, creating an acidic atmosphere and acts as a dopant for polyaniline so many ions are formed SO_4^{2-} that can increase its electrical conductivity [7]. Visual comparison of steel plates without and with inhibitors ready for use presented in Figure 1 (a) to (e).

Figure 1 (a). Uncoated Plate

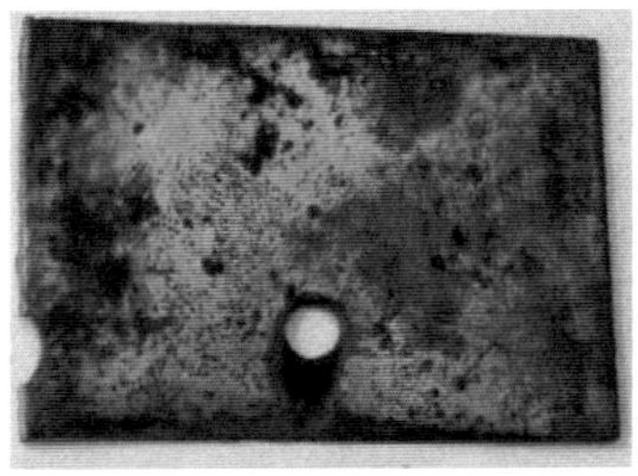

Figure 2 (b). Plate 1

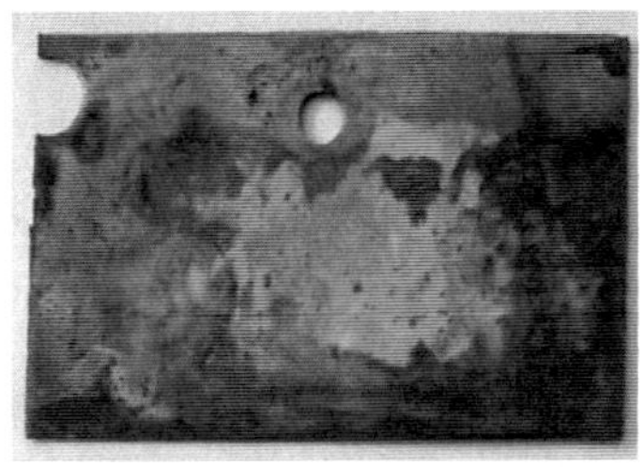

Figure 3 (c). Plate 2

Figure 4 (d). Plate 3

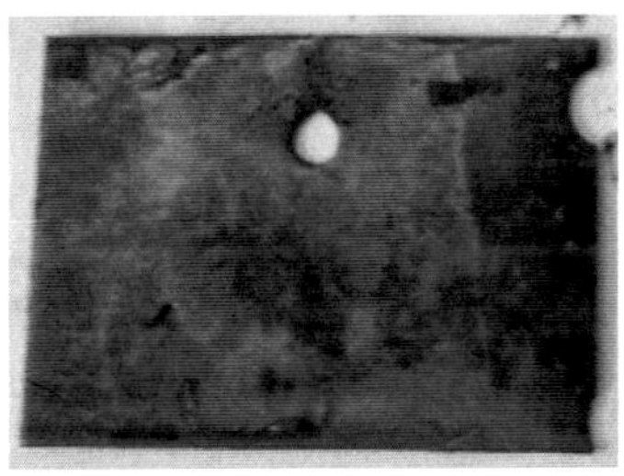

Figure 5 (e). Plate 4

Based on observations, the samples that were the most severely corroded were Plat samples without inhibitors. This is because the uncoated plate, salt, and acid directly damage the surface. Additives in the form of organic compounds have an effect if the material adsorbs on the metal surface and does not have an affinity for water, resulting in a strong inhibitory effect. If organic additives adsorb metal surfaces and have an affinity for water, it will speed up the electrodeposition process. Therefore, the addition of organic compounds is more likely to have an influence in terms of accelerating the formation of deposits resulting in thicker deposits.

There are two factors influence the conductivity of PAni + TiO_2 composites. First, mobility in the composite depends on the concentration of H^+ and Cl^- which decreases dramatically with a decrease in PAni Concentration. The second factor is the length of the chain PAni. At lower PAni concentrations, conduction requires higher activation energy than higher concentrations, which can be attributed to an increase in energy barriers between different sites doing content as the content decreases [8].

Corrosion protection by layers depends on three aspects: (a) water uptake from the layers, (b) water transport in layers and (c) water accessibility to the layer/substrate interface [9]. One way to combat corrosion in an acidic environment is the application of corrosion inhibitors or coatings. This is evident from the results of research that, specimens uncoated has a corrosion rate of 0.007359 mpy. Corrosion is also inhibited by physical barriers in the presence of polymers on metal surfaces that inhibit access by corrosion of species [10]. Some reports suggest that hybrid organic-inorganic composite coatings can increase the corrosion resistance of metal substrates such as iron, aluminum, and magnesium. The percentage of inhibition efficiency is presented in Figure 2.

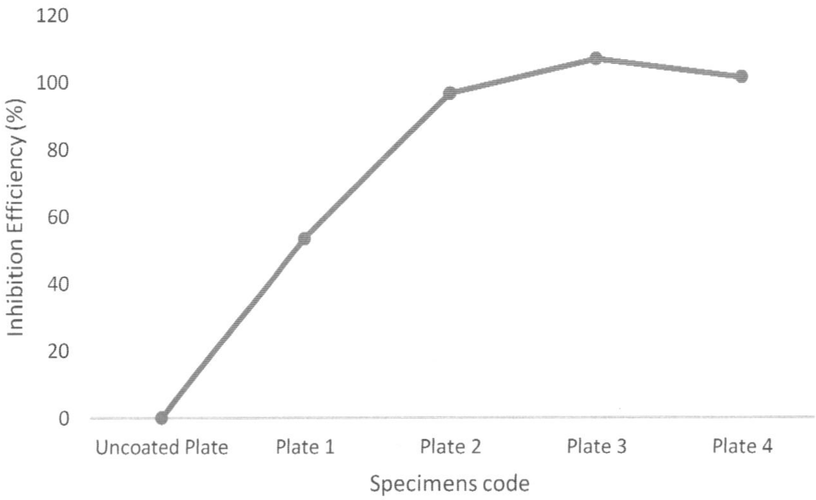

Figure 2. Inhibition Efficiency of Plates

After obtaining the Inhibition Efficiency of plates, each plate was characterized using SEM to see the morphology of the plate surface due to corrosion. PAni has various structures that can be granular, or nanofibers, nanotubes, nanospheres, microspheres, or flakes. the success of PANI / TiO_2 synthesis is characterized by granular and surface debris, in line with abaci's research [5] . Metalic cations interact with PAni imine nitrogen atoms and change the morphology and structure of PAni into compact groups that can greatly reduce the corrosion rate [11]. The PAni redox reaction can be increased significantly by adding metal oxide nanostructured materials. In the electrochemical method, PAni is obtained in the form of ordered thin films on the electrode surface. The morphology of specimens/plates after corrosion is shown in Figure 3 (a) to (e).

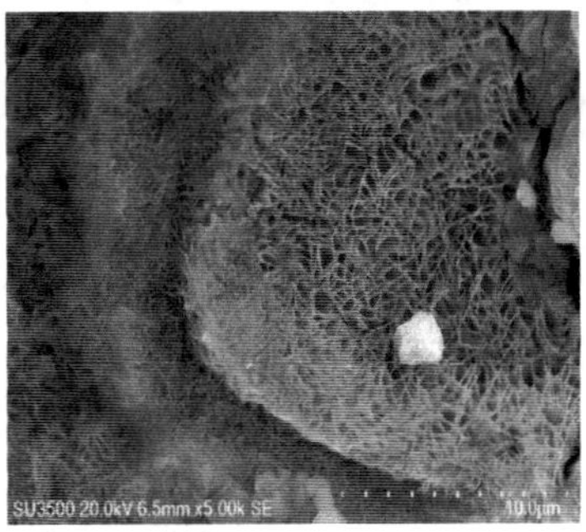

Figure 3 (a). Uncoated Plate

Figure 3 (b). Plate 1

Figure 3 (c). Plate 2

Figure 3 (d). Plate 3

Figure 3 (e). Plate 4

Based on observations, the surface morphology of specimens without inhibitors appeared to have many holes compared to specimens with inhibitors. Corrosion that occurs is homogeneous corrosion. Homogeneous corrosion is caused by chemical reactions that attack uniformly across metal surfaces or in large areas of metal. The attacked metal will become thinner and brittle [12].

4. Conclusions

PAni-TiO_2 formed on the surface of specimens are granular and flakes. Based on visual observations conducted samples undergoing corrosion in the form of uniform corrosion. The addition of PAni-TiO_2 on specimens with the electrodeposition method was effective for inhibiting the corrosion rate. Uncoated specimens have a corrosion rate of 0.007359 mpy. The highest inhibition efficiency was produced on plate 3 with a composition of 10^{-2} M PAni-10^{-3} TiO_2 with a corrosion rate 0.003560 mpy . Uncoated specimens have a corrosion rate of 0.007359 mpy. Visually, PAni-TiO_2 coated specimens in the surface morphology are denser so they are more protected from corrosion. Further research needs to be done to improve the efficiency of PAni-TiO_2 coating inhibition

Acknowledgments

The authors gratefully acknowledge the Ministry of Research Technology and Higher Education of The Republic of Indonesia for providing financial support through the beginner lecturer research scheme.

References

[1] J. Trethewey, K.R., Chambedain,1991 *Korosi Untuk Mahasiswa dan Rekayasawan*
[2] K. J. Pattireuw, F. A. Rauf, R. Lumintang, T. Mesin, U. Sam, and R. Manado, 2013 "Analisis Laju Korosi Pada Baja Karbon Dengan Menggunakan Air Laut Dan H2 SO4,"
[3] Y. Yi, G. Liu, Z. Jin, and D. Feng, 2013 vol. **8**, pp. 3540–3550
[4] A. Permana 2012 *Institut Teknologi Sepuluh Nopember*
[5] Abaci, Serdar; Nessark, Belkacem 2014"Characterization and corrosion protection properties of composite material (PANI+TiO2) coatings on A304"
[6] S. A. Kumar, A. Sankar, and S. Rameshkumar 2013 vol. **2**, no. 9
[7] M. N. Chomari, D. H. Kusumawati, and U. N. Surabaya 2012 vol. **1**, no. 1, pp. 29–32
[8] M. Ibrahim, M. Bassil, T. Khoury, G. E. H. Moussa, M.E. Tahchi and P. Miele 2012 *Mater. Chem. Phys.*, vol. **133**, no. 2–3, pp. 1040–1049
[9] C. Peng, K. Chang, C. Weng, M. Lai, and C. Hsu 2013 *Electrochim. Acta*, vol. **95**, pp. 192–199
[10] P. Li, T. C. Tan, and J. Y. U. Lee 1997 vol. **88** no. March, pp. 237–242
[11] S. Sathiyanarayanan, V. Karpakam, K. Kamaraj, S. Muthukrishnan, and G. Venkatachari 2010 *Surf. Coat. Technol.*, vol. **204**, no. 9–10, pp. 1426–1431
[12] E. Sulistyaningsih and N. Lestari 2018 vol. **3**, no. 2, pp. 52–57

IOP Conf. Series: Materials Science and Engineering **807** (2020) 012047 doi:10.1088/1757-899X/807/1/012047

Design of Ergonomic Dust Collector Room for Coating Process Using Quality Function Deployment

Andrean Emaputra[1*], Taufik Hidayat[2], Argaditia Mawadati[1]

[1] Department of Industrial Engineering, Faculty of Industrial Technology, Institut Sains & Teknologi AKPRIND, Yogyakarta, Indonesia
[2] Department of Mechanical Engineering, Faculty of Industrial Technology, Institut Sains & Teknologi AKPRIND, Yogyakarta, Indonesia

Corresponding e-mail: andrean.emaputra@akprind.ac.id

Abstract. A coating process is an act of giving layers on a surface, for example, painting process, water-proofing process. The coating process needs more attention from enterprises because the processes could release dangerous particles that can be inhaled by workers. The dangerous particles could make the workers suffer from coughs, sneezing, allergic rhinitis, bronchitis, asthma, silicosis, and other decreased-lung functions. Therefore, the coating process should be done in ergonomic circumstances to avoid them from the bad effects of the un-ergonomic coating process. This research aims to find the dust collector room design meeting the ergonomic requirements for the processes. This research used Quality Function Deployment (QFD) to discover the employee needs and the requirements dust collector room should have. The employee and enterprises needs were obtained using group discussion among them working in sanitary product manufacturing made from sands, marble particles, and cement in Bantul, Yogyakarta, Indonesia. The three most important needs of the room are that the dust collector room should collect dust rapidly, localize dust spreading, and give better air circulation. The needs form the three important functional requirements the room should have, i.e., split-plastic wall, concrete wall, high and wide room. Therefore, the dust collector room should be designed based on the voice of the stakeholders to get healthy-working environment.

1. Introduction

A final product is made through many processes like conventional and non-conventional machining. One of the most frequent-used processes is a coating process. A coating process is an act of giving layers on a surface, for example, painting process, water-proofing process. The coating process needs more attention from enterprises because the processes could release dangerous particles that can be inhaled by workers. The dangerous particles could make the workers suffer from coughs, sneezing, allergic rhinitis, bronchitis, asthma, silicosis, and other decreased-lung functions. Dust could also bring bad effects to students' health in Malaysia [1].

Many studies have been conducted to reduce the dust spreading. The wettability of the surface has a significant effect on the dust removal completion [2]. Careful process control and a correct sampling phase in assessing the explosion hazard of dust are needed to prevent a dust explosion depended by variables that affect the fines content [3,4]. A dust explosion can propagate although there is very few dust, or no dust at all in the connected pipes so right placement of the explosion isolation device is very important to ensure that barrier is in place to stop the fireball, besides that the explosion isolation systems need to be validated and verified through large-scale testing [5]. A moving granular bed filter was developed and tested to get 82.0% - 99.8% filtering efficiencies [6,7]. Furthermore, a dust alarm system should be introduced to warn people about the dust level [8].

QFD has been used in many research area and countries especially in a room and build designs. The voice of the customer and the specifications' priority of the construction industries are met together to get the clients' satisfaction leveling up the chance of successful housing design in Nigeria, Iran, Taiwan, India, and Malaysia [9–13]. The specification variables of vessel engine room design are safety, vessel type, vessel's main dimensions, the arrangement of the engine room, material type, coating, vessel speed, ventilation, temperature, vessel tank capacity, mechanics of materials and pressure which are identified using QFD and DEMATEL techniques [14]. The QFD has been used for improving public transport quality, i.e., the second-class cabins quality of high-speed rails (HSRs) in China through some suggestions such as the cabins should have comfortable seat, sensory comfort, and proper luggage storage [15], comfort or not crowded bus in Belgrade [16], and the improvement of Kansai International Airport based on passengers' needs [17]. The workstation of the school workshop should meet the Malaysian design standard having the comfort standard, safety standard and broad working space [18]. The QFD is also used for designing prospective lodging industry [19]. It helps hospitals to get better patients' satisfaction and makes the hospital having higher competitive advantages [20,21]. Moreover, it also elaborates technology usage into education, for example, smart class, multimedia [22].

However, the ergonomic requirements of the dust collector room have not been identified. The identification of the room's specifications is very important because the coating process should be done in ergonomic circumstances so the workers could be avoided from the bad effects of the un-ergonomic coating process. This research aims to find the dust collector room design meeting the ergonomic requirements for the coating process using QFD.

2. Method

The research is done through several steps. First, the voice user of the room is identified. The need is identified through group discussion with an enterprises' manager and 5 workers working on coating proses, i.e. adding cement-layers and liquid-water proof to workpieces using a spray gun. The processes are done in a small-medium enterprise making sanitary products like a sink made from sands, marble particles, and cement in Bantul, Yogyakarta, Indonesia. The small-medium enterprise has been established since 2008. They are asked about what performances and features the dust collector room should have. Then, the answers are entered to the QFD or House of Quality (HOQ) in the customer-requirement column as shown in Figure 1 or in the demanded-quality column as shown in Figure 2. After that, the importance levels of the voice of the customers are assessed (use a scale of 1 to 10, with 10 indicating very important items [23]. Finally, the relative weights of the customer requirements are obtained to compare the degree of importance level among them.

Second, the functional requirements of the dust collector are established based on the voice of the user (customer). The functional requirements are investigated to fulfill the voice of the customer and place to the technical-requirement rows (see Figure 1) or to the quality-characteristic row (see Figure 2). Then the correlations between the functional requirements of the room are assessed (++ = strong positive correlation; + = positive correlation; ━ = negative correlation; ▼ = strong negative correlation [23]).

Third, the correlations between the voice of customer and the functional requirements are determined. The corelation clasification are (1) ⊙ = strong relationship = 9; (2) ○ = moderate relationship = 3; (3) △= weak relationship = 1 [23].

Fourth, the importance weights of the functional requirements of the room are calculated. It is calculated by multiply the value of any relationships shown in the column of the technical requirement by the relative weight of the customer requirement [23]. Finally, the relative weights of the functional requirements can be obtained.

Fifth, the specifications of the room derived from the QFD are implemented. Before the implementation, there was no partition for localizing dust spreading so dust goes elsewhere around the plant.

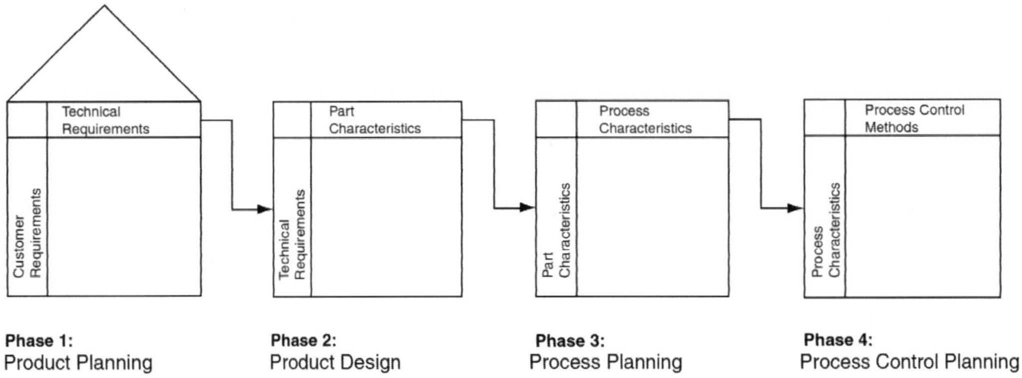

Figure 1. The four phases of the QFD [23].

3. Results and Discussion

The house of quality of the dust collector room identifies 10 needs that should be attached to it. There are demanded requirements having importance level that is more than 8 (see Figure 2). First, the room should absorb dust rapidly so it can reduce the chance of the dust inhalation by workers. It because the dust is dangerous to the humans [1]. Second, the room can prevent dust to spread to other locations, it is very crucial because if dust easily moves to other plant sites the dust could be inhaled by other employees who do not do the coating process. This effort is in line with the dust isolation system that has been introduced [5]. Third, the room can give better air circulation so the worker who does the painting proses could breathe fresh air. Fourth, the workpiece can easily be moved in the room as the workers need to paint other workpieces that have not been painted from the outside. Fifth, the worker can enter to and escape from the room easily without making much dust is going outside since the worker needs allowance for drinking, eating or going to a restroom.

Meanwhile, there are also needs having moderate importance levels. First, the room uses cheap materials because the enterprise could reduce investment cost. Second, the room is easy to be built so that the dust collector room could be made by the existing workers. Third, the room can be used in a long period of time, because it is predicted to be used for the next 4 years. Fourth, the room can decrease the noise level because of the noise resulted by compressors. It shows that the QFD could make a better workstation in an industry like in the lodging industry [19].

Based on the relationship between the demanded requirements and the functional requirements (see Figure 2), the relative weight of the functional requirements of the room can be obtained. There are some functional requirements of the room holding relative weight that is more than 7. First, split-plastic wall (see Figure 4) is very crucial to be attached to the room because it makes the workers and the workpieces go into and outside from the room easily and it prevents dust to scatter outside the room. Second, the concrete floor is also very beneficial for giving easiness movement of the wheeled-table (see Figure 6) and avoiding dust spreading to other sections of the plant. Third, high and wide walls (see Figure 3) makes air circulation goes well, hold a lot of workpieces with many different sizes from small to large. Fourth, fans and blowers (see Figure 3 to Figure 5) are used for pushing and sucking dust from inside the room. While the transparent and thick plastic wall could prevent dust from spreading out and enable the production process to be controlled directly outside the room. Fifth, foam-coated plywood wall is helpful to reduce the noise level.

Sixth, the least importance functional requirement is a mask used by the worker. If the fans and blowers can work properly then the workers do not need to use the mask. Since personal protective equipment is the last effort after controlling dust from the source. This is in line with controlling dust emissions in Beijing through mitigating combustion sources (power, industry, and vehicle exhaust) [24].

The coating process that is done in the ergonomic dust collector room is very beneficial for the enterprise and the workers. The enterprise could reduce a cost for health allowances since the workers can finish the task with minimum health complaints and it can increase the productivity because the workers are able to move the workpieces easily and rapidly. Moreover the workers have higher enthusiasm for doing the job because they are sure that the new dust collector room would bring excellent effect to their health.

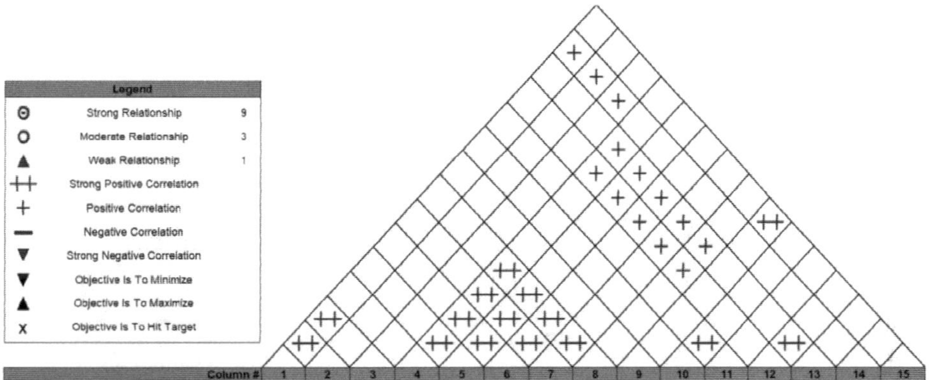

Row #	Max Relationship Value in Row	Relative Weight	Weight / Importance	Demanded Quality (a.k.a. "Customer Requirements" or "Whats") / Quality Characteristics (a.k.a. "Functional Requirements" or "Hows")	blower	fan	dust collector	plywood wall	foam-coated-plywood wall	tarpaulin wall	transparent and thick plastic wall	split-plastic wall	wheeled table	transparent roof (sunlight)	lamp	high wall	wide room	mask used by employee	concrete floor
				Column #	1	2	3	4	5	6	7	8	9	10	11	12	13	14	15
				Direction of Improvement: Minimize (▼), Maximize (▲), or Target (x)	▲	▲	▲	▲	▲	▲	▲	▲	▲	▲	▲	▲	▲	▲	x
1	9	11.9	10.0	the room can absorb dust rapidly	Θ	Θ	O	O	O	O	O	O				▲	▲		O
2	9	11.9	10.0	the room can prevent dust to spread to other locations	Θ	Θ	Θ	Θ	Θ	Θ	Θ	Θ				O	O		Θ
3	9	9.5	8.0	the production process can be seen from outside						Θ	Θ	Θ		Θ	Θ	O	O		
4	9	10.7	9.0	the workpiece can easily be moved in the room							O	Θ	Θ	O	O	Θ	Θ		Θ
5	9	10.7	9.0	the worker can enter to and escape from the room easily without making much dust is going outside							O	Θ	O	O	O	Θ	Θ		Θ
6	9	8.3	7.0	the room uses cheap materials				Θ	Θ	Θ	Θ	Θ		Θ				O	O
7	9	8.3	7.0	the room is easy to be manufactured	Θ	Θ	Θ	Θ	Θ	Θ	Θ	Θ	Θ	Θ	Θ	Θ	Θ		Θ
8	9	8.3	7.0	the room can be used in a long period of time	Θ	Θ	Θ	O	O	▲	▲	▲	O	O	Θ	O	O		Θ
9	9	8.3	7.0	the room can decrease the noise level				Θ	Θ	O	O	O				O	O		
10	9	11.9	10.0	the room can give better air circulation	Θ	Θ	Θ		O			O				Θ	Θ	O	O
				Max Relationship Value in Column	9	9	9	9	9	9	9	9	9	9	9	9	9	3	9
				Weight / Importance	471.4	471.4	400.0	392.9	428.6	411.9	476.2	640.5	228.6	325.0	300.0	501.2	501.2	60.7	546.4
				Relative Weight	7.7	7.7	6.5	6.4	7.0	6.7	7.7	10.4	3.7	5.3	4.9	8.1	8.1	1.0	8.9

Figure 2. The House of Quality of coating process room.

IOP Conf. Series: Materials Science and Engineering **807** (2020) 012047 doi:10.1088/1757-899X/807/1/012047

Figure 3. Plywood wall, tarpaulin wall, and transparent plastic wall are used as a partition for localizing dust.

Figure 4. Split-transparent plastic as a door for easy entering.

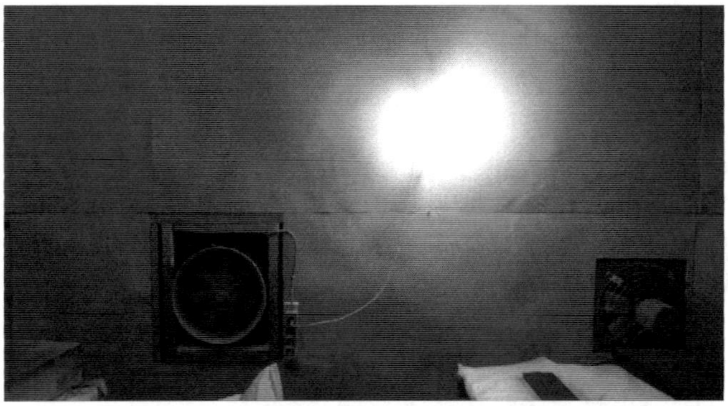

Figure 5. Blowers used for absorbing dust.

Figure 6. Tables equipped by small wheels could make the easy and fast movement of the workpieces.

4. Conclusions

The dust collector room should have abilities to collect dust rapidly, to localize dust spreading, to enable the supervisor to see the production process from the outside, and to transport the workpieces easily. Therefore, the dust collector room should be equipped by blowers, plywood wall, tarpaulin wall, transparent plastic wall, and wheeled tables.

The ergonomic dust collector room gives better insight between the stakeholders of the industry. The top management can level up the productivity and down grade the accident and health cost. The workers have higher time utilization since they can finish the planned job on scheduled by healthier working environment.

Acknowledgments
The authors gratefully acknowledge the Minister of Research, Technology and Higher Education of the Republic of Indonesia for providing financial support and a small-medium enterprise making sanitary products in Bantul, Yogyakarta, Indonesia as a research partner.

References

[1] Norbäck D, Hashim J H, Hashim Z, Cai G H, Sooria V, Ismail S A and Wieslander G 2017 Respiratory symptoms and fractional exhaled nitric oxide (FeNO) among students in Penang, Malaysia in relation to signs of dampness at school and fungal DNA in school dust *Sci. Total Environ.* **577** 148–54

[2] Yang Y, Zhuang D and Ding G 2019 Effect of surface wettability of fins on dust removal by condensate water *Int. J. Heat Mass Transf.* **130** 1260–71

[3] Marmo L, Riccio D and Danzi E 2017 Explosibility of metallic waste dusts *Process Saf. Environ. Prot.* **107** 69–80

[4] Ebadat V 2010 Dust explosion hazard assessment *J. Loss Prev. Process Ind.* **23** 907–12

[5] Taveau J 2017 Dust explosion propagation and isolation *J. Loss Prev. Process Ind.* **48** 320–30

[6] Wenzel B M, Porciúncula C B, Marcilio N R, Menegolla H B, Dornemann G M, Godinho M and Martins C B 2014 Filtration of dust in an intermittent moving granular bed filter : Performance and modeling *Sep. Purif. Technol.* **133** 108–19

[7] Chen Y S, Hsu C J, Hsiau S S and Ma S M 2017 Clean coal technology for removal dust using moving granular bed filter *Energy* **120** 441–9

[8] Schweitzer M D, Calzadilla A S, Salamo O, Sharifi A, Kumar N, Holt G, Campos M and Mirsaeidi M 2018 Lung health in era of climate change and dust storms *Environ. Res.* **163** 36–42

[9] John R, Smith A, Chotipanich S and Pitt M 2014 Awareness and Effectiveness of Quality Function Deployment (QFD) in Design and Build Projects in Nigeria *J. Facil. Manag.* **12** 72–88

[10] Moghimi V, Jusan M B M, Izadpanahi P and Mahdinejad J 2017 Incorporating User Values into Housing Design through Indirect User Participation using MEC-QFD Model *J. Build. Eng.* **9** 76–83

[11] Juan Y K, Hsing N P and Hsu Y H 2019 Applying the Kano two-dimensional model and quality function deployment to develop sustainable planning strategies for public housing in Taiwan *J. Hous. Built Environ.* **34** 265–82

[12] Paul V K and Seth V 2017 Benchmarking and objective selection of technologies for housing in India using quality function deployment *J. Constr. Dev. Ctries.* **22** 63–78

[13] Haron N A, Abdul-Rahman H, Chen W and Wood L C 2015 Quality function deployment modelling to enhance industrialised building system adoption in housing projects *Total Qual. Manag.* **26** 703–18

[14] Cebi S, Ozkok M and Demirci E 2014 Evaluation of Design Parameters for Vessel Engine Room by Using A Modified QFD Technique *J. Mult. Log. Soft Comput.* **23** 559–87

[15] Chin K S, Yang Q, Chan C Y P, Tsui K L and Li Y lai 2019 Identifying passengers' needs in cabin interiors of high-speed rails in China using quality function deployment for improving passenger satisfaction *Transp. Res. Part A Policy Pract.* **119** 326–42

[16] Bajčetić S, Tica S, Živanović P, Milovanović B and Đorojević A 2018 Analysis of Public Transport Users' Satisfaction Using Quality Function Deployment: Belgrade Case Study *Transport* **33** 609–18

[17] Bulut E, Duru O and Huang S T 2018 A multidimensional QFD design for the service quality assessment of Kansai International Airport, Japan *Total Qual. Manag. Bus. Excell.* **29** 202–24

[18] Hashim A M and Dawal S Z M 2012 Kano Model and QFD Integration Approach for

IOP Conf. Series: Materials Science and Engineering **807** (2020) 012047 doi:10.1088/1757-899X/807/1/012047

Ergonomic Design Improvement *Procedia - Soc. Behav. Sci.* **57** 22–32
[19] Kurtulmuşoğlu F B and Pakdil F 2017 Combined analysis of service expectations and perceptions in lodging industry through quality function deployment *Total Qual. Manag. Bus. Excell.* **28** 1393–413
[20] Priyono A and Yulita A 2017 Integrating Kano Model and Quality Function Deployment for designing service in hospital front office *Intang. Cap.* **13** 923–45
[21] Wood L C, Wang C, Abdul-Rahman H and Jamal Abdul-Nasir N S 2016 Green hospital design: Integrating quality function deployment and end-user demands *J. Clean. Prod.* **112** 903–13
[22] Sagnak M, Ada N, Kazancoglu Y and Tayaksi C 2017 Quality function deployment application for improving quality of education in business schools *J. Educ. Bus.* **92** 230–7
[23] Crowson R D 2006 *Product Design and Factory Development* (Florida)
[24] Meng J, Liu J, Fan S, Kang C, Yi K, Cheng Y, Shen X and Tao S 2016 Potential health benefits of controlling dust emissions in Beijing *Environ. Pollut.* **213** 850–9

AUTHOR INDEX